FUNDAMENTAL CONCEPTS OF ELEMENTARY MATHEMATICS

LAWRENCE A. TRIVIERI
Mohawk Valley Community College

HARPER & ROW, PUBLISHERS
New York, Hagerstown, San Francisco, London

Sponsoring Editor: George J. Telecki
Project Editor: Cynthia Hausdorff
Designer: T. R. Funderburk
Production Supervisor: Will C. Jomarrón
Compositor: Santype International Limited
Printer and Binder: Halliday Lithograph Corporation
Art Studio: Vantage Art, Inc.

FUNDAMENTAL CONCEPTS OF ELEMENTARY MATHEMATICS

Library of Congress Cataloging in Publication Data

Trivieri, Lawrence A
Fundamental concepts of elementary mathematics.

Includes index.
1. Mathematics—1961-   I. Title.
QA39.2.T733   510   76-17010
ISBN 0-06-046675-8

CONTENTS

PREFACE

This text has been written primarily to meet the mathematical needs of prospective and in-service elementary school teachers. It also may be used by liberal arts and nonscience students who desire a better understanding of the basic structure and concepts of elementary mathematics.

This book provides the student with a knowledge of the fundamental concepts of elementary mathematics which he or she can build upon and adapt to a particular curriculum or program. The treatment of the mathematical concepts that are basic to elementary school programs is precise but not overwhelming. A mathematical systems approach has been used to give a "big picture" effect to the development of an entire concept.

The first chapter is a brief introduction to various types of reasoning, mathematical systems, and proofs, and answers the question "What is mathematics?" Chapter 2 treats the algebra of propositions, and this material is used in Chapter 3 on the algebra of sets. Both chapters may be treated in depth or only slightly, depending upon the class preparation and instructor's interests. It has been my experience that material relegated to an appendix is considered to be unimportant by most students, and so these chapters have been placed at the beginning of this book. Chapter 4 treats relations and functions and completes what the author considers to be "tool" chapters which are used throughout the remainder of the text.

Chapters 5 through 10 are concerned with the development of the system of real numbers, starting with whole numbers and proceeding, successively, through integers, rational numbers, and real numbers through decimal representations. Following the system of whole numbers is a chapter dealing with various systems of numeration, including bases other than ten. Following the system of integers is a chapter on the basic concepts of number theory.

Chapter 2 treats the system of matrices. This material has been included for several reasons. Primarily, I wish to show that the extension of properties for real numbers is not automatic or universal. For instance, if a and b represent real numbers, then $ab = 0$ implies either $a = 0$ or $b = 0$. However, if A and B are matrices and $AB = \mathbf{0}$, it is not necessarily true that either $A = \mathbf{0}$ or $B = \mathbf{0}$. Other exceptions also are treated.

Chapter 12 deals with other mathematical systems, including groups, rings, fields, modular arithmetic, and movements of a square. Here an attempt is made to demonstrate that objects of a set do not have to be numbers and that operations do not have to be arithmetic.

Chapter 13 has a fairly extensive introduction to informal geometry of size, shape, and measure. This is followed by an introduction to coordinate geometry in Chapter 14. Chapter 15 deals with equations, inequalities, and linear programming and provides the student with an opportunity to further integrate algebra and geometry. The text concludes with two fairly brief chapters on probability and statistics.

There are numerous illustrative examples throughout the text. These are used to motivate a concept or definition and also as applications for certain concepts and definitions. There is an abundance of exercises for almost all sections, and each chapter also contains a summary and a set of review exercises.

This text is designed to be used in a two-semester or three-quarter course sequence. There are enough topics included to allow for great flexibility to meet the needs of a particular group of students.

Proofs of some of the simpler and basic theorems are provided in the text; others are left as exercises. If an instructor desires, all proofs may be omitted and, for that reason, the exercises involving proofs of theorems are given at the end of exercise sets. Hence an instructor can make the course as formal or as informal as he or she wishes.

In developing the topics and preparing the exercise sets, I was guided primarily by extensive experience in teaching this course both at the two year and four-year college level. Students' comments relative to the course made while enrolled in the course, during student-teaching experiences, and also during their early years of full-time teaching were carefully considered. I am deeply indebted to these students for their comments.

The author wishes to express his appreciation to Cummings Publishing Company for graciously granting permission to use material contained in "Elementary Functions" (*Elementary Functions: A Study of Precalculus Mathematics*, Lawrence A. Trivieri, Cummings Publishing Company, Inc., Menlo Park, California, 1972). I also wish to express my sincere appreciation to the several reviewers of the original manuscript for their constructive comments and suggestions. Special gratitude is expressed to Mr. George J. Telecki, mathematics editor, and Mrs. Cynthia Hausdorff, project editor, in the College Department at Harper & Row, and to all those who have assisted in the production of this text.

L. A. T.

Utica, New York

CHAPTER I
INTRODUCTION

In this chapter I will attempt to answer the question "What is mathematics?" and will discuss briefly kinds of reasoning and will introduce the concept of a mathematical system. In the remaining chapters I will discuss various mathematical systems and their applications.

1.1 WHAT IS MATHEMATICS?

Mathematics is different things to different people. If you were to ask someone the question "What is mathematics?" the answer you receive would depend upon the person asked. An elementary schoolchild might respond that mathematics is a subject that helps to tell how many things there are or which one we are talking about. A secondary school student might say that mathematics is the thing that enables one to perform various calculations or to solve problems. Someone else might respond that mathematics is that which is used in balancing a checkbook, preparing a budget, or figuring income taxes—all of which are common experiences for most adults. As seen by these people, then, mathematics may be thought of as a *tool*.

But mathematics is more than a tool. A mathematician might say mathematics is an art. There is much aesthetic beauty and satisfaction associated with the discovery of new mathematical concepts and the proof and development of new theories. Indeed, research mathematicians look upon mathematics as an *art*.

1

Returning to schoolchildren, mathematics may be looked upon as being fun and games—an enjoyable experience. Through proper motivation, schoolchildren set out to play their "games" by learning the rules associated with them. Sometimes these games are played by single individuals, and sometimes they are played by teams of students, and the children soon learn, through competition, that there are certain do's and don'ts associated with these games. These rules include, among others, those for the basic operations of arithmetic. As students become more experienced in playing the games with proper guidance and direction they may even learn to create their own new games.

Mathematics is also considered by some to be a *language*, and indeed it is! A vocabulary, grammar, and sentence structure used for expressing various mathematical concepts are associated with mathematics. The more fluently one learns to converse in mathematics, the better one is able to communicate with others in everyday life experiences. There are, however, some symbols used which are peculiar to mathematics, and we will encounter some of them in this text.

Someone else may answer that mathematics is a *science* and, again, this answer would be correct. It has been said that mathematics is the queen of the sciences insofar as it is the ultimate science. It is the science dealing with preciseness, rigor, and logical reasoning. It is the science involving, for the most part, the use of deductive reasoning, which will be discussed in the next section.

We now have that mathematics is a tool, an art, a science, a language, and a game. You may have an answer that differs from those given and probably depends upon the experiences you had in working with mathematics.

Now that we have some ideas about what mathematics is, let's look at its classifications. Mathematics can be classified as *pure* or *applied*. Pure mathematics, basically, is the study of mathematics for its own sake, whereas applied mathematics is mathematics applied to the solution of problems in various fields of study, including the biological and behavioral sciences as well as the physical sciences and technologies.

Mathematics has many parts, some of which overlap. Sometimes two or more of these parts cover the same subject but from different points of view. Simply classified, these parts may be listed as arithmetic, algebra, geometry, trigonometry, calculus, probability, and statistics. Arithmetic basically is that part of mathematics concerned with numbers and their operations. The concepts and methods associated with arithmetic form the foundations of all mathematics.

Algebra is the part of mathematics that generalizes arithmetic. In algebra symbols other than numerals are used. For instance, the letters a and b can be used to represent arbitrary numbers, whereas in arithmetic the Hindu-Arabic symbols 2 and 3 are used to denote specific numbers. A knowledge of arithmetic and algebra is essential to the study of the other parts of mathematics.

Geometry is that part of mathematics dealing with space figures, spatial relations, and the relations between figures. The basic elements of geometry are points, lines, and planes. All figures in geometry are defined in terms of these elements.

Trigonometry is that part of mathematics involving relationships between and among angles and sides of triangles. These relationships usually include ratios of the measurements of the sides of a triangle. Calculus, in a very simple sense, is that part of mathematics dealing with the analysis of a rate of change between variables and the area under a curve.

Probability and statistics are closely related. Probability involves the likelihood that something will happen, such as predicting weather, determining insurance rates, drawing a particular card or hand of cards from a well-shuffled deck of cards, or rolling a seven with a pair of dice. Statistics is that part of mathematics dealing with the collection and analysis of data and the inferring of certain conclusions from these data.

In this text we will discuss all these parts of mathematics with the exceptions of calculus, and trigonometry.

1.2 TYPES OF REASONING

One type of reasoning is that based upon *intuition*. Webster's dictionary defines intuition as "the power of knowing or the knowledge obtained without recourse to reference or reasoning." Intuitive reasoning, then, depends upon knowledge we gain through experience and the use of our senses, such as the sense of smell or of sight. We know from experience that we cannot always trust our senses and, therefore, we cannot depend upon this type of reasoning in mathematics. Consider, for example, the line segments $\overline{MN}$ and $\overline{PQ}$ in Figure 1.1 The line segment $\overline{MN}$ appears to be longer than the line segment $\overline{PQ}$. However, if we measure them, we find that they are the same length.

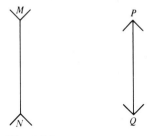

Figure 1.1

A second type of reasoning is *inductive reasoning*. The student is probably most familiar with the process of inductive reasoning from studying the experimental sciences such as chemistry and physics. Here what is called the scientific method is employed. The scientist performs a particular experiment several times, each time under the same set of conditions. If the results of the experiments are the same each time, he reasons that the results will be the same the next time the experiment is performed under the same set of conditions. In this type of reasoning, a general conclusion is drawn from particular observations. However, the results conceivably could vary the next time the experiment is performed under the same set of conditions.

Inductive reasoning is not new in mathematics. Prior to the Greek contributions to mathematics, inductive reasoning was in vogue. Some mathematics was introduced through trial-and-error processes, giving only approximate results. Later these results were refined or corrected. Inductive reasoning is used in mathematics today mainly in the creation of new mathematics. A research mathematician may start with a conjecture based upon several observations. A conjecture is a statement believed to be true but which has not been proven true. Some conjectures remain unproven for long periods of time. For instance, there is the famous Golbach conjectute about prime numbers which dates back to 1742.* To date the conjecture has not been proved or disproved, although outstanding mathematicians have worked on it.

In this text we will emphasize the use of what is known as *deductive reasoning*. The student is probably most familiar with the process of deductive reasoning from his study of Euclidean plane geometry at the high school level. The study of geometry involves the relationships that exist between and among certain figures such as lines, angles, circles, triangles, rectangles, and parallelograms. First we start with certain basic ideas that are understood as common knowledge. There is no need to establish definitions for them. Such ideas are called *undefined ideas* or *undefined terms*. An example from plane geometry is the undefined term *point*. Generally no attempt is made to define this word, yet the term is used in discussions with meaning and understanding.

Often, however, it becomes necessary to define certain terms used in a particular discussion. Hence we also have *definitions*. A definition is basically an agreement on the use of a particular word or phrase. The word or phrase is defined or expressed in terms of other words which have been introduced and agreed upon. Consider, for example, a definition for *rectangle*. A rectangle is defined as a parallelogram having one right angle. The meaning of rectangle would become clear if all the words used in the

---

* The Goldbach conjecture states that any even number (except 2) can be represented as the sum of two prime numbers.

definition had been previously defined and understood. If the word parallelogram had not been defined earlier, then the word rectangle would be without meaning. It is important to note that a definition is accepted; it is not subject to argument.

In addition to the undefined terms and the definitions involved in deductive reasoning, we have postulates or axioms. A *postulate* or an *axiom* is a basic assumption accepted without proof. These may involve relationships between certain undefined terms or properties of the undefined terms. For instance, we have the postulate that "two distinct points determine one and only one straight line." This statement is accepted without proof and appears to be self-evident. Nevertheless, a postulate does not have to be self-evident. One example of a postulate which did not appear to be self-evident is the *parallel postulate*. The controversy centering around this postulate during the nineteenth century led to the development of non-Euclidean geometries.

Using various rules from logic, we can now examine undefined terms, definitions, and postulates and attempt to draw conclusions by valid reasoning. Such conclusions are called *theorems*. An example of a theorem would be the statement "If two sides of a triangle are congruent, then the angles opposite these sides are also congruent." Once a theorem has been proved, we can use it to establish other valid conclusions.

In the foregoing I have described a deductive system which consists of
1. A collection of undefined ideas or undefined terms;
2. A collection of definitions;
3. A collection of postulates or axioms;
4. A collection of theorems established by using various rules of logic.

EXERCISES

For each of the following, read the question carefully and write down your first response. After attempting all of the questions, go back and answer each carefully by applying a suitable type of reasoning.

1. What is fifteen divided by a third?
2. How many 13¢ stamps are there in a strip of 39 13¢ stamps?
3. If you take 3 books from a shelf containing 7 books, how many books do you have?
4. A boy on a motorbike travels 20 miles at 30 miles/hour and back the same 20 miles at 40 miles/hour. What is the average speed for the round trip?
5. Items A and B together cost $2.20. Item A costs $2 more than item B. How much does item B cost?
6. If a is greater than b and the product ab is negative, is a negative?
7. If the temperature on a particular day is 20°F, would it have been twice as warm if the temperature had been 40°F?

8. Why isn't it legal in the United States for a man to marry his widow's sister?
9. What pattern is involved in an infinite sequence of numbers whose first four terms are 1, 2, 3, 5? Write the next three terms in the sequence.
10. If the square of a number divided by the number is equal to the cube of the number divided by the square of the number, what is the number?

1.3 MATHEMATICAL SYSTEMS AND PROOFS

Throughout this text we will be studying various mathematical systems such as the systems of whole numbers, of integers, and of rational numbers; and the systems, or algebra, of propositions and sets. A *mathematical system* consists basically of a set of elements (such as numbers or propositions), at least one operation (such as addition, multiplication, disjunction, or intersection), and a set of axioms or postulates relating to the elements and operations (such as the commutative and associative properties of addition). Generally there is at least one relation involved also (such as the relation "is less than" relative to the whole numbers or "is a subset of" relative to sets). Using a deductive reasoning process, we start with undefined terms, definitions, and axioms and, through a logical argument, arrive at the proof of a theorem.

Proofs may be either direct or indirect. In a *direct proof* one starts with a combination of known axioms, definitions, and previously proved theorems and attempts to arrive at a particular conclusion, using various laws of logic and what is known as a valid argument.

Sometimes it is easier to prove a theorem by what is known as an *indirect proof.* Here one assumes that what is to be proved is false. Then, through a series of steps involving a valid argument, one arrives at a contradiction. This may be a contradiction of a definition, a known axiom, or a known theorem.

At this point the student must be cautioned about proofs. To prove that a statement is true one must prove that it is true in *every possible case.* Therefore, if the statement holds true for one or two particular cases, one may not generalize from those particular cases and conclude that the statement is true in all cases. However, to prove that a statement is false, a single example that the statement is false will suffice. Such an example is called a *counterexample.*

CHAPTER 2
THE ALGEBRA OF PROPOSITIONS

In the previous chapter we learned that a mathematical system consists of a set of elements, at least one operation, and some axioms or postulates relating to the elements and operation(s). In this chapter we will discuss the system or algebra of propositions and an application to switching networks. In the next chapter we will study a related system—the algebra of sets.

2.1 PROPOSITIONS AND PROPOSITIONAL FORMS

We have described the process of deductive reasoning and indicated that various rules of logic are used in arriving at conclusions. In this section we shall introduce some of the terminology used in logic. A statement is a sentence that either asserts, asks, commands, or exclaims. We will now consider a special type of statement called a proposition.

DEFINITION 2.1 A *proposition* is a statement that is either true or false but not both.

A sentence that asserts something is known as a declarative sentence; a proposition must be a declarative sentence. However, not all declarative sentences are propositions. The declarative sentence "She has blue eyes" is not a proposition because it cannot be classified as being either true or false since the truth of the sentence depends upon who "she" is. It becomes a proposition when a specific name is substituted for "she." Then the statement becomes a proposition, and it can be meaningfully classified as being true or false. The declarative statement "x is greater than 3" is not a proposition, because the value assigned to x determines whether the sentence is true or false.

>   **EXAMPLE**   The following are examples of propositions.
>   The month of February has 34 days.
>   Richard M. Nixon was inaugurated as President of the United States on January 20, 1969.
> $3 \leq 4$.

DEFINITION 2.2 A *propositional form* is a statement containing an unknown element or variable which becomes a proposition when the unknown element or variable is replaced by a specific element.

>   **EXAMPLE**   The statement "x is greater than 3" is an example of a propositional form. Without a specific value for the variable x the statement cannot be classified as being true or false. However, if $x = 2$, the propositional form becomes the proposition "2 is greater than 3," which is false.

The values of the variable that make the propositional form a true proposition are called its *truth values*. Hence 5 is a truth value for the propositional form in the above example; it is true that "5 is greater than 3."

DEFINITION 2.3 A *simple proposition* is a proposition containing only one subject and one predicate. Such propositions are generally denoted by the use of lowercase letters such as p, q, and r.

The propositions given in the first example are all simple propositions.

EXERCISES

1. Classify each of the following as a proposition, a propositional form, or neither. If it is a proposition, further classify it as being true or false.
   a. She is beautiful.
   b. Some houses are made of brick.
   c. The football coach is a swell guy.
   d. Every natural number (counting number) is even.
   e. Meat is a protein.
   f. $2x + 3 = 5$.

g. Walter Hickel was secretary of the interior during the Nixon administration.

h. Thirty days in jail or a $100 fine.

i. 4 is exactly divisible by 5.

j. Every month of the year has at least 28 days.

2. Find one truth value for each of the following propositional forms.

a. $3x - 4 = 2$.

b. y is less than 7 but greater than 0.

c. $x < x + 2$, and x is a positive number.

d. $x = 3y$ if $y = 3$ or $y = 5$.

e. On April 1, 1969, _____ was a U.S. senator from New York State.

2.2 CONNECTIVES

It is possible to form new propositions from given propositions by the use of words called *connectives*. We shall now introduce some of the more common connectives used in logic and indicate the agreements among logicians and mathematicians on their usage.

DEFINITION 2.4 The *negation* (or negative) of a proposition, p, is the proposition "it is not true that p." It is denoted by "not p" and is symbolized by $\sim p$.

The usage of this connective is as follows:

If p is true, then $\sim p$ is false.

If p is false, then $\sim p$ is true.

This is summarized in Table 2.1, which is called a *truth table*. The symbol T is used to indicate that the proposition is true; F is used for false.

Table 2.1 Truth table for negation

p	$\sim p$
T	F
F	T

EXAMPLE The following are examples of negations of propositions:

p: $3 \leq 4$ $\sim p$: It is not true that $3 \leq 4$

q: $2 + 2 = 5$ $\sim q$: $2 + 2 \neq 5$ (read 2 plus 2 does not equal 5)

DEFINITION 2.5 The *conjunction* of the propositions p, q is the proposition "p and q," symbolized $p \wedge q$.

The usage of this connective, summarized in truth table 2.2, is as follows:
If p and q are both true, $p \wedge q$ is true.
If at least one of the propositions p, q is false, $p \wedge q$ is false.

Table 2.2 Truth table for conjunction

p	q	$p \wedge q$
T	T	T
T	F	F
F	T	F
F	F	F

EXAMPLE The following are examples of the conjunction of propositions:

p: Sugar is sweet	(T)
q: Roses are red	(T)
$p \wedge q$: Sugar is sweet, *and* roses are red	(T)
r: $2 + 2 = 5$	(F)
s: $3 \leq 4$	(T)
$r \wedge s$: $2 + 2 = 5$, *and* $3 \leq 4$	(F)
m: February has 34 days	(F)
n: There are 6 pints in 1 gallon	(F)
$m \wedge n$: February has 34 days, *and* there are 6 pints in 1 gallon	(F)

DEFINITION 2.6 The *disjunction* of the propositions p, q is the proposition p or q, symbolized $p \vee q$.

The word *or* is used in this definition in the inclusive sense, as in legal language, to mean and/or. The usage of this connective, summarized by truth table 2.3, is as follows:
If at least one of the propositions p, q is true, $p \vee q$ is true.
If both p and q are false, then $p \vee q$ is false.

Table 2.3 Truth table for disjunction

p	q	$p \vee q$
T	T	T
T	F	T
F	T	T
F	F	F

EXAMPLE The following are examples of the disjunction of propositions:

p: Sugar is sweet	(T)
q: Roses are red	(T)
$p \vee q$: Sugar is sweet, *or* roses are red	(T)
r: $2 + 2 = 5$	(F)
s: $3 \leq 4$	(T)
$r \vee s$: $2 + 2 = 5$, *or* $3 \leq 4$	(T)
m: February has 34 days	(F)
n: There are 6 pints in 1 gallon	(F)
$m \vee n$: February has 34 days, *or* there are 6 pints in 1 gallon	(F)

DEFINITION 2.7 A *conditional* is a proposition of the form "if p, then q,"
where p and q are propositions symbolized $p \rightarrow q$.

The proposition p is called the *hypothesis* or the *antecedent*, and the
proposition q is called the *conclusion* or the *consequent*. The usage of this
connective, summarized by truth table 2.4, is as follows:

If p is true, $p \rightarrow q$ is true only if q is true.
If p is false, $p \rightarrow q$ is true whether q is true or false.

Table 2.4 Truth table for conditional

p	q	$p \rightarrow q$
T	T	T
T	F	F
F	T	T
F	F	T

EXAMPLE The following are examples of conditionals:

p: $2 + 2 = 5$	(F)
q: $3 \leq 4$	(T)
$p \rightarrow q$: If $2 + 2 = 5$, then $3 \leq 4$	(T)
r: Sugar is sweet	(T)
s: $6 - 4 = 7$	(F)
$r \rightarrow s$: If sugar is sweet, then $6 - 4 = 7$	(F)
m: $3 + 6 < 4$	(F)
n: $7 + 3 = 12$	(F)
$m \rightarrow n$: If $3 + 6 < 4$, then $7 + 3 = 12$.	(T)

Observe that in a true conditional a true hypothesis must be followed by a true conclusion. This is especially important when we attempt to use conditionals in proofs. Also observe that if the hypothesis is false the conditional is true, whether the conclusion is true or false. This presents an ambiguity that cannot be tolerated in mathematics, in which we use only conditionals having true hypotheses, except in the case of a proof by contradiction. (Recall that a proof by contradiction is an indirect proof in which one assumes the contrary of what is to be proved and then shows a contradiction to one of the hypotheses or to something else known to be true.)

What happens to the conditional $p \rightarrow q$ if we interchange the role of the propositions p and q? If the conditional $p \rightarrow q$ is true, will the conditional $q \rightarrow p$ also be true?

DEFINITION 2.8 The *converse* of the conditional $p \rightarrow q$ is the conditional $q \rightarrow p$. It is formed by interchanging the roles of the propositions p and q.

Even when the conditional $p \rightarrow q$ is true, the converse $q \rightarrow p$ is not necessarily true. Consider, for instance, the proposition p as false and q as true. The conditional $p \rightarrow q$ is true. However, the converse conditional $q \rightarrow p$ is false, since a true hypothesis q is followed by a false conclusion p.

In addition to forming the converse of a conditional, we can also form the inverse and the contrapositive conditionals.

DEFINITION 2.9 The *inverse* of the conditional $p \rightarrow q$ is the conditional $\sim p \rightarrow \sim q$ formed by negating both the hypothesis and the conclusion of the original conditional.

DEFINITION 2.10 The *contrapositive* of the conditional $p \rightarrow q$ is the conditional $\sim q \rightarrow \sim p$.

Observe that the contrapositive of the conditional $p \rightarrow q$ can be formed in one of two different ways:

1. Negate both the hypothesis and conclusion and interchange them;

2. Interchange the hypothesis and conclusion and then negate them.

The relationships among a conditional, its converse, its inverse, and its contrapositive are summarized by truth table 2.5.

Table 2.5 Truth table for conditional, converse, inverse, and contrapositive

p	q	CONDITIONAL $p \rightarrow q$	CONVERSE $q \rightarrow p$	INVERSE $\sim p \rightarrow \sim q$	CONTRAPOSITIVE $\sim q \rightarrow \sim p$
T	T	T	T	T	T
T	F	F	T	T	F
F	T	T	F	F	T
F	F	T	T	T	T

EXAMPLE The following is an example of a true conditional together with its converse, inverse, and contrapositive:

Conditional	If two sides of a triangle are congruent, then the angles opposite these sides are congruent.
Converse	If two angles of a triangle are congruent, then the sides opposite these angles are congruent.
Inverse	If two sides of a triangle are *not* congruent, then the angles opposite these sides are *not* congruent.
Contrapositive	If two angles of a triangle are *not* congruent, then the sides opposite these angles are *not* congruent.

There are times when it is desirable to consider both a conditional and its converse at the same time. Hence we form the conjunction of the two.

DEFINITION 2.11 A *biconditional* is the conjunction of a conditional $p \to q$ and its converse conditional $q \to p$, symbolized by $(p \to q) \land (q \to p)$ or more simply $p \leftrightarrow q$.

The usage for the biconditional, summarized by truth table 2.6, depends upon the usage for the conditional and conjunction.

Table 2.6 Truth table for biconditional

p	q	$p \to q$	$q \to p$	$(p \to q) \land (q \to p)$ $p \leftrightarrow q$
T	T	T	T	T
T	F	F	T	F
F	T	T	F	F
F	F	T	T	T

From Table 2.6 we see that $p \leftrightarrow q$ is true when both p and q are true or when both p and q are false; otherwise, it is false. This connective is also read "if and only if." Its use is especially important in stating definitions. (Many authors use the abbreviation "iff" for "if and only if.")

EXERCISES

1. Write the negation for each of the following propositions.
   a. A dog has four legs.
   b. Vitamins are not metals.
   c. Integers are real numbers.
   d. A house is a building.
   e. $3 + 4 \neq 9$.

2. Write the conjunction for each of the following pairs of propositions.
   a. Red is a color; February is a month of the year.
   b. Five is not a rational number; $p \to q$.
   c. $\sim p$; $2 + 2 = 5$.
   d. $p \vee \sim q$; $\sim(p \to q)$.
3. Write the disjunction for each pair of propositions given in Exercise 2.
4. Write the conditional for each pair of propositions given in Exercise 2 using the first proposition of a pair as the hypothesis and the second as the conclusion.
5. For each of the following conditionals, form (1) the converse, (2) the inverse, and (3) the contrapositive.
   a. If you like to play the piano, then you like music.
   b. $w \to u$.
   c. If the negation of a proposition is true, then the proposition is false.
   d. $\sim p \to q$.
   e. $r \to \sim s$.
6. If a person is convicted of a crime and the law prescribes a punishment of "30 days in jail or a $100 fine," how many different choices of penalties does the judge have? What are they?
7. Laura, captain of the basketball team, said, "My team will win the game this week or it will win the game next week." Later, it was determined that her statement was false. What actually happened?
8. Let r, s, and t represent three propositions. Construct truth tables for each of the following:
   a. $r \vee \sim s$.
   b. $\sim s \to t$.
   c. $r \vee (s \to t)$.
   d. $s \leftrightarrow t$.
   e. $\sim r \wedge (\sim s \vee \sim t)$.
9. Which of the propositions in Exercise 8 have identical truth tables?
10. Consider these two propositions: "It will rain today, or I will wash the car," which has the form "$p \vee q$"; "If it does not rain today, then I will wash the car," which has the form "$\sim p \to q$." Construct truth tables for both propositions, and verify that the truth tables are identical.
11. Write the negation of each of the following propositions.
    a. Brian has brown eyes or brown hair.
    b. Brian has brown eyes and brown hair.
    c. Brian does not have brown eyes or has brown hair.
    d. Brian does not have brown eyes and has brown hair.
    e. Brian has brown eyes and does not have brown hair.
    f. Brian has brown eyes or does not have brown hair.
12. Let p be the proposition "Brian has brown eyes," and let q be the proposition "Brian has brown hair." Write in symbolic form the negation of each of the propositions given in Exercise 11.

13. Let r be the proposition "Brian does not have brown eyes," and let s be the proposition "Brian has brown hair." Write in symbolic form the negation of each of the propositions given in Exercise 11.

14. Let p be the proposition "Food prices are rising," and let q be the proposition "College enrollments are decreasing." Write the following statements out in words.

a. $p \vee q$. b. $p \wedge q$.
c. $p \vee \sim q$. d. $\sim p \vee q$.
e. $p \wedge \sim q$. f. $\sim p \wedge q$.
g. $\sim p \wedge p$. h. $\sim p \wedge \sim q$.
i. $\sim p \vee \sim q$. j. $\sim(p \wedge q)$.
k. $\sim(p \vee q)$. l. $\sim(p \vee \sim q)$.
m. $\sim(\sim p \wedge q)$. n. $\sim(\sim p \vee \sim q)$.

2.3 SIMPLE CIRCUITS

Before continuing our discussion of propositions, let's look at an application of what we have discussed so far to some simple circuits. Suppose that we have a single switch, p, connected to wires between two terminals A and B as illustrated in Figure 2.1. (Assume an appropriate power supply.) There

A •———[Switch, p]———• B

Figure 2.1

are two possible states for the switch. Either it is open, in which case current will not flow from A to B, or it is closed, in which case current will flow from A to B. These two states are illustrated in Figure 2.2.

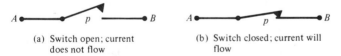

   (a) Switch open; current   (b) Switch closed; current will
       does not flow   flow

Figure 2.2

Further, suppose we refer to the state of the switch p by assigning values 0 and 1 according to whether the switch is open or closed, respectively. Then, if p has a truth value of 0, the negation of p, $\sim p$, must have a truth value of 1 and vice versa. This is illustrated in the truth table given in Table 2.7.

Table 2.7

p	$\sim p$
1	0
0	1

Next we could consider a circuit consisting of two switches. This leads to two cases as follows:

1. The two switches are connected in series as illustrated in Figure 2.3.

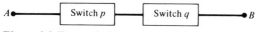

Figure 2.3 Two switches connected in series

Clearly, switch *p* can be open or closed and likewise with switch *q*. We have four possibilities as indicated in Figure 2.4.

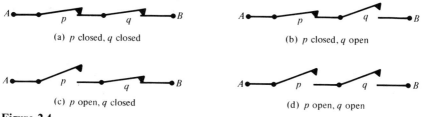

Figure 2.4

Now, for current to flow from terminal *A* to terminal *B*, both switches must be closed; that is, *p* must be closed *and* *q* must be closed. Otherwise current will not flow from *A* to *B*. We show these four cases in Table 2.8.

Table 2.8

p	*q*	*p* and *q* in series
1	1	1
1	0	0
0	1	0
0	0	0

Hence we observe that for the two switches connected in series, both must be closed for current to flow from *A* to *B*; otherwise the circuit is not completed.

2. The two switches are connected in parallel as illustrated in Figure 2.5. Again, depending upon the state of each switch, we have four possibilities to consider as illustrated in Figure 2.6.

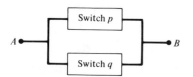

Figure 2.5 Two switches connected in parallel.

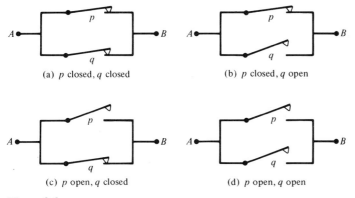

(a) p closed, q closed   (b) p closed, q open

(c) p open, q closed   (d) p open, q open

Figure 2.6

Now for current to flow from terminal A to terminal B, it must be able to flow through either the upper branch of the circuit or the lower branch. Hence at least one switch must be closed to complete the circuit. This is illustrated in Table 2.9.

Table 2.9

p	q	p and q connected in parallel
1	1	1
1	0	1
0	1	1
0	0	0

Hence we observe that for two switches connected in parallel, at least one switch must be closed in order for current to flow from terminal A to terminal B. That is, the circuit will not be completed only when both switches are open.

We will design more elaborate circuits later in this chapter. The point to be made here, however, is relative to the analogues existing between the connectives used in logic and the state of a switch and type of circuits. We will emphasize these corresponding analogues by listing the respective truth tables side-by-side in Figure 2.7. Observe that T corresponds directly with 1 while F corresponds directly with 0.

Notice that we really have two mathematical systems here whose structures are identical. For one system we have a set of propositions together with connectives and the laws pertaining to their use. For the other system we have a set of switches together with circuits containing them and the laws pertaining to these networks. Examination of the two systems reveals that negation, conjunction, and disjunction, relative to statements, correspond respectively with the state of a switch, switches connected in series circuit, and switches connected in parallel circuit.

PROPOSITIONS AND CONNECTIVES	SWITCHES AND CIRCUITS

a. Negation

p	$\sim p$
T	F
F	T

a. State

p	$\sim p$
1	0
0	1

b. Conjunction

p	q	$p \wedge q$
T	T	T
T	F	F
F	T	F
F	F	F

b. Switches in Series Circuit

p	q	p and q
1	1	1
1	0	0
0	1	0
0	0	0

c. Disjunction

p	q	$p \vee q$
T	T	T
T	F	T
F	T	T
F	F	F

c. Switches in Parallel Circuit

p	q	p or q
1	1	1
1	0	1
0	1	1
0	0	0

Figure 2.7

EXERCISES

Fill in the blanks so that the resulting statements are true.
1. The two states for a single switch are _____ and _____.
2. If two switches are connected together, they may be connected in _____ or in _____.
3. The opposite state of a switch corresponds to the _____ of a proposition.
4. The conjunction of propositions corresponds to two switches being connected in _____.
5. The disjunction of propositions corresponds to two switches being connected in _____.

2.4 TAUTOLOGY AND CONTRADICTION

So far we have discussed propositions, propositional forms, and connectives. We have illustrated how we can obtain new propositions by combining simple propositions with various connectives.

DEFINITION 2.12 A *composite proposition* is a proposition formed by combining two or more simple propositions with various connectives.

EXAMPLE The following are examples of composite propositions:

$\sim(p \vee q)$, the negation of the disjunction of p and q.

$p \wedge \sim q$, the conjunction of p and not q.

$\sim(p \vee \sim p)$, the negation of the disjunction of p and not p.

$(p \rightarrow q) \leftrightarrow (\sim q \rightarrow \sim p)$, the biconditional of a conditional with its contrapositive.

For each of the above composite propositions, truth tables can be established. The truth values depend upon the truth values for each of the simple propositions. In some cases, however, the composite propositions are always true regardless of the truth values of the simple propositions.

DEFINITION 2.13 A *tautology* is a composite proposition that is always true.

EXAMPLE Consider the truth table for the last part of the example above.

Solution (see Table 2.10).

Table 2.10

p	q	$p \rightarrow q$	$\sim q \rightarrow \sim p$	$(p \rightarrow q) \leftrightarrow (\sim q \rightarrow \sim p)$
T	T	T	T	T
T	F	F	F	T
F	T	T	T	T
F	F	T	T	T

Since all entries in the last column of the truth table are T's, the composite proposition $(p \rightarrow q) \leftrightarrow (\sim q \rightarrow \sim p)$ is a tautology. ∎

DEFINITION 2.14 A *contradiction* is a composite proposition that is always false.

EXAMPLE Consider which of the composite propositions $(p \vee q) \rightarrow p$, $\sim[(p \wedge \sim p) \rightarrow q]$ are contradictions.

Solution (see Table 2.11).

Table 2.11

(1) p	(2) q	(3) $p \vee q$	(4) $(p \vee q) \to p$	(5) $p \wedge \sim p$	(6) $(p \wedge \sim p) \to q$	(7) $\sim[(p \wedge \sim p) \to q]$
T	T	T	T	F	T	F
T	F	T	T	F	T	F
F	T	T	F	F	T	F
F	F	F	T	F	T	F

Since column (4) contains both T and F values, the composite proposition $(p \vee q) \to p$ is not a contradiction. Note that it is not a tautology either. Since all of the entries in column (7) are F, the composite proposition $\sim[(p \wedge \sim p) \to q]$ is a contradiction. Likewise, the composite proposition $p \wedge \sim p$ is a contradiction. Also observe that the composite proposition given in (6) is a tautology. ■

EXERCISES

For each of the following exercises determine whether the given proposition is a tautology, a contradiction, or neither.

1. $p \vee \sim p$.
2. $p \wedge \sim p$.
3. $p \to p$.
4. $p \to \sim p$.
5. $(p \vee q) \to \sim q$.
6. $(p \wedge q) \to p$.
7. $(p \wedge q) \to \sim p$.
8. $(p \vee \sim q) \wedge q$.
9. $p \vee (q \wedge \sim q)$.
10. $(p \vee q) \to \sim r$.
11. $(\sim p \wedge r) \vee (q \to r)$.
12. $\sim (p \vee q)$.
13. $\sim (p \wedge \sim p)$.
14. $(p \to q) \leftrightarrow (\sim q \to \sim p)$.
15. $\sim[(p \to q) \wedge p] \to q$.
16. $[(p \to q) \wedge \sim q] \to \sim p$.
17. $\sim[(p \to q) \to (\sim p \vee q)]$.
18. $\sim (p \wedge q) \to (\sim p \wedge \sim q)$.
19. $\sim (p \wedge q) \to (\sim p \vee \sim q)$.
20. $(p \to q) \leftrightarrow (q \to p)$.

2.5 IMPLICATION AND EQUIVALENCE

Up to this point we have been concerned with propositions and ways of forming new propositions by using the various connectives introduced. In this section we will consider two relations between given propositions. The first of these is the relation of implication.

DEFINITION 2.15 Consider two propositions p and q. Then the *implication* p implies q, denoted $p \Rightarrow q$, is the relation such that q is true whenever p is true. Specifically, we say that $p \Rightarrow q$ if and only if $p \to q$ is a tautology.

It should be noted that the conditional and the implication are not to be confused. The conditional is a connective between two propositions leading to a new proposition, whereas an implication is a relation between two propositions. From Definition 2.15 we observe that in order to determine whether the proposition p implies the proposition q we must determine whether the conditional $p \to q$ is a tautology.

EXAMPLE Determine whether the proposition $p \to q$ implies the proposition $\sim p \vee q$.

Solution Construct the truth table for p, q, $p \to q$, $\sim p \vee q$, and $(p \to q) \to (\sim p \vee q)$ as shown in Table 2.12.

Table 2.12

p	q	$p \to q$	$\sim p \vee q$	$(p \to q) \to (\sim p \vee q)$
T	T	T	T	T
T	F	F	F	T
F	T	T	T	T
F	F	T	T	T

Since $(p \to q) \to (\sim p \vee q)$ is a tautology, $p \to q$ does imply $\sim p \vee q$. In other words, $\sim p \vee q$ is true whenever $p \to q$ is true. ∎

EXAMPLE Determine whether the proposition $\sim (p \wedge q)$ implies the proposition $\sim p \wedge \sim q$.

Solution Construct the truth table for p, q, $\sim (p \wedge q)$, $\sim p \wedge \sim q$, and $\sim (p \wedge q) \to (\sim p \wedge \sim q)$ as in Table 2.13.

Table 2.13

p	q	$\sim (p \wedge q)$	$\sim p \wedge \sim q$	$\sim (p \wedge q) \to (\sim p \wedge \sim q)$
T	T	F	F	T
T	F	T	F	F
F	T	T	F	F
F	F	T	T	T

Since $\sim (p \wedge q) \to (\sim p \wedge \sim q)$ is not a tautology, we observe that $\sim (p \wedge q)$ does not imply $\sim p \wedge \sim q$. ∎

The second relation we are concerned with in this section is equivalence.

DEFINITION 2.16 Consider the two propositions p and q. We say that p is *equivalent* to q, denoted by $p \Leftrightarrow q$, if and only if $p \leftrightarrow q$ is a tautology.

Again, note that the biconditional and the equivalence (the relation existing between propositions that are equivalent) are not to be confused. The biconditional is a connective between two propositions leading to a new proposition, while an equivalence is a relation between two propositions. From Definition 2.16 we observe that in order to determine whether the propositions p and q are equivalent, the biconditional $p \leftrightarrow q$ must be a tautology.

EXAMPLE Determine whether $p \rightarrow q$ and $\sim q \rightarrow \sim p$ are equivalent.

Solution Construct the truth tables for p, q, $p \rightarrow q$, $\sim q \rightarrow \sim p$, and $(p \rightarrow q) \leftrightarrow (\sim q \rightarrow \sim p)$ as in Table 2.14.

Table 2.14

p	q	$p \rightarrow q$	$\sim q \rightarrow \sim p$	$(p \rightarrow q) \leftrightarrow (\sim q \rightarrow \sim p)$
T	T	T	T	T
T	F	F	F	T
F	T	T	T	T
F	F	T	T	T

Since $(p \rightarrow q) \leftrightarrow (\sim q \rightarrow \sim p)$ is a tautology, the composite propositions $p \rightarrow q$ and $\sim q \rightarrow \sim p$ are equivalent. Observe that $\sim q \rightarrow \sim p$ is the contrapositive of the conditional $p \rightarrow q$. Hence we have established that a conditional and its contrapositive are equivalent. To show that a conditional and its converse are not equivalent, consider the next example. ■

EXAMPLE Determine whether $p \rightarrow q$ and $q \rightarrow p$ are equivalent.

Solution Construct the truth table for p, q, $p \rightarrow q$, $q \rightarrow p$, and $(p \rightarrow q) \leftrightarrow (q \rightarrow p)$ as in Table 2.15.

Table 2.15

p	q	$p \rightarrow q$	$q \rightarrow p$	$(p \rightarrow q) \leftrightarrow (q \rightarrow p)$
T	T	T	T	T
T	F	F	T	F
F	T	T	F	F
F	F	T	T	T

Since $(p \rightarrow q) \leftrightarrow (q \rightarrow p)$ is not a tautology, the conditional $p \rightarrow q$ is not equivalent to its converse. ■

If two propositions p and q are equivalent, we may also symbolize this by $p \equiv q$. Either symbol may be used, and we will use them interchangeably throughout this text. It should also be noted that $p \equiv q$ if and only if p and q have the identical truth values for *every possible case*.

EXERCISES

1. Consider these two statements: (1) "It will rain today, or I will wash the car," which has the form $p \lor q$; (2) "If it does not rain today, then I will wash the car," which has the form $\sim p \to q$. Construct truth tables for both statements, and verify that they are equivalent.

2. Write each of the following disjunctions as equivalent conditionals.
   a. $y = 3$ or $y = 5$.
   b. $\sim p$ or q.
   c. p or $\sim q$.
   d. $\sim p$ or $\sim q$.
   e. I will wash the car, or it will rain today.

3. Write each of the following conditionals as equivalent disjunctions.
   a. If today is not Tuesday, then it will rain.
   b. If $b = 6$, then $b \ne 4$.
   c. If $a \ne 5$, then $a = 7$.
   d. If q, then p.
   e. If $\sim p$, then q.
   f. If $\sim q$, then $\sim p$.

4. For each part of this exercise, determine if the first statement implies the second statement.
   a. $p \lor q$; $q \lor p$.
   b. $p \land q$; $q \land p$.
   c. $p \to q$; $q \to p$.
   d. $p \to q$; $\sim p \lor q$.
   e. $\sim(p \lor q)$; $\sim p \lor \sim q$.
   f. $\sim(p \lor q)$; $\sim p \land \sim q$.
   g. $\sim(p \land q)$; $\sim p \land \sim q$.
   h. $\sim(p \land q)$; $\sim p \lor \sim q$.
   i. $p \to q$; $\sim q \to \sim p$.
   j. $\sim(p \to \sim q)$; $p \land q$.

5. For each part of Exercise 3, determine if the pairs of statements are equivalent.

6. Rewrite each of the following composite propositions in equivalent form using only the connectives $\sim$ and $\lor$.
   a. $p \to q$.
   b. $p \land q$.
   c. $p \leftrightarrow q$.
   d. $q \to \sim p$.
   e. $(p \land q) \to p$.
   f. $p \to (p \land q)$.

2.6 ALTERNATE FORMS OF THE CONDITIONAL AND BICONDITIONAL

When one encounters the conditional in a discussion it is not always in the form "if p, then q." Consider, for instance, the following proposition:

A "A car is economical to operate only if it is a compact."

Basically proposition A states that "If a car is not a compact, then it is not economical to operate," which is in the form "if $\sim q$, then $\sim p$." But "if $\sim q$, then $\sim p$" is the contrapositive of "if p, then q," and hence proposition A is equivalent to proposition

B "If a car is economical to operate, then it is a compact."

Therefore, we have that the conditional "if p, then q" is equivalent to "p, only if q."

Moreover, in mathematics, we encounter the terms *necessary condition* and *sufficient condition*. Now consider the following proposition:

C "A car is economical to operate is a sufficient condition that it is a compact."

Stated in other words, proposition C means that "for a car to be a compact, it is sufficient that it be economical to operate." To say that "p is a sufficient condition for q" is to say that "q happens whenever p happens." Therefore, we have that the conditional "if p, then q" is equivalent to "p is a sufficient condition for q."

Now consider the following proposition:

D "A car is a compact is a necessary condition that it is economical to operate."

Stated in other words, proposition D means that "a car is not economical to operate unless it is a compact."

It should be observed that "if p is a sufficient condition for q," then "q is a necessary condition for p," and conversely. Hence, for the conditional $p \rightarrow q$, the four forms given below are all equivalent:

1. if p, then q.
2. p only if q.
3. p is a sufficient condition for q.
4. q is a necessary condition for p.

EXAMPLE   Consider the conditional "If I study hard, then I will pass this course." Rewrite the conditional in equivalent forms.

Solution   The conditional is already in the form "if p, then q," where p is "I study hard" and q is "I will pass this course." We now write the three equivalent forms given above as

a. p, only if q, or "I study hard only if I pass this course."
b. p is a sufficient condition for q, or "I study hard is a sufficient condition that I will pass this course."
c. q is a necessary condition for p, or "I will pass this course is a necessary condition that I study hard." ■

EXAMPLE   Write an equivalent proposition for "The sun is shining only if it is bright."

Solution   If we let p denote "the sun is shining" and q denote "it is bright," then the given proposition is in the form "p, only if q." In the equivalent form "if p, then q," we would have "If the sun is shining, then it is bright."   ■

Recall that the biconditional is the conjunction of a conditional with its converse. If the conditional is in the form $p \rightarrow q$, then its converse is $q \rightarrow p$. Considering the equivalent forms for the conditional and the equivalent forms for the converse, we conclude that the equivalent forms for the biconditional $p \leftrightarrow q$ are

1. p if and only if q.
2. q if and only if p.
3. p is a necessary and sufficient condition for q.
4. q is a necessary and sufficient condition for p.

EXERCISES

1. Rewrite each of the following in "If ..., then ..." form.
   a. Passing this course is a necessary condition for receiving a grade of B.
   b. $2 + 2 = 5$ is a sufficient condition for $3 \times 2 = 4$.
   c. An angle is a right angle only if its measure is $90°$.
   d. Opposite sides being congruent is a necessary condition that a quadrilateral be a square.
   e. $\sim q$, only if w.
   f. $u \rightarrow w$ is a sufficient condition for $\sim u \vee w$.
   g. Today is Wednesday is a sufficient condition that yesterday was Tuesday.
   h. The judge is fair is a necessary condition that he is impartial.
2. Which of the following, if any, are equivalent to the biconditional $p \leftrightarrow q$?
   a. p, only if q.
   b. p if and only if q.
   c. q, only if p.
   d. q, only if p and if q, then p.
   e. p, only if q and q, only if p.
   f. $\sim q$, if and only $\sim p$.
   g. $\sim p$, only if $\sim q$ and q, only if p.
   h. $p \rightarrow q$ and $\sim p \rightarrow \sim q$.
   i. $p \rightarrow q$ and $\sim p \vee q$.
   j. $p \rightarrow q$ and $\sim q \vee p$.

2.7 RULES FOR THE ALGEBRA OF PROPOSITIONS

In any mathematical system there are various rules or properties associated with the operations introduced. For the algebra of propositions there are some basic properties we will now list. Each of these properties may be proved by constructing appropriate truth tables for the equivalent propositions stated.

Let p, q, and r represent arbitrary propositions, t represent an arbitrary tautology, and c represent an arbitrary contradiction. Then

1. $p \wedge q \equiv q \wedge p$	Commutative property for conjunction
2. $p \vee q \equiv q \vee p$	Commutative property for disjunction
3. $(p \wedge q) \wedge r \equiv p \wedge (q \wedge r)$	Associative property for conjunction
4. $(p \vee q) \vee r \equiv p \vee (q \vee r)$	Associative property for disjunction
5. $p \vee (q \wedge r) \equiv (p \vee q) \wedge (p \vee r)$	Distributive property for disjunction over conjunction
6. $p \wedge (q \vee r) \equiv (p \wedge q) \vee (p \wedge r)$	Distributive property for conjunction over disjunction
7. $p \vee p \equiv p$	Idempotent property for disjunction
8. $p \wedge p \equiv p$	Idempotent property for conjunction
9. a. $p \vee c \equiv p$, and b. $p \vee t \equiv t$	Identity properties for disjunction
10. a. $p \wedge c \equiv c$, and b. $p \wedge t \equiv p$	Identity properties for conjunction
11. a. $p \wedge \sim p \equiv c$, and b. $p \vee \sim p \equiv t$ c. $\sim t \equiv c$ d. $\sim c \equiv t$ e. $\sim \sim p \equiv p$	Complement properties
12. a. $\sim (p \vee q) \equiv \sim p \wedge \sim q$ b. $\sim (p \wedge q) \equiv \sim p \vee \sim q$	De Morgan's laws

These properties should enable the student to simplify composite propositions.

EXAMPLE Simplify the composite proposition $\sim p \vee \sim (\sim p \wedge q)$ by using the properties of this section.

Solution

PROPOSITION	REASON
$\sim p \vee \sim (\sim p \wedge q) \equiv \sim p \vee (p \vee \sim q)$	Property 12(b)
$\equiv (\sim p \vee p) \vee \sim q$	Property 4
$\equiv t \vee \sim q$	Property 11(b)
$\equiv t$	Property 9(b) ■

EXAMPLE Simplify the composite proposition $\sim(p \vee q) \vee (\sim p \wedge q)$.
Solution

PROPOSITION	REASON
$\sim(p \vee q) \vee (\sim p \wedge q) \equiv (\sim p \wedge \sim q) \vee (\sim p \wedge q)$	Property 12(a)
$\equiv \sim p \wedge (\sim q \vee q)$	Property 6
$\equiv \sim p \wedge t$	Property 11(b)
$\equiv \sim p$	Property 10(b) ∎

EXERCISES

1. Simplify each of the following propositions by using the rules for the algebra of propositions given in this section.

a. $p \vee p$.

b. $p \vee \sim p$.

c. $p \wedge p$.

d. $p \wedge \sim p$.

e. $p \vee c$.

f. $p \vee t$.

g. $p \wedge c$.

h. $p \wedge t$.

i. $(p \vee q) \wedge (p \vee r)$.

j. $(p \wedge q) \vee (r \wedge q)$.

k. $\sim q \vee (\sim q \wedge r)$.

l. $(p \wedge q) \vee \sim q$.

m. $\sim(p \wedge q) \vee (p \wedge \sim q)$.

n. $[(p \wedge q) \vee (p \wedge \sim r)] \wedge r$.

o. $(p \wedge q) \vee [(p \wedge r) \vee (p \wedge \sim r)]$.

2. Verify each of the following rules of this section by constructing appropriate truth tables.

a. Rule 5.

b. Rule 6.

c. Rule 9(a).

d. Rule 10(b).

e. Rule 12(a).

f. Rule 12(b).

2.8 VALID ARGUMENTS AND FALLACIES

You probably have heard the statement "Let's settle this argument once and for all times" or read in the newspapers that the police were summoned to a particular address to quell a disturbance resulting from an argument. What is an argument, and when is an argument settled?

As used in logic and mathematics, an *argument* is a sequence of a finite number of propositions, denoted by $s_1, s_2, s_3, \ldots, s_n$, which lead to another proposition, denoted by s_{n+1}. The propositions $s_1, s_2, s_3, \ldots, s_n$ are called the *premises*, and the proposition s_{n+1} is called the *conclusion*.

DEFINITION 2.17 An argument is said to be *valid* if and only if the conjunction of the premises implies the conclusion.

From the above definition we see that if all the premises in an argument are true or are accepted as being true, then the conclusion must also be accepted if the argument is valid.

DEFINITION 2.18 A *fallacy* is an argument that is not a valid argument.

To determine whether an argument is valid or a fallacy, we can construct a truth table to show whether the conjunction of the premises implies the conclusion. Stated in other words, we would have to show that the associated conditional is a tautology.

EXAMPLE Determine whether the following argument is valid.

> If Joe saves his money, then he is rich.
> Joe is poor.
> ∴ Joe does not save his money.

Solution The propositions listed above the line segment are the premises and are accepted as being true. The proposition listed below the line segment is the conclusion. Let p be the proposition "Joe saves his money," and let q be the proposition "Joe is rich." The above argument can then be symbolized as

$$p \rightarrow q\,.$$
$$\frac{\sim q}{\therefore \ \sim p}$$

The symbol ∴ is read "therefore." We now construct the truth table for $[(p \rightarrow q) \wedge \sim q] \rightarrow \sim p$ as in Table 2.16.

Table 2.16

p	q	$p \rightarrow q$	$(p \rightarrow q) \wedge \sim q$	$[(p \rightarrow q) \wedge \sim q] \rightarrow \sim p$
T	T	T	F	T
T	F	F	F	T
F	T	T	F	T
F	F	T	T	T

Since the last column of the above truth table contains only T's, the composite proposition $[(p \rightarrow q) \wedge \sim q] \rightarrow \sim p$ is a tautology and the argument is valid. ∎

EXAMPLE Check the validity of the following argument.

> If Bette has blue eyes, then she is intelligent.
> Bette is intelligent.
> ∴ Bette has blue eyes.

Solution Let p be the proposition "Bette has blue eyes," and q be the proposition "Bette is intelligent." Then we symbolize the above argument as follows:

$$p \rightarrow q$$
$$\frac{q}{\therefore \ p}$$

Next we form the truth table for $[(p \to q) \land q] \to p$ as in Table 2.17

Table 2.17

p	q	$p \to q$	$(p \to q) \land q$	$[(p \to q) \land q] \to p$
T	T	T	T	T
T	F	F	F	T
F	T	T	T	F
F	F	T	F	T

Since the last column of the above truth table contains an F, the proposition $[(p \to q) \land q] \to p$ is not a tautology and the argument is not valid; it is a fallacy. ■

When examining a valid argument, it may appear to be absurd, especially when you look at the conclusion only. It must be noted that the conclusion in a valid argument is not to be interpreted as being true. Consider, for instance, the next example.

EXAMPLE Check the validity of the following argument.

$$\begin{array}{l} \text{If } 2 + 2 = 3, \text{ then } 3 \times 4 = 1. \\ \underline{2 + 2 = 3} \\ \therefore \ 3 \times 4 = 1 \end{array}$$

Solution Clearly we would not accept the proposition "$3 \times 4 = 1$" as being true (modern mathematics notwithstanding). But let's proceed. Let p be the proposition "$2 + 2 = 3$" and q be the proposition "$3 \times 4 = 1$." Then we have

$$\begin{array}{l} p \to q \\ \underline{p} \\ \therefore \ q \end{array}$$

Constructing the truth table for $[(p \to q) \land p] \to q$, we have Table 2.18.

Table 2.18

p	q	$p \to q$	$(p \to q) \land p$	$[(p \to q) \land p] \to q$
T	T	T	T	T
T	F	F	F	T
F	T	T	F	T
F	F	T	F	T

Since the last column of the truth table contains only T's, the composite proposition $[(p \to q) \land p] \to q$ is a tautology and the given argument is valid.

■

EXERCISES

For Exercises 1–10 determine whether each of the arguments is valid or
is a fallacy.

1. $p \to q$
   $\sim p$
   ∴ $\sim q$

2. $p \to q$
   $\sim q$
   ∴ $\sim p$

3. $p \lor q$
   p
   ∴ q

4. $p \land q$
   $\sim p$
   ∴ q

5. $p \to q$
   $q \to r$
   ∴ $r \to p$

6. $p \to q$
   $q \to p$
   ∴ $p \land q$

7. p
   q
   ∴ $p \to q$

8. $p \to q$
   $\sim p \lor q$
   ∴ $p \to \sim q$

9. $p \to q$
   $q \to r$
   $p \land q$
   ∴ r

10. $p \lor \sim q$
    q
    $q \to r$
    ∴ $p \to r$

For each of the following exercises, determine whether the argument is valid
or is a fallacy.

11. If Michael is a student, he is bright.
    Michael is not a student.
    ∴ Michael is not bright.

12. If I study, then I will pass this course.
    If I pass this course, then I will get married.
    I study and I pass this course.
    ∴ I will get married.

13. If Daryl quits school, then he will get a job.
    Daryl does not get a job.
    ∴ Daryl does not quit school.

14. If tall boys date girls, they play basketball.
    Short boys do not play basketball.
    Joe is a short boy.
    ∴ Joe does not date girls.

15. If gasoline prices rise, then I will ride the bus.
    Gasoline prices do not rise if and only if I ride the bus.
    ∴ I ride the bus.

16. Inflation is curbed or the gross national product does not rise.
    Inflation is curbed or price controls are imposed.
    If price controls are imposed, then the gross national product rises.
    ∴ Price controls are imposed or the gross national product rises.

17. New cars are selling, and John is out of work.
    If new cars are selling, then John has a job.
    ∴ New cars are selling and John has a job.

18. Ice cream is cold, or classes have not been canceled.
    Church attendance is up, or ice cream is warm.
    ∴ If classes have been canceled, then church attendance is up.

19. Students study, and professors publish.
    Professors publish.
    ∴ Students study.

20. If the president of the college is a leader, the faculty is happy.
    Professor Geo. M. Etry is a member of the faculty and is unhappy.
    ∴ The president of the college is not a leader.

2.9 SOME BASIC LAWS OF LOGIC

In the previous section we discussed valid arguments and fallacies. It was observed that all arguments can be symbolized as conditionals. Then we test the conditional to determine whether it is a tautology. If the conditional is a tautology, we have a valid argument; otherwise we have a fallacy. In this section we will discuss some of the basic laws used in logic, all of which are examples of valid arguments.

LAW OF THE EXCLUDED MIDDLE For every proposition p, the proposition p or not p, symbolized $p \vee \sim p$, is true.

The truth table for the law of the excluded middle is given in Table 2.19. Observe that the last column contains only T truth values.

Table 2.19 Truth table for law of excluded middle

p	$\sim p$	$p \vee \sim p$
T	F	T
F	T	T

EXAMPLE   Consider the statement "Today is Tuesday, or today is not Tuesday." If we let p represent the proposition "Today is Tuesday," then $\sim p$ represents the proposition "Today is not Tuesday." The truth table for this composite proposition is shown in Table 2.19. Hence the composite proposition "Today is Tuesday, or today is not Tuesday" is a true statement.

LAW OF CONTRADICTION For every proposition p, the proposition p and not p, symbolized $p \wedge \sim p$, is false. Hence $\sim(p \wedge \sim p)$ is true.

Considering the truth table (Table 2.20) for the law of contradiction, we see that the composite proposition is also a tautology.

Table 2.20 Truth table for law of contradiction

p	$\sim p$	$p \wedge \sim p$	$\sim(p \wedge \sim p)$
T	F	F	T
F	T	F	T

EXAMPLE   Consider the statement "It is not true that $2 + 2 = 4$ and $2 + 2 \neq 4$." By the law of contradiction this composite proposition is true. If we let r be the proposition "$2 + 2 = 4$," then $\sim r$ will represent the proposition "$2 + 2 \neq 4$." The truth table for this composite proposition is shown in Table 2.20. Hence the composite proposition "It is not true that $2 + 2 = 4$ and $2 + 2 \neq 4$" is a tautology.

LAW OF DETACHMENT If both the conditional $p \to q$ and the hypothesis p are true, then the conclusion q is also true. Symbolically we have

$$p \to q$$
$$\underline{p}$$
$$\therefore q$$

According to the law of detachment, if we accept the propositions which are above the line segment as being true, then we must accept the proposition which is below the line segment as also true. Rewriting the law of detachment by combining the simple propositions with appropriate connectives, we have $[(p \to q) \wedge p] \to q$.

Observe that the composite proposition is a conditional whose hypothesis is the conjunction of $p \to q$ and p and whose conclusion is q. The truth table (Table 2.21) for this law shows that it is another tautology.

Table 2.21 Truth table for law of detachment

p	q	$p \to q$	$(p \to q) \wedge p$	$[(p \to q) \wedge p] \to q$
T	T	T	T	T
T	F	F	F	T
F	T	T	F	T
F	F	T	F	T

EXAMPLE A theorem in Euclidean geometry states, "If two sides of a triangle are congruent, then the angles opposite these sides are also congruent." In attempting to prove that two angles of a triangle are congruent, we usually accept that the two sides opposite these angles are congruent. The basic structure of the proof would include the following analysis.

Analysis Let p be the proposition "Two sides of a triangle are congruent" and q be the proposition "The angles opposite these sides are congruent." From this we form the conditional $p \to q$, which is the composite proposition "If two sides of a triangle are congruent, then the angles opposite these sides are congruent." We accept the composite proposition $p \to q$ as true since it is a theorem from Euclidean geometry. If we can establish that p is true, then the conjunction of $p \to q$ and p is also true. Observe that p is the hypothesis of the conditional; by examining the truth table for the conditional (Table 2.4), we see that a true conditional that has a true hypothesis must also have a true conclusion. Therefore, the proposition q is true. In other words, "The angles opposite these sides are congruent." ∎

LAW OF CONJUNCTIVE INFERENCE If the proposition p and the proposition q are accepted as true, then the conjunction $p \wedge q$ is also true. Symbolically we have

$$p$$
$$\underline{q}$$
$$p \wedge q$$

EXAMPLE Let x be the proposition "The object a belongs to the set A" and y be the proposition "The object b belongs to the set B." If we accept both the proposition x and the proposition y as true, then by examination of the truth table for conjunction (Table 2.2) we must accept the conjunction $x \wedge y$ as true. Hence our conclusion would be "The object a belongs to the set A and the object b belongs to the set B."

LAW OF CONJUNCTIVE SIMPLIFICATION If the conjunction $p \wedge q$ is accepted as true, then we must accept the proposition p and the proposition q as true. Symbolically, we have

$$\frac{p \wedge q}{p} \quad \text{and} \quad \frac{p \wedge q}{q}$$

or using connectives we have $(p \wedge q) \to p$ and $(p \wedge q) \to q$.

EXAMPLE Let α be the statement "B-52 airplanes are U.S. Air Force heavy bombers" and β be the statement "Nike is a U.S. Army missile."

If we accept the conjunction $\alpha \wedge \beta$ of the two propositions as true, we can conclude that the following propositions are true:

α: B-52 airplanes are U.S. Air Force heavy bombers.

β: Nike is a U.S. Army missile.

These conclusions are established immediately by an examination of the truth table for conjunction (Table 2.2), since the conjunction of two propositions is true only if each proposition is true.

The next law depends upon the definition of the contrapositive of a conditional.

LAW OF CONTRAPOSITIVE INFERENCE If the conditional $p \rightarrow q$ is accepted as true and the proposition q is known to be false, then the proposition p is also false. Symbolically we have

$$p \rightarrow q$$

$$\frac{\sim q}{\sim p}$$

or using connectives we have $[(p \rightarrow q) \wedge \sim q] \rightarrow \sim p$.

The student should construct the truth table for this law and verify that it is another example of a tautology.

EXAMPLE Let p be the proposition "Today is Tuesday" and q be the proposition "Tomorrow is Wednesday." The conditional $p \rightarrow q$ is the composite proposition "If today is Tuesday, then tomorrow is Wednesday," which we accept as true. Further, suppose that we also accept the proposition "Tomorrow is not Wednesday," which is the negation of q or $\sim q$ as true. This gives a true conditional $p \rightarrow q$ and a false conclusion q. Examining the truth table for a conditional (Table 2.4), we must accept the hypothesis p as false or, equivalently, $\sim p$ as true. Hence our conclusion is "Today is not Tuesday."

We have given a few basic laws of logic that serve as examples of tautologies and valid arguments. These laws plus others are used in our deductive system of mathematics to arrive at valid conclusions in what we call proofs.

EXERCISES

1. Construct the truth table for the law of conjunctive inference, and verify that it is an example of a tautology.

2. Construct the truth table for the law of conjunctive simplification, and verify that it is an example of a tautology.
3. Construct the truth table for the law of contrapositive inference, and verify that it is an example of a tautology.
4. For each of the following accept the given propositions as true. Use basic laws of logic to draw valid conclusions, and indicate the laws of logic used in each case.
   a. If $2 + 2 = 3$, then $4 \times 5 = 6$; $2 + 2 = 3$.
   b. If today is Wednesday, then tomorrow is Thursday; tomorrow is not Thursday.
   c. John Smith is 5'8" tall; Mary Sunsweet has blue eyes.
   d. John Smith is 5'8" tall, and Mary Sunsweet has blue eyes.
   e. $3 \times 10 = 30$; $4 + 5 = 9$.
5. Draw appropriate conclusions from each of the following sets of hypotheses.
   a. $p \rightarrow q$; p.
   b. $q \rightarrow p$; $\sim p$.
   c. $p \wedge q$.
   d. $p \vee q$.
   e. $\sim(p \wedge q)$; q.
   f. $p \rightarrow q$; $q \rightarrow r$.
   g. $p \vee \sim q$; q.
   h. $p \rightarrow q$; $\sim q \vee r$; $\sim r$.
   i. $p \rightarrow q$; $\sim p$.
   j. $p \rightarrow q$; $q \rightarrow r$; p.

2.10 QUANTIFIERS

If p is a proposition, then the negation of p is denoted by the symbol $\sim p$. If p is the proposition "$x < 2$," then the negation of p can be written either as "It is not true that $x < 2$" or, simply, "$x \nless 2$" (which is read "x is not less than 2"). However, in mathematics we often wish to negate statements of the following form:

For all college students, studying mathematics is fun.

There exists a real number y such that $y + 2 = 5$.

In the first statement the phrase "for all" refers to every college student. To negate this statement, one asserts that there is at least one college student for whom studying mathematics is not fun. Observe that the negation of the first statement is *not* the statement "For all college students studying mathematics is not fun." In the second statement the phrase "there exists" means that at least one real number y exists such that $y + 2 = 5$. To negate this statement, one asserts that there is no real number y such that $y + 2 = 5$.

The phrases "for all" and "there exists" are called quantifiers and are generally abbreviated by the following symbols:

$$\forall \text{ (for all)} \qquad \exists \text{ (there exists)}$$

The symbol $\forall$ is called the *universal quantifier*; the symbol $\exists$ is called the *existential quantifier*.

The word *some*, which means *at least one and perhaps all*, causes problems in mathematics. It is difficult to negate propositions involving this word. Consider the proposition

> *p*: Some women have blue eyes

This statement asserts that at least one and perhaps all women have blue eyes. To negate the statement means to assert that no women have blue eyes. Hence the negation of *p* is

> ~*p*: All women do not have blue eyes

Observe that the statement "All women do not have blue eyes" is different in structure and meaning from the statement "Not all women have blue eyes."

EXAMPLE The following is a list of a few propositions involving the words "all," "some," and "there exists" together with their negations:

> *p*: All mathematics texts are easy to read.
> ~*p*: Some mathematics texts are not easy to read.

> *q*: Some buildings are tall.
> ~*q*: All buildings are not tall; no buildings are tall.

> *r*: All university presidents are scholars.
> ~*r*: There exists a university president who is not a scholar.

> *s*: There exists an angle which is obtuse.
> ~*s*: All angles are not obtuse.

> *t*: All *Apollo 13* astronauts have black hair.
> ~*t*: At least one *Apollo 13* astronaut does not have black hair.

Observe that the negation of *t* is *not* "All *Apollo 13* astronauts do not have black hair," which means that no *Apollo 13* astronaut has black hair. Some of them may have black hair.

EXERCISES

1. Which of the following statements is (or are) equivalent to "It is not true that all houses are made of brick"?
   a. No houses are made of brick.
   b. Some houses are not made of brick.
   c. All houses are not made of brick.
   d. Some houses are made of brick.
   e. Some houses are made of brick, and some houses are not made of brick.

2. Negate each of the following propositions.
   a. All men are tall.
   b. Some foods are high in caloric value.
   c. No bills are paid.
   d. Every triangle is isosceles.
   e. All numbers are natural numbers.
   f. All animals are cats.
   g. All numbers are even.
   h. There exists a person who is Chinese.
   i. There exists a real number m such that $m + 2 = 4$.
   j. There exists a real number p such that p is odd and less than 7.
3. Let U be the set of all natural numbers which are less than 10. Determine which of the statements below are true and which are false.
   a. All elements of U are greater than 6.
   b. There exists an element in U which is even.
   c. There exists an element d in U such that $d + 6 = 13$.
   d. For all elements c in U, if $c > 8$, then c is odd.
   e. For all e in the set U, there exists f in U such that $e < f$.

2.11 MORE COMPLICATED NETWORKS

In Section 2.3 we considered some simple circuits as an application of the connectives used with propositions. In this section we will consider an extension of this application.

Consider the circuit shown in Figure 2.8 which consists of the switches p, q, and r. Under what conditions will current flow from A to B? First we observe that q and r are connected in parallel and, hence, at least one of them must be closed. Next we observe that the parallel circuit portion of the network is connected to p in series. Therefore, in order for current to flow from A to B, p must be closed and either q or r must be closed. We will verify this by constructing the appropriate truth table (Table 2.22), using "and" for series and "or" for parallel.

From the table we see that current will flow from A to B for cases 1, 2, and 3 only. In case 1, all three switches are closed. In case 2, p is closed and q is closed. Finally, in case 3, p is closed and r is closed. This confirms our observations made earlier.

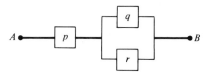

Figure 2.8

Table 2.22

p	q	r	q or r	p and $(q$ or $r)$
1	1	1	1	1
1	1	0	1	1
1	0	1	1	1
1	0	0	0	0
0	1	1	1	0
0	1	0	1	0
0	0	1	1	0
0	0	0	0	0

Now let's reexamine the circuit given above which is in the form "p and $(q$ or $r)$." Suppose that we consider the form "$(p$ and $q)$ or r." Do we get the same network?

"p and q" means that p is connected to q in series; the "or" refers to parallel. Hence our new network would be diagramed as in Figure 2.9 and the corresponding truth table would be as shown in Table 2.23.

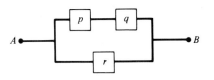

Figure 2.9

Table 2.23

p	q	r	p and q	$(p$ and $q)$ or r
1	1	1	1	1
1	1	0	1	1
1	0	1	0	1
1	0	0	0	0
0	1	1	0	1
0	1	0	0	0
0	0	1	0	1
0	0	0	0	0

Examination of the truth table reveals that current will flow from A to B if p and q are closed, if all three switches are closed, and when only r is closed. Comparing the truth table values in the last column of this truth table with the last column of the previous truth table, we see that they are not the same. Hence the two networks are not equivalent.

EXAMPLE Construct a network which corresponds to the composite proposition

$$(p \wedge q) \vee (p \wedge r)$$

Solution For the propositions p, q, and r we will use the switches p, q, and r. For conjunction we will use a series circuit, and for disjunction we will use a parallel circuit. Our network, then, is as diagramed in Figure 2.10. ■

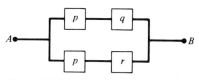

Figure 2.10

EXAMPLE Consider the network given in Figure 2.11. Design an equivalent network that is simpler, if possible.

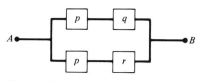

Figure 2.11

Solution From the previous example we know that the given network corresponds to the composite proposition $(p \wedge q) \vee (p \wedge r)$. However, $(p \wedge q) \vee (p \wedge r) \equiv p \wedge (q \vee r)$ by Property 6 of Section 2.7. The corresponding network for $p \wedge (q \vee r)$ is given by Figure 2.12 which is therefore equivalent to the given network. ■

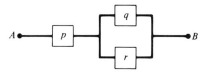

Figure 2.12

In designing circuits it is possible to connect two switches so that they are both closed or both open simultaneously. When this happens we say the switches are equivalent and, in a network, they are represented by the same letter as we illustrated in the example above. It is also possible to connect two switches so that when one switch is open the other is closed, and conversely. In this case, if one switch is represented by p, then the other will be represented by $\sim p$.

EXAMPLE Determine the composite proposition which corresponds to the network shown in Figure 2.13.

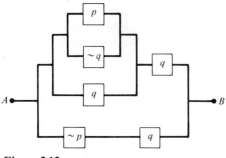

Figure 2.13

Solution The basic circuit is in parallel. The lower branch is $\sim p$ in series with q which yields the proposition $\sim p \wedge q$. The upper branch is in series with one part in parallel. From this we get $[(p \vee \sim q) \vee q] \wedge q$. The complete network, then, corresponds to the composite proposition $\{[(p \vee \sim q) \vee q] \wedge q\} \vee (\sim p \wedge q)$. ■

EXAMPLE Design if possible an equivalent circuit in simple form for that given in the above example.

Solution We will consider the composite proposition obtained in the solution to the previous example and will attempt to simplify it. We have

$$\{[(p \vee \sim q) \vee q] \wedge q\} \vee (\sim p \wedge q) \equiv \{[p \vee (\sim q \vee q)] \wedge q\} \vee (\sim p \wedge q)$$
$$\text{(by Property 4, Section 2.7)}$$

$$\equiv [(p \vee t) \wedge q] \vee (\sim p \wedge q)$$
$$\text{(by Property 11(b), Section 2.7)}$$

$$\equiv (t \wedge q) \vee (\sim p \wedge q)$$
$$\text{(by Property 9(b), Section 2.7)}$$

$$\equiv q \vee (\sim p \wedge q)$$
$$\text{(by Property 10(b), Section 2.7)}$$

The corresponding network, then, would be as shown in Figure 2.14, which is a simpler but equivalent version of that given. ■

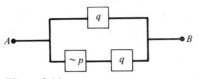

Figure 2.14

EXERCISES

1. For each of the following networks, determine the composite proposition
which corresponds to it.

a.

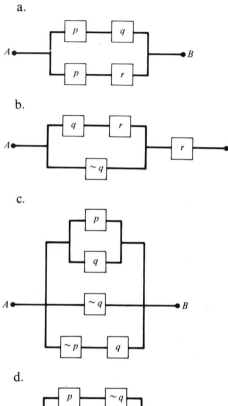

b.

c.

d.

2. For each of the following composite propositions, draw a network which
corresponds to it.

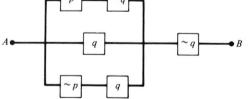

a. $p \lor p$.   b. $p \lor (p \land q)$.

c. $p \land (p \lor q)$.   d. $(p \land q) \lor (q \land r)$.

e. $[(p \lor q) \lor {\sim}q \lor ({\sim}p \land {\sim}q)] \land q$.

f. $[(p \land q) \land {\sim}q] \lor [({\sim}p \land {\sim}q) \land q]$.

g. $q \lor ({\sim}p \land {\sim}q) \lor (p \land q)$.

3. For each part of Exercise 2 determine a simpler network equivalent to the one corresponding to the composite proposition given.
4. For a certain committee of three members the votes are recorded electronically by pressing a button to close a switch. Design a network such that current will flow if and only if a majority of the committee vote in favor of an issue.

SUMMARY

In this chapter we discussed the system or algebra of propositions. We defined and discussed propositions and propositional forms and some of the more common connectives used in logic together with the agreement on their usage. The connectives discussed were negation, conjunction, disjunction, conditional, and biconditional. Truth tables for these connectives also were examined. Also discussed were the converse, inverse, and the contrapositive of a conditional as well as the alternate forms of the conditional and the biconditional.

Composite propositions, their truth tables, tautologies, and contradictions also were introduced and discussed. The relations of implication and equivalence between propositions also were treated. Various properties associated with the connectives, validity of arguments, and laws of logic were introduced and discussed as were quantifiers.

As an application to the material introduced we considered some simple and more complicated switching circuits networks and showed the relationships existing between the algebra of propositions and the system of networks.

In our discussions the following symbols were introduced:

p, q	The propositions p and q
$\sim p$	The negation of the proposition p
$p \wedge q$	The conjunction of the propositions p and q
$p \vee q$	The disjunction of the propositions p and q
$p \rightarrow q$	The conditional if p, then q
$p \leftrightarrow q$	The biconditional which is the conjunction of the conditional $p \rightarrow q$ and its converse $q \rightarrow p$
iff	Abbreviation for if and only if
$p \Rightarrow q$	The relation p implies q
$p \Leftrightarrow q$	The relation p is equivalent to q
$p \equiv q$	Another symbol for $p \Leftrightarrow q$
t	An arbitrary tautology
c	An arbitrary contradiction
$\therefore$	Abbreviation for *therefore*
$\forall$	The universal quantifier *all*
$\exists$	The existential quantifier *there exists*

REVIEW EXERCISES FOR CHAPTER 2

1. Define each of the following terms.
   a. A proposition.
   b. A propositional form.
   c. A simple proposition.
   d. A composite proposition.
   e. The negation of a proposition.
   f. The conjunction of propositions.
   g. The disjunction of propositions.
   h. A conditional.
   i. The converse of a conditional.
   j. The inverse of a conditional.
   k. The contrapositive of a conditional.
   l. The biconditional.
   m. An implication.
   n. An equivalence.
   o. A tautology.
   p. A contradiction.
   q. A valid argument.
   r. A fallacy.
2. Consider the propositions p: Rover is a dog; q: 3 is a natural number; r: $2 + 3 = 0$. Write out each of the propositions indicated in the following.
   a. The negation of p.
   b. p or r.
   c. $q \to r$.
   d. $(p$ and $q)$ or $($not q and $r)$.
   e. $r \to \sim p$.
   f. $(q \to r)$ and $(r \to p)$.
   g. The inverse of $r \to \sim p$.
   h. The contrapositive of $q \to r$.
   i. $(p$ or $q) \to r$.
   j. The converse of $(p$ or $q) \to r$.
3. Which of the composite propositions in Exercise 2, if any, are tautologies? Answer the question by constructing appropriate truth tables.
4. Which of the composite propositions in Exercise 2, if any, are contradictions? Answer the question by constructing appropriate truth tables. (See Example 3.)
5. Negate each of the following propositions.
   a. All women are blondes.
   b. Every quadrilateral is a rectangle.
   c. Some quadrilaterals are rectangles.
   d. There exists y in the set X such that $y = 3$.
   e. All professors are not intelligent.
   f. Apples are candy, and roses are yellow.
   g. Men are mortal, or dogs have three legs.
   h. If today in Monday, then tomorrow is Tuesday. (Hint: First write the conditional as an equivalent disjunction, and then negate.)
6. Which of the following, if any, are equivalent to the conditional "If p, then q"?
   a. $p \Rightarrow q$.
   b. $q \to p$.
   c. $p \to q$.
   d. $\sim p \vee q$.
   e. p, only if q.
   f. q, only if p.
   g. p is a necessary condition for q.
   h. p is a sufficient condition for q.

    i. q is a necessary condition for p.

    j. q is a sufficient condition for p.

    k. p is a necessary and sufficient condition for q.

    l. q is a necessary and sufficient condition for p.

    m. $q \Rightarrow p$.   n. $\sim p \to \sim q$.

    o. $\sim q \to \sim p$.   p. $q \wedge \sim p$.

7. Which of the following, if any, are equivalent to the biconditional "p if and only if q"?

    a. $p \to q$.   b. $q \to p$.

    c. $p \leftrightarrow q$.   d. $q \leftrightarrow p$.

    e. $p \Rightarrow q$.   f. $p \equiv q$.

    g. $q \Leftrightarrow p$.

    h. p is a necessary condition for q.

    i. p is a sufficient condition for q.

    j. q is a necessary condition for p.

    k. q is a sufficient condition for p.

    l. p is a necessary and sufficient condition for q.

    m. q is a necessary and sufficient condition for p.

    n. $p \wedge q$.   o. $(p \to q) \wedge (q \to p)$.

    p. $(\sim p \vee q) \wedge (q \to p)$.   q. $(\sim p \vee q) \wedge (p \vee \sim q)$.

CHAPTER 3
THE ALGEBRA OF SETS

Chapter 2 introduced the algebra of propositions and discussed some of the logic used in mathematics. This chapter will introduce the algebra of sets and will show the relationship between sets and their operations with propositions and their connectives. It also will examine their structures by comparing the properties of the connectives with the corresponding properties of the operations on sets.

3.1 SETS

During the latter part of the last century Georg Cantor decided to use the concept of a set to describe a whole collection, class, or group of objects (just as one refers to a set of books, a collection of stamps, or a group of students). The word *set* refers to any well-defined collection of objects, such as a set of books on a shelf, a set of all people who voted for Hubert Humphrey during the 1968 presidential election, or a set of real numbers.

DEFINITION 3.1 A set is said to be *well-defined* if there exists a method or rule for determining whether an element belongs to the given set. The objects of a set, all of which may or may not be of the same type, are called its *members* or *elements*. Sets are usually denoted by the use of capital letters,

45

while elements of a set are usually denoted by lowercase letters written between braces, { }. Thus we write $A = \{3, 4, 5, 6\}$ or $B = \{a, e, i, o, u\}$.

If an object a is an element of the set A, we write $a \in A$, which is read "a is an element of the set A." If an object a is not an element of the set A, we write $a \notin A$, which is read "a is not an element of the set A." For example, $2 \in \{1, 2, 3\}$ and $m \notin \{a, e, i, o, u\}$.

We can define sets in various ways. Suppose we have the set of all positive integers which are less than 10. We can represent this set as follows:

Let A be the set of all positive integers which are less than 10. The set is clearly defined by the symbol A, and this symbol may now be used to refer to the set.

Let $A = \{1, 2, 3, 4, 5, 6, 7, 8, 9\}$, where { } is a symbol to denote a set. This is known as the *roster method* of listing a set.

Let $A = \{1, 2, 3, \ldots, 9\}$. Here we have introduced the first few elements, followed by "..." indicating that there are other elements, and finally writing the last element. This is especially useful as a representation of a set which has many elements.

Let $A = \{a \mid a$ is a positive integer and $a < 10\}$, which is read "A is the set of all elements of the form or type a such that a is a positive integer and a is less than 10." This is known as the *set-builder method* of denoting a set.

In the set-builder notation for the set A the symbol a is a place-holder which can be replaced by any member of the set. Hence a can be equal to any one of the first nine positive integers.

DEFINITION 3.2 A *variable* is a symbol or a place-holder that can be replaced by any member of a given set. The given set is called the *universal set* or the *universe* of the variable. Each member of the set is called a *value* of the variable.

EXAMPLE   Suppose that u is a variable whose universe is the set $\{u \mid u$ is a natural number, $u > 7$, and $u < 9\}$. We see then that u is a variable that represents a natural number which is greater than 7 and also less than 9. The only natural number satisfying these conditions is 8 and, hence, the only value that u can take is 8.

DEFINITION 3.3 If the universe of the variable is· a set with only one element, then the variable can have only one value. In this case we say that the variable is a *constant*.

Sets may be classified according to the number of their elements. The set $\{1, 2\}$ contains two elements, while the number of elements in the set $\{1, 2, 3, \ldots\}$ cannot be counted.

DEFINITION 3.4 A *finite set* is a set having a number of elements than can be expressed by a whole number. An *infinite set* is a set that is not finite.

The set {1, 2, 3, ..., 39} is finite because it contains 39 elements. However, the set {2, 4, 6, ...}, which is the set of all the even-counting numbers, is infinite because it does not contain a finite number of elements. A finite set may contain only one element such as the set {7}. Moreover, a well-defined set may contain no elements. Consider the set of all real numbers greater than 10 and less than ⁻50. There are no elements in this set.

DEFINITION 3.5 The *empty set* or the *null set*, denoted by $\varnothing$, is defined as the set that has no elements.

EXAMPLE Describe the set B of all the natural numbers that are less than or equal to 12 and are also greater than 4 using the roster method and the set-builder notation.

Solution The roster method lists or enumerates all the objects in the set. Hence $B = \{5, 6, 7, 8, 9, 10, 11, 12\}$. According to the set-builder notation, $B = \{a | a$ is a natural number and a is greater than 4 and less than or equal to 12}. If we let N denote the set of all natural numbers, we could also write $B = \{a | a \in N, 4 < a \leq 12\}$. ∎

EXAMPLE Identify each of the following sets as finite or infinite sets.
$R = \{r | r$ is a real number}
$S = \{s | s$ is an integer, $^-3 < s \leq 4\}$
$T = \{t | t \in R$ and $t \in S\}$ where R and S are the sets described above

Solution The set R is an infinite set since R is the set of *all* real numbers and there are infinitely many real numbers. The set S is finite since the objects of the set S are integers and only those integers greater than ⁻3 and less than or equal to 4. Using the roster method, we note that $S = \{^-2, ^-1, 0, 1, 2, 3, 4\}$. There are exactly seven elements in the set S. In describing T, we want *all* the objects of the form t, such that t is a real number (since $t \in R$) and t is an integer greater than ⁻3 and less than or equal to 4 (since $t \in S$). Observe that we are using the connective of conjunction to describe t. In order to belong to the set, t must be a real number *and* an integer greater than ⁻3 and less than or equal to 4. We know that all integers are real numbers, but not every real number is an integer. We conclude that $T = \{^-2, ^-1, 0, 1, 2, 3, 4\}$ and therefore the set T is finite. ∎

EXAMPLE Identify each of the following statements as true or false.
a. $2 \in \{1, 2, 3, 4\}$.
b. $7 \notin \{b | b \in N, b < 7\}$.

c. The set $C = \{0\}$ is the null set.

d. The set $D = \{\emptyset\}$ is the null set.

Solution

a. True, since 2 is an element of the set containing 1, 2, 3, and 4

b. True, 7 is not an element of the set $\{b \mid b \in N,\ b < 7\}$ since b must be a natural number that is *less than* 7

c. False, the set C contains the element 0 and, therefore, is not the empty or null set

d. False, since the set D contains the null set which is an object; the fact that the null set contains no objects does not mean that D contains no objects; in fact, we say that "D is the set that contains a set that has no objects" ∎

EXERCISES

1. The word "set" has been used to refer to any well-defined collection of objects.
   a. Give two examples of sets that are finite.
   b. Give two examples of sets that are infinite.
   c. Give two examples of empty sets.

2. An example of a set that is not well defined and whose membership would be difficult to determine is the set of all handsome college professors. Give two additional examples of sets which are not well defined.

3. Which of the following sets are well defined?
   a. The set of the first seven natural numbers.
   b. The set of all the great movies of this century.
   c. The set of all whole numbers greater than 3 and less than 20.
   d. The set of all the letters of the English alphabet that are vowels.
   e. The set of all the great politicians who held elective office during the period 1900–1963 inclusive.

4. For each set listed in Exercise 3 that was rejected as being well defined give the reason for your decision.

5. Define each of the following.
   a. The null or empty set.
   b. The universal set.

6. Is $\{\emptyset\}$ the same as $\emptyset$? Why? If your answer is no, explain the difference.

7. In general is $\{A\} = A$, where A is an arbitrary set? Why? If your answer is no, explain the difference between them.

8. Rewrite each of the following sets by enumerating all of the elements.
   a. $A = \{x \mid x \text{ is a positive integer less than } 10\}$.
   b. $B = \{y \mid y \text{ is a natural number such that } y + 2 = 4\}$.
   c. $C = \{z \mid z \text{ is an even integer such that } z^2 - 5z - 6 = 0\}$.
   d. $D = \{a \mid a \text{ is a real number greater than or equal to zero and also less than zero}\}$.

9. Rewrite each of the following sets by describing membership requirements for inclusion. (Example: If $W = \{1, 3, 5\}$, then W is $\{x \mid x$ is an odd natural number that is less than 7$\}$.)

a. $\{1, 2, 3, 4, 5, 6\}$.
b. $\{$Monday, Tuesday, Wednesday, Thursday, Friday, Saturday, Sunday$\}$.
c. $\{a, b, c, d, e\}$.
d. $\{$April, June, September, November$\}$.
e. $\{3, 6, 9, 12, 15, \ldots\}$.

3.2 SUBSETS AND PROPER SUBSETS

Let us consider two arbitrary finite sets A and B. We could examine these sets relative to the number of elements in each. One set may have fewer elements than the other, or both sets may have the same number of elements. If the latter case is true, the elements of the first set may or may not be the same as the elements of the second set. However, finite sets may be compared by examining their elements.

A convenient way of representing sets and relationships between sets is by using Venn diagrams, named for the English logician John Venn (1834–1883). The universal set is represented by a rectangle; subsets of the universal set are represented by circles within the rectangle. In Figure 3.1 we have a diagrammatic representation which illustrates the relationship that set A is contained in set B, while set B is contained in the universal set U.

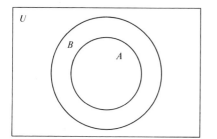

Figure 3.1

Notice that the set A is represented by a circle which is contained completely within the circle representing the set B. Since A and B are subsets of U, both circles are contained within the rectangle.

DEFINITION 3.6 The set A is a *subset* of the set B and is denoted by $A \subseteq B$ if and only if every element of A is also an element of B. If A is not a subset of B, we write $A \nsubseteq B$.

In the above definition we observe that $A \subseteq B$ if and only if $a \in A$ implies $a \in B$ for *every* element in A. To determine if one set is a subset of another set, we simply compare elements; there is no operation being performed on the sets. Hence "is a subset of" is a relation between sets.

Recall in the previous chapter that we encountered the relation *implication* between two propositions. Now let p and q be propositions. If p implies q, we write this $p \Rightarrow q$. Also let P and Q be sets. If P is a subset of Q, we write this $P \subseteq Q$. Hence we note the similarities between the relation "is a subset of" between sets and "implication" between propositions.

EXAMPLE Consider $M = \{1, 2, 3\}$ and $R = \{\{1, 2, 3, 4\}$. $M \subseteq R$ since *every* element of M is also an element of R.

EXAMPLE Consider the sets $M = \{1, 2, 3\}$ and $R = \{1, 2, 3, 4\}$. $R \nsubseteq M$ since it is not true that *every* element of R is also an element of M. We note that $4 \notin M$ but $4 \in R$.

In the first example we concluded that $M \subseteq R$, which means that *every* element of M is also an element of R. However, the set R has the element 4 which is not in M.

DEFINITION 3.7 The set A is a *proper subset* of the set B, denoted by $A \subset B$, if and only if A is a subset of B and, further, B has at least one element that is not in A. If A is not a proper subset of B, we write $A \not\subset B$.

In reference to Definitions 3.6 and 3.7 we should note that the null set is a subset of every set and is also a proper subset of every nonempty set.

EXAMPLE Consider $A = \{a, b, c\}$, $B = \{a, b, d\}$, $D = \{a, b, c, d\}$, and $E = \{b, a, c\}$.

$A \nsubseteq B$ since $c \in A$ but $c \notin B$.
$A \subseteq D$ since *every* element in A is also in D.
$A \subset D$ since $A \subseteq D$ and $d \in D$ but $d \notin A$.
$A \subseteq E$ since *every* element of A is also in E.
$A \not\subset E$ since $A \subseteq E$, but E does not contain elements which are not in A.
$B \nsubseteq A$ since $d \in B$ but $d \notin A$.
$B \subseteq D$ since *every* element of B is also in D.
$B \subset D$ since $B \subseteq D$ and $c \in D$ but $c \notin B$.
$B \nsubseteq E$ since $d \in B$ but $d \notin E$.
$D \nsubseteq A$ since $d \in D$ but $d \notin A$.
$D \nsubseteq B$ since $c \in D$ but $c \notin B$.
$D \nsubseteq E$ since $d \in D$ but $d \notin E$.
$E \subseteq A$ since *every* element of E is also in A.

$E \not\subseteq A$ since $E \subseteq A$, but A does not contain elements which are not in E.
$E \not\subseteq B$ since $c \in E$ but $c \notin B$.
$E \subseteq D$ since *every* element in E is also in D.
$E \subset D$ since $E \subseteq D$ and $d \in D$ but $d \notin E$.
$A \subseteq A$ since *every* element of A is also in A.
$B \subseteq B$ since *every* element of B is also in B.
$D \subseteq D$ since *every* element of D is also in D.
$E \subseteq E$ since *every* element of E is also in E.

In examining the above example, it should be noted that $A \subseteq A$, $B \subseteq B$, $D \subseteq D$, and $E \subseteq E$. In fact *every* set is a subset of itself. However, *no* set is a proper subset of itself. We also should note that $A \subseteq E$ and $E \subseteq A$. Closer examination reveals that the sets A and E are the same except for the order in which their elements are listed.

DEFINITION 3.8 Two sets, A and B, are *equal* and are denoted by $A = B$ if and only if each set is a subset of the other.

It follows from this definition that two sets are equal if and only if they have the same elements.

EXAMPLE If $S = \{a, e, i, o, u\}$ and $T = \{u, o, i, a, e\}$, then $S = T$ since $S \subseteq T$ (*every* element of S is also an element of T) and $T \subseteq S$ (*every* element of T is also an element of S).

DEFINITION 3.9 Two sets, A and B, are *equivalent* and are denoted by $A \sim B$ if and only if every member of either set can be paired with one and only one member of the other set. We say that there exists a *one-to-one correspondence* between the members of the two sets. Equivalence of sets does not, however, imply their equality.

EXAMPLE If $E = \{a, b, c\}$ and $F = \{1, 2, 3\}$, then $E \sim F$ since every element of E can be paired with one and only one element of F and every element of F can be paired with one and only one element of E. Consider the following pairings:

$$a \leftrightarrow 1 \qquad a \leftrightarrow 1$$
$$b \leftrightarrow 2 \quad \text{or} \quad b \underset{\times}{\overset{}{}} 2$$
$$c \leftrightarrow 3 \qquad c 3$$

There are other ways of establishing this one-to-one correspondence. The student is encouraged to determine them.

EXAMPLE If $G = \{1, 2, 3, 4, \ldots\}$ and $H = \{2, 4, 6, 8, 10, \ldots\}$, then $G \sim H$ since every element of G can be paired with that element of H which is twice its value and every element of H can be paired with that element of G which is one-half its value.

EXERCISES

1. List all the subsets for each of the following sets.
   a. $\{a, b, c, d\}$.   b. $\{1, 2, 3\}$
   c. $\{^-2, 0, 3\}$.   d. $\{a, e, i, o, u\}$.
   e. $\{\varnothing\}$.   f. $\{5, 7, ?, \#\}$.
2. Which of the following statements are true if $A = \{a, b, c, d\}$?
   a. $b \in A$.   b. $2 \notin A$.
   c. $c \subseteq A$.   d. $\{d\} \subseteq A$.
   e. $\{a\} \in A$.   f. $\varnothing \in A$.
   g. $\varnothing \subseteq A$.   h. $\varnothing \subset A$.
   i. $\{\varnothing\} \subseteq A$.   j. $\{a, d, c, b\} \subseteq A$.
3. List all of the proper subsets for each set given in Exercise 1.
4. Consider the set $A = \{a, b, 1, 2, 3\}$. Which of the following sets are subsets of A?
   a. $\{a, 1, 3\}$,   b. $\{\ \}$.
   c. $\{1, 2, 3, a, b\}$ d. $\{\varnothing\}$.
   e. $\{a, c, 2, 1\}$.   f. $\{1\}$.
   g. $\{a\}$.   h. $\varnothing$.
   i. $\{a, b, \{1, 2\}\}$.   j. $\{b, a, c\}$.
5. Which of the sets listed in Exercise 4 are proper subsets of A?
6. If $P = Q$, which of the following statements are true?
   a. $P \subseteq Q$.   b. $P \subset Q$.
   c. $Q \subseteq P$.   d. $Q \subset P$.
   e. $Q = P$.   f. $P \subseteq Q$ and $Q \subseteq P$.
7. Define each of the following.
   a. M is a subset of P.
   b. P is a proper subset of M.
   c. Equal sets.
   d. Equivalent sets.
8. Let A and B be the subsets of U where U is the set whose elements are all of the letters of the English alphabet.
   a. Exhibit sets A and B such that $A \subseteq B$.
   b. Exhibit sets A and B such that $A \nsubseteq B$.
   c. Exhibit sets A and B such that $A \subseteq B$ and $B \subseteq A$.
   d. Exhibit sets A and B such that $A \sim B$.

3.3 UNION AND INTERSECTION

Section 3.2 was concerned primarily with comparing sets. This section shall form new sets from given sets by combining them in various ways. Given any two arbitrary sets, we can form a third set by using only the elements which are common to the two given sets. Or we could form a new set by using all of the elements contained in at least one of the two given sets.

DEFINITION 3.10 The *union* of two sets, A and B, denoted by $A \cup B$ (read "A union B") is the set of all elements that are either in A or in B or in both A and B.

In this definition notice that the connective of disjunction is used in describing membership requirements. Hence if $a \in A \cup B$, at least one of the following conditions must be true:

1. a is an element of A (that is, $a \in A$).
2. a is an element of B (that is, $a \in B$).

Further observe that if $a \in A$, then $a \in A \cup B$ no matter what the set B is. Similarly, if $a \in B$, then $a \in A \cup B$ no matter what the set A is.

EXAMPLE Consider the following examples of the union of sets.

Let $R = \{a, b, c\}$ and $S = \{a, e, i, o, u\}$. Then $R \cup S = \{a, b, c, e, i, o, u\}$. (See Figure 3.2.)

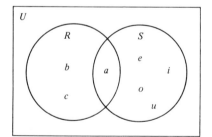

Figure 3.2

If $A = \{1, 2, 3\}$ and $B = \emptyset$, then $A \cup B = \{1, 2, 3\} = A$.

We observe that the union of the set A with the null set yielded the set A.

THEOREM 3.1 Let A be an arbitrary set and B be the null set. Then $A \cup B = A$ or, equivalently, $A \cup \emptyset = A$.

Proof To prove that $A \cup \emptyset = A$, we must show that $A \cup \emptyset \subseteq A$ and that $A \subseteq A \cup \emptyset$. To show that $A \cup \emptyset \subseteq A$, we will let $p \in A \cup \emptyset$, where p is an arbitrary element of $A \cup \emptyset$, and attempt to prove that $p \in A$. Now if $p \in A \cup \emptyset$, either $p \in A$ or $p \in \emptyset$ (using the definition for union of two sets). If $p \in A$, we are done. However, if $p \in \emptyset$, then $p \in A$ since the null set is a subset of every set. Hence if $p \in A \cup \emptyset$, then $p \in A$, and we have shown that $A \cup \emptyset \subseteq A$.

To show that $A \subseteq A \cup \emptyset$, we will let $q \in A$, where q is an arbitrary element of A, and attempt to prove that $q \in A \cup \emptyset$. But if $q \in A$, then $q \in A \cup \emptyset$ by the definition of union of two sets. Hence we have shown that $A \cup \emptyset \subseteq A$.

THEOREM 3.2 Let A and B be arbitrary sets. If $A \subseteq B$, then $A \cup B = B$.

Proof Again, to prove that $A \cup B = B$, it is sufficient to prove that $A \cup B \subseteq B$ and $B \subseteq A \cup B$.

To prove that $A \cup B \subseteq B$, we will let $x \in A \cup B$, where x is an arbitrary element of $A \cup B$, and attempt to prove that $x \in B$. If $x \in A \cup B$, then $x \in A$ or $x \in B$ by the definition of union of two sets. If $x \in B$, we are done. If $x \in A$, then it follows that $x \in B$ since $A \subseteq B$ and, by definition, every element of A is also an element of B. Hence if $x \in A \cup B$, we have shown that $x \in B$, and, therefore, $A \cup B \subseteq B$.

To prove that $B \subseteq A \cup B$, we will let $y \in B$, where y is an arbitrary element of B, and attempt to prove that $y \in A \cup B$. But if $y \in B$, then $y \in A \cup B$ by the definition of union of two sets. Hence we have shown that $B \subseteq A \cup B$.

When we formed the union of the two sets A and B, observe that we used all of the distinct elements that belonged to at least one of the sets. We could also examine the set formed by using only those elements which are common to the two sets A and B.

DEFINITION 3.11 The *intersection* of the two sets A and B, denoted by $A \cap B$ (read "A intersection B"), is the set of all elements that are common to A and B.

EXAMPLE Let $X = \{1, 2, 3, 4, 5, 6, 7, 8\}$ and $Y = \{1, 3, 5, 9\}$. Then $X \cap Y = \{1, 3, 5\}$. (See Figure 3.3.)

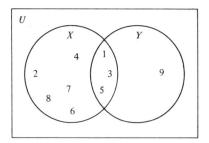

Figure 3.3

DEFINITION 3.12 Two sets, A and B, are said to be *disjoint* if and only if they have no common elements. Hence the intersection of two disjoint sets is the null set.

EXAMPLE Let $A = \{1, 3, 5, 7, \ldots\}$ and $B = \{2, 4, 6, 8, \ldots\}$. Then the sets A and B are disjoint since they have no elements in common. Observe also that $A \cap B = \emptyset$

EXAMPLE If $A = \{1, 2, 3\}$ and $B = \emptyset$, then $A \cap B = \emptyset$.

EXAMPLE If $C = \{1, 2\}$ and $D = \{1, 2, 3, 4\}$, then $C \cap D = \{1, 2\} = C$.

In examining the last two examples, we observe that the intersection of the set A with the null set yielded the null set and that if $C \subseteq D$, then $C \cap D = C$.

THEOREM 3.3 Let A be an arbitrary set and $B = \emptyset$. Then $A \cap B = \emptyset$ or, equivalently, $A \cap \emptyset = \emptyset$.

Proof To prove that $A \cap \emptyset = \emptyset$, it will be sufficient to prove that $A \cap \emptyset \subseteq \emptyset$ and $\emptyset \subseteq A \cap \emptyset$. To prove that $A \cap \emptyset \subseteq \emptyset$, we will let $r \in A \cap \emptyset$, where r is an arbitrary element of $A \cap \emptyset$, and attempt to prove that $r \in \emptyset$. Now if $r \in A \cap \emptyset$, then $r = A$ and $r \in \emptyset$ by the definition of intersection of two sets. But by definition there are no elements in the null set and, hence there are no elements belonging to both A and $\emptyset$. Therefore we have shown that $A \cap \emptyset \subseteq \emptyset$. We ask the reader to complete the proof by showing that $\emptyset \subseteq A \cap \emptyset$.

THEOREM 3.4 Let A and B be arbitrary sets. If $A \subseteq B$, then $A \cap B = A$.
Proof I leave the entire proof as an exercise for the student to complete.

The previous section noted the similarities between the relation "is a subset of" on sets and the "implication" of propositions. This section defines the union of sets using the connective of disjunction and the intersection of sets using the connective of conjunction. What similarities, if any, exist between these operations on sets and the connectives in terms of which they are defined? Let us consider two sets P and Q and an arbitrary element x. We will denote that $x \in P$ by a truth value T and $x \notin P$ by a truth value F and likewise for x and Q. Now consider truth tables 3.1 and 3.2 for union and intersection.

Table 3.1 Truth table for union of sets		
$x \in P$	$x \in Q$	$x \in P \cup Q$
T	T	T
T	F	T
F	T	T
F	F	F

Table 3.2 Truth table for intersection of sets		
$x \in P$	$x \in Q$	$x \in P \cap Q$
T	T	T
T	F	F
F	T	F
F	F	F

Table 3.1, except for the headings for each column, is identical to the truth table for disjunction. In a similar manner Table 3.2, except for the headings for each column, is identical to the truth table for conjunction. We seem to be constructing a mathematical system in this chapter similar in structure to the system or algebra of propositions constructed in Chapter 2 and, indeed, we are!

In fact, consider a universal set U and an arbitrary proposition p relating to the elements of U. Now consider the set P that contains all of those elements of U for which p is true. Clearly $P \subseteq U$. In similar manner let q be another proposition relating to the elements of U and let Q be the set of all elements of U for which q is true. Again, $Q \subseteq U$. It should be clear, then, whenever we have propositions p and q, we can relate them to the sets P and Q and conversely. Also, whenever we have the connectives of disjunction and conjunction involving p and q, we can relate them to the operations union and intersection of P and Q respectively. Combining these similarities with those encountered in the previous section, we can list them as in Table 3.3.

Table 3.3 Similarities between propositions and sets

Propositions	Sets
p	P
q	Q
$p \Rightarrow q$	$P \subseteq Q$
$q \Rightarrow p$	$Q \subseteq P$
$p \Leftrightarrow q$	$P = Q$
$p \vee q$	$P \cup Q$
$p \wedge q$	$P \cap Q$

We will extend Table 3.3 as we continue throughout the remainder of this chapter.

Having remarked upon the similarities as given in Table 3.3, the reader may be anticipating that we also would have properties relating to union and intersection of sets that are analogous to the properties relating to disjunction and conjunction of propositions respectively. Actually we do have such properties and will now discuss them.

3.31 COMMUTATIVE PROPERTIES

Let A and B be two arbitrary sets; then

1. $A \cup B = B \cup A$ Commutative property for union of sets
2. $A \cap B = B \cap A$ Commutative property for intersection of sets

To prove that $A \cup B = B \cup A$, we use the definition for the equality of sets. Hence we must prove that $A \cup B \subseteq B \cup A$ and $B \cup A \subseteq A \cup B$. We leave the proofs for both parts of the above property as exercises.

3.32 ASSOCIATIVE PROPERTIES

Let A, B, and C be three arbitrary sets; then

1. $A \cup (B \cup C) = (A \cup B) \cup C$ Associative property for union of sets
2. $A \cap (B \cap C) = (A \cap B) \cap C$ Associative property for intersection of sets

This property states that the order in which we consider the sets A, B, and C is immaterial in forming the union and intersection. Hence we could form the union (intersection) of A and B and then form the union (intersection) of this new set with C. Or we could form the union (intersection) of B and C and then form the union (intersection) of this new set with A. The results will be the same in either case. (Compare this with the associative properties for the connectives of disjunction and conjunction.)

EXAMPLE Let $P = \{2, 3, 4\}$, $R = \{1, 2, 3, 4\}$, and $S = \{2, 4, 6, 7\}$. Verify that
  a. $P \cup (R \cup S) = (P \cup R) \cup S$
  b. $P \cap (R \cap S) = (P \cap R) \cap S$

Solution
a. $P \cup (R \cup S) = \{2, 3, 4\} \cup [\{1, 2, 3, 4\} \cup \{2, 4, 6, 7\}]$
$\qquad\qquad = \{2, 3, 4\} \cup \{1, 2, 3, 4, 6, 7\}$
$\qquad\qquad = \{1, 2, 3, 4, 6, 7\}$

$(P \cup R) \cup S = [\{2, 3, 4\} \cup \{1, 2, 3, 4\}] \cup \{2, 4, 6, 7\}$
$\qquad\qquad \{1, 2, 3, 4\} \cup \{2, 4, 6, 7\}$
$\qquad\qquad \{1, 2, 3, 4, 6, 7\}$

Hence $P \cup (R \cup S) = (P \cup R) \cup S$.

b. $P \cap (R \cap S) = \{2, 3, 4\} \cap [\{1, 2, 3, 4\} \cap \{2, 4, 6, 7\}]$
$\qquad\qquad = \{2, 3, 4\} \cap \{2, 4\}$
$\qquad\qquad = \{2, 4\}$

$(P \cap R) \cap S = [\{2, 3, 4\} \cap \{1, 2, 3, 4\}] \cap \{2, 4, 6, 7\}$
$\qquad\qquad = \{2, 4, 3\} \cap \{2, 4, 6, 7\}$
$\qquad\qquad = \{2, 4\}$

Hence $P \cap (R \cap S) = (P \cap R) \cap S$. ∎

EXAMPLE Given $A = \{a, b, c\}$, $B = \{b, c, d, e\}$, $C = \{a, c, d, e\}$, form

a. $A \cup (B \cap C)$
b. $(A \cup B) \cap (A \cup C)$
c. $A \cap (B \cup C)$
d. $(A \cap B) \cup (A \cap C)$

Solution

a. $A \cup (B \cap C) = \{a, b, c\} \cup [\{b, c, d, e\} \cap \{a, c, d, e\}]$
$\qquad = \{a, b, c\} \cup \{c, d, e\}$
$\qquad = \{a, b, c, d, e\}$

b. $(A \cup B) \cap (A \cup C) = [\{a, b, c\} \cup \{b, c, d, e\}] \cap [\{a, b, c\} \cup \{a, c, d, e\}]$
$\qquad = \{a, b, c, d, e\} \cap \{a, b, c, d, e\}$
$\qquad = \{a, b, c, d, e\}$

c. $A \cap (B \cup C) = \{a, b, c\} \cap [\{b, c, d, e\} \cup \{a, c, d, e\}]$
$\qquad = \{a, b, c\} \cap \{a, b, c, d, e\}$
$\qquad = \{a, b, c\}$

d. $(A \cap B) \cup (A \cap C) = [\{a, b, c\} \cap \{b, c, d, e\}] \cup [\{a, b, c\} \cap \{a, c, d, e\}]$
$\qquad = \{b, c\} \cup \{a, c\}$
$\qquad = \{a, b, c\}$ ∎

For the particular sets A, B, and C given in the example above we observe that $A \cup (B \cap C) = (A \cup B) \cap (A \cup C)$ and that $A \cap (B \cup C) = (A \cap B) \cup (A \cap C)$. This is also true in general.

3.33 DISTRIBUTIVE LAWS

If A, B, and C are arbitrary sets, then

1. $A \cup (B \cap C) = (A \cup B) \cap (A \cup C)$ — Distributive property of union over intersection on sets

2. $A \cap (B \cup C) = (A \cap B) \cup (A \cap C)$ — Distributive property of intersection over union on sets

The entire proof of this property is left as an exercise.

Earlier in this section I commented that if U is a universal set and p is an arbitrary proposition relating to the elements of U such that P is the set containing all those elements of U for which p is true, then $P \subseteq U$. Clearly P could contain all of the elements of U which means that p would be true for all elements of U and $P = U$. In this case the proposition p would be a tautology. On the other hand, p may be true for no elements of U and $P = \varnothing$. Stated equivalently, p would be false for every element in U. In this case the proposition p would be a contradiction. The following properties, similar to those for disjunction and conjunction, are now given.

3.34 IDEMPOTENT PROPERTIES

If P is an arbitrary set, then
1. $P \cup P = P$
2. $P \cap P = P$

3.35 IDENTITY PROPERTIES

If P is an arbitrary set and U is the universal set containing P, then
1. $P \cup \emptyset = P$
2. $P \cup U = U$
3. $P \cap \emptyset = \emptyset$
4. $P \cap U = P$

The proofs of properties 3.34 and 3.35 are left as exercises.

EXERCISES

1. Let $U = \{1, 2, 3, 4, 5, 6, 7, 8, 9, 10\}$ and let $A = \{6, 8, 9\}$, $B = \{1, 3, 7, 8, 9\}$, and $C = \{2, 6, 8, 9\}$. Form:
   a. $A \cup B$.   b. $A \cap B$.
   c. $B \cup C$.   d. $B \cap C$.
   e. $A \cup (B \cup C)$.   f. $A \cup (B \cap C)$.
   g. $A \cap (B \cap C)$.   h. $A \cap (B \cup C)$.
   i. $(A \cup B) \cap C$.   j. $(A \cap B) \cup C$.

2. Let $U = \{1, 2, 3, 4, 5, 6, 7, 8, 9, 10, 11, 12\}$ and let $P = \{1, 2, 3, 4, 5, 6\}$, $Q = \{4, 5, 6, 7, 8, 9\}$, and $R = \{1, 3, 6, 9, 12\}$. Form:
   a. $P \cup Q$.   b. $P \cup R$.
   c. $Q \cup R$.   d. $Q \cap R$.
   e. $P \cup (Q \cup R)$.   f. $P \cup (Q \cap R)$.
   g. $P \cap (Q \cap R)$.   h. $P \cap (Q \cup R)$.
   i. $(P \cup Q) \cap R$.   j. $(P \cap Q) \cup R$.

3. Which of the following statements are always true for all subsets of U?
   a. $(A \cup B) = A$.   b. $(A \cup B) \subseteq A$.
   c. $(A \cup B) \subset B$.   d. $(A \cap B) \subset A$.
   e. $(A \cap B) \subseteq B$.   f. $A \subseteq (A \cup B)$.
   g. $B \subset (A \cup B)$.   h. $A \subseteq (A \cap B)$.
   i. $A \subset (A \cap B)$.   j. $\emptyset \subset B$.

4. Let $U = \{1, 2, 3, 4, 5, 6, 7, 8, 9, 10, 11, 12, 13, 14, 15\}$ and let $A = \{x | x \in U, x \text{ is even}\}$, $B = \{y | y \in U, y \text{ is odd}\}$, $C = \{z | z \in U, z \text{ is a}$

multiple of 5}, $D = \{u | u \in U, u$ is a divisor of 24}, and $E = \{w | w \in U,$
$2 \leq w < 7\}$. Form:

a. $A \cup C$.

b. $B \cap E$.

c. $(A \cup B) \cap C$.

d. $(A \cup C) \cap (B \cup D)$.

e. $[B \cap (C \cup E)] \cap D$.

f. $[(B \cap D) \cup (A \cap E)] \cap (C \cup D)$.

g. $[(A \cap B) \cup (\emptyset \cap D)] \cap U$.

5. Complete the proof of Theorem 3.3.
6. Prove Theorem 3.4.
7. Prove that the operation of union of sets is a commutative operation.
8. Prove that the operation of intersection of sets is a commutative operation.
9. Prove that the operation of union of sets is an associative operation.
10. Prove that the operation of intersection of sets is an associative operation.
11. Prove the distributive property of union over intersection on sets.
12. Prove the distributive property of intersection over union on sets.
13. Prove both parts of Property 3.34 of this section.
14. Prove all four parts of Property 3.35 of this section.

3.4 ADDITIONAL OPERATIONS ON SETS

This section will continue to observe similarities between the algebra of propositions and the algebra of sets by considering additional operations on sets.

Suppose that we have two sets, A and B, such that A is a subset of B. Then every element of A is also an element of B. We also can form a set of all the elements of B which are not elements of A.

DEFINITION 3.13 If $A \subseteq B$, then the *complement* of A relative to B, denoted by A', is the set of all elements of B that are not in A. Symbolically we have $A \subseteq B$ and $A' = \{x | x \in B$ and $x \notin A\}$.

EXAMPLE Let $A = \{1, 2, 3, 6\}$ and $B = \{1, 2, 3, 4, 5, 6, 7, 8\}$. Then $A' = \{4, 5, 7, 8\}$. (See Figure 3.4.)

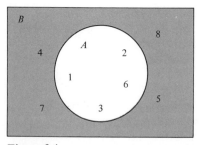

Figure 3.4

Observe in the above example that the set A' contains all of those elements of B which are not in A. Also the sets A and A' are disjoint and $A \cup A' = B$. Now if we consider the set B to be our universal set in this problem and let x be an arbitrary element of B, then $x \in A$ or $x \in A'$ but $x \notin A \cap A'$. Further, let p be the proposition "$x \in A$"; the set of all elements in B for which p is true will be denoted by P. Also let Q be the set of all those elements in B for which p is false. Clearly, then, P and Q are other names for A and A' respectively. Hence the operation of forming the complement of a set corresponds to the negating of a proposition. We will now list properties associated with the complement of a set that are analogous to the properties of the negation of a proposition.

3.41 COMPLEMENT PROPERTIES

If P is an arbitrary set and U is the universal set containing P, then
1. $P \cap P' = \varnothing$
2. $P \cup P' = U$
3. $U' = \varnothing$
4. $\varnothing' = U$
5. $(P')' = P$

3.42 DE MORGAN'S LAWS

If P and Q are arbitrary sets, then
1. $(P \cup Q)' = P' \cap Q'$
2. $(P \cap Q)' = P' \cup Q'$

EXAMPLE Verify the complement properties 1, 2, and 3 above for the sets $U = \{1, 2, 3, 4, 5, 6\}$ and $P = \{2, 4, 5\}$.
Solution
For (1) we must show that $P \cap P' = \varnothing$. Since $P' = \{1, 3, 6\}$, we have $P \cap P' = \{2, 4, 5\} \cap \{1, 3, 6\} = \varnothing$.
For (2) we must show that $P \cup P' = U$. Since $P' = \{1, 3, 6\}$, we have $P \cup P' = \{2, 4, 5\} \cup \{1, 3, 6\} = \{1, 2, 3, 4, 5, 6\} = U$.
For (3) we must show that $(P')' = P$. Since $P' = \{1, 3, 6\}$, we have $(P')' = \{1, 3, 6\}' = \{2, 4, 5\} = P$. ■

EXAMPLE Using the sets $P = \{2, 4, 6, 8\}$, $Q = \{1, 2, 3, 4, 5\}$, and $U = \{1, 2, 3, 4, 5, 6, 7, 8, 9\}$, verify De Morgan's laws.
Solution
a. We wish to show that $(P \cup Q)' = P' \cap Q'$. Since $P \cup Q = \{2, 4, 6, 8\} \cup \{1, 2, 3, 4, 5\} = \{1, 2, 3, 4, 5, 6, 8\}$, we have $(P \cup Q)' = \{1, 2, 3, 4, 5, 6, 8\}' = \{7, 9\}$. Since $P' = \{2, 4, 6, 8\}' = \{1, 3, 5, 7, 9\}$ and $Q' = \{1, 2, 3, 4, 5\}' = \{6, 7, 8, 9\}$, we have $P' \cap Q' = \{1, 3, 5, 7, 9\} \cap \{6, 7, 8, 9\} = \{7, 9\}$. Hence $(P \cup Q)' = P' \cap Q'$.

b. We wish to show that $(P \cap Q)' = P' \cup Q'$. Since $P \cap Q = \{2, 4, 6, 8\}$ $\cap \{1, 2, 3, 4, 5\} = \{2, 4\}$, we have $(P \cap Q)' = \{1, 3, 5, 6, 7, 8, 9\}$. Since $P' = \{2, 4, 6, 8\}' = \{1, 3, 5, 7, 9\}$ and $Q' = \{1, 2, 3, 4, 5\}' = \{6, 7, 8, 9\}$, we have $P' \cup Q' = \{1, 3, 5, 7, 9\} \cup \{6, 7, 8, 9\} = \{1, 3, 5, 6, 7, 8, 9\}$. Hence $(P \cap Q)' = P' \cup Q'$. ∎

In Definition 3.13 we defined the complement of a set A relative to a set B with the condition $A \subseteq B$. Now consider two arbitrary sets, A and B, such that neither is necessarily a subset of the other. We can form a new set which contains only those elements of one of the given sets which are not found in the other. There would be two such sets. (Why?)

DEFINITION 3.14 Given two sets A and B. The *difference set* $A - B$ is the set of all elements that are in A but not in B. Symbolically we have $A - B = \{x \mid x \in A \text{ and } x \notin B\}$.

EXAMPLE   Let $A = \{2, 4, 8, 10, 11\}$ and $B = \{1, 2, 3, 4\}$; then $A - B = \{8, 10, 11\}$ and $B - A = \{1, 3\}$. (See Figure 3.5.)

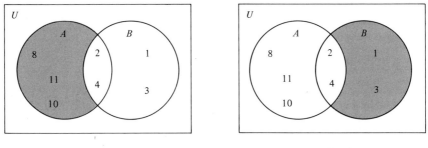

(a) $A - B$ (b) $B - A$

Figure 3.5

There is no direct correspondence between forming the difference of two sets and using a connective with propositions. However, to form the difference of two sets, $A - B$ we want all of the elements x that are in A but not in B. Equivalently, this can be stated as all of those elements x such that $x \in A$ and $x \notin B$. But $x \notin B$ means that $x \in B'$. Hence we are looking for all x such that $x \in A \cap B'$. Therefore $A - B = A \cap B'$. (See Figure 3.5(a).)

EXAMPLE   Using the sets $A = \{1, 2, 3, 4\}$, $B = \{2, 4, 5, 6, 7\}$, and $U = \{1, 2, 3, 4, 5, 6, 7\}$, verify that (a) $A - B = A \cap B'$ and (b) $B - A = B \cap A'$.

Solution
a. $A - B = \{1, 2, 3, 4\} - \{2, 4, 5, 6, 7\} = \{1, 3\}$ and $A \cap B' = \{1, 2, 3, 4\} \cap \{1, 3\} = \{1, 3\}$. Hence $A - B = A \cap B'$.

b. $B - A = \{2, 4, 5, 6, 7\} - \{1, 2, 3, 4\} = \{5, 6, 7\}$ and $B \cap A' = \{2, 4, 5, 6, 7\} \cap \{5, 6, 7\} = \{5, 6, 7\}$. Hence $B - A = B \cap A'$. ■

In comparing two sets we observed the following cases:

1. All of the elements of the first set may also be elements of the second set, in which case the first set is a subset of the second set.

2. None of the elements of the first set are found in the second set, in which case the two sets are disjoint.

It is also possible to exhibit two sets that are not disjoint but such that neither is a subset of the other. Such sets are called overlapping sets.

DEFINITION 3.15 Two sets, A and B, are said to be *overlapping sets* if and only if they have at least one element in common but neither is a subset of the other.

Symbolically, the sets A and B are overlapping if $A \cap B \neq \varnothing$, $A \not\subseteq B$, and $B \not\subseteq A$. The sets given in the example following Definition 3.14 are examples of overlapping sets.

If we have two given sets, we also can form a new set by combining the elements of the given sets such that the elements of our new set would be different from the elements of the given sets. This can be done by pairing each element of one of the given sets, in turn, with every element of the other set. Hence the elements of the new set would be a pair of elements.

DEFINITION 3.16 A pair of objects, a and b, one of which is called the *first component* and the other the *second component*, is called an *ordered pair*. The two objects, separated by a comma, are written within parentheses such as (a, b) or (b, a).

In the above definition the symbol (a, b) refers to an ordered pair of objects a and b, such that a is the first component and b is the second component. The components of an ordered pair may be different or they may be alike. Hence (a, a), $(2, b)$, $(3, 5)$, and $(4, 4)$ are examples of ordered pairs.

We will now consider the construction of a new set whose elements are ordered pairs, all of whose first and second components are elements of two given sets.

DEFINITION 3.17 The *cartesian product* of two sets, A and B, denoted by $A \times B$, is the set of all ordered pairs of the form or type (a, b) such that the first component, a, is an element of A and the second component, b, is an element of B. Symbolically we have $A \times B = \{(a, b) | a \in A \text{ and } b \in B\}$.

Observe that the cartesian product of two given sets is formed by pairing each element of one set, in turn, with every element of the other set. Hence in the above definition if set A had m elements and set B had n elements, the set $A \times B$ would have mn elements.

EXAMPLE   If $C = \{1, 2, 3\}$ and $D = \{a, b\}$, then $C \times D = \{(1, a), (1, b), (2, a), (2, b), (3, a), (3, b)\}$. The graph of $C \times D$ appears in Figure 3.6 as a set of six points.

EXAMPLE   If $C = \{1, 2, 3\}$ and $D = \{a, b\}$, then $D \times C = \{(a, 1), (a, 2), (a, 3), (b, 1), (b, 2), (b, 3)\}$. The graph of $D \times C$ appears in Figure 3.7 as a set of six points.

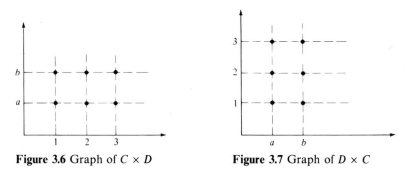

Figure 3.6 Graph of $C \times D$ **Figure 3.7** Graph of $D \times C$

In the two examples above we see that the sets $C \times D$ and $D \times C$ have the same number of elements and, hence, are equivalent. Observe, however, that they are not equal. In general $A \times B \neq B \times A$.

Also, if the two given sets A and B are equal, then $A \times B = A \times A$ which is the cartesian product of the set A with itself.

EXERCISES

1. Given $A = \{1, 2, 3, 4\}$, $B = \{3, 4, 5, 6\}$, $C = \{4, 5, 6, 7\}$, and $D = \{1, 2, 3, 4, 5, 6, 7\}$.
   a. Form $A \cup B$.
   b. Form $B \cap C$.
   c. Form the complement of A relative to D.
   d. Form $A - B$.
   e. Form $C \times A$.
   f. Show that $A \cup (B \cap C) = (A \cup B) \cap (A \cup C)$.
2. Let the universe set $U = \{1, 2, 3, 4, 5, 6, 7, 8, 9\}$, $A = \{1, 2, 3, 4\}$, $B = \{5, 6, 7\}$, $C = \{1, 3, 5, 8\}$, and $D = \{2, 4, 6, 8, 9\}$.
   a. Form A' and B'.
   b. Form $C \cup (A' \cap D)$.
   c. Form $(B \cup C)'$.
   d. Which pair of the above subsets of U are disjoint?

3. If A, B, and C are three distinct and nonempty subsets of some universe set U, display each of the following by use of Venn diagrams.
   a. $A \cap B \neq \emptyset$, $A \cap C \neq \emptyset$, and $B \cap C \neq \emptyset$.
   b. $A \subseteq B$, $B \subseteq C$.
   c. $A \subseteq B$, $B \nsubseteq C$.
   d. $(A \cup B) \cap C = \emptyset$.
   e. $A \cup (B \cap C) = A$.
4. Define each of the following.
   a. Two sets R and S are disjoint.
   b. Two sets W and T are overlapping.
   c. The union of two sets.
   d. The intersection of two sets.
   e. The cartesian product $E \times F$ of the two sets E and F.
   f. The complement of A if $A \subseteq U$.
5. Consider an arbitrary set S and the null set $\emptyset$.
   a. Is $\emptyset \subseteq S$ (that is, is every element of $\emptyset$ also an element of S)? Give a reason for your answer.
   b. Is $S \subseteq \emptyset$? (Remember, S is an arbitrary set.) Give a reason for your answer.
   c. Is $S \cup \emptyset = \emptyset$? Why?
   d. Is $S \cap \emptyset = \emptyset$? Why?
   e. Can $S \cap \emptyset = S$? If yes, explain your answer.
   f. Is $\emptyset \subset S$? Give a reason for your answer.
6. Given $U = \{0, 2, 1, 3, 4, 5, 6\}$. Write the complement of each of the following sets.
   a. $A = \{1, 3, 6\}$.
   b. $B = \{6, 4, 3, 0\}$.
   c. $C = \{2, 4, 1, 6\}$.
   d. $D = \emptyset$.
   e. $E = U$.
   f. $F = B \cup C$.
   g. $G = A \cap B$.
7. Given $A = \emptyset$ and $A \subseteq U$, what is $(A')'$?
8. If the intersection set of two sets M and N is not the empty set, are the sets disjoint? Give a reason for your answer.
9. If the intersection set of two sets M and N is not the empty set, are the sets necessarily overlapping? Give a reason for your answer. Could they be overlapping? Why?
10. Using Venn diagrams show that
    a. The set $\{1, 2, 3\}$ is a subset of $\{1, 2, 3, 4, 5\}$.
    b. The set of all blue-eyed girls is a proper subset of the set of all girls.
    c. The set of all blue-eyed girls is a subset of the set of all girls which, in turn, is a subset of the set of all female persons.
    d. The set of all college males is a subset of the set of all people *and* that the set of all college males and the set of all females are disjoint.

11. Given $A = \{1, 2, 3, 4\}$, $B = \{3, 4, 5, 6\}$, and $U = \{1, 2, 3, 4, 5, 6, 7, 8\}$, using the roster method identify each of the following sets.
    a. $A \cup B$.
    b. B'.
    c. $A \cap B'$.
    d. $(A \cup B)'$.
    e. A'.
    f. $A \cap B$.
    g. $A' \cap B'$.
    h. $A' \cup B'$.
    i. $(A \cap B)'$.
    j. $A' \cup B$.

12. Using the sets $R = \{1, 2, 3\}$, $S = \{2, 4, 5, 6\}$, and $T = \{1, 3, 5\}$, show that
    a. $(R \cup S) \cup T = R \cup (S \cup T)$.
    b. $(R \cap S) \cap T = R \cap (S \cap T)$.
    c. $R \cup (S \cap T) = (R \cup S) \cap (R \cup T)$.
    d. $R \cap (S \cup T) = (R \cap S) \cup (R \cap T)$.

13. Express in roster form each of the following.
    a. $P = \{x \mid x \in A \wedge x \in B\}$, if $A = \{1, 2, 3\}$ and $B = \{2, 3, 4, 5\}$.
    b. $Q = \{y \mid y \in A \vee y \in B\}$, if $A = \{1, 2, 3\}$ and $B = \{2, 3, 4, 5\}$.
    c. $T = \{z \mid z \in P \wedge z \in Q\}$, where P and Q are as defined above.
    d. $W = \{q \mid q \in P \vee q \in Q\}$, where P and Q are as defined above.

14. Given the sets $A = \{a, b, c\}$, $B = \{1, 2, 3, 4\}$, $C = \{b, a, c\}$, $D = \{m, n, p, q\}$, $E = \{0, 1, 2\}$, $F = A \cup C$, $G = B \cap C$, and $H = D \cap E$. Which pairs of the given sets are equivalent?

15. For each pair of equivalent sets identified in the previous exercise, establish a one-to-one correspondence between their elements.

3.5 AN APPLICATION

This section will illustrate the usefulness of some of the concepts discussed so far in this chapter. We are all familiar with various surveys that are taken on various issues. Suppose that we consider a hypothetical survey taken among residents of a particular state in this country. They were asked, among other questions, the three questions listed below:
    1. "Do you favor the way in which the president of the United States is doing his job?"
    2. "Do you favor the way in which the governor of your state is doing his job?"
    3. "Do you favor the way in which your congressional representative is doing his job?"

The results of this survey were tabulated as follows:
    59% favor the way in which the president is doing his job.
    55% favor the way in which the governor is doing his job.
    62% favor the way in which the representative is doing his job.
    35% favor the way in which the president and the governor are doing their jobs.

30% favor the way in which the governor and the representative are doing their jobs.

36% favor the way in which the president and the representative are doing their jobs.

20% favor the way in which the president, the governor, and the representative are doing their jobs.

5% did not respond.

Assuming that 500 people were surveyed, we may now attempt to answer the following questions:

1. How many people favor the way in which the president *or* the representative is doing his job?

2. How many people favor the way in which the president *and* the governor are doing their jobs?

3. How many people favor the way in which the governor is doing his job *but do not* favor the way in which the representative is doing his job?

4. How many people favor the way in which the president *and* the governor are doing their jobs *but do not* favor the way in which the representative is doing his job?

5. How many people favor the way in which *only one* of the three elected officials is doing his job?

To answer the first question, we note that 59 percent of the people favor the way in which the president is doing his job; therefore 59 percent × 500 = 295. Similarly, 62 percent of the people favor the way in which the representative is doing his job; therefore 62 percent × 500 = 310. Since we want the number of people who are satisfied with the president *or* with their representative, we may be inclined to add the 310 to the 295. However, this would yield a total of 605 people, which is 105 people more than we have in our survey. Surely you noted that some of the people who favor the way in which the president is doing his job also favor the way in which their representative is doing his job since we are given that 36 percent favor both. Hence we must subtract from the 605, 36 percent of 500 or 180 people representing those people who were counted twice. Therefore there are 605 − 180 or 425 people who favor the way in which the president is doing his job *or* the way in which their representative is doing his job.

The answer to the other questions also can be determined through a careful analysis of the given data. Fortunately, we can employ the concepts introduced in this chapter to aid us. First, consider the 500 people surveyed to be our universal set. Let P represent the set of all people favoring the way in which the president is doing his job and G and R represent similar sets for the governor and representative. Examining the data given, we observe that $P \cap G \neq \emptyset$, $P \cap R \neq \emptyset$, $G \cap R \neq \emptyset$, and $P \cap G \cap R \neq \emptyset$. With these relationships we form a Venn diagram as in Figure 3.8.

We will now start with the 20 percent of the people who favor the way in which the president, the governor, and their representatives are doing

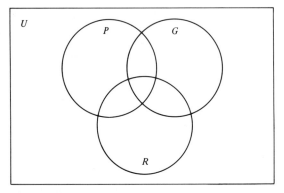

Figure 3.8

their jobs and will record 20 percent $\times$ 500 = 100 in the subset representing $P \cap G \cap R$ as in Figure 3.9.

Next we note that 36 percent favor the way in which the president is doing his job *and* also the way in which the representative is doing his job.

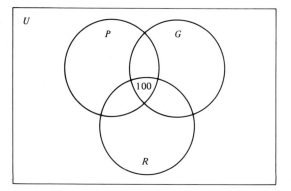

Figure 3.9

That's 36 percent $\times$ 500 = 180 people. We now record this figure in the subset representing $P \cap R$. But observe that we already have 100 included there and only add 80 more as in Figure 3.10.

Continuing now we note that 30 percent favor the way in which both the governor and the representative are doing their jobs. That's 30 percent $\times$ 500 = 150 people for $G \cap R$. Since we already havė 100 in $G \cap R$, we need only add 50 more as in Figure 3.11.

Next we have 35 percent who favor the way in which both the president and the governor are doing their jobs. That's 35 percent $\times$ 500 = 175 people for $P \cap G$. Since we already have 100 in $P \cap G$, we need only add 75 more as in Figure 3.12.

Next we have 62 percent favoring the way in which their representative

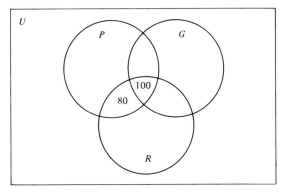

Figure 3.10

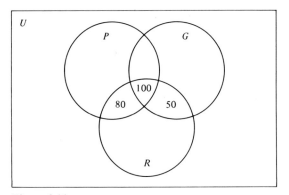

Figure 3.11

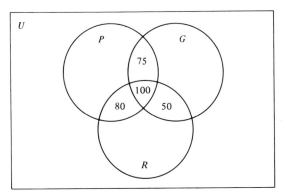

Figure 3.12

is doing his job. That's 62 percent × 500 = 310 people for *R*. But we already have 80 + 100 + 50 = 230 in *R*. Hence we add 80 more as in Figure 3.13.

Next we have 55 percent who favor the way in which the governor is doing his job. That's 55 percent × 500 = 275 people for *G*. But we already

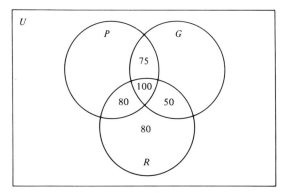

Figure 3.13

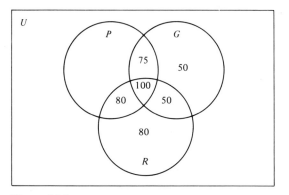

Figure 3.14

have $75 + 100 + 50 = 225$ in G. Hence we add 50 more as in Figure 3.14.

Finally, we have 59 percent who favor the way in which the president is doing his job. That's 59 percent $\times 500 = 295$ people for P. But we already have $80 + 100 + 75 = 255$ people for P. Hence we add 40 more as in Figure 3.15.

We now observe that the number of people accounted for in $P \cup G \cup R$ is $40 + 80 + 100 + 75 + 50 + 50 + 80 = 475$. Since U contains 500, there are $500 - 475 = 25$ people in U but not in $P \cup G \cup R$. We illustrate that fact as in Figure 3.16.

The 25 just included correspond to those who did not respond. (Note that 25 is 5 percent of 500.) From this last diagram we can now answer the questions posed earlier.

    1. The number of people favoring the way in which the president is doing his job *or* the way in which the representative is doing his job would be the number in $P \cup R$. P contains $40 + 80 + 100 + 75 = 295$, and R contains $80 + 100 + 50 + 80 = 310$. But there are $80 + 100 = 180$ in

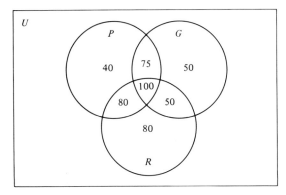

Figure 3.15

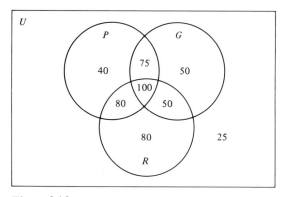

Figure 3.16

$P \cap R$, and these have been counted twice. Hence we have $295 + 310 - 180 = 425$ for our answer.

2. The number of people who favor the way in which the president is doing his job *and* the way in which the governor is doing his job would be the number in $P \cap G$, which is $75 + 100 = 175$.

3. The number of people who favor the way in which the governor is doing his job *but do not* favor the way in which the representative is doing his job would be the number of people in $G - R$, which is $75 + 50 = 125$.

4. The number of people who favor the way in which the president *and* the governor are doing their jobs *but do not* favor the way in which the representative is doing his job would be $(75 + 100) - 100 = 75$, corresponding to $(P \cap G) - R$.

5. The number of people who think that *only one* of the three elected officials is doing his job would be those listed in $P - (G \cup R)$, $G - (P \cup R)$, and $R - (P \cup G)$ or $40 + 50 + 80 = 170$.

EXERCISES

1. The manager of Burger Queen reported that during a given period there were 110 orders of french fries and 160 orders of hamburgers. She further stated that 180 people purchased at least one of the items and that 20 people bought only french fries. How many people bought only hamburgers if no one bought more than one of each item?

2. A car dealer advertised a year-end clearance sale of 80 cars. Some of these cars have some of the following accessories: radio, vinyl roof, and radial tires. The inventory for these cars show

   37 cars with radio
   24 cars with vinyl roof
   23 cars with radial tires
   9 cars with radio and vinyl roof
   12 cars with radio and radial tires
   9 cars with vinyl roof and radial tires
   6 cars with radio, vinyl roof and radial tires

   a. How many cars have at least one accessory?
   b. How many cars have at least two accessories?
   c. How many cars have exactly two accessories?
   d. How many cars have at most two accessories?
   e. How many cars have no accessories?

3. At a ballpark concession stand a supply of 200 hot dogs was prepared in advance of a game as follows:

   56 hot dogs with mustard only
   82 hot dogs with sauerkraut only
   63 hot dogs with relish only
   20 hot dogs with mustard and relish
   15 hot dogs with mustard and sauerkraut
   10 hot dogs with relish and sauerkraut
   5 hot dogs with all three

   a. How many plain hot dogs are there?
   b. How many hot dogs have exactly two of the three choices?
   c. How many hot dogs have at least two of the three choices?
   d. How many hot dogs have at most two of the three choices?
   e. How many hot dogs have at most one of the choices?

3.6 COMPARISON OF THE ALGEBRA OF SETS WITH THE ALGEBRA OF PROPOSITIONS

Let p, q, and r denote arbitrary propositions and P, Q, and R denote their respective truth sets. Let U be the universal set which contains P, Q, and R; t be an arbitrary tautology; and c be an arbitrary contradiction. The following similarities between the algebra of propositions and the algebra of sets can be summarized as follows:

	Propositions	Sets
1. Elements	a. p, q, r	a. P, Q, R
	b. t	b. U
	c. c	c. $\emptyset$
	d. $\sim p$	d. P'
2. Operations	a. $p \vee q$	a. $P \cup Q$
	b. $p \wedge q$	b. $P \cap Q$
	c. $p \rightarrow q \equiv \sim p \vee q$	c. $P' \cup Q$
	d. $p \leftrightarrow q$	d. $(P' \cup Q) \cap (Q' \cup P)$
	e. $p \wedge \sim q$	e. $P - Q$
3. Relations	a. $p \Rightarrow q$	a. $P \subseteq Q$
	b. $p \Leftrightarrow q$	b. $P = Q$
4. Commutative properties	a. $p \vee q \equiv q \vee p$	a. $P \cup Q = Q \cup P$
	b. $p \wedge q \equiv q \wedge p$	b. $P \cap Q = Q \cap P$
5. Associative properties	a. $(p \vee q) \vee r$ $\equiv p \vee (q \vee r)$	a. $(P \cup Q) \cup R$ $= P \cup (Q \cap R)$
	b. $(p \wedge q) \wedge r$ $\equiv p \wedge (q \wedge r)$	b. $(P \cap Q) \cap R$ $= P \cap (Q \cap R)$
6. Distributive properties	a. $p \vee (q \wedge r)$ $\equiv (p \vee q) \wedge (p \vee r)$	a. $P \cup (Q \cap R)$ $= (P \cup Q) \cap (P \cup R)$
	b. $p \wedge (q \vee r)$ $\equiv (p \wedge q) \vee (p \wedge r)$	b. $P \cap (Q \cup R)$ $= (P \cap Q) \cup (P \cap R)$
7. Idempotent properties	a. $p \vee p \equiv p$	a. $P \cup P = P$
	b. $p \wedge p \equiv p$	b. $P \cap P = P$
8. Identity	a. $p \vee c \equiv p$	a. $P \cup \emptyset = P$
	b. $p \vee t \equiv t$	b. $P \cup U = U$
	c. $p \wedge c \equiv c$	c. $P \cap \emptyset = \emptyset$
	d. $p \wedge t \equiv p$	d. $P \cap U = P$
9. Complement properties	a. $p \wedge \sim p \equiv c$	a. $P \cap P' = \emptyset$
	b. $p \vee \sim p \equiv t$	b. $P \cup P' = U$
	c. $\sim t \equiv c$	c. $U' = \emptyset$
	d. $\sim c \equiv t$	d. $\emptyset' = U$
	e. $\sim \sim p \equiv p$	e. $(P')' = P$
10. De Morgan's laws	a. $\sim (p \vee q)$ $\equiv \sim p \wedge \sim q$	a. $(P \cup Q)' = P' \cap Q'$
	b. $\sim (p \wedge q)$ $\equiv \sim p \vee \sim q$	b. $(P \cap Q)' = P' \cup Q'$

SUMMARY

In this chapter we discussed sets and operations on sets with emphasis placed on well-defined sets. Various ways of writing sets, including the set-builder notation and the roster form, were described. A distinction was made between finite sets and infinite sets, and the null or empty set was introduced. Definitions for subset, proper subset, equal sets, equivalent sets, disjoint sets, overlapping sets, and the universal set were given. Relative to operations on sets, we defined and discussed union, intersection, complement, difference, and cartesian product of sets.

Special emphasis was placed on showing the similarities between the algebra of sets and the algebra of propositions. These similarities were discussed and subsequently summarized in Section 3.6. An application involving the use of Venn diagrams and the operations on sets also was introduced.

The following symbols were introduced during our discussions.

SYMBOL	INTERPRETATION
$a \in A$	The object a belongs to the set A
$b \notin B$	The object b does not belong to the set B
$\{\ \}$, or $\varnothing$	The null or empty set (either symbol may be used)
$A \subseteq B$	The set A is a subset of the set B
$A \nsubseteq B$	The set A is not a subset of the set B
$A \subset B$	The set A is a proper subset of the set B
$A \not\subset B$	The set A is not a proper subset of the set B
$A = B$	The sets A and B are equal
$A \sim B$	The sets A and B are equivalent
$A \cup B$	The union of the sets A and B
$A \cap B$	The intersection of the sets A and B
A'	The complement of the set A relative to U, if $A \subseteq U$
U	The universal set
$A - B$	A difference set of A and B
(a, b)	The ordered pair of objects a and b
$A \times B$	The cartesian product of sets A and B

REVIEW EXERCISES FOR CHAPTER 3

1. Define each of the following.
   a. A well-defined set.
   b. A finite set.
   c. An infinite set.
   d. The null or empty set.
   e. A is a subset of B.
   f. M is a proper subset of P.

    g. Equal sets.

    h. Disjoint sets.

    i. Overlapping sets.

    j. Equivalent sets.

    k. The union of two sets.

    l. The intersection of two sets.

    m. The cartesian product of two sets.

    n. The difference of two sets.

    o. The complement of A relative to U, if $A \subseteq U$.

    p. The universal set.

    q. A variable.

    r. A constant.

    s. An ordered pair.

2. Discuss three ways by which we can represent well-defined sets.
3. Give an example of two nonempty sets which are equivalent.
4. Give an example of two sets A and B such that $A \subseteq B$.
5. Give an example of two sets C and D such that $C \subset D$.
6. Give an example of two sets E and F that are disjoint.
7. Give an example of two sets G and H that are overlapping.
8. Give an example of two sets J and K such that $J \cap K \neq \varnothing$
9. Give an example of two sets L and M such that $L \cup M = M$.
10. Identify the following sets as being finite or infinite, and give a reason for your answer.

    a. The set of all positive integers.

    b. The set of all integers.

    c. The set of all hairs on your head right now.

    d. The set of all people in the world at a particular time.

    e. The set of all past U.S. Presidents.

    f. The set of all college graduates between June 1, 1919, and June 30, 1970.

    g. The set of all cheerful South Koreans on January 1, 1959.

    h. The set of all real numbers greater than $^-29$ and less than $^+37$.

    i. The set of all baseballs hit out of Yankee Stadium during World Series games.

    j. The set of all leaves along the curbside in a particular city on a particular fall day.

11. Write each of the following sets using the roster method.

    a. The set of all natural numbers greater than 20 and less than 29.

    b. The set of all integers greater than or equal to $^-4$ and less than $^+3$.

    c. The set of all vowels in the English alphabet.

    d. The set of all the symbols of the Greek alphabet (use your dictionary if necessary).

    e. The set of all the symbols used in the Babylonian number system (consult the mathematics section of your library and look under history of mathematics or early civilization number systems).

f. The set of all the *basic* Roman numerals used to represent the first 1000 natural numbers. (Hint: In our Hindu-Arabic system of numeration, only ten basic numerals are used.)

12. Write each of the following sets using the set-builder notation.
    a. The set of all even natural numbers.
    b. The set of all solutions of the equation $x^2 + 5x + 6 = 0$.
    c. The set of all elements in the union of the sets P and Q.
    d. The set of all elements in the cartesian product $M \times N$.
    e. The set of all elements in the set $R - S$.
    f. The set of all natural numbers that are odd but less than the number y.

13. Let $A = \{x | x$ is a natural number$\}$, $B = \{y | y = 2n, n$ is a natural number$\}$, $C = \{z | z = 2m + 1, m$ is a natural number$\}$, $D = \{u | u$ is a natural number, $u < 9\}$, $E = \{2, 4, 6, 8, 10, 12, 14, 16\}$, and $F = \{1, 3, 9, 10, 13, 20, 21\}$. Identify the following statements as being true or false, and give a reason for your answer.
    a. $E \subseteq B$.   b. $A \sim B$.
    c. $A = C$.   d. $F \subset C$.
    e. $D \sim E$.   f. $E \sim F$.
    g. $B \cap C = \emptyset$ h. B and D are overlapping sets.
    i. $A = B \cup C$.   j. E and F are overlapping sets.

14. Let $U = \{1, 2, 3, 4, 5, 6, 7, 8, 9, 10, 11, 12\}$, $A = \{1, 3, 5, 7\}$, $B = \{2, 5, 8, 12\}$, $C = \{9, 10, 11, 12\}$, and $D = \{4, 5, 7, 11\}$. Form each of the following sets.
    a. A'.   b. B'.
    c. $C' \cup D$.   d. $(A \cap B)'$.
    e. $B' \cap C'$.   f. $C \cap D'$.
    g. $A \cap C$.   h. $B \cup (C \cup D)$.
    i. $B' \cup (C \cap D)'$.   j. $A' \cap (B' \cup C')$.

15. Let $A = \{1, 2, 3, 4, 5\}$, $B = \{3, 4, 5, 6, 7\}$, $C = \{2, 3\}$, and $D = \{1, 5, 7\}$. Form each of the following sets.
    a. $A - B$.   b. $B - A$.
    c. $C \times D$.   d. $D \times B$.
    e. The complement of C relative to A, if possible.
    f. The complement of D relative to B, if possible.
    g. $C - D$.   h. $B - D$.
    i. $D \times C$.   j. $A \times B$.

CHAPTER 4
RELATIONS AND FUNCTIONS

This chapter will introduce some concepts that will be used throughout the rest of the text. Among these is the concept of a relation.

4.1 RELATIONS

Section 3.4 defined the cartesian product $A \times B$ of the two sets A and B as the set of all those ordered pairs (a, b) such that the first component a is an element of A and the second component b is an element of B. We also observed that this cartesian product was formed by pairing each element of A, in turn, with every element of B. We will now define nonempty subsets of cartesian products, which we will call *relations*.

DEFINITION 4.1 The *relation C* on the set $A \times B$ is a nonempty subset of $A \times B$. Hence $C \subseteq A \times B$ and $C \neq \emptyset$. If $A = B$, the relation C is said to be on the set $A \times A$ or simply on the set A.

In the definition of a relation it is not necessary that the elements of set A be of the same form or type as the elements of set B. Suppose, for example, that A is the set of all adult males in a given community and that B is the set of all adult females in the same community. Then the cartesian product $A \times B$ would be the set wherein each male in set A is paired, in turn, with every female in set B.

Assume that our community is a typical community. Then some of the adult males in A would be married to some of the adult females in B

and, in our set $A \times B$, we would have some ordered pairs representing husbands and wives. Let C, a subset of $A \times B$, be the nonempty set of ordered pairs whose first components are adult males in A, whose second components are adult females in B, and such that in each of the ordered pairs of C the first component is the husband of the second component. Then $C = \{(m, f) \mid m \in A, f \in B\}$ represents the relation "is the husband of" on the set $A \times B$. This relation is a set of ordered pairs whose components are people. In a relation it is possible that the components of the ordered pairs may be numbers, triangles, lines, or almost anything. Other examples of relations are

1. The relation "is greater than" on the set of all real numbers is the set of all ordered pairs (a, b) for which a and b are real numbers and $a > b$.
2. The relation "is congruent to" on the set of all triangles is the set of all ordered pairs (x, y) for which x and y are triangles and x is congruent to y.
3. The relation "is the same color as" on the set of all flowers is the set of ordered pairs (p, q) such that p and q are flowers and p is the same color as q.
4. The relation "is a subset of" on the set of all sets is the set of all ordered pairs (A, B) for which A and B are sets and $A \subseteq B$.
5. The relation "implies" on the set of all propositions is the set of all ordered pairs (p, q) for which p and q are propositions and p implies q.
6. The relation "is the brother of" on the set of all male humans is the set of all ordered pairs (u, v) for which u and v are human males and u is the brother of v.
7. The relation "is the sister of" on the set $S \times T$, where S is the set of all human females and T is the set of all human beings, is the set of all ordered pairs (m, n) for which m is a female human and n is a human being and m is the sister of n.
8. The relation "is perpendicular to" on the set of all lines in a plane is the set of all ordered pairs (r, s) for which r and s are lines in the same plane and r is perpendicular to s.

DEFINITION 4.2 The set of all the distinct first components of the ordered pairs in a relation is called the *domain* of the relation. The set of all the distinct second components of the ordered pairs is called the *range* of the relation.

If R is a relation we will symbolize its domain by D_R; in a similar manner the range of R will be denoted by R_R.

EXAMPLE Consider the sets $S = \{1, 2, 3\}$ and $R = \{(1, 1), (1, 3), (2, 3), (3, 2)\}$. R is a relation on S, since R is a nonempty set of ordered pairs all of whose components are elements of S. The domain of R is the *set* whose elements are 1, 2, and 3; that is, $D_R = \{1, 2, 3\}$. The range of R is the *set* whose elements are 1, 2, and 3; that is, $R_R = \{1, 2, 3\}$. In this case $D_R = R_R$. However, this is *not* a requirement for a relation. The graphs of $S \times S$ and R are given in Figures 4.1 and 4.2, respectively. Observe that $R \subseteq S \times S$.

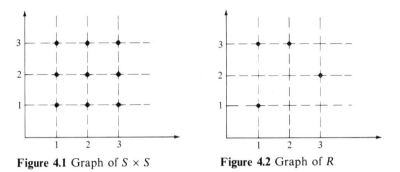

Figure 4.1 Graph of $S \times S$ **Figure 4.2** Graph of R

Suppose that we are given a relation on a given set with all of the ordered pairs enumerated. Suppose, further, that as we were transcribing these ordered pairs we consistently reversed the order of the components in each of the ordered pairs. Obviously we do not have the same relation unless the first and second components in each of the ordered pairs were equal. The result, however, is still a relation.

DEFINITION 4.3 By the *inverse* of the relation R, we mean a new relation denoted by R^*, in which the elements of R^* are the ordered pairs of R with the components reversed. If $(x, y) \in R$, then $(y, x) \in R^*$.

EXAMPLE If $R = \{(1, 2), (2, 3), (3, 4)\}$, then $R^* = \{(2, 1), (3, 2), (4, 3)\}$. Verify that the domain and the range of R become, respectively, the range and domain of R^*. The graphs of R and R^* are given in Figures 4.3 and 4.4, respectively.

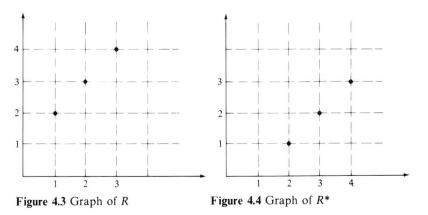

Figure 4.3 Graph of R **Figure 4.4** Graph of R^*

EXERCISES

1. We gave the relation "is the same color as" on the set of all flowers as an example of a nonmathematical relation. Give five additional examples of nonmathematical relations excluding those given in the text.
2. For each of the relations given in Exercise 1, specify its domain and also its range.
3. We gave the relation "is greater than" on the set of all real numbers as an example of a mathematical relation. Give five additional examples of mathematical relations excluding those given in the text.
4. For each of the relations given in Exercise 3, specify the domain and the range.
5. Consider each of the following relations.

$\{(-3, 2), (2, -3), (1, 0), (-5, 2), (4, 7)\}$

$\{(-2, 3), (0, 0), (4, -7), (\sqrt{3}, 2), (-1, -1)\}$

$\{(-5, 2), (-3, 2), (0, 2), (1, 2), (\sqrt{2}, 2), (5, 2), (9, 2)\}$

$\{(a, b) | a = 2b + 3, b \in M\}$ such that $M = \{-5, -4, -3, -2, -1, 0, 1, 2, 3\}$

$\{(x, y) | x, y \in A, x \leq y\}$ such that $A = \{-3, 1, 0, \sqrt{2}, 5\sqrt{3}\}$

   a. For each of the relations given above write the domain and the range.
   b. Graph each of the above relations.
   c. Write the inverse relation for each of the above relations.
   d. Graph each of the inverse relations formed in part c.

4.2 SPECIAL RELATIONS

Let $A = \{1, 2, 3\}$ and $B = \{a, b, c\}$. Then $A \times B = \{(1, a), (1, b), (1, c), (2, a), (2, b), (2, c), (3, a), (3, b), (3, c)\}$. By use of Definition 4.1 it should be obvious that several different relations may now be formed on the set $A \times B$. Hence the definition of a relation represents a very general concept.

Among the various relations considered in mathematics are some very special types called functions, which will be discussed in the next section. Another special type of relation that will be used later in this text is that of an equivalence relation. As we will learn in this section, an equivalence relation will be defined in terms of other special types of relations which we will now consider.

If we are given several relations on a given set S, we may observe that

   1. For *every* element of the set S the element is found paired with itself in some of the given relations.
   2. For *some but not all* elements of the set S the elements may be found paired with themselves in the ordered pairs of the relations.
   3. For *no* elements of the set S is the element paired with itself in any of the ordered pairs in the relation.

Before introducing our next definition, we will note that if R is a relation and if $(x, y) \in R$, we say that x is in the relation R to y, and we write this as xRy. Observe that $xRy \Leftrightarrow (x, y) \in R$. If $(x, y) \notin R$, then we write $x\mathcal{R}y$ which is read "x is not in the relation R to y."

DEFINITION 4.4 A relation R on the set S is said to be *reflexive* if and only if R is a relation on S and xRx for *every* $x \in S$.

EXAMPLE Let $S = \{1, 2, 3, 4\}$ and $R = \{(1, 1), (1, 2), (2, 2), (2, 4), (3, 3), (4, 4)\}$. R is reflexive on S since R is a relation on S (since $R \neq \varnothing$ and $R \subseteq S \times S$), for 1, 2, 3, $4 \in S$ we have $(1, 1), (2, 2), (3, 3), (4, 4) \in R$, and there are no other elements in S.

EXAMPLE Let $S = \{1, 2, 3, 4\}$ and $T = \{(1, 1), (2, 3), (3, 3)\}$. Although T is a relation on S (verify!), T is not a reflexive relation on S since $2 \in S$ but $(2, 2) \notin T$.

If we are given several relations on a given set S, we may also observe that
1. For *every* $(a, b) \in R$ where R is a relation on S we also have $(b, a) \in R$.
2. For *some but not all* $(a, b) \in T$ where T is a relation on S we have $(b, a) \in T$.
3. For *no* $(a, b) \in Q$ where Q is a relation on S is $(b, a) \in Q$.

DEFINITION 4.5 A relation R on the set S is said to be *symmetric* if and only if yRx *whenever* xRy.

Relative to this definition it must be emphasized that the ordered pair (x, y) does not have to be an element of R. However, if $(x, y) \in R$, then (y, x) also must be an element of R for R to be symmetric.

EXAMPLE Let $S = \{1, 2, 3\}$ and $A = \{(1, 2), (2, 2), (2, 1)\}$. A is a symmetric relation on S since A is a relation on S (verify!) and, further, for $(1, 2) \in A$ we have $(2, 1) \in A$, for $(2, 2) \in A$ we have $(2, 2) \in A$, for $(2, 1) \in A$ we have $(1, 2) \in A$, and there are no other pairs in A.

EXAMPLE Let $S = \{1, 2, 3, 4\}$ and $C = \{(1, 2), (1, 4), (2, 3), (2, 1), (3, 2)\}$. C is not a symmetric relation on S since $(1, 4) \in C$ but $(4, 1) \notin C$. (Verify that C is a relation on S.)

EXAMPLE Let $S = \{1, 2, 3\}$ and $B = \{(1, 2), (2, 1), (3, 3), (3, 4), (4, 3)\}$. B is *not* a symmetric relation on S. Although for $(1, 2), (2, 1), (3, 3), (3, 4), (4, 3) \in B$ we *do* have $(2, 1), (1, 2), (3, 3), (4, 3), (3, 4) \in B$, B is not a relation on S. $B \nsubseteq S \times S$ since $(3, 4), (4, 3) \in B$ but $4 \notin S$.

Relations may be further classified according to the following definition.

DEFINITION 4.6 A relation R on the set S is said to be *transitive* if and only if R is a relation on S and xRz whenever xRy and yRz.

Relative to the above definition it should be noted that the ordered pairs (x, y) and (y, z) do not have to be elements of R. However, if they are, then (x, z) must also be an element of R for it to be a transitive relation. Also, note that y appears twice in the definition—as the second component of one ordered pair and as the first component of the other.

EXAMPLE Let $S = \{a, b, c\}$ and $E = \{(a, b), (b, c), (c, c), (a, c), (b, b), (c, b)\}$. E is a relation on S since E is a nonempty subset of $S \times S$. Also, $(a, b), (b, c) \in E$ and $(a, c) \in E$; $(a, b), (b, b) \in E$ and $(a, b) \in E$; $(b, c), (c, b) \in E$ and $(b, b) \in E$; $(b, c), (c, c) \in E$ and $(b, c) \in E$; $(b, b), (b, c) \in E$ and $(b, c) \in E$; $(c, b), (b, b) \in E$ and $(c, b) \in E$; $(c, c), (c, b) \in E$ and $(c, b) \in E$; $(a, c), (c, b) \in E$ and $(a, b) \in E$; $(a, c), (c, c) \in E$ and $(a, c) \in E$; $(c, b), (b, c) \in E$ and $(c, c) \in E$. We have exhausted all possible pairings of the form (x, y) and (y, z) and observed that $(x, z) \in E$ in every case. Hence E is a transitive relation on S.

EXAMPLE Let $S = \{1, 2, 3\}$ and $F = \{(1, 2), (3, 2)\}$. F is a relation on S since F is a nonempty subset of $S \times S$. Now observe that $(1, 2) \in F$ but that there is no ordered pair in F whose first component is 2. Since there is no ordered pair to consider along with $(1, 2)$ we go on to the next ordered pair $(3, 2)$. Again we see that there is no ordered pair in F whose first component is 2. Hence F is a transitive relation on S since since we cannot exhibit two ordered pairs in F, say (x, y) and (y, z) such that $(x, z) \notin F$. We say in this case that the definition for a transitive relation is satisfied vacuously.

EXAMPLE Let $S = \{1, 2, 3, 4\}$ and $G = \{(1, 2), (2, 3), (1, 3), (3, 4)\}$. G is a relation on S since G is a nonempty subset of $S \times S$. Further, observe that $(1, 2), (2, 3) \in G$ and $(1, 3) \in G$, while $(1, 3), (3, 4) \in G$ but $(1, 4) \notin G$. Therefore, G is not a transitive relation on S.

You will note that in this section we defined a relation R on the set S to be symmetric after we defined a relation R on the set S to be reflexive. This should not be interpreted in any manner to mean that if a relation R on the set S is not reflexive it cannot be symmetric. Similarly, a relation on a given set does not have to be symmetric for it to be transitive. However, we can exhibit relations that are reflexive and symmetric, reflexive and transitive, or symmetric and transitive on a given set. We can also exhibit relations on a given set that are reflexive, symmetric, and transitive.

DEFINITION 4.7 A relation R on the set S that is reflexive, symmetric, and transitive on S is said to be an *equivalence* relation on the set S.

The relation "is equal to" on the set of all real numbers and the relation "is the same color as" on the set of all flowers are examples of equivalence relations. We will be concerned with equivalence relations in this text because of their usefulness in mathematics. An equivalence relation on a given set divides or partitions that set into classes or subsets all of whose elements are equivalent in some manner. Such classes or subsets are called *equivalence classes*.

For example, the equivalence relation "is the same color as" on the set of all flowers divides the set of all flowers into equivalence classes by color. All of the flowers in any particular class are equivalent to each other in regard to color.

In the study of the real numbers one encounters equivalence classes. For example, the integer $+2$ can be defined as the equivalence class of all elements of the form or type $a - b$ where a and b are counting numbers and $a = b + 2$. Hence the elements $6 - 4$ and $5 - 3$ belong to the same equivalence class and have the same value, 2.

The rational number $\frac{1}{2}$ can be defined as the equivalence class of all elements of the form or type m/n where m and n are integers, $n \neq 0$, and $n = 2m$. Hence the elements $(+2)/(+4)$ and $(-6)/(-12)$ belong to the same equivalence class and have the same value, $\frac{1}{2}$.

EXERCISES

1. For each of the following indicate whether the set R is a relation on the set S. Give a reason for each of your answers.
   a. $S = \{a, b, c\}$, $R = \{(a, a), (a, c), (b, b), (b, c)\}$.
   b. $S = \{1, 2, 3, 4\}$, $R = \{(1, 2), (2, 4), (4, 3), (5, 2)\}$.
   c. $S = \{x \mid x \in Re, \ x \leq 50\}$, $R = \{(-2, 0), (\frac{1}{2}, -3), (\sqrt{2}, \sqrt{3}), (\pi, -17), (\sqrt{3/2}, -\frac{1}{2})\}$.
   d. $S = \{y \mid y \text{ is an integer}, \ -2 < y < 7\}$, $R = \{(1, 0), (2, -2)\}$.
2. Which of the following relations R on the set S are reflexive?
   a. $S = \{1, 2, 3\}$, $R = \{(1, 2), (2, 2), (3, 2), (3, 3), (2, 1), (1, 1)\}$.
   b. $S = \{a \mid a \text{ is an integer}\}$, $R = \{(2, 2), (-4, -4), (0, 1)\}$.
   c. $S = \{b \mid b \text{ is a natural number}\}$, $R = \{(0, \frac{1}{2}), (-2, 5), (4, 2), (\frac{3}{2}, \frac{2}{5})\}$.
   d. $S = \{c \mid c \text{ is a natural number} < 6\}$, $R = \{(2, 3), (3, 3), (5, 6)\}$.
3. Which of the following relations R on the set S are symmetric?
   a. $S = \{1, 2, 3, 4, 6\}$, $R = \{(1, 2), (2, 1), (3, 3), (4, 5), (5, 4), (6, 6)\}$.
   b. $S = \{a, b, c, d, \ldots, z\}$. $R = \{(e, g), (h, m), (m, n), (g, e), (m, h), (n, m)\}$.
   c. $S = \{x \mid x \in Re\}$, $R = \{(-2, 1), (0, \frac{1}{2}), (1, -2)\}$; $Re = $ the set of real numbers.
   d. $S = \{x \mid x \text{ is an integer}\}$, $R = \{(-5, 4), (3, 0), (1, 7), (6, 0), (-17, 39)\}$.
4. Which of the following relations R on the set S are transitive?
   a. $S = (3, 5, 7, 9)$, $R = \{(3, 3), (5, 7), (7, 9), (9, 3)\}$.
   b. $S = \{a, b, c\}$, $R = \{(a, b), (b, c), (b, a), (a, c), (a, a), (b, b)\}$.

   c. $S = \{1, 2, 3, 6, 9\}$, $R = \{(1, 1), (2, 2), (9, 9), (6, 6), (3, 3)\}$.

   d. $S = \{x \mid x \in Re\}$, $R = \{(0, 2), (4, 7), (7, 4)\}$.

5. Let $S = \{1, 2, 3, 4, 5\}$. Exhibit a relation R on the set S which is

   a. Reflexive and symmetric but not transitive.

   b. Symmetric and transitive but not reflexive.

   c. Reflexive, not symmetric, and transitive.

   d. Not reflexive, not symmetric, and not transitive.

6. Consider the following definitions:

   (1) A relation R on the set S is said to be *nonreflexive* if and only if R is a relation on S and for *some, but not all,* $x \in S$, xRx.

   (2) A relation R on the set S is said to be *irreflexive* or *antireflexive* if and only if R is a relation on S and for *every* $x \in S$, $x\mathcal{R}x$. (Recall that $x\mathcal{R}X$ means "x is not in the relation R to x.")

   a. Which of the relations in Exercise 2 are nonreflexive?

   b. Which of the relations in Exercise 2 are irreflexive?

7. Consider the following definitions:

   (1) A relation R on the set S is said to be *nonsymmetric* if and only if R is a relation on S and xRy implies yRx for *some but not all* $(x, y) \in R$.

   (2) A relation R on the set S is said to be *asymmetric* if and only if R is a relation on S and $y\mathcal{R}x$ for *every* xRy.

   a. Which of the relations in Exercise 3 are nonsymmetric?

   b. Which of the relations in Exercise 3 are asymmetric?

8. Consider the following definitions:

   (1) A relation R on the set S is said to be *nontransitive* if and only if R is a relation on S and for some elements x, y, $z \in S$, xRy, and yRz imply xRz while for other triples x, y, and z, xRy and yRz and $x\mathcal{R}x$.

   (2) A relation R on the set S is said to be *intransitive* if and only if R is a relation on S and *whenever* xRy and yRz we have $x\mathcal{R}z$.

   a. Which of the relations in Exercise 4 are nontransitive?

   b. Which of the relations in Exercise 4 are intransitive?

4.3 FUNCTIONS AS SPECIAL TYPES OF RELATIONS

The specification for a relation may permit the inclusion of two ordered pairs within the set such that the first components are equal and the second components are unequal. For instance, the relation $T = \{(x, y) \mid x^2 + y^2 = 25\}$ has ordered pairs such as $(-3, -4)$ and $(-3, 4)$. Hence the set may contain ordered pairs with equal first components but unequal second components. However, for the relation $R = \{(x, y) \mid y = x + 2\}$, we see that no two ordered pairs have the same first component. In other words, if two ordered pairs within this relation have equal first components, they also have equal second components. Such relations are called *functions*.

DEFINITION 4.8 A *function* is a relation such that each element in the domain is paired with exactly one element in the range.

The *domain* of the function is the set of all the distinct first components of its ordered pairs. The *range* of the function is the set of all the distinct second components of its ordered pairs.

Formally, F is a function if and only if F is a nonempty subset of some cartesian product and, further, if (x_1, y_1) and (x_2, y_2) are elements of F, then $y_1 = y_2$ *whenever* $x_1 = x_2$. If we have a relation that has *distinct* first components, the relation is a function according to this definition.

DEFINITION 4.9 If F is a function and u and v are variables whose universal sets are D_F and R_F, respectively, and if $(u, v) \in F$, then v is called the *image* of u under F.

We use the symbol $F(u)$ to denote the image of u under F. $F(u)$, which is read "F of u", is the second component of the ordered pair. Hence $v = F(u)$ if and only if $(u, v) \in F$. In particular, if $a \in D_F$ and $b = F(a)$, we define $F(a)$ to be the *value* of $F(u)$ at a.

DEFINITION 4.10 The variable u, whose universe is the domain of F, is defined to be an *independent variable*. Similarly, the variable v, whose universe is the range of F, is defined to be the *dependent variable*.

EXAMPLE Consider the set $F = \{(u, v) \mid v = F(u) = u + 3\}$. F is a function since F is a relation on the set of all real numbers such that each

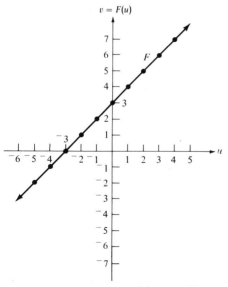

Figure 4.5 Graph of $v = F(u) = u + 3$

value of u is paired with exactly one value of v. The domain of F is the set of all real numbers. It is the universe for the independent variable u. The range of F is also the set of all real numbers. It is the universe for the dependent variable v. The rule of association between the independent variable and the dependent variable is given by the equation $v = u + 3$. The graph of F is the oblique line given in Figure 4.5.

EXAMPLE $G = \{(x, y) \mid y = G(x) = 4\}$ is also a function, since G is a relation on the set of all real numbers such that each value of x is paired with exactly one value of y. The domain of G is the set of all real numbers. The range of G is the set containing only the real number 4. In other words, the second components of the ordered pairs of G are all equal to 4. Such a function is called a *constant valued function*. The graph of G is the horizontal line given in Figure 4.6.

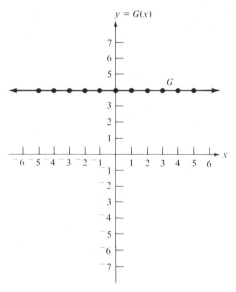

Figure 4.6 Graph of $y = G(x) = 4$

DEFINITION 4.11 A function is a *constant valued function* if and only if all of the second components of its ordered pairs are equal.

We may have a relation on a given set that is not a function but whose inverse is a function. For instance, $B = \{(2, 3), (4, 6), (2, 4), (3, 2)\}$ is a relation on the set of all real numbers, but it is not a function since $(2, 3), (2, 4) \in B$ and the first components are equal but $3 \neq 4$. However, $B^* = \{(3, 2), (6, 4), (4, 2), (2, 3)\}$ is a function since all of its first components are distinct.

Given any relation on a set, we can always form its inverse. The following possibilities exist:

The given relation is a function, and its inverse relation is a function.
The given relation is a function, and its inverse relation is not a function.
The given relation is not a function, and its inverse relation is a function.
The given relation is not a function, and its inverse relation is not a function.

DEFINITION 4.12 If a function has an inverse relation that is also a function, then the function is said to be *one-to-one*.

EXAMPLE $H = \{(1, 2), (2, 3), (3, 4), (7, 9)\}$ is a relation on the set of all real numbers. H is also a function since each element in the domain of H is paired with exactly one element in its range. (Observe that the first components of the ordered pairs in H are all distinct.) Forming the inverse of H, we have $H^* = \{(2, 1), (3, 2), (4, 3), (9, 7)\}$, which is also a function. Since H is a function and H^* is also a function, H is one-to-one. The graphs of H and H^* are given in Figure 4.7(a) and (b).

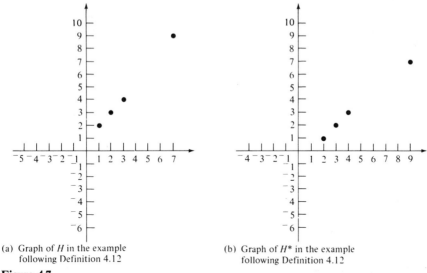

(a) Graph of H in the example following Definition 4.12

(b) Graph of H^* in the example following Definition 4.12

Figure 4.7

EXAMPLE $Q = \{(3, 1), (3, 2), (3, 4), (3, 10), (3, 12)\}$ is a relation on the set of all real numbers. Q is not a function since $(3, 1), (3, 2) \in Q$, $3 = 3$ but $1 \neq 2$. However, $Q^* = \{(1, 3), (2, 3), (4, 3), (10, 3), (12, 3)\}$ is a function since all of its first components are distinct. Observe that all of the second components of the ordered pairs of Q^* are equal. Hence Q^* is a constant valued function. The graphs of Q and Q^* are given in Figure 4.8(a) and (b).

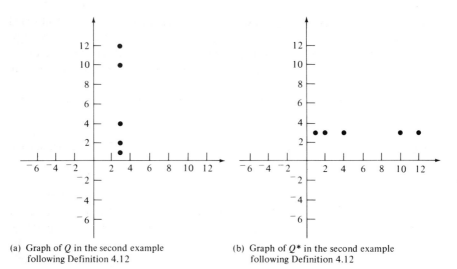

(a) Graph of Q in the second example
following Definition 4.12

(b) Graph of Q^* in the second example
following Definition 4.12

Figure 4.8

EXAMPLE $X = \{(x, y) | y^2 = 4x\}$ is a relation on the set of all real numbers. Is X a function? Let (x_1, y_1) and (x_2, y_2) be two arbitrary elements of X. Suppose that $x_1 = x_2$. Then X is a function if $y_1 = y_2$. But $y_1 = \pm 2\sqrt{x_1}$ and $y_2 = \pm 2\sqrt{x_2}$. From these equalities it follows that $y_1 = y_2$ or $y_1 = -y_2$. For instance, the ordered pairs $(1, -2)$ and $(1, 2)$

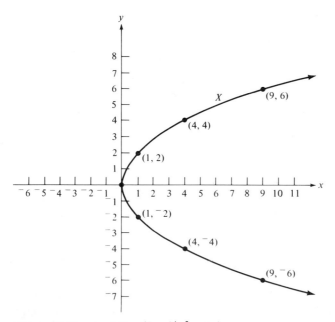

Figure 4.9 Graph of $X = \{(x, y) | y^2 = 4x\}$

are elements of X. The first components are equal, but the second components are unequal. Therefore it does not necessarily follow that the second components of two arbitrary ordered pairs of X are equal whenever the first components are equal; X is not a function. The graph of X is given in Figure 4.9.

Up to this point we have discussed functions of a single independent variable. Functions of two independent variables also exist.

DEFINITION 4.13 A *function of two independent variables* is a set of ordered *triples*, the first two elements of which are values of the independent variables and the third elements of which are the correspondents of the pair of independent variables under the function. A function of two independent variables, then, is a function whose domain is a subset of the cartesian product of some two sets.

As an example of a function of two independent variables, consider $L = \{(x, y; z) \mid z = L(x, y) = x + 3y - 4\}$, where $(x, y) \in D_L = Re \times Re$, where Re is being used to denote the set of all real numbers.

DEFINITION 4.14 A function whose domain is a set of ordered triples is called a *function of three independent variables*. Similarly, if the domain of the function is a set of ordered $n$tuples, the function is called a *function of n-independent variables*.

It is not my intent to elaborate upon functions of two or more independent variables in this text. In Section 4.7, however, I will introduce binary operations as functions of two independent variables.

EXERCISES

1. Consider each of the following relations.
    (1) $\{(-3, 2), (0, 4), (5, 5), (7, -3), (\sqrt{2}, \sqrt{3})\}$
    (2) $\{(-5, 2), (-3, 2), (1, 2), (4, 2)\}$
    (3) $\{(2, -3), (2, -1), (2, 0), (2, 5), (2, 7)\}$
    (4) $\{(1, 2), (2, 3), (3, 4), (4, 5), (5, 6), (6, 7)\}$
    (5) $\{(-4, 0), (-3, 4), (-3, 1), (0, 2), (1, -4), (5, 0)\}$
    (6) $\{(x, y) \mid x, y \in Re, y = x^2\}$
    (7) $\{(x, y) \mid x, y \in Re, x = y^2\}$
    (8) $\{(x, y) \mid x, y \in Re, 2x + 3y = 5\}$
    (9) $\{(x, y) \mid x, y \in Re, y = 2\}$
    (10) $\{(x, y) \mid x, y \in Re, x = -3\}$
    (11) $\{(x, y) \mid x, y \in Re, x^2 + y^2 = 25\}$
    (12) $\{(x, y) \mid x, y \in Re, xy = 1\}$

a. Which of the above relations are functions? Give a reason for your responses.
b. Which of the functions identified in part (a) are one-to-one? Why?
c. Graph each of the above relations.
d. Examine each of the graphs in part (c). For the relations you classified as functions, can you construct lines perpendicular to the axis of the domain through each point in the domain and have the lines intersect the graph only once? (If the answer is no, either your graph is incorrect or your answer that the relation is a function is incorrect or both are incorrect.)

2. For each of the relations given in Exercise 1 form the inverse relation.
3. Which of the relations given in Exercise 1 have inverses that are functions? Give a reason for each of your answers.
4. Which of the relations given in Exercise 1 are one-to-one functions? Why?
5. Give an example of a nonmathematical function that is not one-to-one.
6. Give an example of a nonmathematical relation that is not a function but such that its inverse relation is a function.
7. Give an example of a nonmathematical relation that is not a function and such that its inverse relation is not a function.

4.4 FUNCTION MACHINES

The previous section discussed the concept of a function as a special type of relation. It was observed that for each function there was a pairing of elements—one in the domain of the function with exactly one in its range. The rule of association or the specification of the function was used to determine this pairing of elements. It is the recipe or the set of instructions we follow to determine how the dependent variable is related to the independent variable.

For instance, if we have a function, F, with specification $y = F(x) = x + 3$, then the independent variable is denoted by x. We next observe that any real value can be assigned to x and the dependent variable, denoted by y, also will be a real value. In particular, if $x = 2$, then $y = 2 + 3$ or 5 and the ordered pair $(2, 5) \in F$. If $x = 4$, then $y = 4 + 3$ or 7 and the ordered pair $(4, 7) \in F$. The specification of this function enables us to determine the value of y for each value assigned to x—simply add 3 to x to get the value for y. Is it possible to invent a machine to "produce" values of y for corresponding values of x? If the answer is yes, is it possible that the machine could be so designed that when it receives a value of x, called the input, it would automatically add 3 to it to produce the y value or output?

In this section we will "construct" such machines. Further, we will combine two or more of these simple machines to produce a more complex machine. Such machines will be called *function machines* and are included in many texts at the elementary school level. Our emphasis throughout this section will be on determining the domain of a particular function.

Assuming that you are now ready and willing to do so, we will construct a machine to generate values of y associated with the function F specified by $y = F(x) = x + 3$. In examining the specification, we observed that to obtain the y values, one simply adds 3 to the respective x values. Hence it would appear that we should attempt to construct an adder machine— but not just any adder machine. We want a machine that will add 3—not 4 or 0 or -2. Further, we will design our machine to look like an adder and also identify it as a 3-adder machine as in Figure 4.10.

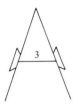

Figure 4.10 A 3-adder machine

Did you notice the two protusions on the sides of the machine? If so, good; if not, look again. The one on the left is for the input or the values of x that will be "fed" into the machine. The one on the right is for the output or the corresponding values of y. Any value of x that is fed into the machine will have 3 added to it and "transformed" into $x + 3$ as denoted in Figure 4.11.

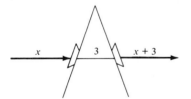

Figure 4.11 A 3-adder machine with input and output

Now if we feed in an x value of 2, we will get an output (or y value) of $2 + 3$ or 5. If we feed in an x value of -3, we will get an output of $-3 + 3 = 0$. If we feed in an x value of 0, we will get an output of $0 + 3$ or 3. Since the domain of the function under consideration is the set of all real numbers (Do you agree?) our machine will accept as input data any real value of x to which will be added 3 to produce a y (or $x + 3$) value.

Now that we have constructed a 3-adder machine we could easily construct a whole family of adder machines, each one of which would add to the given input data a unique (distinct) real number to produce the output

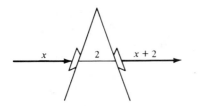

(a) A 2–adder machine

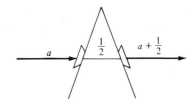

(b) A $\frac{1}{2}$-adder machine

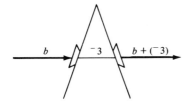

(c) A $^-$3–adder machine

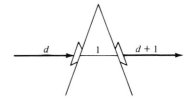

(d) A 1–adder machine

Figure 4.12 Some adder machines with input and output

data. Some of these machines are illustrated in Figure 4.12. How many such machines can be constructed (theoretically)?

Returning to our function F with specification $y = F(x) = x + 3$, we determined that the domain of F is the set of all real numbers. The range of F is also the set of all real numbers. (Verify!) Further, observe that F is one-to-one since for each element b in the range there is exactly one value, $b - 3$, in the domain. Hence the inverse function F exists, and we will denote it by the function G with the specification $y = G(x) = x - 3$. To produce the ordered pairs of this function, observe that for each x value we must subtract 3 to obtain the corresponding y values. Putting our genius to work, we will now design a 3-subtracter machine as illustrated in Figure 4.13.

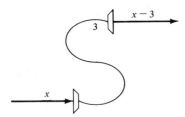

Figure 4.13 A 3-subtracter machine with input and output

Before we rush off to design a whole family of subtracter machines (How many would there be?) we ask, "Is it necessary to do so?" Clearly, subtracting 3 from x is equivalent to adding -3 to x. Hence instead of using a 3-subtracter, we could use the -3-adder already designed. Similarly, instead of using a 5-subtracter, we could use the -5-adder. In general, for each a-subtracter, we would use an $-a$-adder.

EXAMPLE Consider the function A with the specification $y = A(x)$ $= x + 2$. Assuming that we have already designed an appropriate function machine to produce the function values for A, determine (a) what type of function machine is required and (b) the output or function values for the input x values of 2, -3, 0, $\frac{1}{2}$, and $\sqrt{3}$.

Solution

a. Since the specification for the function A is $A(x) = x + 2$, we require a 2-adder as illustrated in Figure 4.14.

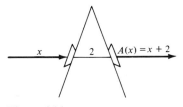

Figure 4.14

b. The output associated with such a function machine is designated as y or $A(x)$ or $x + 2$. Hence we have

INPUT	OUTPUT	RESULTS
$x = 2$	$A(x) = 2 + 2 = 4$	$A(2) = 4$
$x = -3$	$A(x) = -3 + 2 = -1$	$A(-3) = -1$
$x = 0$	$A(x) = 0 + 2 = 2$	$A(0) = 2$
$x = \frac{1}{2}$	$A(x) = \frac{1}{2} + 2 = 2\frac{1}{2}$	$A(\frac{1}{2}) = 2\frac{1}{2}$
$x = \sqrt{3}$	$A(x) = \sqrt{3} + 2 = 2 + \sqrt{3}$	$A(\sqrt{3}) = 2 + \sqrt{3}$ ∎

Having "designed" a family of adder machines that will also subtract, we can now design multiplier machines that will also divide. Recall, for instance, that dividing by 2 is equivalent to multiplying by $\frac{1}{2}$. Some of these machines are illustrated in Figure 4.15.

Now consider the function H with specification $\beta = H(\alpha) = 3\alpha + 4$. To design an appropriate function machine to produce the function values, β, we observe that we must multiply α by 3 (hence a 3-multiplier machine is required) and then add 4 to the result (hence a 4-adder machine is needed). To produce the necessary function values, β, we therefore would need a 3-multiplier machine and a 4-adder machine, *but* they must be appropriately connected. Since the 4 is to be added to the result of multiplying α by 3, it would be necessary to "feed" the α values into the 3-multiplier machine first and then feed the 3α values into the 4-adder as

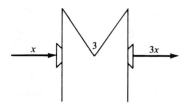

(a) A 3–multiplier machine

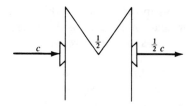

(b) A $\frac{1}{2}$-multiplier machine

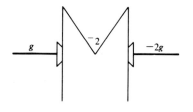

(c) $A^{-}2$-multiplier machine

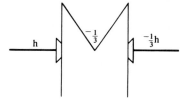

(d) $A^{-}\frac{1}{3}$-multiplier machine

Figure 4.15 Some multiplier machines with input and output

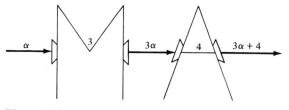

Figure 4.16

illustrated in Figure 4.16. Observe that the output of the 3-multiplier machine becomes the input for the 4-adder machine. Some representative function values are obtained as shown in the facing table.

Now that we have acquired the basic know-how for the construction of function machines that will perform simple arithmetic operations of addition, multiplication, subtraction, and division, let us try our hand at designing some machines of a more sophisticated nature. The first one will be the squaring machine, which will be associated with the function R, having the specification $q = R(p) = p^2$. The input data for this machine will be a p value, and the output value will be a p^2 value. That is, the output value

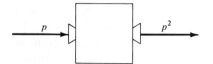

Figure 4.17 The squaring machine with input and output

INPUT FOR 3-MULTIPLIER	OUTPUT FOR 3-MULTIPLIER / INPUT FOR 4-ADDER	OUTPUT FOR 4-ADDER	RESULTS
$\alpha = -2$	$3\alpha = 3(-2) = -6$	$H(\alpha) = 3\alpha + 4$ $= -6 + 4 = -2$	$H(-2) = -2$
$\alpha = 0$	$3\alpha = 3(0) = 0$	$H(\alpha) = 3\alpha + 4$ $= 0 + 4 = 4$	$H(0) = 4$
$\alpha = 1$	$3\alpha = 3(1) = 3$	$H(\alpha) = 3\alpha + 4$ $= 3 + 4 = 7$	$H(1) = 7$
$\alpha = \sqrt{2}$	$3\alpha = 3(\sqrt{2}) = 3\sqrt{2}$	$H(\alpha) = 3\alpha + 4$ $= 3\sqrt{2} + 4$	$H(\sqrt{2}) = 3\sqrt{2} + 4$
$\alpha = \frac{2}{3}$	$3\alpha = 3(\frac{2}{3}) = 2$	$H(\alpha) = 3\alpha + 4$ $= 2 + 4 = 6$	$H(\frac{2}{3}) = 6$

will be a value of p multiplied by itself. This is no ordinary multiplier machine because the multiplier does not remain constant—it varies as the p value varies. The squaring machine is illustrated in Figure 4.17.

EXAMPLE Using the squaring machine, compute the values of $R(p) = p^2$ if $p = 2, -1, 4, 0, \sqrt{2}, -\sqrt{3}, 7,$ and $-6.$
Solution

INPUT	OUTPUT	RESULTS
$p = 2$	$R(p) = 2^2 = 4$	$R(2) = 4$
$p = -1$	$R(p) = (-1)^2 = 1$	$R(-1) = 1$
$p = 4$	$R(p) = 4^2 = 16$	$R(4) = 16$
$p = 0$	$R(p) = 0^2 = 0$	$R(0) = 0$
$p = \sqrt{2}$	$R(p) = (\sqrt{2})^2 = 2$	$R(\sqrt{2}) = 2$
$p = -\sqrt{3}$	$R(p) = (-\sqrt{3})^2 = 3$	$R(-\sqrt{3}) = 3$
$p = 7$	$R(p) = 7^2 = 49$	$R(7) = 49$
$p = -6$	$R(p) = (-6)^2 = 36$	$R(-6) = 36$ ■

Did you observe that the output values in the above example were all nonnegative? Is this true for all output values? Why?

From the squaring machine we can advance to the cubing machine as illustrated in Figure 4.18. In a similar manner we could, if necessary, design machines to raise any real number to a positive whole number power such as x^4 or y^5 or u^6 etc. However, in this section we will not go beyond the cuber.

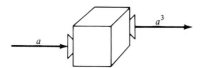

Figure 4.18 The cubing machine with input and output.

EXAMPLE Combine as necessary some of the machines already designed in this section to compute values of $R(b) = 4b^3 + 5$. Then compute $R(b)$ for $b = -3, -2, -\frac{1}{2}, 0, 2, \sqrt{3},$ and 5 using the "computer."

Solution The specification for the function R is given by the equation $R(b) = 4b^3 + 5$. Hence to obtain $R(b)$ for any given value of b in the domain of R, we perform the following operations with the appropriate machines:

OPERATION	MACHINE REQUIRED
Cube b (compute b^3)	Cuber
Multiply b^3 by 4	4-multiplier
Add 5 to $4b^3$	5-adder

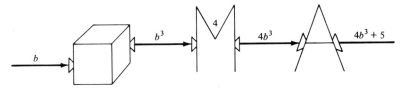

Figure 4.19

The required machines, connected in appropriate order, are illustrated in Figure 4.19. Our computed values are as indicated in the table on page 98. ■

With the squarer, cuber, and other positive integral power machines, the input values can be any real numbers and the output values will also be real numbers. The next machine we will construct will be the square-rooter, and we will note that caution will have to be exercised when working with the input values.

The square-rooter machine is illustrated in Figure 4.20. The square-rooter machine will only accept for input values those real numbers which are nonnegative since the square root of a negative real number is not a real number. We should not expect the machine to do anything which we cannot do (except possibly to work faster and more accurately) and, hence, we will not use negative values for the input.

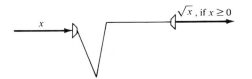

Figure 4.20 The square-rooter machine with input and output

In a similar manner we could, if necessary, construct any nth-rooter machine such that n is a positive whole number greater than or equal to 2. If n is even (such as 2, 4, 6, 8, 10, etc.), the input values must be nonnegative since an even root of any negative real number is not a real number. If n is odd, there is no problem with the input values. A typical nth-rooter machine is illustrated in Figure 4.21.

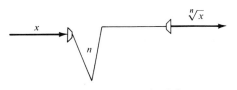

Note: 1. n is a positive whole number ≥ 2
      2. If n is even, the x value must be nonnegative

Figure 4.21 A typical nth-rooter machine with input and output

INPUT FOR CUBER	OUTPUT FOR CUBER / INPUT FOR 4-MULTIPLIER	OUTPUT FOR 4-MULTIPLIER / INPUT FOR 4-ADDER	OUTPUT FOR 5-ADDER	RESULTS
$b = -3$	$b^3 = (-3)^3 = -27$	$4b^3 = 4(-27) = -108$	$R(b) = 4b^3$ $+5 = -108$ $+5 = -103$	$R(-3) = -103$
$b = -2$	$b^3 = (-2)^3 = -8$	$4b^3 = 4(-8) = -32$	$R(b) = 4b^3$ $+5 = -32$ $+5 = -27$	$R(-2) = -27$
$b = -\frac{1}{2}$	$b^3 = (-\frac{1}{2})^3 = -\frac{1}{8}$	$4b^3 = 4(-\frac{1}{8}) = -\frac{1}{2}$	$R(b) = 4b^3$ $+5 = -\frac{1}{2}$ $+5 = 4\frac{1}{2}$	$R(-\frac{1}{2}) = 4\frac{1}{2}$
$b = 0$	$b^3 = 0^3 = 0$	$4b^3 = 4(0) = 0$	$R(b) = 4b^3$ $+5 = 0$ $+5 = 5$	$R(0) = 5$
$b = 2$	$b^3 = 2^3 = 8$	$4b^3 = 4(8) = 32$	$R(b) = 4b^3$ $+5 = 32$ $+5 = 37$	$R(2) = 37$
$b = \sqrt{3}$	$b^3 = (\sqrt{3})^3 = 3\sqrt{3}$	$4b^3 = 4(3\sqrt{3}) = 12\sqrt{3}$	$R(b) = 4b^3$ $+5 = 12\sqrt{3}$ $+5 = 5 + 12\sqrt{3}$	$R(\sqrt{3}) = 5 + 12\sqrt{3}$
$b = 5$	$b^3 = 5^3 = 125$	$4b^3 = 4(125) = 500$	$R(b) = 4b^3$ $+5 = 500$ $+5 = 505$	$R(5) = 505$

EXAMPLE Design an appropriate "computer" by arranging in order machines designed in this section to compute the function values for the function T with specification $y = T(x) = \sqrt{3x^2 - 4}$. Then, if possible, compute $T(2)$, $T(-3)$, $T(0)$, $T(\frac{1}{2})$, $T(5)$, and $T(1)$.

Solution The specification for the function T is given by the equation $y = T(x) = \sqrt{3x^2 - 4}$. Hence, to obtain the y value for given x value in the domain of T, we perform the following operations in order with the indicated machines:

OPERATION	MACHINE REQUIRED
Square x	Squarer
Multiply x^2 by 3	3-multiplier
Subtract 4 from $3x^2$	-4-adder
Take the square root of $3x^2 - 4$	Square-rooter

The required "computer" is illustrated in Figure 4.22.

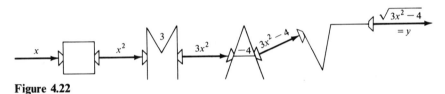

Figure 4.22

Using the above "computer", we will now attempt to computer $T(2)$, $T(-3)$, $T(0)$, $T(\frac{1}{2})$, $T(5)$, and $T(1)$ as shown on page 100.

Observe that $T(0)$, $T(\frac{1}{2})$, and $T(1)$ are not defined on the set of real numbers. That is because 0, $\frac{1}{2}$, and 1 are not in the domain of T. What is the domain of T? Clearly, if x is any real number, then x^2 is a real number since the square of every real number is a real number. Also, $3x^2$ is a real number since the product of two real numbers is a real number. Further, $3x^2 - 4$ is a real number since if 4 is subtracted from a real number, the result is a real number. However, $\sqrt{3x^2 - 4}$ is a real number only if $3x^2 - 4 \geq 0$. Therefore, the permissible values we can assign to x are those such that $3x^2 - 4 \geq 0$, which means that $x \geq 2/\sqrt{3}$ or $x \leq -2/\sqrt{3}$. (verify!) The domain of T, then, is the set of all those real numbers which are greater than or equal to $2/\sqrt{3}$ or which are less than or equal to $-2/\sqrt{3}$. Only those values can be used as input values for our "computer." Knowing this in advance, we could install a "scanner" outside the square-rooter to scan its input values. If we attempt to feed into the square-rooter a value of x which is greater than $-2/\sqrt{3}$ and also less than $2/\sqrt{3}$, the scanner would "reject" this value. Observe that in this example $x = 0$ would be rejected by the scanner. But $x = 0$ would be accepted as an input value for the squarer, its output value, $0^2 = 0$, would be accepted as an input value

INPUT FOR SQUARER	OUTPUT FOR SQUARER	INPUT FOR 3-MULTIPLIER	OUTPUT FOR 3-MULTIPLIER	INPUT FOR −4-ADDER	OUTPUT FOR −4-ADDER	INPUT FOR SQUARE-ROOTER	OUTPUT FOR SQUARE-ROOTER = RESULTS
$x = 2$	$x^2 = 2^2$	$= 4$	$3x^2 = 3(4)$	$= 12$	$3x^2 - 4$	$= 12 - 4$ $= 8$	$\sqrt{3x^2 - 4} = \sqrt{8} = 2\sqrt{2}$ $T(2) = 2\sqrt{2}$
$x = -3$	$x^2 = (-3)^2$	$= 9$	$3x^2 = 3(9)$	$= 27$	$3x^2 - 4$	$= 27 - 4$ $= 23$	$\sqrt{3x^2 - 4} = \sqrt{23}$ $T(-3) = \sqrt{23}$
$x = 0$	$x^2 = 0^2$	$= 0$	$3x^2 = 3(0)$	$= 0$	$3x^2 - 4$	$= 0 - 4$ $= -4$	$\sqrt{3x^2 - 4} = \sqrt{-4} \notin Re$ $T(0)$ is not defined
$x = \frac{1}{2}$	$x^2 = (\frac{1}{2})^2$	$= \frac{1}{4}$	$3x^2 = 3(\frac{1}{4})$	$= \frac{3}{4}$	$3x^2 - 4$	$= (\frac{3}{4}) -$ $= -\frac{13}{4}$	$\sqrt{3x^2 - 4} = \sqrt{-13/4} \notin Re$ $T(\frac{1}{2})$ is not defined
$x = 5$	$x^2 = 5^2$	$= 25$	$3x^2 = 3(25)$	$= 75$	$3x^2 - 4$	$= 75 - 4$ $= 71$	$\sqrt{3x^2 - 4} = \sqrt{71}$ $T(5) = \sqrt{71}$
$x = 1$	$x^2 = 1^2$	$= 1$	$3x^2 = 3(1)$	$= 3$	$3x^2 - 4$	$= 3 - 4$ $= -1$	$\sqrt{3x^2 - 4} = \sqrt{-1} \notin Re$ $T(1)$ is not defined

for the 3-multiplier, this new output value, $3(0) = 0$, would be accepted as an input value for the -4-adder, and its output value, $0 - 4 = -4$, would then be rejected by the scanner in front of the square rooter. Would it not be more practical to place the scanner in front of the squarer component of our computer to reject all those values of x which are greater than $-2/\sqrt{3}$ and also less than $2/\sqrt{3}$? Surely it would be! But we must "program" our computer in advance to let the scanner know what to reject. That means that we must examine the domain of the function first. Knowing what the domain of the function is, we would not attempt to compute function values for those elements which are not in the domain. This would save us time and effort whether we actually perform the operations or program the computer to do it for us. ■

EXAMPLE A group of students have been designing "computers" to solve various problems suggested to them. One such problem suggested by Michelle follows: "If 9 is subtracted from 4 times an integer and the result is divided by 3, then the answer may or may not be an integer. Can you design a machine to perform the operations indicated and compute five integer values that can be used such that the answer is an integer?" Several members of the group designed a machine quickly and computed the five integer values. One such solution follows.

Solution Reading the problem carefully, we note that 9 is to be subtracted from 4 times an integer. Hence, before the subtraction takes place, the integer must be multiplied by 4. The first operation, then, is to multiply the integer by 4 using a 4-multiplier. After this operation is performed, 9 will be subtracted from the result of the first operation using -9-adder. Reading further we note that the result of this operation is to be divided by 3. This can be accomplished by using a $(\frac{1}{3})$-multiplier. After this operation is performed, we examine the result to determine whether it is an integer. Our computer could be designed as illustrated in Figure 4.23.

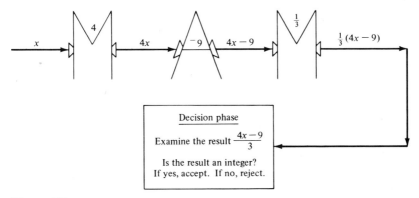

Figure 4.23

We now feed in integer values for the input:

x	$4x$	$4x-9$	$\dfrac{4x-9}{3}$	Examine to Accept or Reject
1	4	-5	$-\frac{5}{3}$	x
2	8	-1	$-\frac{1}{3}$	x
3	12	3	1	x
0	0	-9	-3	x
-1	-4	-13	$-\frac{13}{3}$	x
-2	-8	-17	$-\frac{17}{3}$	x
-3	-12	-21	-7	x
6	24	15	5	x
21	84	75	25	x

Can you find 5 additional integer values of x such that the results will be integers? ■

We will now consider very briefly the design of machines that could be used to compute function values for more complicated function specifications.

Suppose that we have a function, R, with the specification $R(a) = 3a^2 - 2a + 7$. What kind of machine would be needed to compute various values of $R(a)$? We start as before and observe that for a given value a in the domain of R we square a to obtain a^2 using a squarer machine and then multiply a^2 by 3 using 3-multiplier. The next step requires that we subtract $2a$ from the results obtained so far. But first we would have to multiply a by 2 using a 2-multiplier. That means that we would have to leave or store the $3a^2$ value somewhere while the $2a$ value is being computed. Once we have the $2a$ value, we could retrieve the $3a^2$ value and proceed to form the difference, $3a^2 - 2a$. The next step would be to add 7 to the result using a 7-adder. There are two new operations introduced here— storing and retrieving—which we had not encountered earlier. We are now on the threshold of designing a minicomputer. Rather than illustrate the design of such machines, we will indicate the sequence of instructions that will be fed into our computer. This will be given in the so-called flow chart in Figure 4.24.

It is not my intent to proceed further with the design of minicomputers in this text. My intent has been to make the student aware that we do have machines (slide rules, hand calculators, desk calculators, high-speed digital computers, and analogue computers) that can be used to perform certain operations or sequences of operations. However, these machines will only do what we instruct (or program) them to do. Therein lies our secret of success! If we can determine what has to be done and, equally important,

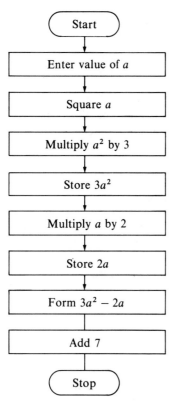

Figure 4.24 Flow chart for computing $3a^2 - 2a + 7$ for one value of a

what cannot be done (such as dividing by zero or extracting the square root of a negative number), we may be able to solve many different types of problems.

EXERCISES

For each of the functions with specifications given below, design an appropriate function machine that could be used to find function values for the particular function. Also compute the function values as directed.

1. $A(x) = 3x^2 + 4$
   Compute $A(-4)$, $A(-\sqrt{2})$, $A(0)$, $A(1)$, $A(\sqrt{3})$, and $A(6)$.

2. $B(u) = \frac{1}{2}u^3 - 5$
   Compute $B(-5)$, $B(-3)$, $B(-\frac{1}{2})$, $B(0)$, $B(\frac{1}{3})$, and $B(\frac{7}{4})$.

3. $D(v) = -3\sqrt{v} + 4$
   Compute $D(0)$, $D(1)$, $D(3)$, $D(4)$, $D(8)$, $D(16)$, and $D(24)$.
   Does $D(-2)$ exist? Why (why not)? What about $D(-\frac{3}{2})$? $D(-5)$?

4. $E(t) = \sqrt{2t + 4}$

Compute $E(t)$ for 4 distinct positive values of t and for 4 distinct negative values of t. Also compute $E(0)$.

5. $F(s) = \sqrt[3]{s - 3}/-5$

Compute $F(-3)$, $F(-2)$, $F(0)$, $F(3)$, $F(5)$, $F(24)$, and $F(30)$.

Each of the following equations is the specification of a function. If appropriate function machines were designed to compute the respective function values and a "scanner" was placed in front of each machine, what values of the independent variable, if any, would be "rejected" by the scanner?

6. $H(a) = (2a - 3)/5$.

7. $J(b) = (3b - 2)^{1/2}$.

8. $K(c) = -2\sqrt{c} + 5$.

9. $L(d) = \frac{1}{3}\sqrt{4 - d}$.

10. $M(f) = \sqrt[3]{3f^2 - 4}$.

For each of the following construct an appropriate flow chart for computing a single function value for the function with given specification.

11. $A(x) = 5x^2 - 2x + 7$.

12. $B(y) = \frac{2}{3}y^3 - 5y^2 + 4y$.

13. $C(u) = 4u^2 - 5u + 2 - (3/u)$.

14. $D(w) = (3w^2 - 2)(w^3 + 5)$.

15. $E(a) = (3a^2 - 2)/(a + 4)$.

16. $F(b) = \sqrt[3]{4 - 3b + 5b^2}$.

17. $H(c) = (c - 2)(3c + 1)/(c^2 + 2c - 3)$.

4.5 MAPPINGS

The mathematician who is primarily interested in the area of mathematics known as algebra defines a function as a mapping. A mapping consists of three things:

1. A set called the *domain.*

2. A set called the *image set.*

3. A rule of correspondence describing the relationship between each element of the domain and the corresponding element of the image set; this correspondence may be given in the form of an equation, or it may be represented in tabular form.

DEFINITION 4.15 A mapping, f, from a set A into a set B is said to be a or rule that associates a unique element $b \in B$ with each element $a \in A$. Such a mapping may be denoted by $f: A \rightarrow B$, where the correspondence is written $f(a) = b$. The element b is called the image or correspondent of a under the mapping f.

Before examining various examples of mappings, we will emphasize the following implications of our definition:

1. If $a \in A$, then there is exactly one element $b \in B$ such that $b = f(a)$.

2. Every element in the image set B does not have to be used under the mapping f; there may exist some $b \in B$ such that there is no $a \in A$ with the property that $b = f(a)$. The subset of B that is used under the mapping f is called the *range* of the mapping, denoted by the symbol R_f. Hence the range of a mapping is a subset of the image set of the mapping.

3. It is possible that two or more a's from the set A will be paired with the same $b \in B$. For example, if a_1 and a_2 belong to the set A with $a_1 \neq a_2$, then $f(a_1)$ may equal $f(a_2)$.

4. There is nothing sacred about our choice of symbols such as A, B, and f. Indeed, it is possible to denote a mapping from the set C into the set D, $\alpha: C \rightarrow D$; or a mapping from the set X into the set Y, $g: X \rightarrow Y$.

EXAMPLE Let $A = \{1, 2, 3, 4\}$ and $B = \{a, b, c\}$. Define f by the following rule of association:

$$
\begin{array}{c}
f \\
1 \rightarrow a \\
2 \rightarrow b \\
3 \rightarrow b \\
4 \rightarrow c
\end{array}
$$

which may also be written $f(1) = a$, $f(2) = b$, $f(3) = b$, and $f(4) = c$.

Clearly, f is a mapping from A into B, written $f: A \rightarrow B$. Observe that $D_f = \{1, 2, 3, 4\} = A$, $R_f = \{a, b, c\}$ and that the range of the mapping is the entire image set B. Also observe that the element b in the image set is associated or paired with the elements 2 and 3 of the domain. This does not contradict the uniqueness part of the definition of a mapping. The mapping $f: A \rightarrow B$ also could be denoted by the diagram in Figure 4.25.

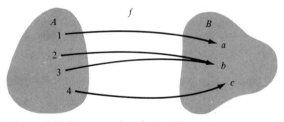

Figure 4.25 The mapping $f: A \rightarrow B$

Note that if the arrows in Figure 4.25 were reversed, a different rule of association results. This association from the set B into the set A does not yield a mapping, however, since there are two arrows from a single element, b, of the set B going to two different elements of the set A. (See Figure 4.26.)

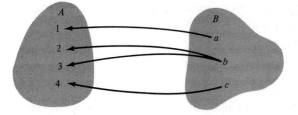

Figure 4.26 No mapping from B into A

EXAMPLE Let $X = \{-2, -1, 0, 1, 2\}$ and $Y = \{y \mid y \in Z, -2 \le y \le 5\}$, where Z is the set of all integers. Define h by the equation $h(x) = x^2$ for $x \in X$. So $h(-2) = 4$, $h(-1) = 1$, $h(0) = 0$, $h(1) = 1$, and $h(2) = 4$, which is a mapping from X into Y as shown in Figure 4.27.

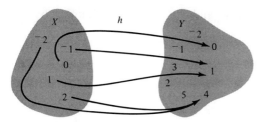

Figure 4.27 The mapping $h: X \to Y$

Observe that $R_h = \{0, 1, 4\}$. In this case the entire image set Y was not used for the range of the mapping h. The range of the mapping h is a proper subset of Y.

DEFINITION 4.16 A mapping, f, from a set A into a set B is said to be a mapping from A *onto* B, denoted by $f: A \xrightarrow{\text{onto}} B$, if every element of the set B is the image of some element of A.

It follows, then, that if f is a mapping from A onto B, then the range of f and the image set B are equal. Hence *every function is a mapping from its domain onto its range.*

EXAMPLE Let $M = \{a, b, c, d, e\}$ and $N = \{\alpha, \beta, \gamma, \delta, \varepsilon\}$. Define g as given in Figure 4.28.

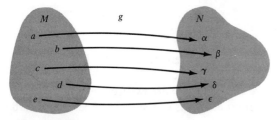

Figure 4.28 The mapping $g: M \to N$

Now the correspondence $g(a) = \alpha$, $g(b) = \beta$, $g(c) = \gamma$, $g(d) = \delta$, $g(e) = \varepsilon$ is a mapping from M into N. Since the entire set N is used under this correspondence, g is also a mapping from the set M *onto* the set N.

EXERCISES

1. Let $A = \{1, 2, 3, 4, 5\}$ and $B = \{a, b, c, d\}$. Define f as given below:

$$f$$
$$1 \rightarrow a$$
$$2 \rightarrow b$$
$$3 \rightarrow b$$
$$4 \rightarrow c$$
$$5 \rightarrow d$$

    a. Verify that Definition 4.15 is satisfied for $f: A \rightarrow B$.
    b. Is $f: A \xrightarrow{\text{onto}} B$? Why?
    c. If the arrows were reversed, would a new mapping be defined? Why?
    d. Compute $f(1)$, $f(3)$, and $f(5)$.
2. Let $C = Re$ (where Re denotes the set of all real numbers) and $D = \{x \mid x \in Re,\ x \geq 0\}$. Define g as $g(x) = x^2$ for all $x \in C$.
    a. Verify that Definition 4.15 is satisfied for $g: C \rightarrow D$.
    b. Is $g: C \xrightarrow{\text{onto}} D$? Why?
    c. Compute $g(-2)$, $g(0)$, $g(\frac{1}{2})$, $g(1)$, and $g(3)$.
3. Let $E = Re$ and $F = Re$. Define h as $h(x) = \pm x$ for all $x \in E$. Prove or disprove that h is a mapping from E into F.
4. Let $G = \{-2, -1, 0, 1, 2, 3\}$ and $H = \{a, b, c\}$.
    a. Define a mapping k from G into H such that k is not onto H.
    b. Define a mapping l from G into H such that l is onto H.
    c. Define a mapping m from H into G such that m is not onto G.
    d. Define a mapping n from H into G such that n is onto G.
5. If temperature readings were taken every hour of the day on the hour and the data recorded in tabular form with the hours listed in the left column and the temperature readings listed on the right, would this constitute a mapping from the set of elements in the left column into the set of elements in the right column? Why?
6. The following table displays the batting averages for five baseball players. Does this represent a mapping from the set of players into the set of averages? Why?

PLAYERS	AVERAGES
1	0.386
2	0.374
3	0.372
4	0.361
5	0.302

7. Test scores for a group of six students are shown below.

STUDENTS	SCORES
1	73
2	82
3	73
4	77
5	92
6	89

a. Does this represent a mapping from the set of students into the set of scores? Why?

b. Does this represent a mapping from the set of scores into the set of students? Why?

4.6 INVERSE OF A FUNCTION

Observe that if the arrows in Figure 4.28 were reversed, each element of the set N would correspond to a unique element in the set M. Whenever no element of N is used more than once the function is said to be one-to-one.

DEFINITION 4.17 A mapping g from a set M into a set N is said to be *one-to-one*, denoted $1 : 1$, if distinct (unequal) elements of M have distinct (unequal) images in N; that is, if $a \neq b$, then $g(a) \neq g(b)$.

Two facts should be clear from the above definition:

1. If g is a one-to-one mapping from M into N, then each element of M is associated with a unique element of N and each element of N that is used under this correspondence is associated with a unique element of M.

2. A $1 : 1$ mapping does not have to be onto.

The mapping $g: M \to N$ in the third example of Section 4.5 is one-to-one and $g: M \xrightarrow{\text{onto}} N$. Consider now a mapping which is $1 : 1$ but not onto.

EXAMPLE Let $P = \{1, 2, 3\}$ and $Q = \{a, b, c, d, e\}$. Define q as given in Figure 4.29. The correspondence $q(1) = a$, $q(2) = b$, $q(3) = d$ is clearly a mapping from P into Q. Note that not all of Q is used under the mapping and, therefore, q is not a mapping from P onto Q. However, the mapping is one-to-one. Reversing the arrows in Figure 4.29 gives a new correspondence; those elements in the set Q which are used are paired with unique elements in the set P. The domain of the new correspondence is a proper subset of Q.

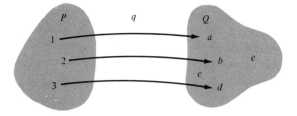

Figure 4.29 The mapping $q: P \to Q$

EXAMPLE Let $R = \{1, 2, 3, 4\}$ and $S = \{a, b, c, d\}$. Define α as given in Figure 4.30. The correspondence $\alpha(1) = a$, $\alpha(2) = b$, $\alpha(3) = c$, $\alpha(4) = d$ is clearly a $1:1$ mapping from R onto S. Further, reversing the arrows in Figure 4.30 reveals that each element of S is associated with a unique element of R.

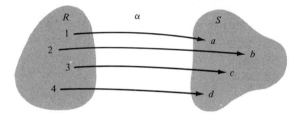

Figure 4.30 The mapping $\alpha: R \to S$

Observe that by reversing the arrows in Figure 4.30, a new mapping exists from S onto R. (See Figure 4.31.) If we name this new mapping α^{-1}, we now have the correspondence $\alpha^{-1}(a) = 1$, $\alpha^{-1}(b) = 2$, $\alpha^{-1}(c) = 3$, and $\alpha^{-1}(d) = 4$, which is a mapping from S onto R.

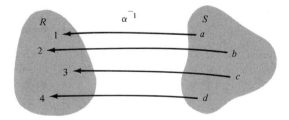

Figure 4.31 A new mapping $\alpha^{-1}: S \to R$

DEFINITION 4.18 Let A and B be sets, and let g be a one-to-one mapping from A onto B. Then the mapping g has an *inverse*, denoted by g^{-1}, which is also a mapping such that if $g(a) = b$, then $g^{-1}(b) = a$.

EXERCISES

1. Let $P = \{a, b, c, d, e\}$ and $Q = \{1, 2, 3\}$. Define f as given below:

$$
\begin{array}{c}
f \\
a \to 1 \\
b \to 2 \\
c \to 3 \\
d \to 2 \\
e \to 1
\end{array}
$$

   a. Verify that f is a mapping from $P \to Q$.
   b. Compute $f(a), f(b), f(c), f(d)$, and $f(e)$.
   c. Is f a mapping from P onto Q? Why?
   d. Is f a one-to-one mapping? Why?
   e. Does $f: P \to Q$ possess an inverse mapping? Why?

2. Let P and Q be the sets as defined in Exercise 1. Define g as given below:

$$
\begin{array}{c}
g \\
1 \to a \\
1 \to e \\
2 \to b \\
2 \to d \\
3 \to c
\end{array}
$$

   a. Prove or disprove that g is a mapping from Q into P.
   b. If g is a mapping from Q into P, is g onto? Why?
   c. If g is a mapping from Q into P, is g $1:1$? Why?

3. Let $A = Re$ (the set of all real numbers) and $B = Re$. Define h as given below:

$$
h(x) = \begin{cases} x, & \text{if } x > 0 \\ 0, & \text{if } x = 0 \\ -x, & \text{if } x < 0 \end{cases}
$$

Caution: Only one rule of association is given above, not three different rules. However, the specification for h is given according to whether x is positive, zero, or negative. For instance, $h(3) = 3$, $h(\frac{1}{2}) = \frac{1}{2}$, $h(-2) = 2$, and $h(-\sqrt{3}) = \sqrt{3}$.

   a. Verify that h is a mapping from A into B.
   b. What is the domain of h?
   c. What is the range of h?
   d. Is h a mapping from A onto B? Why?
   e. Is h a one-to-one mapping? Why?
   f. Does the mapping $h: A \to B$ possess an inverse mapping? Why?

g. Compute (1) $h(-3)$; (2) $h(-1.5)$; (3) $h(0)$; (4) $h(2.1)$; (5) $h(-3/2)$; (6) $h(x^2)$; (7) $h(3x)$; (8) $h(x+2)$; (9) $h(\sqrt{x})$; (10) $h(-x^2)$.

4. Let $A = Re$ and $B = \{y|y \in Re, y \geq 0\}$. Define h as given in Exercise 3 above. Now answer (a) to (f) of Exercise 3 for this new set B.

5. Give an example (other than those in the text) of a mapping $p: Re \to Re$ such that p is one-to-one but not onto Re.

6. Give an example (other than those in the text) of a mapping $q: Re \to Re$ such that q has an inverse mapping.

7. Give an example (other than those in the text) of a mapping $s: S \to T$ such that s is onto T but not one-to-one. The sets S and T are to be defined by the student as appropriate subsets of Re satisfying the conditions for s.

8. Let $R = Re$ and $W = \{b|b \in Re, b > 0\}$. Define a mapping $\alpha: R \to W$ by the rule of association $\alpha(a) = 3^a$ for all $a \in R$.
   a. Verify that α is a mapping from R into W.
   b. Is α a mapping from R onto W? Why?
   c. Is α a one-to-one mapping? Why?
   d. Does the mapping $\alpha: R \to W$ possess an inverse mapping? Why?
   e. Compute (1) $\alpha(-3)$; (2) $\alpha(\frac{1}{2})$; (3) $\alpha(0)$; (4) $\alpha(-\frac{1}{3})$; (5) $\alpha(5)$; (6) $\alpha(-2)$; (7) $\alpha(-a)$; (8) $\alpha(a^2)$; (9) $\alpha(1/a)$ if $a \neq 0$.

9. Given t is a mapping that possesses an inverse, and if $t(0) = 2$, $t(2) = 4$, $t(-2) = 1$, and $t(5) = 0$, compute
   a. $t^{-1}(0)$.
   b. $t^{-1}(2)$.
   c. $t^{-1}(1)$.
   d. $t^{-1}(4)$.

10. If w is a mapping such that $w(-3) = 2$, $w(-2) = 5$, $w(0) = 2$, $w(\frac{1}{2}) = -3$, and $w(4) = 4$, does w possess an inverse mapping? Why?

4.7 A COMPARISON OF THE DEFINITIONS FOR A FUNCTION

In Section 4.5 we defined a function as a mapping from one set into another set, and in Section 4.3 we defined a function as a special type of a relation. In either case the reader probably has noticed that there is a pairing of objects and that the pairing is ordered. In this section we will compare the definitions by reconsidering examples given in the earlier sections of this chapter.

EXAMPLE Consider the mapping of the first example in Section 4.4. We were given the sets $A = \{1, 2, 3, 4\}$ and $B = \{a, b, c\}$ and defined the mapping $f: A \to B$ by the following rule of association:

$$f$$
$$1 \rightarrow a$$
$$2 \rightarrow b$$
$$3 \rightarrow b$$
$$4 \rightarrow c$$

To rewrite this function as a set of ordered pairs, we observe that $1 \in A$ is paired with $a \in B$, $2 \in A$ is paired with $b \in B$, $3 \in A$ is paired with $b \in B$, and $4 \in A$ is paired with $c \in B$. Since the mapping is from A into B, we can abbreviate the rule of association by forming the ordered pairs $(1, a)$, $(2, b)$, $(3, b)$, and $(4, c)$, respectively. Now we may write the function f as the set containing these ordered pairs. Hence $f = \{(1, a), (2, b), (3, b), (4, c)\}$.

EXAMPLE   Consider the mapping of the third example in Section 4.5. We were given the sets $M = \{a, b, c, d, e\}$ and $N = \{\alpha, \beta, \gamma, \delta, \varepsilon\}$ and defined the mapping $g: M \rightarrow N$ as given in Figure 4.28. Since the mapping is from M into N, we can abbreviate the rule of association for the function g by forming the ordered pairs (a, α), (b, β), (c, γ), (d, δ), and (e, ε), respectively. Hence $g = \{(a, \alpha), (b, \beta), (c, \gamma), (d, \delta), (e, \varepsilon)\}$.

EXAMPLE   Consider the function F of the first example in Section 4.3 in which $F = \{(u, v) \,|\, v = F(u) = u + 3\}$. The domain of F is the set of all real numbers. To rewrite the function F as a mapping we will introduce the two sets R and S and consider $F: R \rightarrow S$. The set R is the domain of F, and S is the image set that must contain the range of F. But the domain of F is Re; hence $R = Re$. The range of F is also Re and, hence, $Re \subseteq S$. Clearly, we can let $S = Re$. Therefore we have $F: Re \rightarrow Re$.

Further, since F is given as a set of ordered pairs of the form or type (u, v) such that $v = u + 3$, we know that the u's in the set R will be paired with unique v's in the set S. Since there are infinitely many values of u, we will not be able to list them all. However, we may show the association between the u's and v's as illustrated in Figure 4.32.

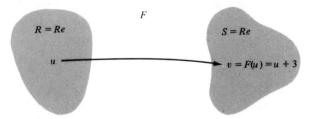

Figure 4.32 The mapping $F: R \rightarrow S$ such that $v = F(u) = u + 3$

Using the rule of association given, we see that if $u = 4$, then the $v \in S$ associated with 4 will be $4 + 3$ or 7. Similarly, if $u = -5$, $v = -2$, etc.

EXERCISES

1. At a particular weather station the temperature readings are recorded every hour on the hour using a 24-hour clock system (for example, 0200 hours means 2:00 A.M., 1200 hours means noon, and 1630 hours means 4:30 P.M.). The temperature recorded is a function of the time when it was recorded. Some of the data recorded over a 24-hour period of time is as follows:

$T(0100) = 47°$ $T(0400) = 48°$ $T(0700) = 51°$ $T(1000) = 53°$
$T(1200) = 53°$ $T(1400) = 53°$ $T(1500) = 54°$ $T(2300) = 49°$

All temperatures recorded are in degrees Fahrenheit. The symbol $T(0400) = 48°$, for instance, means that the temperature was 48° at 4:00 A.M.

a. Verify that T is a function from the set of hours into the set of temperature readings in degrees Fahrenheit.
b. What was the temperature recorded at 3:00 P.M. that day?
c. Based on the information given above, what was the temperature range from 7:00 A.M. to 3:00 P.M. that day?
d. Represent the function T as an appropriate mapping.
e. Represent the function T as an appropriate set of ordered pairs.
f. Is the function T one-to-one? Why?
g. Graph that portion of the function T given above.

2. Karen's age changes once a year on her birthday. Her age at any particular instant is a function, A, of time. For this problem, assume that Karen was born on December 20, 1960.

a. How old, in years, was Karen on May 23, 1968?
b. When will Karen become 42 years old?
c. Represent the function A as an appropriate mapping.
d. Represent the function A as an appropriate set of ordered pairs.
e. Is the function A a 1 : 1 mapping? Why?
f. Graph the function A from the period December 20, 1960, to December 20, 1972.

3. First class postal rates for letters mailed in the United States are computed on the basis of weight as follows: 13 cents for each ounce or fraction of an ounce. Although there are other restrictions, the cost of mailing a letter in the United States is a function M of weight.

a. Represent the function M as an appropriate mapping.
b. Represent the function M as an appropriate set of ordered pairs.
c. Is the function M one-to-one? Why?
d. Compute the costs of mailing letters in the United States with the following weights: (1) 2.1 oz; (2) 0.7 oz; (3) 5.6 oz; (4) 4.5 oz; (5) 1 lb; (6) 2 lb 3 oz; (7) 3 lb 5 oz; (8) 16 lb; (9) 25 lb 2 oz; and (10) 3.216 oz.

4. In physics and engineering the following symbols are encountered together with their respective interpretations:

SYMBOL	INTERPRETATION
R	Resistance
X	Reactance
L	Inductance
X_L	Inductive reactance
C	Capacitance
X_C	Capacitive reactance
Z	Impedance
V	Voltage
I	Current

Let A be the set of all the symbols given above, and let B be the set of all the interpretations given above. Define f as a rule of association that pairs each symbol on the left with its respective interpretation on the right.

a. Verify that f is a mapping from A into B.

b. Is the mapping f from A onto B? Why?

c. Is the mapping $f: A \to B$ one-to-one? Why?

d. If $f: A \to B$ has an inverse mapping, describe it.

e. Represent the function f as a special set of ordered pairs.

f. Compute each of the following: (1) $f(Z)$; (2) $f(C)$; (3) $f(X_L)$; (4) $f(V)$; (5) $f(R)$.

5. Let G be the set of some of the Greek alphabet symbols given below; their English equivalents are also given.

SYMBOL	TRANSLITERATION	SYMBOL	TRANSLITERATION
α (alpha)	a	μ (mu)	m
β (beta)	b	ν (nu)	n
γ (gamma)	g	ξ (xi)	x
δ (delta)	d	o (omicron)	o
ε (epsilon)	e	π (pi)	p
ζ (zeta)	z	ρ (rho)	r
ι (iota)	i	σ (sigma)	s
κ (kappa)	k	τ (tau)	t
λ (lambda)	l	υ (upsilon)	y

Let E be the set of all the symbols of the English alphabet.

a. Are the sets G and E equivalent? Why?

b. Is it possible to define a mapping g from G into E? If so, do it; if not, indicate why not.

c. Is it possible to define a mapping h from G into E such that h is onto? Why?

d. Is it possible to define a mapping k from E into G such that k is one-to-one? Why?

e. Is it possible to define a mapping m from E into G such that m is onto? If so, do it; if not, indicate why not.

f. Is it possible to define a mapping p from E into G such that p would have an inverse mapping? If so, do it; if not, indicate why not.

4.8 BINARY OPERATIONS

Another type of function is a function of two independent variables. As defined in Section 4.3, a function of two independent variables is a set of ordered *triples* the first two elements of which are values of the independent variables and the third elements of which are the correspondents of the pair of independent variables under the function. Such a function has a domain that is a subset of some cartesian product of two sets. Ordinary addition and multiplication are examples of functions of two independent variables.

DEFINITION 4.19 The concept of *ordinary addition*, denoted by $+$, is defined to be the mapping $+: Re \times Re \to Re$, satisfying the correspondence that $+((a, b)) = a + b$ for every pair of elements $(a, b) \in Re \times Re$. Such a function from $Re \times Re \to Re$ is called a *binary operation* on Re.

As a set of ordered pairs, we may write this function as

$$+ = \{((x, y), z) \mid z = +((x, y)) = x + y \text{ for every } (x, y) \in Re \times Re\}$$

Observe that z is the correspondent of the ordered pair of real numbers (x, y). The x's and y's form the independent variables, and the z's represent the dependent variable. Thus we have a set of ordered pairs such that the first component of each ordered pair is an ordered pair. Such an ordered pair may be written as an ordered triple $(x, y; z)$, using the semicolon to separate the independent variables from the dependent variable.

A few observations regarding this function are in order.

1. The domain of $+$ is $Re \times Re$.

2. The range of $+$ is Re.

3. The function $+$ is well defined since, for each pair of real numbers (a, b), the sum $a + b$ is a unique real number.

4. The mapping $+$ is from $Re \times Re$ *onto* Re since the range is Re.

5. The mapping $+$ is not one-to-one; for instance, both ordered pairs $(2, 3)$ and $(1, 4)$ in the domain of $+$ are paired with the real number 5 in the range.

6. The mapping $+$ does not have an inverse since $+$ is not one-to-one.

DEFINITION 4.20 The binary operation of *ordinary multiplication*, denoted by x, is defined to be the mapping $x: Re \times Re \to Re$ satisfying the correspondence that $x((a, b)) = a \times b = ab$ for every pair of elements $(a, b) \in Re \times Re$.

As a set of ordered pairs, we may rewrite this function as

$$x = \{(a, b; c) \mid c = x((a, b)) = ab \text{ for every } (a, b) \in Re \times Re\}$$

Observations similar to those for the binary operation of ordinary addition may be made for the binary operation of ordinary multiplication.

EXERCISES

1. Let $+$ and $\times$ be the binary operations of ordinary addition and ordinary multiplication, respectively. Evaluate each of the following.
   a. $+(-2, 3)$
   b. $+(-4, -6.2)$
   c. $+(17.9, 23.6)$
   d. $+(-59, 72)$
   e. $+(4.6, 9.7)$
   f. $+(\sqrt{2}, -\sqrt{3})$
   g. $+(2\sqrt{5}, -3\sqrt{5})$
   h. $+(\pi, e)$
   i. $+(17\sqrt{2}, -9)$
   j. $+(6.009, 7.146)$
   k. $\times(-6, -9)$
   l. $\times(7, -19)$
   m. $\times(1.2, 2.3)$
   n. $\times(-7.1, 5.2)$
   o. $\times(-8.1, -9.8)$
   p. $\times(-\sqrt{3}, -\sqrt{2})$
   q. $\times(1/\sqrt{3}, \sqrt{3}/3)$
   r. $\times(-\frac{2}{3}, -\frac{7}{9})$
   s. $\times(\frac{16}{23}, 3\frac{1}{2})$
   t. $\times(\frac{19}{4}, \frac{16}{37})$
2. Define $S : Z \times Z \to Z$ such that $S((a, b)) = a - b$ and Z is the set of all integers.
   a. Is S a mapping from $Z \times Z$ onto Z? Give a reason for your answer.
   b. Is S a binary operation? Give a reason for your answer.
   c. Is S one-to-one? Give a reason for your answer.
   d. Does S possess an inverse mapping? Give a reason for your answer.
3. Is the operation of division a binary operation on Re? Give a reason for your answer.
4. Is the operation of division a binary operation on the set $Re - \{0\}$? Give a reason for your answer.

SUMMARY

In this chapter we introduced the concept of a relation as a nonempty subset of some cartesian product. Special types of relations such as reflexive, symmetric, transitive, and equivalence were discussed. Equivalence relations were singled out for their importance in mathematics. Functions were introduced as special types of relations and also as mappings. One-to-one mappings and the inverse of a function also were discussed. The domain and the range of both a relation and a function also were defined. Binary operations as special types of functions were introduced. Function machines were discussed.

The following symbols were introduced during our discussions:

SYMBOL	INTERPRETATION
$f : A \to B$	The mapping f from the set A into the set B
$f(a) = b$	b is the image or the correspondent of a under the mapping f
D_f	The domain of f
R_f	The range of f
$f : A \xrightarrow{\text{onto}} B$	The mapping f from the set A onto the set B
f is $1:1$	The mapping f is one-to-one
f^{-1}	The inverse mapping of the mapping f if it exists
xHy	x is in the relation H to y $(xHy \leftrightarrow (x, y) \in H)$
$R*$	The inverse relation of the relation R (it is not necessarily a function if R is a function, although it may be a function even if R is not a function)
Re	The set of all real numbers
$+$	The binary operation of ordinary addition
$\times$	The binary operation of ordinary multiplication

REVIEW EXERCISES FOR CHAPTER 4

1. Define each of the following.
   a. A mapping g from X into Y.
   b. The domain of a function.
   c. The range of a function.
   d. A mapping that is onto.
   e. A mapping that is one-to-one.
   f. The inverse of a mapping.
   g. A relation on the set $P \times Q$.
   h. A relation on the set T.
   i. The inverse of a relation.
   j. A constant valued function.
   k. A binary operation.
2. Let $R = \{1, 2, 3, 4, 5\}$, $S = \{a, b, c, d\}$, and $T = \{\alpha, \beta, \gamma\}$.
   a. Define a relation on the set $R \times S$ that is not a function.
   b. Define a relation on the set $R \times T$ that is a function.

    c. Define a relation on the set $S \times T$ that is not a function but such that its inverse relation is a function.

    d. Define a mapping f from R into S such that f is not onto.

    e. Define a mapping g from T into S such that g is $1:1$.

    f. Define a mapping h from S into S such that h has an inverse mapping.

    g. If possible, define a mapping k from S into T that is onto and one-to-one. If this is not possible, indicate why not.

    h. If possible, define a mapping l from T into R that is one-to-one. If this is not possible, indicate why not.

3. A professor has 27 students in one of his classes. He has recorded their names in alphabetical order by last name on the first 27 numbered lines on a page in his roll book.

    a. Does there exist a function between the set of the first 27 natural numbers and the set of names of the 27 students? Why?

    b. If such a mapping does exist, is it one-to-one? Why?

    c. At the end of the term the professor assigned grades to the 27 students as follows: The first named student received a grade A, the second B, the third C, the fourth D, the fifth F, the sixth A, the seventh B, and so on. The procedure was repeated after each group of five names of students. Does there exist a function between the set of names of students and the set of grades assigned? Why?

    d. If such a function does exist, in part (c), is it one-to-one? Why?

4. A partial postal ZIP code list for a certain Central New York State area is given below.

LOCALITY	ZIP CODE
Barneveld	13304
Chadwicks	13318
Marcy	13403
Remsen	13458
Utica	13501
Utica	13502

Let A be the set containing the names of all the localities listed in the left column, and let B be the set containing the ZIP code numbers listed in the right column.

    a. Is it possible to define a mapping from A into B? If so, do it; if not, indicate why not.

    b. Is it possible to define a mapping from B into A? If so, do it; if not, indicate why not.

    c. Are the two sets A and B equivalent? Why?

5. Freight that is light and bulky is sometimes shipped by a private parcel service. The rates for shipments depend upon weight and distance traveled.

A hypothetical schedule of rates is given below.

(1) WEIGHT	(2) 100 MILES	(3) 200 MILES	(4) 500 MILES
1 to 23 lb	$6.30	$7.95	$8.90
24 to 62 lb	8.40	9.80	11.10
63 to 100 lb	9.80	11.20	13.25

a. Define an appropriate function α from the set of elements in column 1 into the set of elements in column 2. Is α one-to-one? Why?

b. Define an appropriate function β from the set of elements in column 1 into the set of elements in column 4. Is β one-to-one? Why?

c. Compute: (1) α(12 lb); (2) β(60.5 lb); (3) α(97 oz); (4) β(7 lb 3 oz).

d. For $9.80, how large a package may be shipped by parcel service and for how far?

CHAPTER 5
THE SYSTEM OF WHOLE NUMBERS

The first type of numbers which a student encounters are the *natural* or *counting* numbers. These numbers together with zero constitute the set of *whole* numbers. This chapter will examine the system of whole numbers—that is, the set of whole numbers together with arithmetic operations defined on them and also the properties associated with these operations. It also will introduce the relation of order on the set of whole numbers. We will denote the set of all natural numbers by N and the set of all whole numbers by W. Thus

$$N = \{1, 2, 3, 4, 5, 6, 7, \ldots\} \quad \text{and} \quad W = \{0, 1, 2, 3, 4, 5, 6, \ldots\} = N \cup \{0\}$$

5.1 NUMBERS AND NUMERALS

A number is an abstraction or a concept. You have never seen a number nor will you ever see one. When you think of the number two, you may hold up fingers on your hand such as in Figure 5.1. What is seen in Figure

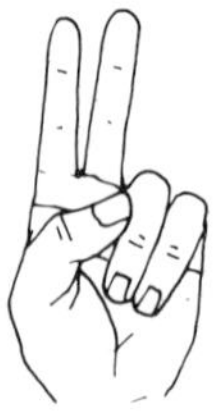

Figure 5.1

120

5.1 is a symbolic representation for the number two. Other symbols for the same number are $||$, $..$, or $//$.

DEFINITION 5.1 A *numeral* is a symbol representing a number.

The most common numerals used today are the Hindu-Arabic numerals you used throughout elementary school and continue to use today. Most of the pages of this text are numbered with them. The Hindu-Arabic numerals are 1, 2, 3, 4, 5, 6, 7, 8, 9, 10, 11, ..., where the symbol "..." means "and so on". These numerals are called the Hindu-Arabic numerals because they had their origin with the Hindus but were introduced to the Western world by the Arabs. The reader is probably also familiar with Roman numerals, the first twelve of which are I, II, III, IV, V, VI, VII, VIII, IX, X, XI, and XII. We will look closer at these and other numeration systems in the next chapter.

Instead of using the Hindu-Arabic numerals to represent numbers, we also could use the English language and refer to the first few counting numbers as one, two, three, four, five, six, seven, eight, nine, ten, eleven, twelve, and so forth. Or using the French language we would have un, deux, trois, quatre, cinq, six, sept, huit, neuf, dix, onze, douze, treize, quatorze, and so forth.

The numerals introduced so far represent what are known as the cardinal numbers.

DEFINITION 5.2 A *cardinal number* is a number that answers the question "How many?"

If we have a set $A = \{\alpha, \beta, \gamma\}$ we would say that the set A has three elements or that the cardinality of the set A is 3. We will introduce the symbol $n(A)$ to denote "the cardinality of the set A" or "the number of elements in the set A." Hence $n(A) = 3$.

In addition to cardinal numbers, we also have ordinal numbers.

DEFINITION 5.3 An *ordinal number* is a number that answers the question "Which one?"

Referring again to the set $A = \{\alpha, \beta, \gamma\}$, we would say that α is the first element listed, β is the second element listed, and γ is the third element listed in the set.

Throughout the balance of this chapter we will be primarily concerned with cardinal numbers.

EXERCISES

For each of the following determine whether the number indicated is a cardinal number or an ordinal number.

1. 23 students in class.
2. The second test.
3. The third term.
4. 12 books.
5. Three dollars.
6. The thirty-second president of the United States.
7. The 4-minute clock.
8. The eighth floor.
9. Nine people.
10. The tenth page of the book.
11. A dozen donuts.
12. A set with seven elements.
13. The fourth seat on the bus.
14. 100 miles.
15. 1 million times.
16. The thousandth character.
17. The first row.
18. 23 new automobiles.
19. The second right turn.
20. The last exercise.

5.2 COUNTING

The most important thing that one learns to do in mathematics is to count. As we will see later in this chapter, addition will be defined in terms of counting and multiplication also will be defined in terms of counting.

The need for counting should be quite obvious today in our attempt to account for various inventories of stock, members of a camping party, students in a class, enrollments in a college, and so forth. In early times a shepherd would probably account for his herd of sheep by establishing a one-to-one correspondence between his set of sheep and a set of stones in a pile. There was no need for written numerals until about the period known as historical times. As the number of objects to be accounted for became larger and larger, it was necessary to expand and improve upon (simplify?) the numeration systems. Let's look at our system of numeration.

When we count using the Hindu-Arabic numerals we have 1, 2, 3, 4, 5, 6, 7, 8, 9, 10, 11, 12, 13, 14, 15, 16, 17, 18, 19, 20, 21, 22, 23, 24, ..., 30, 31, ..., 39, 40, 41, 42, 43, 44, ..., 50, 51, ..., 59, 60, 61, ..., 70, 71, ..., 80, 81, 82, ..., 90, 91, ..., 99, 100, 101, 102, ...

Notice that after 9 we wrote 10 and went from a one-digit numeral to a two-digit numeral. Also notice that from 10 to 19 each of these two-digit numerals begins with 1; from 20 to 29, each of these beings with 2; and so forth. Also notice that after 99 we continue with a three-digit numeral. Why?

To answer this question, let's rearrange these numerals as in Table 5.1.

Table 5.1

1	11	21	31	41	51	61	71	81	91	101
2	12	22	32	42	52	62	72	82	92	102
3	13	23	33	43	53	63	73	83	93	103
4	14	24	34	44	54	64	74	84	94	104
5	15	25	35	45	55	65	75	85	95	105
6	16	26	36	46	56	66	76	86	96	106
7	17	27	37	47	57	67	77	87	97	107
8	18	28	38	48	58	68	78	88	98	108
9	19	29	39	49	59	69	79	89	99	109
10	20	30	40	50	60	70	80	90	100	110, etc.

Now suppose that we let $b = 10$ and list our numerals as in Table 5.2.

Table 5.2

1	$b + 1$	$2b + 1$	$3b + 1$	$4b + 1$	$5b + 1$	$6b + 1$
2	$b + 2$	$2b + 2$	$3b + 2$	$4b + 2$	$5b + 2$	$6b + 2$
3	$b + 3$	$2b + 3$	$3b + 3$	$4b + 3$	$5b + 3$	$6b + 3$
4	$b + 4$	$2b + 4$	$3b + 4$	$4b + 4$	$5b + 4$	$6b + 4$
5	$b + 5$	$2b + 5$	$3b + 5$	$4b + 5$	$5b + 5$	$6b + 5$
6	$b + 6$	$2b + 6$	$3b + 6$	$4b + 6$	$5b + 6$	$6b + 6$
7	$b + 7$	$2b + 7$	$3b + 7$	$4b + 7$	$5b + 7$	$6b + 7$
8	$b + 8$	$2b + 8$	$3b + 8$	$4b + 8$	$5b + 8$	$6b + 8$
9	$b + 9$	$2b + 9$	$3b + 9$	$4b + 9$	$5b + 9$	$6b + 9$
b	$b + b$	$2b + b$	$3b + b$	$4b + b$	$5b + b$	$6b + b$
or	$2b$	$3b$	$4b$	$5b$	$6b$	$7b$

$7b + 1$	$8b + 1$	$9b + 1$	$bb + 1$	$bb + b + 1$
$7b + 2$	$8b + 2$	$9b + 2$	$bb + 2$	$bb + b + 2$
$7b + 3$	$8b + 3$	$9b + 3$	$bb + 3$	$bb + b + 3$
$7b + 4$	$8b + 4$	$9b + 4$	$bb + 4$	$bb + b + 4$
$7b + 5$	$8b + 5$	$9b + 5$	$bb + 5$	$bb + b + 5$
$7b + 6$	$8b + 6$	$9b + 6$	$bb + 6$	$bb + b + 6$
$7b + 7$	$8b + 7$	$9b + 7$	$bb + 7$	$bb + b + 7$
$7b + 8$	$8b + 8$	$9b + 8$	$bb + 8$	$bb + b + 8$
$7b + 9$	$8b + 9$	$9b + 9$	$bb + 9$	$bb + b + 9$
$7b + b$	$8b + b$	$9b + b$	$bb + b$	$bb + b + b$, etc.
or $8b$	$9b$	bb		

The alert reader may have observed that the symbol b corresponds to the base used in this numeration system where $b = 10$. What happens if we have a system with fewer digits? Let's consider a system with the digits 0, 1, 2, 3, 4, and 5 only. Then we would count as in Table 5.3.

Table 5.3

1	$b+1$	$2b+1$	$3b+1$	$4b+1$	$5b+1$
2	$b+2$	$2b+2$	$3b+2$	$4b+2$	$5b+2$
3	$b+3$	$2b+3$	$3b+3$	$4b+3$	$5b+3$
4	$b+4$	$2b+4$	$3b+4$	$4b+4$	$5b+4$
5	$b+5$	$2b+5$	$3b+5$	$4b+5$	$5b+5$
b	$b+b$	$2b+b$	$3b+b$	$4b+b$	$5b+b$
or	$2b$	$3b$	$4b$	$5b$	bb

$bb+1$	$bb+b+1$
$bb+2$	$bb+b+2$
$bb+3$	$bb+b+3$
$bb+4$	$bb+b+4$
$bb+5$	$bb+b+5$
$bb+b$	$bb+b+b$, etc.

Did you detect that for this system $b = 6$? With $b = 6$, observe that we arrived at the digit bb faster than with $b = 10$, whereas if $b = 12$ we would arrive at the digit bb much later. Why?

When $b = 10$ the digits 1 and 0 have *place value*. That is, it makes a difference whether we write 10 or 01 as we attempted to illustrate above. More specifically, consider the numeral 27. Counting from 1 to 27, we would have Table 5.4.

Table 5.4

1	11	21
2	12	22
3	13	23
4	14	24
5	15	25
6	16	26
7	17	27
8	18	
9	19	
10	20	

Now this time let $T = 10$, and rewrite the above numerals as in Table 5.5

Table 5.5

1	$T+1$	$2T+1$
2	$T+2$	$2T+2$
3	$T+3$	$2T+3$
4	$T+4$	$2T+4$
5	$T+5$	$2T+5$
6	$T+6$	$2T+6$
7	$T+7$	$2T+7$
8	$T+8$	
9	$T+9$	
T	$T+T$	
or	$2T$	

Hence $27 = 2T + 7 = 2(10) + 7$, whereas by following a similar procedure, $72 = 7T + 2 = 7(10) + 2$ and, clearly, $7(10) + 2 \neq 2(10) + 7$. Hence $27 \neq 72$ although both numerals involve the identical digits.

Next consider all the numerals from 1 to 123 as in Table 5.6.

Table 5.6

1	11	21	31	41	51	61	71	81	91	101	111	121
2	12	22	32	42	52	62	72	82	92	102	112	122
3	13	23	33	43	53	63	73	83	93	103	113	123
4	14	24	34	44	54	64	74	84	94	104	114	
5	15	25	35	45	55	65	75	85	95	105	115	
6	16	26	36	46	56	66	76	86	96	106	116	
7	17	27	37	47	57	67	77	87	97	107	117	
8	18	28	38	48	58	68	78	88	98	108	118	
9	19	29	39	49	59	69	79	89	99	109	119	
10	20	30	40	50	60	70	80	90	100	110	120	

Again let $T = 10$, and rewrite the above numerals as in Table 5.7.

Hence $123 = TT + 2T + 3$. Now what meaning if any can we assign to the symbol TT? Since $T = 10$, then TT means 10×10 or 100. Therefore we have $123 = (TT) + 2(T) + 3 = 100 + 2(10) + 3$. Later we will observe that 10×10 can be written as 10^2 and, therefore, $123 = (1 \times 10^2) + (2 \times 10) + 3$.

Table 5.7

1	$T + 1$	$2T + 1$	$3T + 1$	$4T + 1$	$5T + 1$	$6T + 1$
2	$T + 2$	$2T + 2$	$3T + 2$	$4T + 2$	$5T + 2$	$6T + 2$
3	$T + 3$	$2T + 3$	$3T + 3$	$4T + 3$	$5T + 3$	$6T + 3$
4	$T + 4$	$2T + 4$	$3T + 4$	$4T + 4$	$5T + 4$	$6T + 4$
5	$T + 5$	$2T + 5$	$3T + 5$	$4T + 5$	$5T + 5$	$6T + 5$
6	$T + 6$	$2T + 6$	$3T + 6$	$4T + 6$	$5T + 6$	$6T + 6$
7	$T + 7$	$2T + 7$	$3T + 7$	$4T + 7$	$5T + 7$	$6T + 7$
8	$T + 8$	$2T + 8$	$3T + 8$	$4T + 8$	$5T + 8$	$6T + 8$
9	$T + 9$	$2T + 9$	$3T + 9$	$4T + 9$	$5T + 9$	$6T + 9$
T	$T + T$	$2T + T$	$3T + T$	$4T + T$	$5T + T$	$6T + T$
or	$2T$	$3T$	$4T$	$5T$	$6T$	$7T$

$7T + 1$	$8T + 1$	$9T + 1$	$TT + 1$	$TT + T + 1$	$TT + 2T + 1$
$7T + 2$	$8T + 2$	$9T + 2$	$TT + 2$	$TT + T + 2$	$TT + 2T + 2$
$7T + 3$	$8T + 3$	$9T + 3$	$TT + 3$	$TT + T + 3$	$TT + 2T + 3$
$7T + 4$	$8T + 4$	$9T + 4$	$TT + 4$	$TT + T + 4$	
$7T + 5$	$8T + 5$	$9T + 5$	$TT + 5$	$TT + T + 5$	
$7T + 6$	$8T + 6$	$9T + 6$	$TT + 6$	$TT + T + 6$	
$7T + 7$	$8T + 7$	$9T + 7$	$TT + 7$	$TT + T + 7$	
$7T + 8$	$8T + 8$	$9T + 8$	$TT + 8$	$TT + T + 8$	
$7T + 9$	$8T + 9$	$9T + 9$	$TT + 9$	$TT + T + 9$	
$7T + T$	$8T + T$	$9T + T$	$TT + T$	$TT + T + T$	
or $8T$	$9T$	TT		$TT + 2T$	

In a similar manner 4267 can be shown to be equal to (4×10^3) $+ (2 \times 10^2) + (6 \times 10) + 7$.

We have established so far that our numeration system is in terms of the base used and that the order in which we write the digits in a numeral is important since our system has associated with it place value.

EXERCISES

1. Counting is used to determine whether two sets are equivalent. For instance, any set equivalent to the set $\{1, 2, 3, 4\}$ must contain 4 elements. In general, any set equivalent to the set $\{1, 2, 3, \ldots, n\}$ must contain n elements. For each part of this exercise, determine the number of elements in the given set.

   a. $\{a, c, b\}$.
   b. $\{4, 7, 1, 3, 6\}$.
   c. $\{a, b, c, d, e, f, g\}$.
   d. $\{x \mid x$ is a natural number less than 19$\}$.
   e. $\{x \mid x$ is a natural number greater than 4 and less than 20$\}$.
   f. $\{y \mid y \in N, 2 \leq y \leq 18\}$.
   g. $\{y \mid y \in N, 3 \leq y < 27\}$.
   h. $\{u \mid u$ is a whole number less than or equal to 17$\}$.
   i. $\{u \mid u \in W, w < 26\}$.
   j. $\{u \mid u \in W, w \leq 16, w \neq 3\}$.

2. If a system of numeration consists of the digits 0, 1, 2, 3, and 4 only, write the numeral for each of the following.

   a. 2 more than 4.
   b. 3×3.
   c. $(2 + 4) \times 3$.
   d. $(3 + 4) - (2 \times 3)$.
   e. The tenth whole number.

3. Identify each of the following statements as being true or false.

   a. If $a = b$, then $aa = bb$.
   b. If $c > d$, then $cc = dd$.
   c. If $n(P) = n(Q)$, then $P = Q$.
   d. If $n(A) - n(B) = 0$, then $A = B$.
   e. If $A \sim B$, then $n(A) = n(B)$.
   f. If $A = B$, then $n(A) = n(B)$.
   g. $n(P \cup Q) = n(P) + n(Q)$.
   h. $n(P \cap Q) = n(P) + n(Q)$.
   i. $n(A \cap B) = n(A \cup B) - n(A) - n(B)$.
   j. $n(R \times S) = n(S \times R)$.

5.3 ADDITION OF WHOLE NUMBERS

Consider the sets $A = \{a, b, c\}$ and $B = \{\#, !, ?, -\}$. We observe that $n(A) = 3$ and $n(B) = 4$. Form $A \cup B = \{a, b, c, \#, !, ?, -\}$, and observe that $n(A \cup B) = 7$. Now consider the sets $C = \{a, b, c, d\}$ and $D = \{b, d, e, f, g\}$. We observe that $n(C) = 4$ and $n(D) = 5$. Form $C \cup D = \{a, b, c, d, e, f, g\}$, and observe that $n(C \cup D) = 7$.

For the sets A and B we have $n(A \cup B) = 7 = 3 + 4 = n(A) + n(B)$, and for the sets C and D we have $n(C \cup D) = 7 \neq 4 + 5 = n(C) + n(D)$. This is not surprising since A and B are disjoint sets and none of their elements were used twice in forming the union; hence no elements were counted twice. However, C and D are overlapping sets and two objects (b and d) are contained in both sets; hence they were both counted twice above.

DEFINITION 5.4 Let $a, b \in W$, and let A and B be two *disjoint* sets such that $n(A) = a$ and $n(B) = b$. Then we define the *sum of a and b*, denoted by $a + b$, to be the cardinality of the union of A and B. That is, $a + b = n(A \cup B)$.

In other words, the sum of the two whole numbers a and b is given as the cardinality of the union of two *disjoint* sets whose cardinalities are, respectively, a and b.

EXAMPLE   Determine the sum of 3 and 4.
Solution   Consider two sets P and Q such that P and Q are *disjoint* and such that $n(P) = 3$ and $n(Q) = 4$. Let $P = \{a, b, c\}$ and $Q = \{d, e, g, f\}$. Form $P \cup Q = \{a, b, c, d, e, f, g\}$, and determine that $n(P \cup Q) = 7$. Hence $3 + 4 = 7$.   ∎

In many elementary mathematics textbooks and workbooks the above example would be illustrated as in Figure 5.2. P and Q are represented as disjoint sets and are combined together in forming their union, as indicated by the arrows.

An equivalent representation, using fingers on your hands, would be as indicated in Figure 5.3. The hands, with the appropriate numbers of fingers extended, would correspond to the disjoint sets. (Clearly, a finger on one hand is not on the other hand also!) Now to form the union of these disjoint sets we simply bring the hands together as indicated in Figure 5.4. Counting the extended fingers, we find that there are 7, representing the sum of 3 and 4.

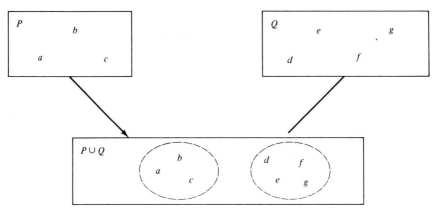

Figure 5.2

One hand with 3
fingers extended

One hand with 4
fingers extended

Figure 5.3

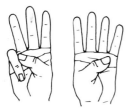

Figure 5.4

Addition, using the fingers on both hands, may call attention to the fact that we have a total of ten fingers on both hands and that our numeration system is a base 10 system.

EXAMPLE Determine the sum of 7 and 4.

Solution Consider two *disjoint* sets A and B such that $n(A) = 7$ and $n(B) = 4$. Let $A = \{a, b, c, d, e, f, g\}$ and $B = \{u, v, x, y\}$. Next form $A \cup B = \{a, b, c, d, e, f, g, u, v, x, y\}$ and determine that $n(A \cup B) = 11$. Hence $7 + 4 = 11$.

Alternate Solution Using "finger" addition as described above, we would hold up one hand containing 4 fingers extended such as indicated in Figure 5.5. Then, starting with the 7, we would continue to count, using the extended fingers and getting: 7, 8, 9, 10, 11. Hence $7 + 4 = 11$. ∎

Figure 5.5 One hand with 4 fingers extended

What happens if we try to add a natural number and 0? If we are counting objects in a set, we always start with 1 and proceed. What, then, is the whole number 0?

DEFINITION 5.5 The *whole number* 0 is defined to be the cardinality of the empty set.

Now let's consider an arbitrary natural number a and let A be a set such that $n(A) = a$. From Definition 5.5 we now have that $n(\varnothing) = 0$. Form $A \cup \varnothing$, and determine $n(A \cup \varnothing)$. But from Chapter 3 we know that $A \cup \varnothing = A$. Hence $n(A \cup \varnothing) = n(A) = a$. Hence $a + 0 = a$ for every $a \in N$.
What about $0 + 0$? Since $0 = n(\varnothing)$, then $0 + 0 = n(\varnothing \cup \varnothing)$. But $\varnothing \cup \varnothing = \varnothing$ and, hence, $n(\varnothing \cup \varnothing) = n(\varnothing) = 0$. Combining this result with the immediately preceding result, we now conclude that if a is an arbitrary whole number, then $a + 0 = a$ for *every* $a \in W$.

EXERCISES

1. Using the definition for the sum of two whole numbers, determine each of the following sums.
   a. $2 + 3$.
   b. $3 + 4$.
   c. $5 + 0$.
   d. $3 + 3$.
   e. $0 + 0$.
   f. $7 + 3$.
   g. $4 + 7$.
   h. $3 + 6$.
   i. $4 + 9$.
   j. $6 + 7$.
2. For each of the following form $A \cup B$ and determine whether or not $n(A \cup B) = n(A) + n(B)$.
   a. $A = \{1, 2, 3\}$, $B = \{4, 5\}$.
   b. $A = \{2, 3, 5, 6, 7\}$, $B = \{1, 2, 3, 4\}$.

c. $A = \{4, 6, 8\}, B = \varnothing$.
d. $A = \{x \mid x \in N, x \leq 6\}, B = \{y \mid y \in N, 6 \leq y < 9\}$.
e. $A = \{x \mid x \in N, x \leq 8\}, B = \{y \mid y \in N, 8 < y \leq 16\}$.
f. $A = P \cup Q, B = P \cap Q$, where $P = \{1, 2, 4, 6, 7\}$ and $Q = \{u \mid u \in N, u$ is odd, $1 < u < 7\}$.
g. $A = R \times S, B = S \times R$, where $R = \{1, 2, 3\}$ and $S = \{a, b\}$.

5.4 MULTIPLICATION OF WHOLE NUMBERS

To motivate the definition we will use for the multiplication of two whole numbers, let's consider the following. Suppose that we have three children and that each child has five nickels. How many nickels do the three children have among them? Clearly, we could arrange the nickels as in Figure 5.6.

First child's nickels   Second child's nickels   Third child's nickels

Figure 5.6

Now by counting we would determine that there are 15 nickels. However, instead of arranging the nickels as we have them in Figure 5.6, we could arrange them as in Figure 5.7.

Again, counting them we would find that there are 15 nickels. Now, suppose that we were to superimpose a coordinate grid over the nickels as in Figure 5.8. Further, if we treat the entries shown on the vertical axis

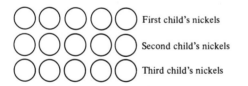

First child's nickels

Second child's nickels

Third child's nickels

Figure 5.7

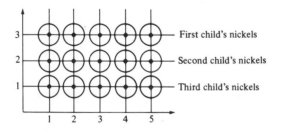

Figure 5.8

as the elements of a set B and treat the entries shown on the horizontal axis as the elements of a set A, we would then have $B = \{1, 2, 3\}$ and $A = \{1, 2, 3, 4, 5\}$. Also, if we formed the cartesian product $A \times B$ we would see that the center of each nickel in Figure 5.8 would coincide with a unique element of $A \times B$.

For this problem we have $n(A) = 5$ (representing the number of nickels per child), $n(B) = 3$ (representing the number of children), and $n(A \times B) = 15$. (Do you agree?) Combining all of this information, we would have that $5 \times 3 = n(A) \times n(B) = n(A \times B) = 15$.

DEFINITION 5.6 Let a and b be arbitrary whole numbers. Consider two sets A and B such that $n(A) = a$ and $n(B) = b$. Then the *product of a and b*, denoted by $a \times b$ or, simply, ab is given by $ab = n(A \times B)$.

The reader should observe that in Definition 5.6 the sets A and B do not have to be disjoint as was the case in our definition for the addition of whole numbers. The sets A and B are arbitrary with the only restrictions being that $n(A) = a$ and $n(B) = b$. It's possible that A is a subset of B or that A and B are overlapping sets. In fact, the elements in A don't even have to resemble the elements in B.

If you examine closely how we arrived at our definition of multiplication of whole numbers, you may have noted that multiplication involves repeated additions. In Figure 5.6 we added the five nickels three times. Hence 3×5 may be thought of as $5 + 5 + 5$. We will now state a theorem without proof.

THEOREM 5.1 If a is a natural number and b is a whole number, then $a \times b = b + b + b + \cdots + b$, where b is used in the sum a times.

EXAMPLE Evaluate 2×6 using (a) Theorem 5.1 and (b) Definition 5.6.
Solution
a. By Theorem 5.1, $2 \times 6 = 6 + 6 = 12$.
b. By Definition 5.6, we exhibit two sets A and B such that $n(A) = 2$ and $n(B) = 6$. Let $A = \{1, 2\}$ and $B = \{a, b, c, d, e, f\}$. Next form $A \times B = \{(1, a), (1, b), (1, c), (1, d), (1, e), (1, f), (2, a), (2, b), (2, c), (2, d), (2, e), 2, f)\}$ and determine that $n(A \times B) = 12$. Hence $2 \times 6 = n(A \times B) = 12$. ■

What happens when $b = 0$ in Theorem 5.1? To answer this question consider the following example.

EXAMPLE   Evaluate 2×0 using (a) Theorem 5.1 and (b) Definition 5.6.
Solution
a. By Theorem 5.1, $2 \times 0 = 0 + 0 = 0$.
b. By Definition 5.6, we exhibit two sets C and D such that $n(C) = 2$ and $n(D) = 0$. Let $C = \{a, b\}$ and $D = \varnothing$. Next form $C \times D = \varnothing$ and determine that $n(C \times D) = 0$. Hence $2 \times 0 = n(C \times D) = 0$.   ∎

THEOREM 5.2   In general, if a is a counting number and $b = 0$, then $a \times b = a \times 0 = 0$.
Proof   The proof of this theorem is left as an exercise for the student.

EXERCISES

1. Using the definition for the product of two whole numbers, determine each of the following products.
   a. 3×5.   b. 2×4.
   c. 4×3.   d. 1×2.
   e. 5×0.   f. 5×3.
   g. 4×4.   h. 2×5.
   i. 0×6.   j. 4×6.
2. Using Theorem 5.1, determine each of the following products.
   a. 2×6.   b. 3×5.
   c. 4×3.   d. 2×8.
   e. 3×0.   f. 3×7.
   g. 4×5.   h. 5×2.
   i. 6×3.   j. 8×9.
3. For each of the following form $A \times B$ and determine whether or not $n(A \times B) = n(A) \cdot n(B)$.
   a. $A = \{1, 2, 3\}$, $B = \{4, 6\}$.
   b. $A = B = \{x \mid x \in W, x \leq 4\}$.
   c. $A = \{a, b, c\}$, $B = \{a, c, e\}$.
   d. $A = P \cup Q$, $B = P \cap Q$, where $P = \{1, 2, 3\}$ and $Q = \{2, 3, 4\}$.
   e. $A = R \cup S$, $B = R \cap S$, where $R = \{x \mid x \in N, \ 6 < x \leq 8\}$ and $S = \{y \mid y \in N, y < 4\}$.
4. Prove Theorem 5.2.

5.5 PROPERTIES FOR ADDITION AND MULTIPLICATION OF WHOLE NUMBERS

So far in this chapter we have introduced the set $W = \{0, 1, 2, 3, 4, 5, 6, \ldots\}$ defined as the set of whole numbers. We have also defined the operations of addition and multiplication on these numbers. Hence we have a nonempty set together with at least one operation defined on its elements. In this section we will examine and discuss properties associated with these operations and will continue the development of the system of whole numbers. We will start with the commutative properties.

5.51 COMMUTATIVE PROPERTIES OF ADDITION AND MULTIPLICATION OF WHOLE NUMBERS

If $a, b \in W$, then:

1. $a + b = b + a$ Commutative property for addition
2. $ab = ba$ Commutative property for multiplication

These properties are easily established by referring to the respective definitions for addition and multiplication of whole numbers. For instance, to determine the sum $a + b$ select two *disjoint* sets A and B such that $n(A) = a$ and $n(B) = b$. Next form the union of A and B and conclude that $a + b = n(A \cup B)$. But from Chapter 3 we know that $A \cup B = B \cup A$. Also, two sets that are equal are equivalent and, hence, have the same cardinality. Therefore, $n(B \cup A) = n(A \cup B)$ and we have $a + b = n(A \cup B) = n(B \cup A)$. But since $A \cap B = \varnothing$, $n(B \cup A) = n(B) + n(A)$. Finally, we have $a + b = n(A \cup B) = n(B \cup A) = n(B) + n(A) = b + a$.

The proof that $ab = ba$ is left as an exercise.

The commutative property for addition of whole numbers allows us to add the two whole numbers a and b by adding b to a or by adding a to b. That is, the order in which we add two whole numbers is not important. Likewise, the order in which we multiply two whole numbers is not important. What happens when we want to add or multiply three or more whole numbers?

5.52 ASSOCIATIVE PROPERTIES FOR ADDITION AND MULTIPLICATION OF WHOLE NUMBERS

If $a, b,$ and c are any whole numbers, then

1. $(a + b) + c = a + (b + c)$ Associative property for addition
2. $(ab)c = a(bc)$ Associative property for multiplication

Again, these properties are easily established by referring to the respective definitions for addition and multiplication of whole numbers together with properties of union and cartesian product of sets. Before we attempt to establish these properties, however, we will state a theorem.

THEOREM 5.3 If $A, B,$ and C are any three sets, then $(A \times B) \times C$ is equivalent to $A \times (B \times C)$.

Proof The proof of this theorem is left as an exercise.

Now to establish that $(ab)c = a(bc)$, we would select any three sets P, Q, and R such that $n(P) = a$, $n(Q) = b$, and $n(R) = c$. Then

$(ab)c = n(P \times Q) \times n(R)$	By Definition 5.6
$n(P \times Q) \times n(R) = n[(P \times Q) \times R]$	By Definition 5.6
$(ab)c = n[(P \times Q) \times R]$	Transitive property for equality*
But $(P \times Q) \times R$ is equivalent to $P \times (Q \times R)$	Theorem 5.3
Therefore $n[(P \times Q) \times R] = n[P \times (Q \times R)]$	Equivalent sets have the same cardinality
$(ab)c = n[P \times (Q \times R)]$	Transitive property for equality
But $n[P \times (Q \times R)] = n(P) \times n(Q \times R)$	Definition 5.6
$n(P) \times n(Q \times R) = a(bc)$	Definition 5.6
Therefore $(ab)c = a(bc)$	Transitive property for equality

Establishing the associative property for addition of whole numbers is left as an exercise.

We can now use the commutative and associative properties of addition and multiplication of whole numbers to prove statements such as $2 + (3 + 4) = (4 + 2) + 3$.

EXAMPLE Using properties of this section, show that $2 + (3 + 4) = (4 + 2) + 3$.

Solution

$2 + (3 + 4) = (3 + 4) + 2$	Property 1, Section 5.51
$(3 + 4) + 2 = (4 + 3) + 2$	Property 1, Section 5.51
$2 + (3 + 4) = (4 + 3) + 2$	Transitive property for equality
$(4 + 3) + 2 = 4 + (3 + 2)$	Property 1, Section 5.52
$2 + (3 + 4) = 4 + (3 + 2)$	Transitive property for equality
$4 + (3 + 2) = 4 + (2 + 3)$	Property 1, Section 5.51
$2 + (3 + 4) = 4 + (2 + 3)$	Transitive property for equality
$4 + (2 + 3) = (4 + 2) + 3$	Property 1, Section 5.52
$2 + (3 + 4) = (4 + 2) + 3$	Transitive property for equality ■

EXAMPLE Using the properties of this section, show that $(2 \times 3) + (4 \times 6) = (6 \times 4) + (3 \times 2)$.

---

* We will use the expression "transitive property for equality" throughout this text as an abbreviation for "the relation is equal to on the set S is a transitive relation on S." (In particular, $S = W$.)

Solution

$(2 \times 3) + (4 \times 6) = (4 \times 6) + (2 \times 3)$	Property 1, Section 5.51
$(4 \times 6) + (2 \times 3) = (6 \times 4) + (3 \times 2)$	Property 1, Section 5.52
$(2 \times 3) + (4 \times 6) = (6 \times 4) + (3 \times 2)$	Transitive property for equality

■

5.53 DISTRIBUTIVE PROPERTIES OF MULTIPLICATION OVER ADDITION ON THE SET W

If a, b, and c are any whole numbers, then

1. $a(b + c) = ab + ac$ Left-hand distributive property
2. $(a + b)c = ac + bc$ Right-hand distributive property

In Property 1, Section 5.53, we observe that in order to evaluate $a(b + c)$ we first add b and c and then multiply a by the sum obtained or we multiply a and b, multiply a and c, and then add the products obtained. Although we have listed the distributive properties in two parts, throughout the remainder of the chapter we will refer to either part as the distributive property of multiplication over addition on the set of whole numbers.

EXAMPLE   Evaluate $2 \times (3 + 4)$ using the property in Section 5.53.
Solution

$$2 \times (3 + 4) = (2 \times 3) + (2 \times 4) = 6 + 8 = 14$$

Observe that if we first added $3 + 4 = 7$ and then multiplied 2 by 7, we would have obtained the same result.   ■

EXAMPLE   Evaluate $(15 \times 16) + (15 \times 4)$ using the property in Section 5.53.
Solution

$$(15 \times 16) + (15 \times 4) = 15 \times (16 + 4) = 15 \times 20 = 300 \quad ■$$

EXAMPLE   Evaluate 23×41 using the property in Section 5.53.
Solution   Since $41 = 40 + 1$, we may write $23 \times 41 = 23 \times (40 + 1) = (23 \times 40) + (23 \times 1) = 920 + 23 = 943$.   ■

5.54 IDENTITY PROPERTIES FOR WHOLE NUMBERS

1. ADDITIVE IDENTITY FOR WHOLE NUMBERS   If a is an arbitrary whole number, then $0 + a = a$ and $a + 0 = a$.
2. MULTIPLICATIVE IDENTITY FOR WHOLE NUMBERS   If a is an arbitrary whole number, then $1 \cdot a = a$ and $a \cdot 1 = a$.

Property 1 in this section can easily be established if we let A be an arbitrary set such that $n(A) = a$. Then $0 + a = n(\emptyset) + n(A) = n(\emptyset \cup A)$ by the definition for addition of whole numbers. But $\emptyset \cup A = A$ and, therefore, $n(\emptyset \cup A) = n(A)$. Hence $0 + a = n(\emptyset \cup A) = n(A) = a$. The second part of this property, $a + 0 = a$, follows immediately by the use of the commutative property for addition of whole numbers. This property identifies the whole number 0 as being the additive identity for the *set* of whole numbers. It states that 0 added to any whole number yields a sum that is that whole number.

To establish property 2 in this section, we let S and T be any two sets such that $n(S) = 1$ and $n(T) = a$. Then we have $1 \cdot a = n(S) \cdot n(T) = n(S \times T)$. But the set $S \times T$ will have exactly a elements since there is only 1 element in S which will get paired, in turn, with each of the a elements in T. Therefore, $n(S \times T) = n(T)$ and $1 \cdot a = n(S \times T) = n(T) = a$. The second part of the property, $a \cdot 1 = a$, follows immediately from the commutative property for multiplication of whole numbers. This property identifies the whole number 1 as being the multiplicative identity for the *set* of whole numbers. It states that the product of 1 and any whole number is always that whole number.

We will conclude this section with four useful theorems.

THEOREM 5.4 If a, b, and c are whole numbers and $a = b$, then $a + c = b + c$.

Proof Let A be a set such that $n(A) = a$. Since we are given that $a = b$, we also conclude that $n(A) = b$. Now consider a set B such that A and B are disjoint and $n(B) = c$. By the definition of addition of whole numbers, we now have that $a + c = n(A \cup B)$ and $b + c = n(A \cup B)$. But $b + c = n(A \cup B)$ implies that $n(A \cup B) = b + c$ by the symmetric property for equality. We now have that $a + c = n(A \cup B)$ and $n(A \cup B) = b + c$. By the transitive property for equality, we conclude that $a + c = b + c$.

THEOREM 5.5 If a, b, c, and d are any whole numbers and $a = b$ and $c = d$, then $a + c = b + d$.

Proof The proof follows immediately by the use of the above theorem and the transitive property for equality. Since we are given that $a = b$, then $a + c = b + c$ by Theorem 5.4. Also, since we are given that $c = d$, then $b + c = b + d$ by the same theorem. From $a + c = b + c$ and $b + c = b + d$, we conclude that $a + c = b + d$ by the transitive property for equality.

THEOREM 5.6 If a, b, and c are any whole numbers and $a = b$, then $ac = bc$.

Proof The proof is similar to that for Theorem 5.4 and is left as an exercise.

THEOREM 5.7 If a, b, c, and d are whole numbers and $a = b$ and $c = d$, then $ac = bd$.

Proof The proof is similar to that for Theorem 5.5 and is left as an exercise.

EXERCISES

1. For each of the following state which property of this section is being used.
   a. $2 + 4 = 4 + 2$.
   b. $3 \times (2 \times 4) = (3 \times 2) \times 4$.
   c. $5 + 0 = 5$.
   d. $5 \times 4 = 4 \times 5$.
   e. $7 \times 1 = 7$.
   f. $0 + 0 = 0$.
   g. $0 \times 1 = 0$.
   h. $(4 + 2) + 3 = 3 + (2 + 4)$.
   i. $(4 + 2) + 3 = 4 + (2 + 3)$.
   j. $8 + [2 + (3 + 4)] = (8 + 2) + (3 + 4)$.
   k. $(p + q) \times r = (p \times r) + (q \times r)$.
   l. $a + (b \times c) = a + (c \times b)$.
   m. $a \times (b + 0) = a \times b$.
   n. $(p \times 1) + (1 \times q) = p + q$.
2. Prove each of the following statements and give a reason for each step of the proof.
   a. $(2 + 3) + 4 = (4 + 2) + 3$.
   b. $2 \times (3 + 4) = (4 \times 2) + (3 \times 2)$.
   c. $(2 + 0) \times (3 \times 1) = 3 \times 2$.
   d. $(4 \times 5) \times 2 = (2 \times 4) \times 5$.
   e. $[(3 \times 4) \times 1] + 0 = 4 \times 3$.
   f. $(1 + 2) + (3 + 4) = 2 + (1 + 4) + 3$.
   g. $(a + b) + (c + d) = (b + d) + (c + a)$.
   h. $(a + b) \times (c + d) = [(a + b) \times c] + [(a + b) \times d]$.
   i. $(p + q) \times (r + s) = [(r + s) \times p] + [q \times (s + r)]$.
   j. $[a \times (b + 0)] \times 1 = b \times a$.
3. Using the distributive property of multiplication over addition of whole numbers, simplify each of the following computations.
   a. 16×21.
   b. 18×11.
   c. 31×17.
   d. $(16 \times 3) + (16 \times 7)$.
   e. $(6 \times 29) + (4 \times 29)$.
4. Use the commutative and associative properties of multiplication of whole numbers to simplify each of the following computations.
   a. $(16 \times 25) \times 4$.
   b. $20 \times (36 \times 5)$.
   c. $4 \times (16 \times 5)$.
   d. $(2 \times 9) \times (6 \times 5)$.
   e. 5×168.
5. Prove that if a, $b \in W$, then $ab = ba$.
6. Prove Theorem 5.3.
7. Prove Theorem 5.6.
8. Prove Theorem 5.7.

5.6 ORDERING OF WHOLE NUMBERS

In this section we will introduce the relation of order associated with the system of whole numbers. Let's consider counting from 1 to 10 and observe that we would get to the whole number 3 before the whole number 7. In fact, we would have to add 4 to 3 to get 7. In a similar manner we would have to add 3 to 5 to get to 8. When counting, we are imposing an order since some counting numbers come before others or, we may say, that some counting numbers are less than others.

DEFINITION 5.7 If a and b are any whole numbers, then a is said to be *less than* b, denoted by $a < b$, if and only if there exists a *natural number* c such that $b = a + c$.

EXAMPLE
a. The whole number 3 is less than the whole number 7 since there does exist a natural number, namely 4, such that $7 = 3 + 4$.
b. $5 < 8$ since there exists $3 \in N$ such that $8 = 5 + 3$.
c. $8 < 17$ since there exists $9 \in N$ such that $17 = 8 + 9$.
d. The whole number 7 is not less than the whole number 5 since there is no natural number c such that $5 = 7 + c$. Since 7 is not less than 5, we write $7 \not< 5$.

DEFINITION 5.8 If a and b are any whole numbers, then a is said to be *greater than* b, denoted by $a > b$, if and only if $b < a$.

From Definition 5.8 we see that if a and b are whole numbers with a greater than b, then b is less than a. Hence the statements $a > b$ and $b < a$ say the same thing.

EXAMPLE
a. $8 > 3$ since $3 < 8$. $3 < 8$ since there exists $5 \in N$ such that $8 = 3 + 5$.
b. $17 > 11$ since $11 < 17$. $11 < 17$ since there exists $6 \in N$ such that $17 = 11 + 6$.

EXAMPLE
a. $7 \not< 7$ since there does not exist a *natural number* c such that $7 = 7 + c$.
b. $7 \not> 7$ since $7 \not< 7$.

In mathematics we frequently combine equality and inequalities such as "less than or equal to" or "greater than or equal to." For instance, if a and b are whole numbers, we may write $a \leq b$ to mean $a < b$ or $a = b$. Similarly, $c \geq d$ means that $c > d$ or $c = d$. Although the connective of disjunction is being used here, it is used in the *exclusive* sense.

5.61 TRICHOTOMY PROPERTY FOR WHOLE NUMBERS

If a and b are any whole numbers, then *exactly* one of the following statements is true:

1. $a < b$.
2. $a = b$.
3. $a > b$.

Hence if we are given that p and q are whole numbers and also that p is not less than q, we may conclude that *either* p is equal to q *or* that p is greater than q, but not both.

EXERCISES

1. For each of the following insert the symbol $<$, $>$, or $=$ between the two whole numbers given to make the resulting statement correct.

  a. 2   5.
  b. $(3 + 2)$ $(1 + 4)$.
  c. 4 $(3 + 0)$.
  d. $2(3 + 4)$ $3(2 + 5)$.
  e. (8×3) (5×5).
  f. $3(4 + 2)$ (2×9).
  g. $5 \times (4 + 2)$ $5 + (6 \times 4)$.
  h. (5×0) $(1 + 2)$.
  i. $(3 + 4)$ (6×1).
  j. $(0 + 0)$ (0×0).

2. Let a represent the first whole number and b represent the second whole number of the pairs of whole numbers given in each of the following. Using the definitions of this section, show why $a < b$ or $a > b$.

  a. 2, 3.
  b. 1, 0.
  c. 14, 9.
  d. 5, 3.
  e. 2, 7.
  f. 0, 3.
  g. $3 + 4, 6 \times 1$.
  h. $5 + 0, 4 \times 0$.
  i. $2(3 + 2), 2(3 \times 2)$.
  j. $3(4 \times 5), 4(2 + 3)$.

3. Is the relation "is greater than" on the set of whole numbers a reflexive relation? Why or why not?

4. Is the relation "is less than" on the set of whole numbers a symmetric relation? Why or why not?

5. Is the relation "is greater than or equal to" on the set of whole numbers a reflexive relation? Why or why not?

6. Is the relation "is less than or equal to" on the set of whole numbers a symmetric relation? Why or why not?

7. Prove that if a, b, and c are any whole numbers with $a < b$ and $b < c$, then $a < c$.

8. What property does the statement in Exercise 7 characterize?

5.7 SUBTRACTION AND DIVISION ON THE SET OF WHOLE NUMBERS

Suppose that we have a bag of marbles that contains 15 marbles and we add to them another 8 marbles. We would then have $15 + 8$ or 23 marbles in the bag. Now suppose that we start with 23 marbles in the bag and take

8 marbles out of the bag. We would now have $23 - 8$ or 15 marbles left in the bag. Taking the 8 marbles out of the bag was the opposite of putting 8 marbles in the bag. Putting marbles in the bag corresponds to the operation of addition, and taking marbles out of the bag corresponds to the operation of subtraction; these opposite operations are called *inverse* operations.

If the bag contains 15 marbles, we could add another 17 marbles to obtain $15 + 17$ or 32 marbles. Starting with the 15 marbles in the bag, could we take 17 marbles out of the bag?

DEFINITION 5.9 If a and b are any two whole numbers, then the *difference* $a - b$ found by subtracting the whole number b from the whole number a is equal to the whole number c if and only if $a = b + c$.

It should be clear from an examination of the above definition that the whole number c will exist and satisfy the equation $a = b + c$ when a is greater than b or when a is equal to b. If $a < b$, then $a - b$ is not defined; that is, $a - b$ is not a whole number.

At this point we should comment on the fact that if a and b are any two whole numbers, then $a + b$ is also a whole number and ab also is a whole number. Stated in other words, the sum of any two whole numbers is always a whole number and the product of any two whole numbers is always a whole number. This is known as *closure*, and we say that the set of whole numbers is closed under the operations of addition and multiplication. However, the set of whole numbers is *not* closed under the operation of subtraction since $a - b$ is not always a whole number.

EXAMPLE
a. $7 - 3 = 4$ since $4 \in W$ and $7 = 3 + 4$
b. $8 - 8 = 0$ since $0 \in W$ and $8 = 8 + 0$.
c. $17 - 10 = 7$ since $7 \in W$ and $17 = 10 + 7$.
d. $0 - 0 = 0$ since $0 \in W$ and $0 = 0 + 0$.
e. $9 - 11 \notin W$ since there is no $c \in W$ such that $9 = 11 + c$.

Just as addition has an inverse operation called subtraction, multiplication has an inverse operation called *division*.

DEFINITION 5.10 If a is a whole number and b is a natural number, then *a divided by b*, denoted $a \div b$ or a/b or $\frac{a}{b}$, is equal to the whole number c if and only if $a = bc$.

It should be clear from Definition 5.10 and the following examples that the set of whole numbers is not closed under the operation of division.

EXAMPLE
a. $18 \div 2 = 9$ since $9 \in W$ and $18 = 2 \times 9$.
b. $27 \div 9 = 3$ since $3 \in W$ and $27 = 9 \times 3$.

EXAMPLE $17 \div 4 = ?$
Solution   If $17 \div 4$ has a solution, then we want a whole number c such that $17 = 4 \times c$. But there is no whole number c such that the product of 4 and c is 17. Hence $17 \div 4$ does not have a solution on the set of whole numbers. ■

EXAMPLE
a. $0 \div 7 = 0$ since $0 \in W$ and $0 = 7 \times 0$.
b. $0 \div 13 = 0$ since $0 \in W$ and $0 = 13 \times 0$.

EXAMPLE $5 \div 0 = ?$
Solution   If $5 \div 0$ has a solution on the set of whole numbers, then we want a whole number d such that $5 = 0 \times d$. But $0 \times d = 0$ for *every* whole number d. Hence no whole number d exists and $5 \div 0$ does not have a solution on the set of whole numbers. ■

EXAMPLE $0 \div 0 = ?$
Solution   If $0 \div 0$ has a solution, then we want a whole number m such that $0 = 0 \times m$. But $0 \times m = 0$ for *every* whole number m. Hence $0 \div 0$ has infinitely many solutions on the set of whole numbers. ■

Considering the last two examples, we find the divisor 0 is not a natural number as is required by Definition 5.10. However, we say that division by 0 is not possible for the following reasons:
1. If $a \neq 0$, then $a \div 0$ is not defined since there is no whole number c such that $a = 0 \times c$.
2. $0 \div 0$ is not defined since there is no *unique* solution since $0 = 0 \times d$ for *every* whole number d.
We will conclude this section with the following theorems.

THEOREM 5.8   If a is a natural number and b and c are whole numbers with $b \geq c$, then $a(b - c) = ab - ac$.
Proof   Since $b \geq c$, $b - c$ is a whole number by the definition of subtraction. Let $b - c = d$; hence $b = c + d$ and $a(b - c) = ad$. Now since $b = c + d$, we have

$ab = a(c + d)$	By Theorem 5.6
$a(c + d) = ac + ad$	Property 1, Section 5.53
$ab = ac + ad$	Transitive property for equality
$ac + ad = ad + ac$	Property 1, Section 5.51
$ab = ad + ac$	Transitive property for equality

Hence $ab - ac = ad$ Definition for subtraction since $ad \in N$ (closure
                             property for multiplication of N)
$ad = ab - ac$ Symmetric property for equality

We now have that $a(b - c) = ad$ and $ad = ab - ac$. Therefore, by the
transitive property for equality we conclude that $a(b - c) = ab - ac$.

In the above theorem a was restricted to be a natural number. However,
it is easy to verify that the theorem would also hold true if $a = 0$. The
above theorem is known as the *distributive property for multiplication over
subtraction on the set of whole numbers* whenever the subtraction can be
formed.

EXAMPLE   Verify Theorem 5.8 with $a = 2$, $b = 8$, and $c = 5$.
Solution

$$a(b - c) = 2(8 - 5) = (2)(3) = 6$$

and

$$ab - ac = (2)(8) - (2)(5) = 16 - 10 = 6$$

Hence $a(b - c) = ab - ac$.   ■

EXAMPLE   Verify Theorem 5.8 with $a = 4$, $b = 10$, and $c = 4$.
Solution

$$a(b - c) = 4(10 - 4) = (4)(6) = 24$$

and

$$ab - ac = (4)(10) - (4)(4) = 40 - 16 = 24$$

Hence $a(b - c) = ab - ac$.   ■

EXAMPLE   Verify Theorem 5.8 with $a = 0$, $b = 7$, and $c = 2$.
Solution

$$a(b - c) = 0(7 - 2) = (0)(5) = 0$$

and

$$ab - ac = (0)(7) - (0)(2) = 0 - 0 = 0$$

Hence $a(b - c) = ab - ac$.   ■

THEOREM 5.9   If a, b, c, and d are any whole numbers with $a \geq c$
and $b \geq d$, then $(a + b) - (c + d) = (a - c) + (b - d)$.

Proof Since $a \geq c$ and $b \geq d$, we can let $a - c = p$ and $b - d = q$, where $p, q \in N$. Hence

$a = c + p$ and $b = d + q$	Definition of subtraction of whole numbers
$a + b = (c + p) + (d + q)$	Theorem 5.5
$(c + p) + (d + q) = c + [p + (d + q)]$	Property 1, Section 5.52
$a + b = c + [p + (d + q)]$	Transitive property for equality
$c + [p + (d + q)] = c + [(p + d) + q]$	Property 1, Section 5.52
$a + b = c + [(p + d) + q]$	Transitive property for equality
$c + [(p + d) + q] = c + [(d + p) + q]$	Property 1, Section 5.51
$a + b = c + [(d + p) + q]$	Transitive property for equality
$c + [(d + p) + q] = c + [d + (p + q)]$	Property 2, Section 5.52
$a + b = c + [d + (p + q)]$	Transitive property for equality
$c + [d + (p + q)] = (c + d) + (p + q)$	Property 1, Section 5.52
$a + b = (c + d) + (p + q)$	Transitive property for equality
$(a + b) - (c + d) = p + q$	Definition for subtraction since $p + q \in N$ (closure property for addition on N)

Also, since $a - c = p$ and $b - d = q$, we have

$(a - c) + (b - d) = p + q$	Theorem 5.5
$p + q = (a - c) + (b - d)$	Symmetric property for equality

We now have $(a + b) - (c + d) = p + q$ and $p + q = (a - c) + (b - d)$. Therefore, by the transitive property for equality, we conclude that

$$(a + b) - (c + d) = (a - c) + (b - d)$$

EXAMPLE Verify Theorem 5.9 with $a = 8$, $b = 7$, $c = 5$, and $d = 3$.
Solution

$$(a + b) - (c + d) = (8 + 7) - (5 + 3) = 15 - 8 = 7$$

and

$$(a - c) + (b - d) = (8 - 5) + (7 - 3) = 3 + 4 = 7$$

Hence $(a + b) - (c + d) = (a - c) + (b - d)$. ■

EXAMPLE Verify Theorem 5.9 with $a = 10$, $b = 9$, $c = 2$, and $d = 5$.
Solution

$$(a + b) - (c + d) = (10 + 9) - (2 + 5) = 19 - 7 = 12$$

and

$$(a - c) + (b - d) = (10 - 2) + (9 - 5) = 8 + 4 = 12$$

Hence $(a + b) - (c + d) = (a - c) + (b - d)$. ■

THEOREM 5.10 If a is a natural number and b and c are any whole numbers with $b \geq c$, then $(a + b) - c = a + (b - c)$.

Proof Since $b \geq c$, we will let $b - c = k$ where $k \in N$. Then

$b = c + k$	Definition of subtraction of whole numbers
$a + b = a + (c + k)$	Theorem 5.4
$a + (c + k) = (a + c) + k$	Property 1, Section 5.52
$a + b = (a + c) + k$	Transitive property for equality
$(a + c) + k = (c + a) + k$	Property 1, Section 5.51
$a + b = (c + a) + k$	Transitive property for equality
$(c + a) + k = c + (a + k)$	Property 1, Section 5.52
$a + b = c + (a + k)$	Transitive property for equality
$(a + b) - c = a + k$	Definition of subtraction since $a + k \in N$ (closure property for addition on N)

Also, since $b - c = k$, we have

$a + (b - c) = a + k$	Theorem 5.4
$a + k = a + (b - c)$	Symmetric property for equality

We now have $(a + b) - c = a + k$ and $a + k = a + (b - c)$. Therefore, by the transitive property for equality, we conclude that $(a + b) - c = a + (b - c)$.

Two observations should be made relative to the above theorem:
1. Although we restricted a to be a natural number, it easily can be verified that the theorem will also hold true if $a = 0$.
2. The theorem is *not* to be compared with the statement $(a - b) - c \stackrel{?}{=} a - (b - c)$ because in general $(a - b) - c \neq a - (b - c)$.

EXAMPLE Using $a = 10$, $b = 6$, and $c = 3$, verify that $(a - b) - c \neq a - (b - c)$.

Solution

$$(a - b) - c = (10 - 6) - 3 = 4 - 3 = 1$$

and

$$a - (b - c) = 10 - (6 - 3) = 10 - 3 = 7$$

Since $1 \neq 7$, $(a - b) - c \neq a - (b - c)$. ∎

EXAMPLE Verify Theorem 5.10 with $a = 3$, $b = 9$, and $c = 3$.

Solution

$$(a + b) - c = (3 + 9) - 3 = 12 - 3 = 9$$

and

$$a + (b - c) = 3 + (9 - 3) = 3 + 6 = 9$$

Hence $(a + b) - c = a + (b - c)$. ∎

EXERCISES

1. Perform each of the following subtractions, and verify your results by using Definition 5.9.
   - a. $8 - 6$.
   - b. $4 - 2$.
   - c. $23 - 16$.
   - d. $4 - 0$.
   - e. $9 - 9$.
   - f. $(2 \times 3) - (1 + 4)$.
   - g. $3(4 + 2) - 2(3 + 3)$.
   - h. $(5 \times 0) - (0 + 0)$.
   - i. $(8 + 2) - 7$.
   - j. $13 - (4 + 5)$.

2. Perform each of the following divisions, and verify your results by using Definition 5.10.
   - a. $10 \div 2$.
   - b. $27 \div 9$.
   - c. $54 \div 27$.
   - d. $16 \div 4$.
   - e. $0 \div 5$.
   - f. $(2 \times 3) \div 3$.
   - g. $(5 \times 0) \div 3$.
   - h. $(8 \div 2) \div 4$.
   - i. $81 \div (27 \div 3)$.
   - j. $(256 \div 16) \div (64 \div 8)$.

3. Using Theorem 5.8, evaluate each of the following.
   - a. 3×9.
   - b. 9×11.
   - c. 6×29.
   - d. 7×39.
   - e. 12×99.

4. Perform the indicated operations on the whole numbers given if possible; otherwise, indicate that no answer is possible among the whole numbers.
   - a. $6 - (4 - 2)$.
   - b. $(5 + 7) \div 3$.
   - c. $9 \div (5 - 3)$.
   - d. $(6 - 6) \div (4 + 2)$.
   - e. $7 - (0 \div 8)$.
   - f. $8 \div (5 - 3)$.
   - g. $(8 \div 5) - 3$.
   - h. $9 - (7 - 5)$.
   - i. $(5 - 5) \div (7 - 7)$.
   - j. $6 - (4 - 7)$.

5. For each of the following indicate whether $a < b$ or $a > b$.
   - a. $a = 5 - 3; b = 6 \div 2$.
   - b. $a = 9 + (5 - 0); b = 4 - (8 \div 2)$.
   - c. $a = (3 \times 2) \div 6; b = (4 + 3) - 2$.
   - d. $a = (18 \div 6) \div 3; b = 18 \div (6 \div 3)$.
   - e. $a = (19 - 7) - 5; b = 19 - (7 - 5)$.

6. Construct examples to prove that subtraction on the set of whole numbers is neither a commutative nor an associative operation.

7. Construct examples to prove that division on the set of whole numbers is neither a commutative nor an associative operation.

8. Construct examples to prove that the set of whole numbers is not closed under the operation of subtraction.

9. Construct examples to prove that the set of whole numbers is not closed under the operation of division.

10. Is the set $S = \{1\}$ closed under the operation of subtraction? Why or why not?

11. Is the set $S = \{1\}$ closed under the operation of division? Why or why not?

5.8 THE NUMBER LINE

In this chapter we have defined and discussed the various arithmetic operations on the whole numbers and also have defined and discussed order
among them. In this section we will strive for a better comprehension of
these concepts by introducing what is known as the number line. A *number
line* is a geometric representation of the set of whole numbers superimposed
upon a line in a particular manner. The line will be represented as in
Figure 5.9 with the arrows at either end denoting the fact that the line
can be extended indefinitely in either direction.

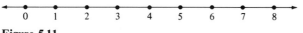

Figure 5.9

Next we locate two points on the line, one representing the whole number
0 and the other representing the whole number 1. This is represented in
Figure 5.10.

Figure 5.10

We have established what is known as a scale or a unit of measure and
can now locate the rest of the whole numbers on the line as indicated in
Figure 5.11.

Figure 5.11

Observe that the whole numbers are placed on the number line using the
concept of order. That is, if a and b are whole numbers with $a > b$, then the
whole number a is placed to the right of the whole number b. Also note
that the whole numbers are equally spaced along the line such that 2 is as
far to the right of 1 as 1 is to the right of 0, 3 is as far to the right of 2
as 2 is to the right of 1, 4 is as far to the right of 3 as 3 is to the right of 2,
and so forth.

We can now represent the whole numbers by the use of *vectors*. For
instance, to represent the whole number 3, we would start at 0 on the number
line and count over three *units*. For emphasis we will illustrate this by
placing the vector above the line as in Figure 5.12.

Instead of starting at 0, we could have started at 1 and counted over
3 units or at 2 and counted over 3 units and so forth. See Figure 5.13.

We will now represent the order relation, operations, and their properties
for whole numbers using the number line.

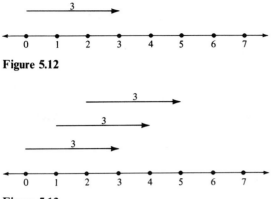

Figure 5.12

Figure 5.13

5.81 ORDER REPRESENTED ON THE NUMBER LINE

To represent the fact that $2 < 4$, observe in Figure 5.14 that we "construct" the vectors representing 2 and 4 with both vectors starting from the same point (in this case, the 0-point) and that the 2-vector is "shorter" than the 4-vector.

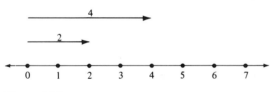

Figure 5.14

It would appear that we would have to add a whole number to 2 to obtain 4 and, indeed, that is correct! Compare this statement with Definition 5.7.

To represent the fact that $7 > 3$, observe in Figure 5.15 that we construct the vectors 7 and 3 with both starting from the same point (in this case, the 1-point) and that the 7-vector is "larger" than the 3-vector since the 3-vector is "shorter" than the 7-vector. Compare this statement with Definition 5.8.

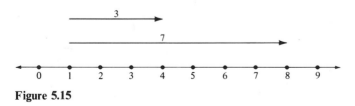

Figure 5.15

5.82 ADDITION OF WHOLE NUMBERS
REPRESENTED ON THE NUMBER LINE

To represent the statement that $3 + 4 = 7$, proceed as follows. In Figure 5.16 start at the 0-point and construct a 3-vector. Since we want to add 4 to 3, we will then start at the end of the 3-vector and construct a 4-vector. Now if we return to and construct a new vector from the 0-point to the end of the 4-vector in the diagram, we would conclude that our new vector is a 7-vector, which represents the sum of $3 + 4$.

A representation of the statement $5 + 3 = 8$ is given in Figure 5.17.

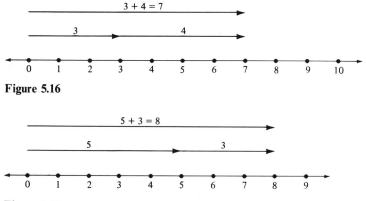

Figure 5.16

Figure 5.17

5.83 COMMUTATIVE PROPERTY FOR ADDITION
OF WHOLE NUMBERS

To represent the statement that $3 + 2 = 2 + 3$, proceed as follows. In Figure 5.18, starting at the 0-point *above* the number line, we construct a 3-vector. Starting at the end of this vector, construct a 2-vector. Now return to and construct a new vector from the 0-point to the end of the 2-vector in the diagram. This represents $3 + 2$. Next, starting at the 0-point *below* the number line, represent $2 + 3$ as indicated. Observe that the $(3 + 2)$-vector and the $(2 + 3)$-vector are the same.

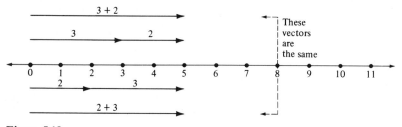

Figure 5.18

5.84 ASSOCIATIVE PROPERTY FOR ADDITION
OF WHOLE NUMBERS

We will now indicate the representation of $(2 + 3) + 4 = 2 + (3 + 4)$ by the diagram in Figure 5.19.

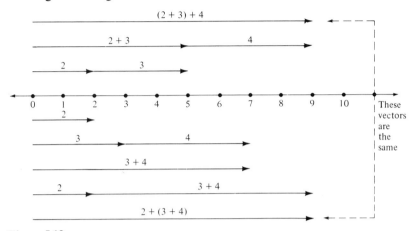

Figure 5.19

5.85 MULTIPLICATION OF WHOLE NUMBERS REPRESENTED
ON THE NUMBER LINE

To represent the statement $3 \times 2 = 6$, recall that 3×2 is equivalent to $2 + 2 + 2$ and proceed as indicated in Figure 5.20.

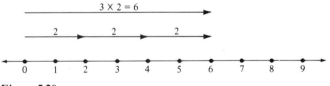

Figure 5.20

5.86 COMMUTATIVE PROPERTY FOR MULTIPLICATION
OF WHOLE NUMBERS

To represent the statement $3 \times 2 = 2 \times 3$, proceed as indicated in Figure 5.21.

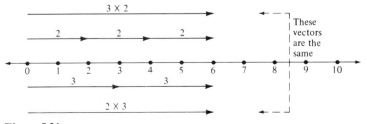

Figure 5.21

5.87 ASSOCIATIVE PROPERTY FOR MULTIPLICATION OF WHOLE NUMBERS

To represent the statement $(2 \times 3) \times 2 = 2 \times (3 \times 2)$, proceed as indicated in Figure 5.22.

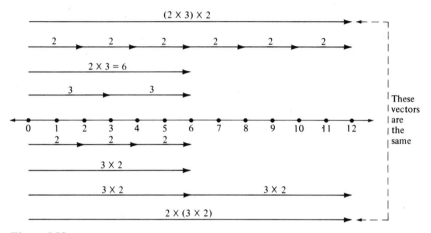

Figure 5.22

5.88 THE DISTRIBUTIVE PROPERTY OF MULTIPLICATION OVER ADDITION ON WHOLE NUMBERS

To represent the statement $2(3 + 2) = (2)(3) + (2)(2)$, proceed as indicated in Figure 5.23.

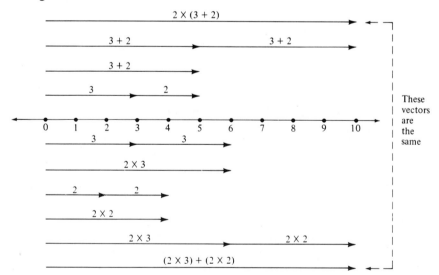

5.89 SUBTRACTION OF WHOLE NUMBERS REPRESENTED ON THE NUMBER LINE

In our discussion of subtraction we indicated that subtraction was the inverse or the opposite of addition. Hence, since we move to the right along the number line for addition, we will adopt the convention of moving to the left along the number line for subtraction. We will now represent the statement $8 - 2 = 6$. Starting at the 0-point, move to the right 8 units to construct the 8-vector. To subtract 2, we will now start at the end of the 8-vector constructed and move to the left 2 units as indicated. The vector labeled $8 - 2 = 6$ is our solution vector. (See Figure 5.24.)

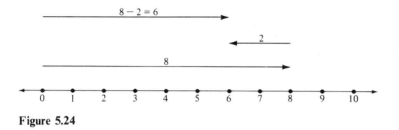

Figure 5.24

5.810 DIVISION REPRESENTED ON THE NUMBER LINE

Division is the opposite or inverse of multiplication. Since multiplication can be interpreted as successive additions, then division also can be interpreted as successive subtractions. To illustrate $2 \times 3 = 6$, we considered $2 \times 3 = 3 + 3$. Hence to illustrate $12 \div 4 = 3$, we will want to determine how many 4's can be subtracted from 12 which, of course, is 3. We will now illustrate the statement $12 \div 4 = 3$ as indicated in Figure 5.25 representing the solution as the 3-vector where 3 represents the number of 4's contained in 12.

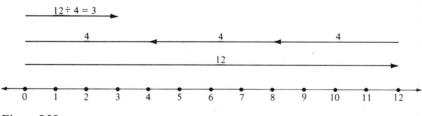

Figure 5.25

Consider Figure 5.26. Observe, that we started with 9, subtracted 4, subtracted another 4, and then could only subtract 1. This illustrates that 9 can not be (exactly) divided by 4; there is a "remainder" (of 1).

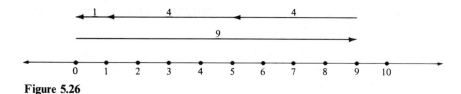

Figure 5.26

EXERCISES

Illustrate each of the following statements using the number line.

1. $5 < 7$.
2. $9 > 6$.
3. $5 + 3 = 3 + 5$.
4. $7 - 4 = 3$.
5. $5 - 3 \neq 3 - 5$.
6. $5 + 2 = 7$.
7. $(3 + 2) + 4 = 3 + (2 + 4)$.
8. $5 \times 3 = 15$.
9. $2 \times 4 = 4 \times 2$.
10. $(2 \times 3) \times 2 = 2 \times (3 \times 2)$.
11. $9 > (4 + 3)$.
12. $(2 + 3) < (3 + 4)$.
13. $2 \times (4 + 3) = (2 \times 4) + (2 \times 3)$.
14. $3 \times (4 - 2) = (3 \times 4) - (3 \times 2)$.
15. $3 \times 4 = 6 + 6$.
16. $5 + 3 = 10 - 2$.
17. $12 \div 3 = 4$.
18. $15 \div 5 = 3$.
19. $10 - (6 - 2) \neq (10 - 6) - 2$.
20. $3 \times (4 + 2) \neq 3 + (4 \times 2)$.

5.9 A LOOK AT THE SYSTEM OF WHOLE NUMBERS

Throughout this chapter we have developed the system of whole numbers. We started with the set of whole numbers denoted by W, defined and discussed the operations of addition, multiplication, subtraction, and division of whole numbers together with their properties, and introduced the relation of order on the set of whole numbers. In this section we will summarize our results.

If a, b, and c are any whole numbers, then

5.91 OPERATIONS

1. $a + b = n(A \cup B)$ such that $n(A) = a$, $n(B) = b$, and $A \cap B = \emptyset$.
2. $ab = n(A \times B)$ such that $n(A) = a$ and $n(B) = b$.
3. $a - b = p$ iff $p \in W$ and $a = b + p$.
4. $a \div b = q$ iff $b \in N$, $q \in W$ and $a = bq$.

5.92 CLOSURE PROPERTIES

1. The set W is closed under addition; that is, if a, $b \in W$, then $a + b \in W$
2. The set W is closed under multiplication; that is, if a, $b \in W$, then $ab \in W$

3. The set W is *not* closed under the operation of subtraction; that is, if $a, b \in W$, then $a - b \in W$ iff $a \geq b$ (for example, $7 - 9 \notin W$).
4. The set W is *not* closed under the operation of division; that is, if $a \in W$, $b \in N$, then $a \div b \in W$ iff $q \in W$ and $a = bq$ (for example, $12 \div 5 \notin W$).

5.93 COMMUTATIVE PROPERTIES

1. FOR ADDITION: If $a, b \in W$, then $a + b = b + a$.
2. FOR MULTIPLICATION: If $a, b \in W$, then $ab = ba$.

5.94 ASSOCIATIVE PROPERTIES

1. FOR ADDITION: If $a, b, c \in W$, then $(a + b) + c = a + (b + c)$.
2. FOR MULTIPLICATION: If $a, b, c \in W$, then $(ab)c = a(bc)$.

5.95 DISTRIBUTIVE PROPERTIES

1. MULTIPLICATION OVER ADDITION: If $a, b, c \in W$, then
   a. $a(b + c) = ab + ac$,
   b. $(a + b)c = ac + bc$.
2. MULTIPLICATION OVER SUBTRACTION: If $a, b, c \in W$ and $b \geq c$, then
   $a(b - c) = ab - ac$.

5.96 IDENTITY PROPERTIES

1. ADDITIVE IDENTITY: If a is an arbitrary whole number, then $a + 0 = 0 + a = a$; 0 is the additive identity on the set W
2. MULTIPLICATIVE IDENTITY: If a is an arbitrary whole number, then $a \cdot 1 = 1 \cdot a = a$; 1 is the multiplicative identity on the set W.

5.97 ORDER PROPERTIES

1. If $a, b \in W$, then $a < b$ iff $\exists\, p \in N$ such that $b = a + p$.
2. If $a, b \in W$, then $a > b$ iff $b < a$.
3. TRICHOTOMY PROPERTY: If $a, b \in W$, then *exactly* one of the following must hold:

$$\text{Either } a < b \quad \text{or} \quad a = b \quad \text{or} \quad a > b.$$

5.98 PROPERTIES OF ZERO

1. $0 \cdot a = 0$ for all $a \in W$.
2. $0 \div a = 0$, if $a \neq 0$.
3. $0 \div 0$ is not defined since it is *indeterminate*; that is, $0 \div 0$ does not have a *unique* solution.
4. $a \div 0$, if $a \neq 0$, is not defined since it is *impossible* to divide a nonzero whole number by 0.

SUMMARY

In this chapter we developed the system of whole numbers. We started with the set of natural numbers, N, and extended it to the set of whole numbers, W, by considering $W = N \cup \{0\}$. The operations of addition and multiplication, together with their various properties, were defined and discussed. With suitable restrictions we also defined and discussed the operations of subtraction and division. The order relation was also introduced and discussed. To emphasize and reinforce the concepts introduced in this chapter, we also discussed the number line as a geometric representation of the set of whole numbers.

The following symbols were introduced during our discussions.

Symbol	Interpretation
N	The set of natural (or counting) numbers
W	The set of whole numbers
$n(A)$	The cardinality of (or the number of objects in) the set A
$a + b$	The sum of the whole numbers a and b
$a \times b$	The product of the whole numbers a and b
$a \cdot b$	Another symbol for $a \times b$
ab	Another symbol for $a \times b$
$a < b$	The whole number a is less than the whole number b
$a \not< b$	The whole number a is not less than the whole number b
$a > b$	The whole number a is greater than the whole number b
$a \not> b$	The whole number a is not greater than the whole number b
$a = b$	The whole number a is equal to the whole number b
$a \leq b$	The whole number a is less than or equal to the whole number b
$a \geq b$	The whole number a is greater than or equal to the whole number b
$a - b$	The difference of the whole numbers a and b, if $a \geq b$
$a \div b$	The quotient of the whole number a by the natural number b whenever the symbol is meaningful

REVIEW EXERCISES FOR CHAPTER 5

Identify each of the following statements as being true or false. If the statement is true, indicate precisely what is meant by it. If the statement is false, correct it so that the resulting statement is true or give a counterexample to substantiate your answer.

1. The set of whole numbers is closed under the operation of addition.
2. The set of whole numbers is closed under the operation of subtraction.
3. The set of whole numbers is closed under the operation of multiplication.

4. The set of whole numbers is closed under the operation of division.
5. The operation of addition is a commutative operation on the set W.
6. The operation of addition is an associative operation on the set W.
7. The operation of subtraction is a commutative operation on the set W.
8. The operation of multiplication is a commutative operation on the set W.
9. The operation of subtraction is an associative operation on the set W.
10. The operation of multiplication is an associative operation on the set W.
11. The operation of division is a commutative operation on the set W.
12. The operation of division is an associative operation on the set W.
13. The operation of addition can be distributed over the operation of multiplication on the set of whole numbers.
14. The operation of multiplication can be distributed over the operation of subtraction on the set of whole numbers.
15. If a and b are arbitrary whole numbers, the statement that either $a < b$ or $a = b$ or $a > b$ is known as the distributive property.
16. If $n(A) = 3$ and $n(B) = 4$, then $n(A \cup B) = 7$.
17. If $n(A) = 3$ and $n(B) = 4$, then $n(A \cap B) = 12$.
18. If $n(P) = 2$ and $n(Q) = 5$, then $n(P \cup Q) \leq 7$.
19. If $a \in W$, then $0 \div a$ is always defined.
20. If $a, b \in W$, then $a - b \in W$ only when $a > b$.
21. If $b \in N$, then $0 \div b$ is always defined.
22. If $b \in N$, then $b \div 0$ may be indeterminate.
23. The set N has an additive identity element.
24. The set N has an additive identity element, or the set W has a multiplicative identity element.
25. Cardinal numbers are used to indicate how many elements are in a set, whereas ordinal numbers refer to the location of the elements in the set.
26. If $A \subseteq B$, $a \in A$ and $b \in B$, then $a \leq b$.
27. If $A \cap B = \varnothing$, then $n(A \times B) = n(A) \times n(B)$.
28. If $A \subset B$, then $n(A) \leq n(B)$.
29. If $4 < 6$, then $(3 \times 6) > (3 \times 4)$.
30. If $4 < 6$, then $7 - 4 < 7 - 6$.

CHAPTER 6
SYSTEMS OF NUMERATION

The previous chapter discussed the system of whole numbers. Before expanding this system to the system of integers, as we will do in the next chapter, we will take a closer look at the system of numeration used today.

It should be obvious that there was no need for representing numerals until after people learned to count, just as there was no need to record the spoken word until after people learned to speak. The very earliest method used to record a particular count involved a tally system employing pebbles, sticks, or other physical objects. To appreciate our system of numeration and to develop a better understanding of it, we will first look at some earlier systems.

6.1 EARLY NUMERATION SYSTEMS

Chapter 5 defined a numeral to be a symbol that represents a number. Numbers themselves usually are the same for all peoples, but the symbols used to represent these numbers vary. For instance, to represent the number fourteen we use the Hindu-Arabic numeral 14, whereas the Romans used the numeral XIV, the Egyptians used the numeral $\cap|\,|\,|$, and the Babylonians used $\lhd \; ^{\vee\;\vee\;\vee}_{\;\;\vee}$.

It is quite probable that the number words associated with numbers preceded the numerals used to represent them. Primitive people kept a record of their possessions by making various marks on the wall of a cave or by notches in a stick or marks on a piece of clay and so forth. As possessions increased or as the tribe increased in members, this tallying method was improved by grouping various sticks or notches and by the introduction of new symbols. For instance, a set of 5 strokes | | | | |, each stroke representing a finger of a hand, was represented by a hand ⍦, and a group of 20 strokes was represented by a whole man ⚲. The 20 strokes probably referred to the 10 fingers and 10 toes on the man.

6.2 EGYPTIAN NUMERATION SYSTEM

The early Egyptian numerals were picture symbols known as hieroglyphics. The Egyptians used the stroke, |, to represent 1, the heel bone, ⌒, to represent 10, the scroll or coil of rope, ○, to represent 100, the lotus flower, ⚲, to represent 1000, the bent finger, ⌐, to represent 10,000, the bourbot fish, ◁, to represent 100,000, and the astonished man, ⚲, to represent 1 million. This system was recorded by the Egyptian scribe Ahmes on papyrus about 1800 B.C. There is no place value associated with this numeration system and, hence, the symbols ⌒⌒| and ⌒|⌒ both represent 21. This system also lacked 0.

The first nine symbols were generally written

one	two	three	four	five	six	seven	eight	nine
|	||	|||	|||	|||	|||	|||	|||	|||
			|	||	|||	|||	|||	|||
					|	||	|||

In a similar manner, the heel bones were grouped as

ten	twenty	thirty	forty	fifty	sixty	seventy	eighty	ninety

The pattern was repeated for grouping coils of rope, lotus flowers, and other symbols.

Clearly, the Egyptian numeration system seems to have been developed on the pattern of a base ten.

EXAMPLE

a. The Egyptian symbol for 2136 would be

$$⚲⚲○⌒⌒⌒|||||||$$

or

$$⚲⚲○⌒⌒⌒|||$$

b. The Egyptian symbol for 42,968 would be

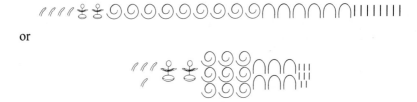

or

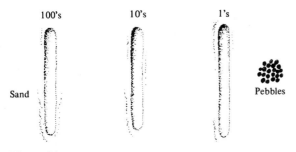

For merchants and traders this counting process was very tedious and, subsequently, methods of calculations were introduced. A simple counting board used for addition and subtraction was constructed by making grooves in the sand and placing pebbles in the grooves accordingly. There would be as many grooves in the sand as were needed with the first groove on the right representing ones, the next groove to the left tens, the next groove to the left hundreds, and so forth. Such a counting board is illustrated in Figure 6.1.

Figure 6.1

If, for instance, an Egyptian wanted to add 236 to 351, he would proceed as follows: 351 would be represented on the counting board as in Figure 6.2, with 1 pebble in the one's groove, 5 pebbles in the ten's groove, and 3 pebbles in the hundred's groove.

Then a small stick or similar object would be placed above those pebbles and additional pebbles would be added to correspond to 236 as illustrated in Figure 6.3.

Figure 6.2

Figure 6.3

The sticks or other objects would then be removed and the pebbles in each groove counted with the result being 587 as illustrated in Figure 6.4.

Figure 6.4

Figure 6.5 illustrates the addition of 465 to 378.

Figure 6.5

Observe that in the one's groove there are $5 + 8 = 13$ pebbles. The Egyptian would count up to 10 of the pebbles, discard 9 of them, and place the tenth one in the ten's groove. The additional 3 pebbles would remain in the one's groove. The counting board would now look like what is represented in Figure 6.6.

Now, working in the ten's groove, the Egyptian would count up to 10 of the pebbles, discard 9 of them, and place the tenth one in the hundred's groove. The additional 4 pebbles would remain in the ten's groove. (See Figure 6.7.)

Counting the number of pebbles in the hundred's groove, it would not

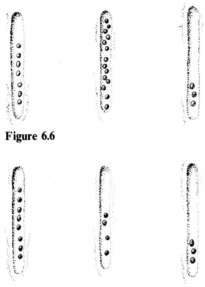

Figure 6.6

Figure 6.7

exceed 9; hence there was nothing more to do but determine the result. He would now determine that $378 + 465 = 843$.

EXAMPLE Using the picture symbols, represent the sum $351 + 236$.
Solution

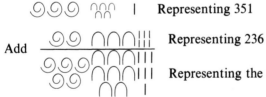

Add

Representing 351

Representing 236

Representing the desired sum, 587 ∎

EXAMPLE Using the picture symbols, represent the sum $378 + 465$.
Solution

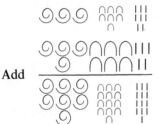

Add

Representing 378

Representing 465

Representing the desired sum *before* simplifying

or

 Representing the desired sum, 843 ■

The Egyptian could also perform subtraction using either the counting board or the picture symbols. Consider the subtraction of 234 from 676, using the counting board. First 676 would be represented as in Figure 6.8.

Figure 6.8

To subtract 234, he would start in the one's groove and remove 4 pebbles, which is possible since 6 pebbles are in that groove. Then he would remove 3 pebbles from the ten's groove and 2 pebbles from the hundred's groove. The result, 442, is illustrated in Figure 6.9.

Figure 6.9

EXAMPLE Using the picture symbols, represent the difference 676 − 234.

Solution

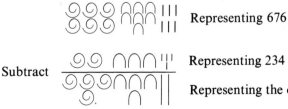

Subtract

Representing 676

Representing 234

Representing the desired result, 442

■

Using the counting board, the Egyptian would subtract 258 from 632 as follows. First 632 would be represented as in Figure 6.10.

Figure 6.10

To subtract 258, he would start in the one's groove and remove 8 pebbles. But there are only 2 pebbles there. Hence he would remove 1 pebble from the ten's groove and place 10 additional pebbles in the one's groove and then remove 8 pebbles from that groove. (See Figure 6.11.)

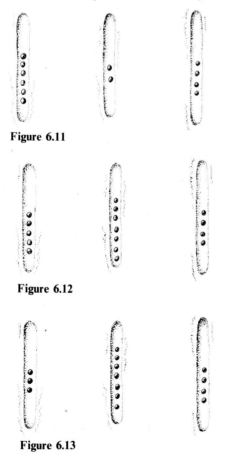

Figure 6.11

Figure 6.12

Figure 6.13

To remove 5 pebbles from the ten's groove, he would first have to remove 1 pebble from the hundred's groove, place 10 additional pebbles in the ten's groove, and then remove 5 pebbles from the same groove. (See Figure 6.12.)

Finally, he would remove 2 pebbles from the hundred's groove. The result would be 374 as illustrated in Figure 6.13.

EXAMPLE   Using picture symbols, represent the difference $632 - 258$.
Solution

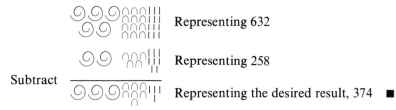

Since the subtraction cannot be performed in this representation, we would rewrite the problem as follows:

Relative to the operation of multiplication, there is evidence that the Egyptians used a modified form of repeated addition to find the product of two numbers. For instance, to find the product of 17 and 26, the Egyptian would start with the product of 1 and 26, as represented by step (a) in the following. Next he would double both the multiplier and the multiplicand as in (b). He would then double parts again as in (c). Doubling again, he would arrive at step (d). Doubling again, he would arrive at step (e). At this point, he would recognize that doubling again would give him a multiplier of 32, which is too large. However, knowing that $17 = 16 + 1$, he would add (a) and (e) to obtain the desired result.

a. | Representing $1 \times 26 = 26$

b. || Representing $2 \times 26 = 52$

c.   Representing $4 \times 26 = 104$

d.   Representing $8 \times 26 = 208$

e.   Representing $16 \times 26 = 416$

a + e.   Representing the sum of $1 \times 26 = 26$ and $16 \times 26 = 416$ for the desired result of $17 \times 26 = 442$

Of course he could have used other combinations of 17 such as $8 + 8 + 1$ or $8 + 4 + 4 + 1$ and so forth. The method described here is known as the *method of doubling and adding*.

EXAMPLE Using the method of doubling and adding, find the product of 23 and 31.

Solution

a.	∩∩∩|	$1 \times 31 = 31$
b.	∩∩∩||	$2 \times 31 = 62$
c.	𝒪∩∩|||	$4 \times 31 = 124$
d.	𝒪𝒪∩∩∩|||	$8 \times 31 = 248$
e.	𝒪𝒪𝒪∩∩∩|||	$16 \times 31 = 496$

Now, since $23 = 16 + 4 + 2 + 1$, we have

$$(a) + (b) + (c) + (e) \qquad ∩∩\|\|\| \;\Big|\; ∩∩∩∩\|\|\|$$

or the desired result, $23 \times 31 = 713$. ■

Division as repeated subtractions also was attempted by the Egyptians, but the procedure is rather cumbersome. In the exercises for this section the student will be asked to consider such a procedure.

EXERCISES

1. Write the Egyptian symbol for each of the following.
   - a. 163.
   - b. 278.
   - c. 683.
   - d. 1123.
   - e. 12,345.
   - f. 86,569.
   - g. 120,032.
   - h. 169,946.
   - i. 2,123,864.
   - j. 3,234,567.
2. Write the numeral that each of the following Egyptian symbols represents.
   - a. 𝒪𝒪∩∩\|\|\|
   - b. ⚘∩𝒪\|\|𝒪
   - c. ∩\|\|⚘⁄
   - d. ꙮ𝒪∩\|\|
   - e. ⁄⁄⚘𝒪𝒪∩∩\|\|\|
   - f. ꙮ⁄⚘⚘𝒪𝒪𝒪∩∩\|\|\|
   - g. ⚘⁄⁄⁄⚘𝒪𝒪∩∩∩\|\|
   - h. ⁄⁄⚖⚖∩∩⚘\|\|\|
   - i. ⚘\|\|\|𝒪⚘∩ꙮ
   - j. ⚘⚘⚘⚘𝒪𝒪𝒪∩\|\|\|
3. Using the Egyptian symbols for the numerals given, find the indicated sums as was done in the second and third examples of this section.
   - a. $247 + 132$.
   - b. $345 + 413$.
   - c. $253 + 358$.
   - d. $1234 + 6785$.
   - e. $7653 + 2968$.
   - f. $37,456 + 56,982$.
   - g. $867 + 1,246 + 23,696$.
   - h. $23,495 + 139,463 + 965,472$.

4. Using the Egyptian symbols for the numerals given, find the indicated differences as was done in the fourth and fifth examples of this section.
   a. $768 - 234$.   b. $12,864 - 9,413$.
   c. $923 - 789$.   d. $18,762 - 12,586$.
   e. $463,579 - 237,987$.   f. $2,169,437 - 1,456,123$.
5. Using the method of doubling and adding described in this section, find the indicated products using Egyptian symbols.
   a. 17×46.   b. 24×53.
   c. 33×92.   d. 52×123.
   e. 87×1234.   f. 123×2468.
6. Consider a procedure that could be used to perform the operation of division using Egyptian symbols.

6.3 BABYLONIAN NUMERATION SYSTEM

The Babylonians used what are known as cuneiform numerals. The word cuneiform is a combination of the Latin words *cuneus*, meaning a wedge, and *forma*, meaning a shape. They used a stylus to impress their symbols in the clay tablets and then baked the tablets in the hot sun or in a kiln. After the tablets were baked dry, they would last for a very long time. Obtaining clay tablets was no problem since clay was very abundant around Babylon.

The symbols used in the Babylonian numeration system were $\triangledown$ for one and $\triangleleft$ for ten. Using these two symbols only, numbers up to 59 could be represented using a pattern similar to the Egyptian numeration system. For instance, the first nine symbols were generally written impressed as

one	two	three	four	five	six	seven	eight	nine
$\triangledown$	$\triangledown\triangledown$	$\triangledown\triangledown\triangledown$	$\triangledown\triangledown\triangledown$	$\triangledown\triangledown\triangledown$	$\triangledown\triangledown\triangledown$	$\triangledown\triangledown\triangledown$	$\triangledown\triangledown\triangledown$	$\triangledown\triangledown\triangledown$
			$\triangledown$	$\triangledown\triangledown$	$\triangledown\triangledown\triangledown$	$\triangledown\triangledown\triangledown$	$\triangledown\triangledown\triangledown$	$\triangledown\triangledown\triangledown$
						$\triangledown$	$\triangledown\triangledown$	$\triangledown\triangledown\triangledown$

The next ten symbols were impressed as

ten	eleven	twelve	thirteen	fourteen
$\triangleleft$	$\triangleleft\triangledown$	$\triangleleft\triangledown\triangledown$	$\triangleleft\triangledown\triangledown\triangledown$	$\triangleleft\,\triangledown\triangledown\triangledown\,\triangledown$

fifteen	sixteen	seventeen	eighteen	nineteen
$\triangleleft\,\triangledown\triangledown\triangledown\,\triangledown\triangledown$	$\triangleleft\,\triangledown\triangledown\triangledown\,\triangledown\triangledown\triangledown$	$\triangleleft\,\triangledown\triangledown\triangledown\,\triangledown\triangledown\triangledown\,\triangledown$	$\triangleleft\,\triangledown\triangledown\triangledown\,\triangledown\triangledown\triangledown\,\triangledown\triangledown$	$\triangleleft\,\triangledown\triangledown\triangledown\,\triangledown\triangledown\triangledown\,\triangledown\triangledown\triangledown$

The pattern for impressing tens is similar to that for ones but only up to 50 and is illustrated as follows:

ten	twenty	thirty	forty	fifty
$\triangleleft$	$\triangleleft\triangleleft$	$\triangleleft\triangleleft\triangleleft$	$\triangleleft\triangleleft\triangleleft$	$\triangleleft\triangleleft\triangleleft$
			$\triangleleft$	$\triangleleft\triangleleft$

These symbols were used in combination to impress numbers up to 59, which was represented as

$$\triangleleft \ \triangleleft \ \triangleleft \underset{\triangledown \ \triangledown \ \triangledown}{\overset{\triangledown \ \triangledown \ \triangledown}{\underset{\triangledown \ \triangledown \ \triangledown}{}}}$$
$$\triangleleft \ \triangleleft$$

To represent 60 they returned to the pattern used to represent the one. Hence 60 was represented by $\triangledown$. Clearly, this created an ambiguity since $\triangledown$ could now represent 1 or 60. However, it is believed that the Babylonians were the first people to use place value in their numeration system. For instance, to represent 61 they would impress $\triangledown$ for 60 followed by a space and then impress $\triangledown$ again for 1 as follows: $\triangledown \quad \triangledown$.

The base for the Babylonian numeration system is a combination of 10 and 60. The 10 was used for grouping and may have been influenced by the Egyptians. The 60 is attributed to the Babylonians because of their extensive work in astronomy. Recall that a degree (as a measure of an angle) is divided into 60 equal parts called minutes and that a minute is divided into 60 equal parts called seconds. Likewise, an hour is divided into 60 minutes and a minute into 60 seconds. Thus, the Babylonian influence is still with us.

The Babylonian numeration system differed from the Egyptian numeration system insofar as a subtraction symbol, $\triangledown^\triangleright$, was used by the Babylonians. For instance, to represent 26, the Babylonians would impress $\triangleleft \ \triangleleft \ \overset{\triangledown \ \triangledown \ \triangledown}{\triangledown \ \triangledown \ \triangledown}$ or, using the subtraction symbol, would impress $\triangleleft \ \triangleleft \ \triangleleft \quad \triangledown^\triangleright \overset{\triangledown \ \triangledown \ \triangledown}{\triangledown}$, meaning $30 - 4$ or 26.

There are no known methods that may have been used by the Babylonians for performing arithmetic operations.

EXAMPLE

a. $\triangledown \ \triangledown \quad \triangleleft \ \triangledown$ Represents $2(60) + (10 + 1)$ or 131
b. $\triangleleft \ \triangledown \quad \triangleleft \ \triangleleft \ \triangledown$ Represents $11(60) + (20 + 1)$ or 681
c. $\triangleleft \ \overset{\triangledown \ \triangledown \ \triangledown}{\triangledown} \quad \overset{\triangleleft \ \triangleleft \ \triangleleft \ \triangledown \ \triangledown}{\triangleleft \ \triangleleft \quad \triangledown}$ Represents $14(60) + (50 + 4)$ or 894

EXAMPLE

a. $\triangleleft \ \triangleleft \triangledown^\triangleright \ \triangledown$ Represents $20 - 1$ or 19
b. $\overset{\triangleleft \ \triangleleft \ \triangleleft}{\triangleleft} \triangledown^\triangleright \triangleleft \overset{\triangledown \ \triangledown \ \triangledown}{\triangledown \ \triangledown \ \triangledown}$ Represents $40 - 16$ or 24

EXAMPLE   Represent 6962 using Babylonian symbols.
Solution   Rewriting, we have

$$6962 = 3600 + 3360 + 2$$
$$= (1 \times 60^2) + (56 \times 60) + (2 \times 1)$$

Hence we have

$$\triangledown \quad \overset{\triangleleft \ \triangleleft \ \triangleleft \ \triangledown \ \triangledown}{\triangleleft \ \triangleleft \ \triangledown \ \triangledown \ \triangledown} \quad \triangledown \ \triangledown \quad \blacksquare$$

EXERCISES

1. Write the Babylonian symbol for each of the following.
   a. 42.   b. 59.
   c. 119.   d. 327.
   e. 592.   f. 1440.
   g. 1923.   h. 2347.
   i. 3615.   j. 5000

2. Write the numeral that each of the following Babylonian symbols represents.
   a. ᐯ ᐯ   b. ᐯ ◄ ◄
   c. ᐯ ᐯ ᐯ ᐯ   d. ◄ ◄ ᐯ ◄ ◄ ◄ ᐯ ᐯ
   e. ᐯᐯᐯ ◄ ◄ ᐯᐯᐯ   f. ᐯ ᐯ ᐯ
   g. ᐯ ᐯ ◄ ◄ ᐯᐯᐯ ◄◄◄ ᐯ ᐯ   h. ◄ ◄ ᐯ▷ ◄ ᐯ
   i. ◄◄◄ ᐯ▷ ◄ ᐯᐯᐯ   j. ᐯ ◄ ᐯ ᐯ▷ ◄ ᐯ

3. Write the Egyptian symbol that corresponds to each Babylonian symbol given in Exercise 2.

4. Perform the indicated operation, and express your results using Babylonian symbols.
   a. ᐯ ◄ + ᐯ   ᐯ ᐯ + ◄ ◄ ᐯ   ◄ ᐯ
   b. ◄ ᐯ   ◄ ᐯ − ᐯ   ᐯ
   c. ᐯ ᐯ ᐯ   ◄ + ◄   ᐯ ᐯ ᐯ
   d. ᐯ ᐯ ᐯ   ◄ − ◄   ᐯ ᐯ ᐯ
   e. ◄ ᐯᐯᐯ ◄◄◄ ᐯ + ◄ ◄ ᐯ ◄◄◄ ᐯ ᐯ ᐯ
   f. ◄ ◄ ᐯᐯᐯ ◄ ᐯ − ◄ ᐯ ᐯ ◄ ◄ ◄ ᐯ ᐯ

5. Express the result of each part of Exercise 4 using Egyptian symbols.

6. Perform the indicated operations, and express your results using Babylonian symbols.
   a. ⚲∩|| + ∩∩||◎◎⚲
   b. ◎◎∩| − ∩∩∩|||
   c. ◎∩ × ⚲◎∩|
   d. ⌐◎∩⚲|∩ + ୧◎◎∩||∩
   e. ⚲◎|◎∩∩|∩ − ◎∩||◎|◎

6.4 GREEK NUMERATION SYSTEM

The Greek numeration system was quite different from the Egyptian and the Babylonian systems. As the system evolved the Greeks used letters of their alphabet for their numerals. There were originally 27 letters in the old Greek alphabet associated with numerals as follows:

α for 1	ι for 10	ρ for 100
β for 2	κ for 20	σ for 200
γ for 3	λ for 30	τ for 300
δ for 4	μ for 40	υ for 400
ε for 5	ν for 50	φ for 500
ϝ for 6	ξ for 60	χ for 600
ζ for 7	ο for 70	ψ for 700
η for 8	π for 80	ω for 800
θ for 9	ϟ for 90	⌐ for 900

There was no place value associated with their system because each numeral had its own unique value regardless of the position in which it appeared. The Greek youngster had a distinct advantage over the youngster today—there was nothing extra for him to learn since he had to learn his alphabet anyway. If by chance the Greek numeral also spelled a word, a bar would be placed above the numeral to distinguish it as such. For numbers larger than 999, the symbol $_\prime$ was used and was placed in front of a given numeral. For example, ρ represents 100 but $_\prime\rho$ represents $100 \times 1000 = 100,000$.

EXAMPLE

a. χμε   Represents $600 + 40 + 5$ or 645
b. ωπθ   Represents $800 + 80 + 9$ or 889
c. $_\prime$βφλη   Represents $(2 \times 1000) + 500 + 30 + 8$ or 2538

Other alphabetic numeration systems are the Hebrew, the Syrian, and the early Arabic.

EXERCISES

1. Write the numeral represented by each of the following Greek symbols.
   a. τξη.   b. ψνβ.
   c. ωλα.   d. ⌐πθ.
   e. ρια.   f. $_\prime$εωμδ.
   g. $_\prime$θφοθ.   h. $_\prime$ϟψμϝ.
   i. $_\prime$γυκη.   j. $_\prime$πχϟβ.
2. Express each of the symbols given in Exercise 1 in equivalent form using Egyptian symbols.
3. Express each of the symbols given in Exercise 1 in equivalent form using Babylonian symbols.
4. Perform the indicated operations, and write your results using Greek symbols.
   a. ψϝ + ξη.   b. ϟθ + μδ + λγ.
   c. σλδ + χνη.   d. τ + ρ + φ + ⌐.

e. $\omega - \sigma$.   f. $\psi\xi\varepsilon - \pi\theta$.

g. $\omega\pi\eta - \Theta\theta$.   h. $_\prime\beta\nu - \sigma_F\beta$.

i. $_\prime\beta + _\prime\delta - \chi\varepsilon$.   j. $_\prime\eta - \omega - \mu_F$.

6.5 ROMAN NUMERATION SYSTEM

The Roman numerals have survived until the present day. The Roman numeration system is similar to the Egyptian numeration system since it is basically a tallying system. The Roman numerals for 1 through 4 are identical with those for the Egyptian numerals. However, to represent 5 the Romans used V which may have been a hieroglyphic for a hand. Using both I and V, the Romans were able to symbolize numbers from 6 to 9. Again, for 10 they used X, which may have been a hieroglyphic for two hands in the form of V and Λ put together. The symbol X was used in a numeral a maximum of four times. For 50 the Romans used L. Using the symbols I, V, and X, the Romans could write all of the numbers from 1 to 49. With the addition of L they could write all of the numbers from 1 to 99. For 100 the Romans used C, possibly because it is the first letter of the Latin word *centum*, which means 100, or possibly because the Egyptian coil of rope ◯ was sometimes written as ◖. Again, a maximum of four Cs were used, with the symbol D being introduced to represent 500. Why D?

The early Romans used the symbol (I) to represent 1000, and it is conjectured that the symbol I) may have been used to represent one-half of (I), or 500. Hence it is possible that D is a simplification of I). The D, like V and L, is used only once in any numeral. The early Roman symbol I for 1000 was later replaced by M, which is the first letter in the Latin word *mille*, which means 1000. Like I, X, and C, M could be used a maximum of four times in any numeral.

EXAMPLE

a. XXII is the Roman numeral for 22.

b. DCCXIIII is the Roman numeral for 714.

c. MMDCLXXXII is the Roman numeral for 2682.

EXAMPLE Represent (a) 465 and (b) 1743 using Roman numerals.

Solution

a. $465 = 400 + 60 + 5$

        $= (4 \times 100) + (50 + 10) + 5$

Using C for 100, L for 50, X for 10, and V for 5, we have CCCCLXV.

b. $1743 = 1000 + 700 + 40 + 3$

$\qquad = 1000 + 500 + 200 + 40 + 3$

$\qquad = 1000 + 500 + (2 \times 100) + (4 \times 10) + (3 \times 1)$

Using M for 1000, D for 500, C for 100, X for 10, and I for 1, we have MDCCXXXXIII. ∎

The Romans also used a subtraction principle in their system of numeration. For example, instead of representing 4 by IIII (as was done by the early Romans), the symbol IV was introduced to denote that 1 was being subtracted from 5. In a similar manner IX denoted 1 being subtracted from 10 and, hence, IX represents 9. Likewise, XL denotes 10 being subtracted from 50 and, hence, XL represents 40.

EXAMPLE
a. XIV represents $10 + (5 - 1)$ or 14.
b. CXLIV represents $100 + (50 - 10) + (5 - 1)$ or 144.
c. CMLXIX represents $(1000 - 100) + 50 + 10 + (10 - 1)$ or 969.

Throughout the years, the *vinculum* (a Latin word for horizontal bar) was used in various manuscripts over certain Roman numerals to denote multiplication by 1000. For example, $\overline{V}$ means 5×1000 or 5000 and $\overline{C}$ means 100×1000 or 100,000. Two such bars over a numeral means multiplication by 1000×1000. Hence $\overline{\overline{V}}$ means $5 \times 1000 \times 1000$ or 5,000,000. Although we still use the vinculum with Roman numerals today, there is no evidence that the Romans actually originated the practice.

EXAMPLE
a. $\overline{C}$MCX means $100,000 + 1,000 + 100 + 10$ or 101,110.
b. $\overline{\text{II}}$ is another symbol for MM or 2000.

EXAMPLE Rewrite 623,482 using Roman numerals
Solution

$623,482 = 623,000 + 400 + 80 + 2$

Hence we have $\overline{\text{DCXXIII}}$CDLXXXII. ∎

The reader who has compared the Roman numeration system with the Egyptian numeration system may be anticipating that arithmetic operations using Roman numerals may be performed in a somewhat analogous manner as using Egyptian numerals and, indeed, this is correct. However, in performing arithmetic operations using Roman numerals it is sometimes easier to use the older Roman forms such as IIII for IV and CCCC for CD. Let's attempt to add CCXLIX to CCCLXVII. As with the Egyptian method, we would line up like symbols after first rewriting in older Roman numeral form as follows:

	CCC	L X	V	II
Add	CC	XXXX	V	IIII
	CCCCC L XXXXX VV IIIIII			

The spaces between different symbols are for convenience

Combining but not simplifying

| Or | D | L L | X | VI |

Simplifying

Or   DCXVI

As the final result

Checking the above addition by converting to equivalent Hindu-Arabic numerals we obtain

	CCCLXVII	or	367
Add	CCXLIX		+ 249
	DCXVI		616	Which checks!

To subtract CCXLIX from CCCLXVII, we would proceed as follows:

	CCC L	X V	II
Subtract	CC	XXXX V	IIII	Rewriting

Recognizing that we can't subtract IIII from II or XXXX from X, we regroup as follows and subtract:

	CCC XXXXX VV IIIIIII
Subtract	CC XXXX V IIII
	C X V III	Or, in compact form, CXVIII

Checking the above subtraction using equivalent Hindu-Arabic numerals we obtain

	CCCLXVII	or	367
Subtract	CCXLIX		− 249
	CXVIII		118	Which checks!

To multiply LII by XXXII we would proceed as follows:

		LII
Multiply		XXXII
	L	II

Multiplying by I and spacing for convenience

| | L | II |

Multiplying by I again

| | D | XX |

Multiplying by X

| | D | XX |

Multiplying by X again

| | D | XX |

Multiplying by X again

DDDLLXXXXXXIIII

Adding the partial products but not simplifying

Or   MDCLXIV

The product in simplified form

Division could be performed using the principle of repeated subtraction. However, the student should first become familiar with basic products such as V times X = L, V times C = D, X times C = M, and others.

EXERCISES

1. What numbers are represented by the following Roman numerals?
   - a. XXXIII.
   - b. LVII.
   - c. MCLXVI.
   - d. DCXLIX.
   - e. MCLIV.
   - f. MMCCLXVII.
   - g. MDCCXLIX.
   - h. MMDCLXIV.
   - i. LXXXVIIII.
   - j. MCCCCXXXXIIII.
   - k. MDCLXVI.
   - l. CCCCLXXXXVIIII.
   - m. CMXCIX.
   - n. $\overline{\overline{\text{VI}}}$.
   - o. $\overline{\text{X}}$MMDL.
   - p. $\overline{\overline{\text{IX}}}$.
   - q. $\overline{\text{IV}}$DX.
   - r. DCDCDC.
   - s. $\overline{\text{MMDCLMMMDCCXLVII}}$.
   - t. $\overline{\overline{\text{MMMDCCLXXVI}}}$DCXLDCCVIIII.

2. Rewrite each of the following using Roman numerals.
   - a. 37.
   - b. 169.
   - c. 909.
   - d. 1464.
   - e. 3976.
   - f. 19,372.
   - g. 39,947.
   - h. 205,690.
   - i. 476,123.
   - j. 675,675.
   - k. 990,009.
   - l. 999,999.
   - m. 1,000,000.
   - n. 1,500,500.
   - o. 3,333,333.
   - p. 9,999,999.
   - q. 99,999,999.
   - r. 123,123,234.
   - s. 575,467,892.
   - t. 3,469,762,999.

3. Perform the following indicated operations, and express your results using Roman numerals.
   - a. CCLV + CXII.
   - b. CCLV − CXII.
   - c. MDCL + DCCLII.
   - d. MDCL − DCCLII.
   - e. MMCVII + MDCXLV + CCI.
   - f. MMCVII + MDCXLV − CCI.
   - g. MMCVII − (MDCXLV + CCI).
   - h. $\overline{\text{X}}$MMDCLIX + $\overline{\text{VII}}$DCCLXX.
   - i. $\overline{\text{XXV}}$CCIV − $\overline{\text{XIV}}$CMXLIV.
   - j. $\overline{\text{D}}$DCCLXVIII + $\overline{\text{VI}}$DCLCMII.

4. Complete the following multiplication table using Roman numerals.

×	I	II	III	IV	V	VI	VII	VIII	IX	X
I										
II										
III										
IV										
V										
VI										
VII										
VIII										
IX										
X										

5. Complete the following multiplication table using Roman numerals.

×	I	V	X	L	C	D	M
I							
V							
X							
L							
C							
D							
M							

6. Perform the indicated operations, and express your results using Roman numerals.

a. Multiply XXII by VI.   b. Multiply LVI by XXII.
c. Multiply MMDCCL by MCXLV.   d. Divide M by L.
e. Divide MMDCXL by XXX.   f. Divide $\overline{X}$D by CCC.

6.6 MAYAN SYSTEM OF NUMERATION

The Mayas of Central America, like the Chinese and Japanese, have a system of numeration where their numerals are written in vertical form. Their system consists of dots, horizontal line segments, and a symbol, $\bigcirc$, for zero. The first stage of grouping is in fives as noted below for the numerals from 0 to 19 inclusive.

0	1	2	3	4	5	6	7	8	9

10	11	12	13	14	15	16	17	18	19

All of the numerals above represent numbers that can be written in the units place. Numbers larger than 19 were then written in a base 20, possibly because they had 20 days in a month. Instead of counting from 20 to 20^2, they continued from 20 to 18(20), possibly because they had 18 months in their year. The system then continued from 18(20) to $18(20)^2$ to $18(20)^3$ and so forth.

EXAMPLE   .

$$1 \times 18(20) = 360$$

$\bigcirc$ Represents $0 \times 20 = 0$

. $1 \times 1 = 1$ or $1 + 0 + 360 = 361$

EXAMPLE   ..

$$2 \times 18(20) = 720$$

.   Represents $1 \times 20 = 20$

— $5 \times 1 = 5$ or $5 + 20 + 720 = 745$

EXAMPLE   .

$$1 \times 18(20)^2 = 7200$$

.. $2 \times 18(20) = 720$

$\cdot\!\!-$ Represents $6 \times 20 = 120$

$\bigcirc$ $0 \times 1 = 0$

     or $0 + 120 + 720 + 7200 = 8040$

EXAMPLE   Represent 869 in Mayan symbols.
Solution

$869 = 720 + 140 + 9$
$\quad = (2 \times 360) + (7 \times 20) + (9 \times 1)$

Hence we have

2×360 ..

7×20 or   ⠒

9×1 ....   as the required symbol ■

EXAMPLE   Represent 9008 in Mayan symbols.
Solution

$9008 = 7200 + 1800 + 8$
$\quad = (1 \times 7200) + (5 \times 360) + (0 \times 20) + (8 \times 1)$

1×7200 .

5×360 —

0×20 or   ⊖

8×1 ⠇   as the required symbol ■

The Mayan system of numeration system, like most other early numeration systems, was not practical for arithmetic calculations as we know them today. These early systems were used primarily for representing numbers.

EXERCISES

1. Write each of the following numerals using Mayan symbols.
   a. 37.
   b. 119.
   c. 345.
   d. 482.
   e. 976.
   f. 1234.
   g. 4699.
   h. 7640.
   i. 83,463.
   j. 275,692.
2. What numbers are represented by the following Mayan symbols?

   a. ⠈
         ..

   b. ⠤
         ⠄

   c. ⠒
         ⠄
         ⊖

   d. —
         ..
         —
         ..

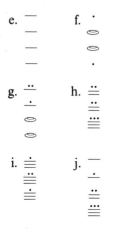

3. Write each number represented by the Mayan symbols in Exercise 2 in equivalent form using Roman numerals.
4. Write each number represented by the Mayan symbols in Exercise 2 in equivalent form using Babylonian numerals.
5. Write each number represented by the Mayan symbols in Exercise 2 in equivalent form using Egyptian numerals.
6. Write each number represented by the given symbols in Mayan symbols.

a. ◡◡∩∩|⚹   b. ◁▽ ▽▽
c. MDCLVII   d. ◁⚹ ◦◦◦ ∩∩|||
e. ◁ ◁ ▽ ▽ ▽▽▷◁ ▽   f. MMDCCLXXIX

6.7 SOME OLDER ALGORITHMS

When performing arithmetic operations we recall some basic addition and multiplication facts that were learned at the elementary grade level. These involve the sums and products of single digit numerals. However, when working with numerals involving two or more digits, we perform the arithmetic operations based upon certain rules or procedures. These procedures are called *algorithms*. We have discussed some of the earlier algorithms used in previous sections of this chapter.

Throughout the ages various computers have been used to perform the more complex arithmetic operations. The counting board used by the Egyptians represented a form of a computer. The abacus was also an early form of the computer and is still in use today in most of Asia. A simple abacus is depicted in Figure 6.14. It is a simple four-wire abacus. To represent various numbers by means of an abacus, beads may be used such that not more than nine beads would be on any one wire. For instance, 3724 would be represented on a four-wire abacus by placing $3 + 7 + 2 + 4$ or 16 beads on the wires as illustrated in Figure 6.15. The

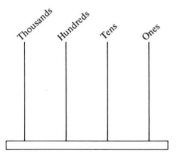

Figure 6.14 A four-wire abacus

beads are all alike but the value associated with a bead depends upon the particular wire on which the bead is placed. To add another number to the one already recorded, additional beads are placed on the wires with the understanding that not more than nine beads can be placed on any one wire. If 2698 were to be added to 3724 already displayed on the abacus, we would proceed as follows.

First add 8 beads to the one's wire. But that would yield $4 + 8 = 12$ beads, which are too many beads for the wire. However, since 10 beads on the one's wire is equivalent to 1 bead on the ten's wire, 10 of these beads would be removed and "carried" over to the next wire to the left in the form of 1 bead. The abacus at that point would appear as in Figure 6.16(a). Now 9 beads would be added to the ten's wire, giving $3 + 9 = 12$ beads. Again, 10 of these beads would be removed and carried over to the next wire to the left in the form of 1 bead. The abacus now appears as in Figure 6.16(b). Next 6 beads would be added to the hundreds wire, giving $8 + 6 = 14$ beads. Again, 10 beads would be removed and carried over to the next wire to the left in the form of 1 bead. The abacus now appears as in Figure 6.16(c). Finally, 2 beads are added to the thousands wire, giving $4 + 2 = 6$ beads. Since 6 is less than 10, we are done and can read the sum of 3724 and 2698 as 6422 represented on the abacus as illustrated in Figure 6.16(d).

Subtraction would be performed in a similar manner using the abacus. However, instead of adding beads one would remove beads, and instead of carrying beads over to the wire on the left, one would remember that 1

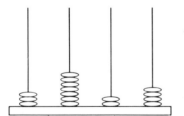

Figure 6.15

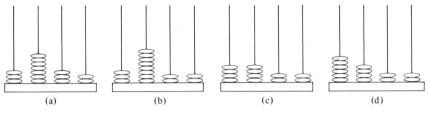

(a) (b) (c) (d)

Figure 6.16

bead on a wire is equivalent to 10 beads on the next wire to the right of it. Of course if 5 or more digits were involved in the numbers being used, additional wires would have to be placed on the abacus.

The Japanese use a modern abacus, known as a *soroban*, which differs from the simple abacus described above insofar as only seven beads are used on any one wire, with two of these beads set apart from the other five. Each of these two beads represents five beads in one and may be of a different color. In addition, the beads are not removed from the wire but are moved along the wire toward the middle. Figure 6.17 represents 2874 being shown on such an abacus.

With sufficient practice the reader could become very proficient in the use of an abacus to perform the operation of addition.

What about multiplication? Using the abacus, multiplication can be performed by considering repeated additions. An earlier method used involved what is known as the *lattice method of multiplication*. This method was known to the Arabians who, in turn, may have learned about it from the Hindus. The method derives its name from a lattice diagram that is constructed with each "cell" divided by a diagonal. For instance, to multiply 238 by 73, the lattice would be arranged as in Figure 6.18, with the multiplicand, 238, arranged across the top and the multiplier, 73, arranged to the right. We now form products of pairs of digits, one from the multiplicand and one from the multiplier. For example, we would have $2 \times 7 = 14$ and would enter 14 in the "rectangular box" in the first row and first column with the 1 above the diagonal and the 4 below the diagonal. In a similar manner, then, we would find the products $3 \times 7 = 21$, $8 \times 7 = 56$, etc. and

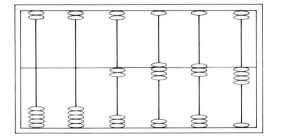

Figure 6.17

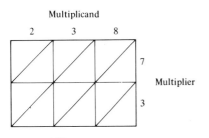

Figure 6.18

enter their products accordingly. These products are displayed in Figure 6.19.

To obtain the product, we now add elements between adjacent diagonals beginning at the lower right. Only one element, 4, is found below the lowest diagonal, and that entry is placed below the lattice diagram as shown in Figure 6.20. Adding the elements between the next two adjacent diagonals, we obtain $6 + 2 + 9 = 17$. We write the unit's digit, 7, as indicated and carry the ten's digit, 1, to be added to the elements between the next two adjacent diagonals and continue in this manner. When all diagonal elements have been summed, we read the product starting from the left and going across the bottom. Hence the product of 238 and 73 is 17,374, which can be easily verified by using the algorithm for multiplication with which you are more familiar.

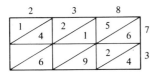

Figure 6.19

Another interesting algorithm is that known as the *peasant method of multiplication*. If we wish to multiply two whole numbers using this method, we would write the two numerals side by side with the larger one to the left of the smaller one. Directly below these we would write two more numerals. The one on the left would be half of that above it, disregarding any remainder, and the one on the right would be double that above it. We

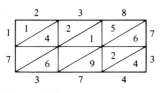

Figure 6.20

would repeat this process until the entry on the left becomes 1. For instance, consider the product of 369 and 17. We would write

$$369 \times 17$$

and proceed as indicated above.

$$369 \times 17$$
$$184 \times 34$$
$$92 \times 68$$
$$46 \times 136$$
$$23 \times 272$$
$$11 \times 544$$
$$5 \times 1088$$
$$2 \times 2176$$
$$1 \times 4352$$

Now we will go back and examine each numeral in the left-hand column. If it is odd, leave it alone; if it is even, cross it out together with the corresponding numeral in the right-hand column. This stage in the process is illustrated below:

$$369 \times 17$$
$$\cancel{184} \times \cancel{34}$$
$$\cancel{92} \times \cancel{68}$$
$$\cancel{46} \times \cancel{136}$$
$$23 \times 272$$
$$11 \times 544$$
$$5 \times 1088$$
$$\cancel{2} \times \cancel{2176}$$
$$1 \times 4352$$

Finally, we add *all* the remaining numerals in the right-hand column that were not crossed out. That sum will be the product of our two numbers. In the problem being illustrated we have $17 + 272 + 544 + 1088 + 4352 = 6273$ as the product of 369 and 17.

A more current version of the lattice method of multiplication is that referred to as Napier's bones. Multiplication using Napier's bones or Napier's rods, as they are also called, involves strips of bone, wood, metal, or even cardboard. For each of the ten digits, strips would be prepared that contain the 1 through 9 multiples of that digit, such as those illustrated in Figure 6.21. Duplicate strips would be necessary to accommodate multiplicands with repeated digits. To perform multiplications using these strips is simply a matter of selecting and aligning appropriate strips to represent the problem. For instance, to multiply 2364 by 349 we would proceed as follows. Arrange strips headed 2, 3, 6, and 4 side by side as shown in Figure 6.22, and determine the products (3×2364), (4×2364), and

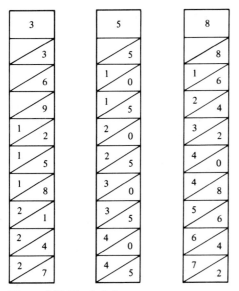

Figure 6.21 Napier's bones

(9 × 2364) as indicated. Notice the similarity between this method and the lattice method of multiplication in determining these products; we add between adjacent diagonals for a particular row. Finally, we would add the partial products and take into consideration the place value of the digits in the multiplier. For instance, 4 × 2364 really is 40 × 2364. The addition is done to the right in Figure 6.22.

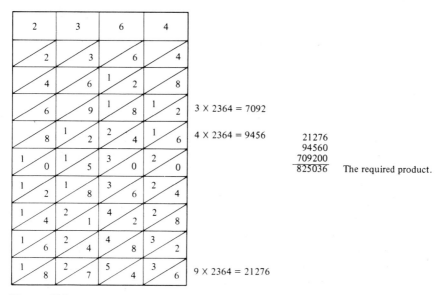

3 X 2364 = 7092

4 X 2364 = 9456

9 X 2364 = 21276

21276
94560
709200
825036   The required product.

Figure 6.22

The algorithms discussed in this chapter may appear to be difficult to work with only insofar as we lack the experience working with them. As we gain the experience, we may conclude that some of them are easier to work with and may even make more sense that the algorithms we were instructed in and use today. To conclude this section, let's look at a very arbitrary operation we will define as follows.

Suppose that on the set of whole numbers we define the operation $\#$ such that if $a,b \in W$; then $a \# b = 2a + 3b$. For example, $2,3 \in W$ and $2 \# 3 = 2(2) + 3(3) = 4 + 9 = 13$. Also, $5,7 \in W$ and $5 \# 7 = 2(5) + 3(7) = 10 + 21 = 31$.

Relative to the operation $\#$ as defined above, we could now ask the following questions:

1. Is the set W closed under the operation $\#$?

2. Is the operation $\#$ a commutative operation on the set W?

3. Is the operation $\#$ an associative operation on the set W?

4. Does there exist an identity element in W for the operation $\#$?

We will now attempt to answer these questions. If W is closed under the operation $\#$, then $a \# b \in W$ for all $a,b \in W$. Now if a and b are arbitrary whole numbers, then $a \# b = 2a + 3b$ by the definition of $\#$. Since $2,3 \in W$, we have that $2a$ and $3b$ are whole numbers because the set W is closed under the operation of ordinary multiplication. Also, since $2a$ and $3b$ are whole numbers, we have that $2a + 3b$ is a whole number because the set W is closed under the operation of ordinary addition. Finally, since a and b are arbitrary whole numbers and $a \# b \in W$, we conclude that the set W is closed under the operation $\#$.

To determine if the operation of $\#$ is commutative on W, we consider p and q as being arbitrary whole numbers and ask if $p \# q = q \# p$? Now $p \# q = 2p + 3q$ and $q \# p = 2q + 3p$, and it appears that $2p + 3q \neq 2q + 3p$. Now recall from Section 1.3 that to prove a statement false requires a single counterexample. In particular, let $p = 2$ and $q = 4$, and determine that $p \# q = 2 \# 4 = 2(2) + 3(4) = 4 + 12 = 16$ and that $q \# p = 4 \# 2 = 2(4) + 3(2) = 8 + 6 = 14$. Since $16 \neq 14$, we have that $p \# q \neq q \# p$ and, hence, the operation $\#$ is not a commutative operation on the set W.

To determine if the operation $\#$ is an associative operation on the set W, we consider m, n, and p as being arbitrary whole numbers and ask if $(m \# n) \# p = m \# (n \# p)$? Now $(m \# n) \# p = 2(m \# n) + 3p = 2(2m + 3n) + 3p = 4m + 6n + 3p$ and $m \# (n \# p) = 2m + 3(n \# p) = 2m + 3(2n + 3p) = 2m + (6n + 9p) = 2m + 6n + 9p$. It appears that $4m + 6n + 3p \neq 2m + 6n + 9p$ and, to support our conjecture, we will let $m = 3$, $n = 2$, and $p = 4$. Then $(m \# n) \# p = (3 \# 2) \# 4 = 2(3 \# 2) + 3(4) = 2[2(3) + 3(2)] + 12 = 2(6 + 6) + 12 = 2(12) + 12 = 24 + 12 = 36$ and $m \# (n \# p) = 3 \# (2 \# 4) = 2(3) + 3(2 \# 4) = 6 + 3[2(2) + 3(4)] = 6 + 3(4 + 12) =$

$6 + 3(16) = 6 + 48 = 54$. Since $36 \neq 54$, we conclude that $(m \# n) \# p \neq m \# (n \# p)$ and that the operation $\#$ is not an associative operation on the set W.

Finally, to determine if there exists an identity element in W for the operation $\#$, we consider $e \in W$ and ask is $a \# e = a$ for *every* $a \in W$? But since $a \# e = 2a + 3e$, we want to know for what value of $e \in W$, if any, is $2a + 3e = a$? Solving this last equation for e, we have $3e = -a$ or that $e = -a/3$. Since $a \in W$, $e = -a/3 \notin W$ and, hence, we conclude that there is no identity element in W for the operation $\#$.

Since this operation is not commutative and not associative and since there is no identity element for the operation on the set W, we may ask, "Why are we interested in it?" The answer lies in the fact that once we know *how* the operation is defined we will be able to work with it whether it is a useful operation or not. Other arbitrary operations will be introduced in the exercises of this section.

EXERCISES

1. Using a 4-wire abacus as depicted in Figure 6.14, perform the indicated operations.
   a. $1234 + 3462$.   b. $3692 + 1893$.
   c. $1069 + 239$.   d. $7687 - 2432$.
   e. $6597 - 3989$.
2. Using the lattice method of multiplication, perform the following operations.
   a. 123×37.   b. 469×89.
   c. 524×237.   d. 797×467.
   e. 2345×6712.
3. Using the peasant method of multiplication, perform the following operations.
   a. 462×23.   b. 546×87.
   c. 783×39.   d. 2378×123.
4. By forming appropriate Napier's bones as depicted in Figure 6.21, determine the following products.
   a. 46×12.   b. 69×72.
   c. 423×34.   d. 527×231.
   e. 2479×312.   f. 3472×469.
   g. 5469×235.
5. Consider the operation @ defined on the set of whole numbers such that if $a, b \in W$, then $a @ b = a + b + 1$.
   a. Compute $2 @ 4$.   b. Compute $4 @ 9$.
   c. Compute $(6 @ 7) @ 2$.   d. Compute $5 @ (8 @ 6)$.
   e. Is the set W closed under the operation @? Why or why not?
   f. Is the operation @ a commutative operation on W? Why or why not?

g. Is the operation @ an associative operation on W? Why or why not?

h. Does there exist an identity element in W for the operation @? If so, what is it?

6. Consider the operation $\triangle$ defined on the set of whole numbers such that if $a, b \in W$, then $a \triangle b = b$.

a. Compute $2 \triangle 3$. b. Compute $4 \triangle 7$.

c. Compute $(6 \triangle 8) \triangle 9$. d. Compute $6 \triangle (8 \triangle 9)$.

e. Is the set W closed under the operation $\triangle$? Why or why not?

f. Is the operation $\triangle$ a commutative operation on W? Why or why not?

g. Is the operation $\triangle$ an associative operation on W? Why or why not?

h. Does there exist an identity element in W for the operation $\triangle$? If so, what is it?

6.8 HINDU-ARABIC SYSTEM OF NUMERATION

As we commented upon earlier in the text, the system of numeration we use today and which is used almost exclusively in civilized countries of the world is known as the Hindu-Arabic system. The *Hindu* part refers to the people of India and is a Persian name for the region of northern India along the Indus River. The Arabs were the first non-Hindu people to use these new numerals. An Arab scholar, al-Khwarizmi, wrote a book about the use of these numerals about the year 825. Partly due to the translation of this book into Latin by the English scholastic philosopher Adelard of Bath during the early part of the twelfth century, the Hindu-Arabic numerals spread throughout the Western world.

The Hindu-Arabic system represents a combination of the best features of the Egyptian and Babylonian systems of numeration. Recall that the Egyptian system used a single base 10, but had no place value associated with it and had symbols that were difficult to reproduce. The Babylonian system had place value, but was a combination of the bases 10 and 60. The Hindu-Arabic system uses the simple base 10 from the Egyptians, the place value from the Babylonians, and its own simplified numerals which are 0, 1, 2, 3, 4, 5, 6, 7, 8, 9.

In Chapter 5 we discussed counting and defined the operations of addition, multiplication, subtraction, and division on the set of whole numbers. In this section we will discuss algorithms used in performing these operations. Consider, now, what happens when we add 32 and 46. Since $46 = (4 \times 10) + (6 \times 1)$ and $32 = (3 \times 10) + (2 \times 1)$, it is possible that we could group the tens together and group the ones together in a similar manner as the Egyptians grouped heel bones together and coils of rope together, etc., or the Babylonians grouped $\triangledown$'s together and $\triangleleft$'s together.

And, indeed, we do! Hence

		Tens	Ones
	46 =	4	6
Add	32 =	3	2
		7	8

and we see that $46 + 32 = 78$.

Now consider the sum of 239 and 487. We would set up the addition problem as follows:

		Hundreds	Tens	Ones
	239 =	2	3	9
Add	487 =	4	8	7
		6	11	16

Adding the ones, we obtain $9 + 7 = 16$. But 10 ones are equivalent to 1 ten. Just as the Egyptians did with their counting boards or the Japanese do with their abacus, we would now write $16 = 10 + 6$, drop the 10 from the one's column, and add 1 to the ten's column, obtaining

		Hundreds	Tens	Ones
	239 =	2	3	9
Add	487 =	4	8	7
		6	12	6

In a similar manner, the 12 in the ten's column could be written as $10 + 2$ and, again, we note that 10 tens are equivalent to 1 hundred. Hence we could drop the 10 from the ten's column and add 1 to the hundred's column, obtaining

		Hundreds	Tens	Ones
	239 =	2	3	9
Add	487 =	4	8	7
		7	2	6

We now have single digits for the sum in each column and can read the sum of 239 and 487 to be 726.

In the above algorithm we added like units such as ones to ones, tens to tens, and hundreds to hundreds. Also, we employed what is known as a carry-over procedure making use of the facts that 10 ones are equivalent to 1 ten, 10 tens are equivalent to 1 hundred, and so forth.

EXAMPLE Using the expanded notation as described above, find the sum of 2398 and 4637.

Solution

	Thousands	Hundreds	Tens	Ones
2398 =	2	3	9	8
Add 4637 =	4	6	3	7
	6	9	12	15

Now since 15 in the one's column is equal to $10 + 5$ ones or 1 ten plus 5 ones, we may use the carry-over procedure and obtain

	Thousands	Hundreds	Tens	Ones
2398 =	2	3	9	8
Add 4637 =	4	6	3	7
	6	9	13	5

Again, 13 in the ten's column is equal to $10 + 3$ tens or 1 hundred plus 3 tens. Using the carry-over procedure we now have

	Thousands	Hundreds	Tens	Ones
2398 =	2	3	9	8
Add 4637 =	4	6	3	7
	6	10	3	5

Further, 10 in the hundred's column is equal to $10 + 0$ hundreds or 1 thousand plus 0 hundreds. Using the carry-over procedure, we now have

	Thousands	Hundreds	Tens	Ones
2398 =	2	3	9	8
Add 4637 =	4	6	3	7
	7	0	3	5

Finally, since each sum in every column is represented by a single digit, we can now read the sum of 2398 and 4637 to be 7035. ∎

Subtraction can be accomodated in a similar manner employing a procedure of "borrowing" instead of carrying-over as we will now illustrate. Let's attempt to subtract 234 from 698.

	Hundreds	Tens	Ones
698 =	6	9	8
Subtract 234 =	2	3	4
	4	6	4

Since the difference obtained in each column is represented by a single digit, we can now determine that $698 - 234 = 464$.

EXAMPLE   Using the expanded notation, subtract 487 from 724.
Solution

	Hundreds	Tens	Ones
$724 =$	7	2	4
Subtract $487 =$	4	8	7

We would first attempt to subtract 7 from 4 ones, but this is not possible. Therefore we would "borrow" 1 of the tens associated with 724, which is equivalent to 10 ones, and add these 10 ones to the 4 ones already present; this leaves 1 ten in the ten's column. We now subtract 7 ones from 14 ones as indicated:

	Hundreds	Tens	Ones
$724 =$	7	1	14
Subtract $487 =$	4	8	7
			7

To continue, we would attempt to subtract 8 tens from 1 ten but, again, this is not possible. Borrowing again, we have the following, which enables us to subtract in the ten's column as indicated:

	Hundreds	Tens	Ones
$724 =$	6	11	14
Subtract $487 =$	4	8	7
		3	7

Finally, subtracting in the hundred's column, we have

	Hundreds	Tens	Ones
$724 =$	6	11	14
Subtract $487 =$	4	8	7
	2	3	7

Since the difference in each column is represented by a single digit, we now determine that $724 - 487 = 237$. ■

To motivate the algorithm used for multiplication of whole numbers, let's consider the product of 239 by 123. This could be written as 239×123. Now, writing $123 = 100 + 20 + 3$, we would have $239 \times (100 + 20 + 3)$. Using the distributive property of multiplication over addition, we would have $(239 \times 100) + (239 \times 20) + (239 \times 3)$, which is equal to 23,900 + 4,780 + 717 or 29,397. Observe that the multiplication was performed by using repeated additions. If we set up this multiplication problem in the customary form used today, we would have

$$
\begin{array}{r}
239 \\
\times \quad 123 \\
\hline
717 \\
478 \\
239 \\
\hline
29397
\end{array}
$$

The product of 239 and 3
The product of 239 and 20
The product of 239 and 100

In the above procedure we multiplied 239 by 3 and obtained the partial product 717. Next we multiplied 239 by 2, obtained 478, and "indented" one place to the left. Really what we did was to multiply 239 by 20 (since the 2 is in the ten's place) and suppressed the 0 in 4780. Likewise, we multiplied 239 by 1, obtained 239, and indented another place to the left. Again, what we really did was to multiply 239 by 100 (since the 1 is in the hundred's place) and to suppress the 00 in 23900. Finally, we added the partial products 717, 4,780, and 23,900 and obtained 29,397 as the product.

The carry-over and borrow terms used in the preceding discussions are sometimes also referred to as "renaming." In fact, many of the modern mathematics programs used at the elementary grade level today refer to "renaming" or "regrouping" or "substituting" very frequently.

Before discussing an algorithm used for division, I want to remark that the algorithms discussed so far depend upon one's ability to perform basic addition, subtraction, and multiplication. That is, the basic addition and multiplication facts must be known. We will now list these in Tables 6.1 and 6.2.

The algorithm discussed for multiplication involved the use of repeated additions. In a similar manner, the algorithm for division involves the use

Table 6.1 Basic addition tables, base ten

+	0	1	2	3	4	5	6	7	8	9
0	0	1	2	3	4	5	6	7	8	9
1	1	2	3	4	5	6	7	8	9	10
2	2	3	4	5	6	7	8	9	10	11
3	3	4	5	6	7	8	9	10	11	12
4	4	5	6	7	8	9	10	11	12	13
5	5	6	7	8	9	10	11	12	13	14
6	6	7	8	9	10	11	12	13	14	15
7	7	8	9	10	11	12	13	14	15	16
8	8	9	10	11	12	13	14	15	16	17
9	9	10	11	12	13	14	15	16	17	18

Table 6.2 Basic multiplication tables, base ten

×	0	1	2	3	4	5	6	7	8	9
0	0	0	0	0	0	0	0	0	0	0
1	0	1	2	3	4	5	6	7	8	9
2	0	2	4	6	8	10	12	14	16	18
3	0	3	6	9	12	15	18	21	24	27
4	0	4	8	12	16	20	24	28	32	36
5	0	5	10	15	20	25	30	35	40	45
6	0	6	12	18	24	30	36	42	48	54
7	0	7	14	21	28	35	42	49	56	63
8	0	8	16	24	32	40	48	56	64	72
9	0	9	18	27	36	45	54	63	72	81

of repeated subtractions. Consider the following division problem given in "short" form:

$$
\begin{array}{r}
357 \\
27\overline{)9639} \\
81 \\
\hline
153 \\
135 \\
\hline
189 \\
189 \\
\hline
\end{array}
$$

Now, to divide 9639 by 27, we are looking for a whole number q such that $9639 = 27 \cdot q$. Stated in other words, we want to know how many 27's are contained in 9639. To find out, we can subtract 27 repeatedly until it is not possible to subtract additional 27's. For instance, there are at least ten 27's contained in 9639 since $10 \times 27 < 9639$. Hence

$$
\begin{array}{r}
9639 \\
- \quad 270 \quad (10 \times 27) \\
\hline
9369
\end{array}
$$

It now appears that there are at least 100 more 27's contained in 9639 since $100 \times 27 < 9369$. Hence

$$
\begin{array}{r}
9639 \\
- \quad 270 \quad (10 \times 27) \\
\hline
9369 \\
-2700 \quad (100 \times 27) \\
\hline
6669
\end{array}
$$

We have now subtracted $10 + 100 = 110$ 27's and have many more contained in 9639. How many more? What about 500 more? Well,

$500 \times 27 = 13,500$, which is greater than 6669. What about 200 more? $200 \times 27 = 5400$, which is less than 6669. Hence we have

```
  9639
-  270   (10 × 27)
──────
  9369
- 2700   (100 × 27)
──────
  6669
- 5400   (200 × 27)
──────
  1269
```

How many more 27's are contained in 1269? What about 50? $50 \times 27 = 1350$, which is greater than 1269. What about 40? $40 \times 27 = 1080$, which is less than 1269. Hence we have

```
  9639
-  270   (10 × 27)
──────
  9369
- 2700   (100 × 27)
──────
  6669
- 5400   (200 × 27)
──────
  1269
- 1080   (40 × 27)
──────
   189
```

How many more 27's are contained in 189? What about 5? $5 \times 27 = 135$, which is less than 189. Hence we have

```
  9639
-  270   (10 × 27)
──────
  9369
- 2700   (100 × 27)
──────
  6669
- 5400   (200 × 27)
──────
  1269
- 1080   (40 × 27)
──────
   189
-  135   (5 × 27)
──────
    54
```

How many more 27's are contained in 54? Clearly, the answer is 2 since $2 \times 27 = 54$. Hence we have

```
   9639
 −  270  (10 × 27)
  ‾‾‾‾‾
   9369
 − 2700  (100 × 27)
  ‾‾‾‾‾
   6669
 − 5400  (200 × 27)
  ‾‾‾‾‾
   1269
 − 1080  (40 × 27)
  ‾‾‾‾‾
    189
 −  135  (5 × 27)
  ‾‾‾‾‾
     54
 −   54  (2 × 27)
  ‾‾‾‾‾
      0
```

All together, then, we have subtracted $10 + 100 + 200 + 40 + 5 + 2$ or 357 27's from 9639 and there was nothing left to subtract. That is, the division is *exact*. The short form given earlier represents a short-cut for the repeated subtractions, and the quotient (the answer in the division problem) of 357 represents the number of 27's contained in 9639. Taking a closer look at the short form for this problem, we should really write the following:

```
        357
     ‾‾‾‾‾‾
  27)9639
   − 8100  (300 × 27)
    ‾‾‾‾‾
     1539
   − 1350  (50 × 27)
    ‾‾‾‾‾
      189
   −  189  (7 × 27)
    ‾‾‾‾‾
        0
```

EXERCISES

1. Using the expanded notation for writing Hindu-Arabic numerals (see the first example), perform the following additions.
   - a. $23 + 35$.   b. $37 + 26$.
   - c. $234 + 378$.   d. $976 + 249$.
   - e. $4789 + 3472$.   f. $9872 + 4036$.
2. Using the expanded notation for writing Hindu-Arabic numerals (see the second example), perform the following subtractions.
   - a. $68 − 42$.   b. $82 − 47$.
   - c. $239 − 147$.   d. $4276 − 979$.
   - e. $7069 − 4907$.   f. $9002 − 3169$.

3. Using the distributive property of multiplication over addition, perform the following multiplications.
   a. 234×32.   b. 425×127.
   c. 6234×234.   d. 5562×405.
   e. $27{,}683 \times 2704$.
4. For each part of Exercise 3 determine the product the "short" way.
5. Using the method of repeated subtractions, determine the quotient in each of the following exercises.
   a. $4977 \div 21$.   b. $12{,}924 \div 36$.
   c. $24{,}720 \div 48$.   d. $231{,}756 \div 372$.
   e. $106{,}124 \div 1{,}234$.   f. $539{,}580 \div 2{,}346$.
6. Perform the indicated divisions in each part of Exercise 5 using the "short" method.

6.9 BASES OTHER THAN TEN

Section 5.1 commented that if b is less than ten, then when counting, we would arrive at bb quicker than when b is equal to ten. Also, if b is greater than ten, we would arrive at bb later than when b is equal to ten. Using Hindu-Arabic numerals we can develop systems of numeration with bases other than ten. In this section we will discuss such systems using b less than ten and b greater than ten. Working with such systems should also help us to better understand our base ten system.

Let's begin with a base five system. In such a system we use only the digits 0, 1, 2, 3, and 4. Before attempting to perform arithmetic operations using this system, we must first learn to count and then learn the basic addition and multiplication facts associated with this system. Counting as indicated in Section 5.1, we have Table 6.3.

Table 6.3

1	11	21	31	41	101	111	121	131	141	201	211	221	231
2	12	22	32	42	102	112	122	132	142	202	212	222	232
3	13	23	33	43	103	113	123	133	143	203	213	223	233
4	14	24	34	44	104	114	124	134	144	204	214	224	234
10	20	30	40	100	110	120	130	140	200	210	220	230	240,

and so on.

Numerals in base five will be written as 234_{five}.

Next we would form the basic addition and basic multiplication tables, using our definitions for ordinary addition and ordinary multiplication on the set of whole numbers, renamed, and our new counting techniques (see Tables 6.4 and 6.5). The student should verify that these tables are correct.

Table 6.4 Basic
addition tables, base
five

+	0	1	2	3	4
0	0	1	2	3	4
1	1	2	3	4	10
2	2	3	4	10	11
3	3	4	10	11	12
4	4	10	11	12	13

Table 6.5 Basic
multiplication tables,
base five

×	0	1	2	3	4
0	0	0	0	0	0
1	0	1	2	3	4
2	0	2	4	11	13
3	0	3	11	14	22
4	0	4	13	22	31

EXAMPLE Using the basic addition table, base five, find the sum of 234_{five} and 322_{five}.

Solution Just as 234_{ten} means $(2 \times 10^2) + (3 \times 10) + (4 \times 1)$ or 2 hundreds + 3 tens + 4 ones, 234_{five} means $(2 \times 5^2) + (3 \times 5) + (4 \times 1)$ or 2 twenty-fives + 3 fives + 4 ones. Hence we have

$$
\begin{array}{lccc}
 & 25\text{'s} & 5\text{'s} & 1\text{'s} \\
234_{\text{five}} = & 2 & 3 & 4 \\
\text{Add} \quad 322_{\text{five}} = & 3 & 2 & 2 \\
\hline
 & 10 & 10 & 11
\end{array}
$$

Now since 10_{five} ones are equal to 1_{five} fives and since 11_{five} $= 10_{\text{five}} + 1_{\text{five}}$, we can carry-over 10_{five} ones to the five's column as 1_{five} five and add this to the 10_{five} fives already there as follows:

$$
\begin{array}{lccc}
 & 25\text{'s} & 5\text{'s} & 1\text{'s} \\
234_{\text{five}} = & 2 & 3 & 4 \\
\text{Add} \quad 322_{\text{five}} = & 3 & 2 & 2 \\
\hline
 & 10 & 11 & 1
\end{array}
$$

In a similar manner we can carry over 10_{five} fives to the twenty-five's column as 1_{five} twenty-five and add this to the 10_{five} twenty-fives already there as follows:

$$
\begin{array}{lccc}
 & 25\text{'s} & 5\text{'s} & 1\text{'s} \\
234_{\text{five}} = & 2 & 3 & 4 \\
\text{Add} \quad 322_{\text{five}} = & 3 & 2 & 2 \\
\hline
 & 11 & 1 & 1
\end{array}
$$

We now write the sum of 234_{five} and 322_{five} as 1111_{five}. ∎

EXAMPLE   Find the sum of 2312_{five} and 1444_{five}.
Solution

$$
\begin{array}{l}
\phantom{\text{Add}\quad}2312_{five} \\
\text{Add}\quad 1444_{five} \\
\hline
\phantom{\text{Add}\quad}4311_{five} \quad \blacksquare
\end{array}
$$

Subtraction in base five is treated as the inverse of addition, base base. For instance, to subtract 231_{five} from 443_{five}, we would have

$$
\begin{array}{rccc}
 & 25\text{'s} & 5\text{'s} & 1\text{'s} \\
443_{five} = & 4 & 4 & 3 \\
\text{Subtract}\quad 231_{five} = & 2 & 3 & 1 \\
\hline
 & 2 & 1 & 2
\end{array}
$$

and, hence, $443_{five} - 231_{five} = 212_{five}$. $\quad \blacksquare$

EXAMPLE   Subtract 4321_{five} from 32433_{five}.
Solution

$$
\begin{array}{l}
\phantom{\text{Subtract}\quad}32433_{five} \\
\text{Subtract}\quad4321_{five} \\
\hline
\phantom{\text{Subtract}\quad}23112_{five}
\end{array}
$$

In this problem, to subtract 4 from 2 in the fourth column from the right is impossible. Hence we borrowed a 1_{five} from the fifth column of the top numeral and wrote it as 10_{five} units to be added in the fourth column giving $10_{five} + 2_{five} = 12_{five}$. Then we subtracted 4_{five} from 12_{five}, obtaining 3_{five}. $\quad \blacksquare$

Multiplication in base five can be done by employing the same algorithm as for multiplication in base ten and using the multiplication table for base five.

EXAMPLE   Multiply 234_{five} by 12_{five}.
Solution

$$
\begin{array}{ll}
\phantom{\text{Multiply}\quad}234_{five} & \\
\text{Multiply}\quad12_{five} & \\
\hline
\phantom{\text{Multiply}\quad}1023 & (2 \times 234)_{five} \\
\phantom{\text{Multiply}\quad}234 & (10 \times 234)_{five} \\
\hline
\phantom{\text{Multiply}\quad}3413_{five} & \blacksquare
\end{array}
$$

EXAMPLE Multiply 4321_{five} by 324_{five}.
Solution

$$
\begin{array}{r}
4321_{five} \\
\text{Multiply} \qquad 324_{five} \\
\hline
33334 \qquad (4 \times 4321)_{five} \\
14142 \qquad (20 \times 4321)_{five} \\
24013 \qquad (300 \times 4321)_{five} \\
\hline
3132104_{five} \quad \blacksquare
\end{array}
$$

Division in base five is treated as repeated subtraction in a manner similar to subtraction in base ten. Consider the following example of dividing 4343_{five} by 23_{five}.

$$
\begin{array}{r}
141_{five} \\
23_{five} \overline{)4343_{five}} \\
23 \qquad (100 \times 23)_{five} \\
\hline
204 \\
202 \qquad (40 \times 23)_{five} \\
\hline
23 \\
23 \qquad (1 \times 23)_{five} \\
\hline
0
\end{array}
$$

EXAMPLE Divide 34243_{five} by 32_{five}.
Solution

$$
\begin{array}{r}
1034_{five} \\
32_{five} \overline{)34243_{five}} \\
32 \qquad (1000 \times 32)_{five} \\
\hline
224 \\
201 \qquad (30 \times 32)_{five} \\
\hline
233 \\
233 \qquad (4 \times 32)_{five} \\
\hline
0 \quad \blacksquare
\end{array}
$$

EXAMPLE Rewrite 3423_{five} in equivalent form using a base ten numeral.
Solution

$$
\begin{aligned}
3423_{five} &= [(3 \times 10^3) + (4 \times 10^2) + (2 \times 10) + (3 \times 1)]_{five} \\
&= [(3 \times 5^3) + (4 \times 5^2) + (2 \times 5) + (3 \times 1)]_{ten} \\
&= [(3 \times 125) + (4 \times 25) + (10) + (3)]_{ten} \\
&= (375 + 100 + 10 + 3)_{ten} \\
&= 488_{ten} \quad \blacksquare
\end{aligned}
$$

EXAMPLE   Rewrite 40342_{five} in equivalent form using a base ten numeral.

Solution

$$40342_{five} = [(4 \times 10^4) + (0 \times 10^3) + (3 \times 10^2) + (4 \times 10) + (2 \times 1)]_{five}$$
$$= [(4 \times 5^4) + (0 \times 5^3) + (3 \times 5^2) + (4 \times 5) + (2 \times 1)]_{ten}$$
$$= [(4 \times 625) + (0 \times 125) + (3 \times 25) + (20) + (2)]_{ten}$$
$$= (2500 + 0 + 75 + 22)_{ten}$$
$$= 2597_{ten} \quad \blacksquare$$

EXAMPLE   Rewrite 234_{ten} in equivalent form using a base five numeral.

Solution   We will first consider the various powers of 5 which are, in base ten, $5^0 = 1$, $5^1 = 5$, $5^2 = 25$, $5^3 = 125$, $5^4 = 625$, and so on. Since $5^4 > 234_{ten}$ and $5^3 < 234_{ten}$, we ask how many 5^3's are contained in 234_{ten}. The answer is 1, and we subtract $1 \times 5^3 = 125_{ten}$ from 234_{ten}, obtaining 109_{ten}. We now ask how many 5^2's are contained in 109_{ten}. The answer is 4_{ten}, and we subtract $4 \times 5^2 = 100_{ten}$ from 109_{ten}, obtaining 9_{ten}. Now we ask how many 5^1's are contained in 9_{ten}. The answer is 1, and we subtract $1 \times 5 = 5_{ten}$ from 9_{ten}, obtaining 4_{ten}. Finally, we ask how many 5^0's are contained in 4_{ten}. The answer is 4, and we subtract $4 \times 5^0 = 4_{ten}$ from 4_{ten}, obtaining 0_{ten}. Hence $234_{ten} = 1414_{five}$. $\quad \blacksquare$

Table 6.6

1	11	21	31	41	51	61	71	81	91
2	12	22	32	42	52	62	72	82	92
3	13	23	33	43	53	63	73	83	93
4	14	24	34	44	54	64	74	84	94
5	15	25	35	45	55	65	75	85	95
6	16	26	36	46	56	66	76	86	96
7	17	27	37	47	57	67	77	87	97
8	18	28	38	48	58	68	78	88	98
9	19	29	39	49	59	69	79	89	99
T	1T	2T	3T	4T	5T	6T	7T	8T	9T
E	1E	2E	3E	4E	5E	6E	7E	8E	9E

T1	E1	101	111	121	131	141	
T2	E2	102	112	122	132	142	
T3	E3	103	113	123	133	143	
T4	E4	104	114	124	134	144	
T5	E5	105	115	125	135	145	
T6	E6	106	116	126	136	146	
T7	E7	107	117	127	137	147	
T8	E8	108	118	128	138	148	
T9	E9	109	119	129	139	149	
TT	ET	10T	11T	12T	13T	14T	
TE	EE	10E	11E	12E	13E	14E	and so on.

Next let's consider a base twelve system. In such a system we use the digits 0, 1, 2, 3, 4, 5, 6, 7, 8, 9 as we do in the base ten system. However, additional symbols are necessary in this system and we will introduce T for ten and E for eleven. We will now learn to count in this system as indicated in Table 6.6.

Numerals in base twelve will be written as $67T9_{\text{twelve}}$.

Next we will construct the basic addition and multiplication tables for base twelve, and the student should verify that they are correct (see Tables 6.7 and 6.8).

Table 6.7 Basic addition tables, base twelve

+	0	1	2	3	4	5	6	7	8	9	T	E
0	0	1	2	3	4	5	6	7	8	9	T	E
1	1	2	3	4	5	6	7	8	9	T	E	10
2	2	3	4	5	6	7	8	9	T	E	10	11
3	3	4	5	6	7	8	9	T	E	10	11	12
4	4	5	6	7	8	9	T	E	10	11	12	13
5	5	6	7	8	9	T	E	10	11	12	13	14
6	6	7	8	9	T	E	10	11	12	13	14	15
7	7	8	9	T	E	10	11	12	13	14	15	16
8	8	9	T	E	10	11	12	13	14	15	16	17
9	9	T	E	10	11	12	13	14	15	16	17	18
T	T	E	10	11	12	13	14	15	16	17	18	19
E	E	10	11	12	13	14	15	16	17	18	19	1T

Table 6.8 Basic multiplication tables, base twelve

×	0	1	2	3	4	5	6	7	8	9	T	E
0	0	0	0	0	0	0	0	0	0	0	0	0
1	0	1	2	3	4	5	6	7	8	9	T	E
2	0	2	4	6	8	T	10	12	14	16	18	1T
3	0	3	6	9	10	13	16	19	20	23	26	29
4	0	4	8	10	14	18	20	24	28	30	34	38
5	0	5	T	13	18	21	26	2E	34	39	42	47
6	0	6	10	16	20	26	30	36	40	46	50	56
7	0	7	12	19	24	2E	36	41	48	53	5T	65
8	0	8	14	20	28	34	40	48	54	60	68	74
9	0	9	16	23	30	39	46	53	60	69	76	83
T	0	T	18	26	34	42	50	5T	68	76	84	92
E	0	E	1T	29	38	47	56	65	74	83	92	T1

EXAMPLE Using the basic addition tables, base twelve, find the sum of $89T7_{\text{twelve}}$ and $795E_{\text{twelve}}$.

Solution Just as 234_{ten} means $(2 \times 10^2) + (3 \times 10) + (4 \times 1)$ or 2 hundreds + 3 tens + 4 ones, $89T7_{\text{twelve}}$ means $(8 \times 12^3) + (9 \times 12^2)$

$+ (10 \times 12) + (7 \times 1)$ or 8 seventeen twenty-eights $+ 9$ one hundred forty-fours $+ 10$ twelves $+ 7$ ones. Hence we have

	1728's	144's	12's	1's
$89T7_{\text{twelve}} =$	8	9	T	7
Add $795E_{\text{twelve}} =$	7	9	5	E
	13	16	13	16

Now since 10_{twelve} ones are equivalent to 1_{twelve} twelve and since $16_{\text{twelve}} = 10_{\text{twelve}} + 6_{\text{twelve}}$, we can carry over 10_{twelve} ones to the twelve's column as 1_{twelve} twelve and add this to the 13_{twelve} twelves already there as follows:

	1728's	144's	12's	1's
$89T7_{\text{twelve}} =$	8	9	T	7
Add $795E_{\text{twelve}} =$	7	9	5	E
	13	16	14	6

In a similar manner we can carry over 10_{twelve} twelves to the one hundred forty-four's column as 1_{twelve} one hundred forty-four and add this to the 16_{twelve} one hundred forty-fours already there as follows:

	1728's	144's	12's	1's
$89T7_{\text{twelve}} =$	8	9	T	7
Add $795E_{\text{twelve}} =$	7	9	5	E
	13	17	4	6

Similarly, we can carry over 10_{twelve} one hundred forty-fours to the seventeen twenty-eight's column as 1_{twelve} seventeen twenty-eight and add this to the 17_{twelve} seventeen twenty-eights already there as follows:

	1728's	144's	12's	1's
$89T7_{\text{twelve}} =$	8	9	T	7
Add $795E_{\text{twelve}} =$	7	9	5	E
	14	7	4	6

We now write the sum of $89T7_{\text{twelve}}$ and $795E_{\text{twelve}}$ as 14746_{twelve}. ∎

EXAMPLE Find the sum of $9TE5_{\text{twelve}}$ and $234T_{\text{twelve}}$.
Solution

$9TE5_{\text{twelve}}$
Add $\quad 234T_{\text{twelve}}$
10243_{twelve} ∎

Subtraction in base twelve is treated as the inverse of addition in base twelve. For instance, to subtract 4935_{twelve} from $9E6T_{twelve}$, we would have

	1728's	144's	12's	1's
$9E6T_{twelve} =$	9	E	6	T
Subtract $4935_{twelve} =$	4	9	3	5
	5	2	3	5

and, hence, $9E6T_{twelve} - 4935_{twelve} = 5235_{twelve}$.

EXAMPLE Subtract $5T69_{twelve}$ from $2E7T8_{twelve}$.
Solution

$$
\begin{array}{r}
2E7T8_{twelve} \\
\text{Subtract} \quad 5T69_{twelve} \\
\hline
2593E_{twelve}
\end{array}
$$

In this problem, to subtract 9 from 8 in the first column on the right is impossible. Hence we borrowed a 1_{twelve} from the second column of the top numeral and wrote it as 10_{twelve} units to be added in the first column giving $10_{twelve} + 8_{twelve} = 18_{twelve}$. Then we subtracted 9_{twelve} from 18_{twelve} and obtained E_{twelve}. A similar procedure was employed when we tried to subtract T_{twelve} from 7_{twelve} in the third column. ∎

For multiplication in base twelve we employ the same algorithm as for multiplication in base ten but use the multiplication table for base twelve.

EXAMPLE Multiply 987_{twelve} by $T2_{twelve}$.
Solution

$$
\begin{array}{r}
897_{twelve} \\
\text{Multiply} \quad T2_{twelve} \\
\hline
1572 \quad (2 \times 897)_{twelve} \\
73ET \quad (T0 \times 897)_{twelve} \\
\hline
75552_{twelve} \quad ∎
\end{array}
$$

EXAMPLE Multiply $9TE5_{twelve}$ by $4E5_{twelve}$.
Solution

$$
\begin{array}{r}
9TE5_{twelve} \\
\text{Multiply} \quad 4E5_{twelve} \\
\hline
41691 \quad (5 \times 9TE5)_{twelve} \\
91057 \quad (E0 \times 9TE5)_{twelve} \\
33798 \quad (400 \times 9TE5)_{twelve} \\
\hline
410E841_{twelve} \quad ∎
\end{array}
$$

Division in base twelve is treated as repeated subtraction in a manner similar to subtraction in base ten. Consider the following example of dividing $45T32_{\text{twelve}}$ by 67_{twelve}.

$$
\begin{array}{r}
822_{\text{twelve}} \\
67_{\text{twelve}} \,\overline{\big)\, 45T32_{\text{twelve}}} \\
448 \qquad (800 \times 67)_{\text{twelve}} \\
\overline{123} \\
112 \qquad (20 \times 67)_{\text{twelve}} \\
\overline{112} \\
112 \qquad (2 \times 67)_{\text{twelve}}
\end{array}
$$

EXAMPLE Divide $T9E7946_{\text{twelve}}$ by 346_{twelve}.
Solution

$$
\begin{array}{r}
32615_{\text{twelve}} \\
346_{\text{twelve}} \,\overline{\big)\, T9E7946_{\text{twelve}}} \\
T16 \qquad (30000 \times 346)_{\text{twelve}} \\
\overline{857} \\
690 \qquad (2000 \times 346)_{\text{twelve}} \\
\overline{1879} \\
1830 \qquad (600 \times 346)_{\text{twelve}} \\
\overline{494} \\
346 \qquad (10 \times 346)_{\text{twelve}} \\
\overline{14T6} \\
14T6 \qquad (5 \times 346)_{\text{twelve}} \quad \blacksquare
\end{array}
$$

EXAMPLE Rewrite $45T7_{\text{twelve}}$ in equivalent form using a base ten numeral.

Solution

$$
\begin{aligned}
45T7_{\text{twelve}} &= [(4 \times 10^3) + (5 \times 10^2) + (T \times 10) + (7 \times 1)]_{\text{twelve}} \\
&= [(4 \times 12^3) + (5 \times 12^2) + (10 \times 12) + (7 \times 1)]_{\text{ten}} \\
&= [(4 \times 1728) + (5 \times 144) + (120) + (7)]_{\text{ten}} \\
&= (6912 + 720 + 120 + 7)_{\text{ten}} \\
&= 7759_{\text{ten}} \quad \blacksquare
\end{aligned}
$$

EXAMPLE Rewrite 4697_{ten} in equivalent form using a base twelve numeral.

Solution We will first consider the various powers of 12 which are, in base ten, $12^0 = 1$, $12^1 = 12$, $12^2 = 144$, $12^3 = 1728$, $12^4 = 20736$, and so on. Since $12^4 > 4697_{\text{ten}}$ and $12^3 < 4697_{\text{ten}}$, we ask how many 12^3's are contained in 4697_{ten}. The answer is 2, and we subtract $2 \times 12^3 = 2 \times 1728 = 3456_{\text{ten}}$ from 4697_{ten}, obtaining 1241_{ten}. We now ask how many 12^2's are contained in 1241_{ten}. The answer is 8_{ten}, and we subtract $8 \times 12^2 = 8 \times 144_{ten} = 1152_{ten}$ from 1241_{ten}, obtaining 89_{ten}. Now we ask how many 12's are contained in 89_{ten}. The answer is 7_{ten}, and we subtract $7 \times 12_{ten} = 84_{ten}$ from 89_{ten}, obtaining 5_{ten}. Finally, we ask how many 12^0's are contained in 5_{ten}. The answer is 5_{ten}, and we subtract 5_{ten} from 5_{ten}, obtaining 0_{ten}. Hence we have $4697_{ten} = 2875_{twelve}$. ∎

Other bases would be discussed in a similar manner. We would first learn to count in the base and then learn the basic addition and multiplication tables in that base. Finally, we would learn to perform arithmetic operations using that base. The last base we will discuss in this chapter is the base two. This base has been very popular in recent years because the high-speed digital computer recognizes only base two. In the base two the only digits used are 0 and 1. To count in base two we proceed as follows:

| 1 | 11 | 101 | 111 | 1001 | 1011 | 1101 | 1111 |
| 10 | 100 | 110 | 1000 | 1010 | 1100 | 1110 | 10000 | and so on. |

Numerals in base two will be written as 101_{two}.

Next we will construct the basic addition and multiplication tables (Tables 6.9 and 6.10) for base two, and the student should verify that they are correct.

Table 6.9
Basic addition tables,
base two

+	0	1
0	0	1
1	1	10

Table 6.10 Basic
multiplication tables,
base two

×	0	1
0	0	0
1	0	1

EXAMPLE Using the basic addition tables, base two, find the sum of 1011_{two} and 1101_{two}.
Solution

	8's	4's	2's	1's
$1011_{two} =$	1	0	1	1
Add $1101_{two} =$	1	1	0	1
	10	1	1	10

Employing the carry-over procedure, we have

$$1011_{two}$$
Add $\quad 1101_{two}$
$$\overline{11000_{two}}$$ ∎

EXAMPLE Find the sum of 110010_{two} and 101001_{two}.
Solution

$$
\begin{array}{r}
110010_{two} \\
\text{Add} \quad 101001_{two} \\
\hline
1011011_{two}
\end{array}
$$ ∎

EXAMPLE Subtract 1101_{two} from 100110_{two}.
Solution

$$
\begin{array}{r}
100110_{two} \\
\text{Subtract} \quad 1101_{two} \\
\hline
11001_{two}
\end{array}
$$ ∎

For multiplication in base two we employ the same algorithm for multiplication as in base ten but use the multiplication table for base two.

EXAMPLE Multiply 1011_{two} by 101_{two}.
Solution

$$
\begin{array}{rl}
1011_{two} & \\
\text{Multiply} \quad 101_{two} & \\
\hline
1011 & (1 \times 1011)_{two} \\
10110 & (100 \times 1011)_{two} \\
\hline
110111_{two} &
\end{array}
$$ ∎

Division in base two is treated as repeated subtraction in a manner similar to subtraction in base ten.

EXAMPLE Divide 11011001_{two} by 111_{two}.
Solution

$$
\begin{array}{rl}
\phantom{111_{two}|}11111_{two} & \\
111_{two}\,|\,\overline{11011001_{two}} & \\
\quad 111 & (10000 \times 111)_{two} \\
\hline
\quad 1101 & \\
\quad 111 & (1000 \times 111)_{two} \\
\hline
\quad 1100 & \\
\quad 111 & (100 \times 111)_{two} \\
\hline
\quad 1010 & \\
\quad 111 & (10 \times 111)_{two} \\
\hline
\quad 111 & \\
\quad 111 & (1 \times 111)_{two} \\
\hline
\end{array}
$$ ∎

EXAMPLE Rewrite 11011_{two} in equivalent form using a base two numeral.

Solution

$$
\begin{aligned}
11011_{two} &= [(1 \times 10^4) + (1 \times 10^3) + (0 \times 10^2) + (1 \times 10) + (1 \times 1)]_{two} \\
&= [(1 \times 2^4) + (1 \times 2^3) + (0 \times 2^2) + (1 \times 2) + (1 \times 1)]_{ten} \\
&= [(1 \times 16) + (1 \times 8) + (0 \times 4) + (2) + (1)]_{ten} \\
&= (16 + 8 + 0 + 2 + 1)_{ten} \\
&= 27_{ten} \quad \blacksquare
\end{aligned}
$$

EXAMPLE Rewrite 326_{ten} in equivalent form using a base two numeral.
Solution We will first consider the various powers of 2 which are, in base ten, $2^0 = 1$, $2^1 = 2$, $2^2 = 4$, $2^3 = 8$, $2^4 = 16$, $2^5 = 32$, $2^6 = 64$, $2^7 = 128$, $2^8 = 256$, $2^9 = 512$, and so on. Since $2^9 > 326_{ten}$ and $2^8 < 326_{ten}$, we ask how many 2^8's are contained in 326_{ten}. The answer is 1, and we subtract $1 \times 2^8 = 1 \times 256_{ten}$ from 326_{ten}, obtaining 70_{ten}. We now ask how many 2^7's are contained in 70_{ten}. The answer is 0 since $2^7 = 128_{ten}$ and $128_{ten} > 70_{ten}$. How many 2^6's are contained in 70_{ten}? The answer is 1, and we subtract $1 \times 2^6 = 1 \times 64 = 64_{ten}$ from 70_{ten}, obtaining 6_{ten}. How many 2^5's are contained in 6_{ten}? The answer is 0 since $2^5 = 32_{ten} > 6_{ten}$. How many 2^4's are contained in 6_{ten}? Again the answer is 0 since $2^4 = 16_{ten} > 6_{ten}$. How many 2^3's are contained in 6_{ten}? The answer is again 0 since $2^3 = 8_{ten} > 6_{ten}$. How many 2^2's are contained in 6_{ten}? The answer is 1, and we subtract $1 \times 2^2 = 1 \times 4 = 4_{ten}$ from 6_{ten}, obtaining 2_{ten}. How many 2's are contained in 2_{ten}? The answer is 1, and we subtract $1 \times 2^1 = 1 \times 2 = 2_{ten}$ from 2_{ten}, obtaining 0. How many 2^0's are contained in 0? The answer is 0. Hence we have $326_{ten} = 101000110_{two}$. $\blacksquare$

EXERCISES

1. Using Hindu-Arabic numerals together with T for ten and E for eleven, express the first twenty counting numbers in
   a. Base two. b. Base five.
   c. Base eight. d. Base nine.
   e. Base eleven. f. Base twelve.
2. Using the basic addition and multiplication tables, base five, given in this section, perform the indicated operations.
   a. $2342_{five} + 324_{five}$. b. $4032_{five} + 3440_{five}$.
   c. $3434_{five} - 2313_{five}$. d. $4032_{five} - 2313_{five}$.
   e. $342_{five} \times 234_{five}$. f. $4312_{five} \times 431_{five}$.
   g. $(214_{five} + 341_{five}) - 444_{five}$.
   h. $(233_{five} + 143_{five}) \times (243_{five} - 214_{five})$.

3. Using the basic addition and multiplication tables, base five, given in this section, determine the quotient and the remainder for each of the following division exercises.

   a. $23412_{five} \div 34_{five}$.
   b. $40302_{five} \div 21_{five}$.
   c. $32234_{five} \div 123_{five}$.
   d. $4321043_{five} \div 203_{five}$.

4. Perform the indicated additions in the base indicated.

   a. $365_{seven} + 456_{seven}$.
   b. $10101_{two} + 11011_{two}$.
   c. $231_{four} + 123_{four}$.
   d. $5431_{six} + 2543_{six}$.
   e. $9TE24_{twelve} + 8E0T7_{twelve}$.
   f. $7854_{nine} + 12675_{nine}$.
   g. $20112_{three} + 10220_{three}$.
   h. $764351_{eight} + 5734_{eight}$.

5. Perform the indicated subtractions in the base indicated.

   a. $111011_{two} - 10110_{two}$.
   b. $87654_{nine} - 6805_{nine}$.
   c. $6T9E42_{twelve} - 3E97ET_{twelve}$.
   d. $14652_{seven} - 5416_{seven}$.
   e. $30201_{four} - 1312_{four}$.

6. Perform the indicated multiplications in the base indicated.

   a. $654_{seven} \times 241_{seven}$.
   b. $3132_{four} \times 203_{four}$.
   c. $8761_{nine} \times 546_{nine}$.
   d. $20E9T_{twelve} \times TE1_{twelve}$.
   e. $3764_{eight} \times 546_{eight}$.
   f. $35425_{six} \times 5042_{six}$.

7. Determine the quotient and the remainder for each of the following division exercises in the indicated base.

   a. $101010_{two} \div 101_{two}$.
   b. $51043_{six} \div 204_{six}$.
   c. $123456_{seven} \div 213_{seven}$.
   d. $237643_{eight} \div 1564_{eight}$.
   e. $4080706_{nine} \div 2004_{nine}$.
   f. $TE98T4E5_{twelve} \div 23E_{twelve}$.

8. Rewrite each of the following in equivalent form as a base ten numeral.

   a. 2346_{seven}
   b. $T0E09_{twelve}$.
   c. 31204_{five}.
   d. 1001110_{two}.
   e. 23456_{eleven}.
   f. 897_{twelve}.
   g. 645123_{eight}.
   h. 20102002_{three}.

9. Rewrite each of the following base ten numerals in equivalent form with the indicated base.

   a. 2316_{ten}, in base four.
   b. 9876_{ten}, in base seven.
   c. 789_{ten}, in base three.
   d. 12986_{ten}, in base twelve.
   e. 4769_{ten}, in base six.
   f. 8076_{ten}, in base two.
   g. 67108_{ten}, in base nine.
   h. 909_{ten}, in base eight.

10. For each of the following determine the value of the base b.

   a. $25_{ten} = 100_b$.
   b. $84_{ten} = 70_b$.
   c. $24_{ten} = 40_b$.
   d. $49_{ten} = 61_b$.
   e. $10_{ten} = 1010_b$.
   f. $21_{ten} = 23_b$.
   g. $108_{ten} = 99_b$.
   h. $164_{ten} = 432_b$.
   i. $84_{ten} = 314_b$.
   j. $117_{ten} = 99_b$.

SUMMARY

This chapter introduced and discussed some early systems of numeration. These included the Egyptian, the Babylonian, the Greek, the Roman, and the Mayan systems. The Hindu-Arabic system of numeration also was discussed and compared in part with the early systems. Older algorithms such as the lattice method of multiplication and the peasant method of multiplication also were introduced. Napier's bones or Napier's rods were discussed as a method used for multiplication. In the discussion of the Hindu-Arabic system of numeration emphasis was placed on the base ten, but bases other than ten also were discussed. These included bases two, five, and twelve.

In our discussion the following symbols were introduced.

SYMBOL	INTERPRETATION
/	The stroke, representing the number 1 in Egyptian and Roman systems of numeration
∩	Egyptian numeral for 10
๑	Egyptian numeral for 100
⚲	Egyptian numeral for 1000
⫽	Egyptian numeral for 10,000
⟁	Egyptian numeral for 100,000
⚲	Egyptian numeral for 1 million
▽	Babylonian numeral for 1 or 60
◁	Babylonian numeral for 10
▽▷	Babylonian symbol for subtraction
V	Roman numeral for 5
X	Roman numeral for 10
L	Roman numeral for 50
C	Roman numeral for 100
D	Roman numeral for 500
M	Roman numeral for 1000
(I)	Early Roman numeral for 1000
$\overline{\text{V}}$	Roman numeral for 5000
$\overline{\overline{\text{V}}}$	Roman numeral for 5 million
⊖	Mayan numeral for 0
.	Mayan numeral for 1
—	Mayan numeral for 5

REVIEW EXERCISES FOR CHAPTER 6

All Hindu-Arabic symbols are base ten unless otherwise indicated.
1. Write the Egyptian symbol for each of the following.
  a. 262.   b. 375.

c. 623. d. 1015.
e. 13,234. f. 23,962.
g. 37,462. h. 169,754.
i. 342,123. j. 1,023,011.

2. Write the Babylonian symbol for each of the Hindu-Arabic symbols listed in Exercise 1.

3. Write the Roman numeral for each of the Hindu-Arabic symbols listed in Exercise 1.

4. Write the Mayan symbol for each of the Hindu-Arabic symbols listed in Exercise 1.

5. Write the Hindu-Arabic numeral for each of the following Egyptian symbols.

a. ⌒∩||

b. ⌒∩⌒∩

c. ⌒∩⌒∩∩

d. ⌒∩∩|||

e. ∩∩|||⌒⌒⌒

f. ⌒∩⌒∩∩∩∩|||

6. Write the Hindu-Arabic numeral for each of the following Babylonian symbols.

a. Y ◁Y

b. ◁ ◁

c. ◁YY ◁Y

d. ◁◁YY ◁◁YYY

e. ◁◁◁ Y ◁Y

f. YY ◁Y Y◁◁◁

7. Write the Hindu-Arabic numeral for each of the following Roman numerals.

a. MDCLXV. b. MMCCCCXXXX.
c. MMDCCLIV. d. CDLXXIX.
e. X̄MIV. f. M̄D̄CCXLIV.

8. Write the Hindu-Arabic numeral for each of the following Mayan symbols.

a.

b.

c.

d.

e.

f.

9. Using the lattice method of multiplication, determine the following products.

a. 247 × 27. b. 523 × 142.

10. Using the peasant method of multiplication, determine the following products.

    a. 246×19.   b. 397×112.

11. Using the expanded notation for writing Hindu-Arabic numerals, perform the indicated operations.

    a. $37 + 29$.   b. $246 + 127$.

    c. $6392 + 2769$.   d. $48 - 29$.

    e. $273 - 126$.   f. $4096 - 1769$.

12. Using the distributive property of multiplication over addition, perform the following multiplications.

    a. 416×23.   b. 719×37.

    c. 896×125.   d. 697×572.

13. Using the method of repeated subtractions, determine the quotient in each of the following.

    a. $3198 \div 26$.   b. $41107 \div 37$.

    c. $164268 \div 169$.   d. $230625 \div 225$.

14. Perform the indicated operations in the base indicated.

    a. $2346_{seven} + 4562_{seven}$.   b. $1010011_{two} - 10101_{two}$.

    c. $234_{five} \times 412_{five}$.   d. $9T8E76_{twelve} \div 23T_{twelve}$.

15. Rewrite each of the following as a base ten Hindu-Arabic numeral.

    a. 23768_{nine}.   b. $9T4E76_{twelve}$.

    c. 1101100111_{two}.   d. 4123214_{five}.

    e. 254315_{seven}.

CHAPTER 7
THE SYSTEM OF INTEGERS

Chapter 5 discussed the system of whole numbers and showed that the set of whole numbers is closed under the operations of addition and multiplication but not under the operations of subtraction and division. That is, if we consider the equations $a = b + x$ and $c = dy$, then the first equation would have a solution x on the set of whole numbers only if $a \geq b$ and the second equation would have a solution y on the set of whole numbers only if d is a natural number and d divides c. These facts result directly from the definitions of subtraction and division on the set W.

This chapter will introduce the system of integers and, on the set of integers, we will be able to determine the solution x of the equation $a = b + x$ for any integers a and b.

7.1 THE INTEGERS

Consider a strip painted on the floor of a room such as that indicated in Figure 7.1. Starting from the line segment marked on the strip, take two steps to the right and mark the spot where you stopped as indicated in Figure 7.2. Now return to the starting point marked on the strip, take two steps to the left, and mark the spot where you stopped as indicated in Figure 7.3.

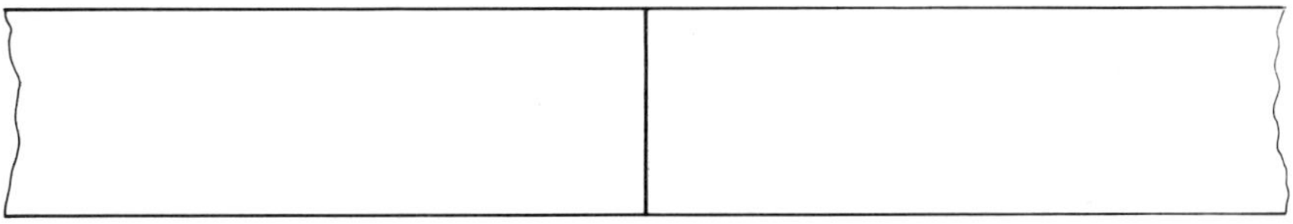

Figure 7.1

Clearly, in each case two steps were taken but in different directions. If for convenience we represent the two steps to the right by the numeral $^+2$ (read positive two), we could represent the two steps to the left as $^-2$ (read negative two). Now return to the starting mark on the strip and move two steps to the right. Turn around, and move two steps to the left. If you assume that all four steps were equally spaced, you would end up back at the starting mark. Do you agree? Hence we have $^+2 + {}^-2 = 0$, indicating that the two steps to the left (denoted by $^-2$) undid the two steps to the right (denoted by $^+2$), returning you to the starting mark on the strip.

Two steps to right

Figure 7.2

Now consider the next event. Suppose that you start the day with no money in your pocket. Two hours later a friend calls upon you and pays back the $5 borrowed from you the day before. You now have $5 in your pocket (assuming that the money was put there). Later in the day you spend $5. How much money is left in your pocket? Again, if the money received is indicated by $^+5$ (in dollars), then the money spent would be denoted by $^-5$ (in dollars). Clearly, if you started the day with no money in your pocket, received $5 and spent $5, you should have no money in your pocket after these two transactions. Hence, in dollars, we would have $^+5 + {}^-5 = 0$.

For a third event consider the following. A particular stock was selling for $56 a share at the start of a business day. During the day the stock

Two steps to left

Figure 7.3

"rose" $3 a share and later "dropped" $3 a share. There were no other fluctuations in the price of a share of that stock during the day. Clearly, at the close of the business day the stock was worth $56 a share, representing a gain of $0 per share. Now if we represent the rise in the price of the stock as $^+3$ (in dollars), then the drop in the price of the stock would be $^-3$ (in dollars), and we have $^+3 + {}^-3 = 0$ for the *gain* in dollars.

As a final example consider temperature readings on a particular day. Suppose that at 7:00 A.M. the temperature reading in a certain locality was 0°F and during the next hour the temperature "dropped" 10°. During the next two hours, suppose that the temperature "rose" 10°. What would the temperature reading be at 10:00 A.M. that day? Did you get 0°F? Again, we could consider the drop in temperature as $^-10$ (in degrees) and the rise in temperature as $^+10$ (in degrees). The *change* in temperature, then, would be $^-10 + {}^+10 = 0$ in degrees.

We could consider other examples to show that if n represents a natural number, then we could introduce a new number, ^-n, such that $n + {}^-n = 0$.

DEFINITION 7.1 If n is a natural number, then we define the *negative* of n, denoted by ^-n, to be a unique new number such that $n + {}^-n = {}^-n + n = 0$.

In the above definition we defined ^-n as the *negative* of n. We also could refer to ^-n as the *opposite* of n or as the *additive inverse* of n. Also, observe that if ^-n is the negative of n, then n is the negative of ^-n. Or if ^-n is the additive inverse of n, then n is the additive inverse of ^-n.

EXAMPLE
a. $^-3$ is the additive inverse of 3.
b. $^-2$ is the negative of 2.
c. 4 is the additive inverse of $^-4$.
d. 1 is the opposite of $^-1$.
e. 0 is the additive inverse of 0.

Combining these new numbers introduced in Definition 7.1 with the whole numbers, we obtain a new set of numbers known as the set of integers.

DEFINITION 7.2 Consider the set $W = \{0, 1, 2, 3, 4, 5, \ldots, n, \ldots\}$ and the set $\{^-1, {}^-2, {}^-3, {}^-4, {}^-5, \ldots, {}^-n, \ldots\}$ such that ^-n is the negative of the natural number n with the property that $^-n + n = n + {}^-n = 0$. Then, by the set of all *integers* denoted by I we shall mean the set

$$\{\ldots, {}^-5, {}^-4, {}^-3, {}^-2, {}^-1, 0, 1, 2, 3, 4, 5, \ldots\}$$

From the above definition we can readily determine that $W \subset I$ and that $N \subset I$. Further, we can write $I = \{\ldots, \,^-5, \,^-4, \,^-3, \,^-2, \,^-1\} \cup \{0\} \cup \{1, 2, 3, 4, 5, \ldots\}$. The set $\{1, 2, 3, 4, 5, \ldots\}$ is called the set of all *positive integers* and is sometimes written $\{^+1, \,^+2, \,^+3, \,^+4, \,^+5, \ldots\}$. The set $\{^-1, \,^-2, \,^-3, \,^-4, \,^-5, \ldots\}$ is called the set of all *negative integers*, and 0 is called the *zero integer*. Hence the set of integers can be considered as the union of three disjoint sets—the set of all positive integers, the set of all negative integers, and the integer 0. Sometimes it is convenient to consider the set $\{0, 1, 2, 3, 4, 5, \ldots\}$ known as the set of all *nonnegative integers*. Also, the set $\{0, \,^-1, \,^-2, \,^-3, \,^-4, \,^-5, \ldots\}$ is called the set of all *nonpositive integers*.

Throughout the remainder of this chapter we will develop the system of integers. In so doing it will be our intent to extend the system of whole numbers and to preserve, if possible, what was already developed. For instance, we will want to define addition and multiplication of integers without having to redefine the respective operations relative to whole numbers.

EXERCISES

1. For each of the following write its additive inverse (a and b represent integers).

    a. 4.

    b. $^-7$.

    c. 6.

    d. $^-9$.

    e. 0.

    f. a.

    g. ^-b.

    h. $2 + 3$.

    i. $4 - 7$.

    j. 3×6.

    k. ab.

    l. $a + b$.

    m. $b - 2$.

    n. $3 + \,^-3$.

    o. $16 \div 4$.

    p. $^-16 \div (8 \div \,^-2)$.

2. Evaluate each of the following expressions (a and b represent integers).

    a. $4 + \,^-4$.

    b. $^-3 + 3$.

    c. $a + \,^-a$.

    d. $(b + a) + \,^-b$.

    e. $^-(a + b) + (a + b)$.

    f. $(b - a) + \,^-(b - a)$.

    g. $^-(6 \times 2) + (6 \times 2)$.

    h. $^-(7 - 9) + (9 - 7)$.

3. What is the largest integer?

4. What is the smallest integer?

5. What is the largest nonpositive integer?

6. What is the largest negative integer?

7. What is the smallest positive integer?

8. What is the smallest nonnegative integer?

9. Which integers if any can be classified as being both nonnegative and nonpositive?

10. Identify each of the following statements as being true or false. $N =$ set of natural numbers, $W =$ set of whole numbers, $I =$ set of integers, $I_p =$ set of positive integers, and $I_n =$ set of negative integers.

a. $N \subseteq W$.

b. $N \subset W$.

c. $W \subseteq I$.

d. $W \subset N$.

e. $W \cap I_n = \varnothing$.

f. $I_n \cup I_p = I$.

g. $I_n \cap I_p \subset W$.

h. $N \subseteq W \subseteq I$.

i. $I_p \cup \{0\} \subseteq W$.

j. $N \cap I = W$.

k. $I - I_p = I_n$.

l. $I_n \cup \varnothing = I_p$.

m. $I_n \cap I_p = 0$.

n. $I_n \cap I_p = \{0\}$.

o. $I_n \cap I_p = \varnothing$.

7.2 ADDITION OF INTEGERS

In this section we will define the sum of two integers as an extension of the addition of two whole numbers and we will assume the following properties:

1. CLOSURE PROPERTY   If $a, b \in I$, then $a + b \in I$.
2. COMMUTATIVE PROPERTY FOR ADDITION   If $a, b \in I$, then $a + b = b + a$.
3. ASSOCIATIVE PROPERTY FOR ADDITION   If $a, b, c \in I$, then $(a + b) + c = a + (b + c)$.
4. ADDITIVE IDENTITY PROPERTY   There exists $0 \in I$ such that $a + 0 = 0 + a = a$ for *every* $a \in I$.
5. ADDITIVE INVERSE PROPERTY   If $a \in I$, then there exists a unique element $^-a \in I$ such that $a + {}^-a = {}^-a + a = 0$.

We will now proceed with the definition for the sum of two integers. Since the integers have been classified as positive, negative, and zero, we must consider the following cases:

1. The sum of two positive integers.
2. The sum of two negative integers.
3. The sum of a positive integer with a negative integer.

The first case can be considered quickly. Since the positive integers are other names for the corresponding natural numbers, we can define the sum of two positive integers as we defined the sum of two natural numbers. That is, if a and b are any positive integers, then $a + b = n(A \cup B)$, where $a = n(A)$, $b = n(B)$, and $A \cap B = \varnothing$.

To motivate the definition for the sum of two negative integers, consider the following examples.

EXAMPLE   Evaluate $(^-4 + {}^-3) + (4 + 3)$.
Solution

$$(^-4 + {}^-3) + (4 + 3) = (^-4 + {}^-3) + (3 + 4) \qquad \text{Commutative property for addition}$$

$(^-4 + {}^-3) + (3 + 4) = {}^-4 + [^-3 + (3 + 4)]$ — Associative property for addition

$(^-4 + {}^-3) + (4 + 3) = {}^-4 + [^-3 + (3 + 4)]$ — Transitive property for equality

$^-4 + [^-3 + (3 + 4)] = {}^-4 + [(^-3 + 3) + 4]$ — Associative property for addition

$(^-4 + {}^-3) + (4 + 3) = {}^-4 + [(^-3 + 3) + 4]$ — Transitive property for equality

$^-4 + [(^-3 + 3) + 4] = {}^-4 + (0 + 4)$ — Additive inverse property on I

$(^-4 + {}^-3) + (4 + 3) = {}^-4 + (0 + 4)$ — Transitive property for equality

$^-4 + (0 + 4) = {}^-4 + 4$ — Additive identity property

$(^-4 + {}^-3) + (4 + 3) = {}^-4 + 4$ — Transitive property for equality

$^-4 + 4 = 0$ — Additive inverse property

$(^-4 + {}^-3) + (4 + 3) = 0$ — Transitive property for equality

■

From the above example we see that $(^-4 + {}^-3) + (4 + 3) = 0$. But $^-4 + {}^-3 \in I$ by the closure property for addition. Since we have the sum of two integers equal to zero, each integer, then, is the additive inverse of the other. In particular $^-4 + {}^-3$ is the unique additive inverse of $(4 + 3)$. Hence we have $^-4 + {}^-3 = {}^-(4 + 3)$.

EXAMPLE If $a, b \in W$, evaluate $(^-a + {}^-b) + (a + b)$.
Solution Since $a, b \in W$, then $a, b, {}^-a, {}^-b \in I$ and we have

$(^-a + {}^-b) + (a + b) = (^-a + {}^-b) + (b + a)$ — Commutative property for addition

$(^-a + {}^-b) + (b + a) = {}^-a + [^-b + (b + a)]$ — Associative property for addition

$(^-a + {}^-b) + (a + b) = {}^-a + [^-b + (b + a)]$ — Transitive property for equality

$^-a + [^-b + (b + a)] = {}^-a + [(^-b + b) + a]$ — Associative property for addition

$(^-a + {}^-b) + (a + b) = {}^-a + [(^-b + b) + a]$ — Transitive property for equality

$^-a + [(^-b + b) + a] = {}^-a + (0 + a)$ — Additive inverse property

$(^-a + {}^-b) + (a + b) = {}^-a + (0 + a)$ — Transitive property for equality

$^-a + (0 + a) = {}^-a + a$	Additive identity property
$(^-a + {}^-b) + (a + b) = {}^-a + a$	Transitive property for equality
$^-a + a = 0$	Additive inverse property
$(^-a + {}^-b) + (a + b) = 0$	Transitive property for equality ∎

From the above example we have $(^-a + {}^-b) + (a + b) = 0$. Since $(^-a + {}^-b)$, $(a + b) \in I$ by the closure property for addition, and since we have the sum of two integers equal to zero, each integer must be the unique additive inverse of the other. In particular, $(^-a + {}^-b)$ must be the unique additive inverse of $(a + b)$ and, hence, we have $(^-a + {}^-b) = {}^-(a + b)$.

EXAMPLE
a. $^-2 + {}^-3 = {}^-(2 + 3) = {}^-5$.
b. $^-4 + {}^-7 = {}^-(4 + 7) = {}^-11$.
c. $^-5 + {}^-8 = {}^-(5 + 8) = {}^-13$.
d. $^-9 + {}^-6 = {}^-(9 + 6) = {}^-15$.

To motivate the definition for the sum of a positive integer with a negative integer, consider the following examples.

EXAMPLE Evaluate $5 + {}^-3$.
Solution

$5 + {}^-3 = (2 + 3) + {}^-3$	Substituting $2 + 3$ for 5
$(2 + 3) + {}^-3 = 2 + (3 + {}^-3)$	Associative property for addition
$5 + {}^-3 = 2 + (3 + {}^-3)$	Transitive property for equality
$2 + (3 + {}^-3) = 2 + 0$	Additive inverse property
$5 + {}^-3 = 2 + 0$	Transitive property for equality
$2 + 0 = 2$	Additive identity property
$5 + {}^-3 = 2$	Transitive property for equality

We now have that $5 + {}^-3 = 2$. Also, we know that $5 - 3 = 2$. Combining these statements, then, we have $5 + {}^-3 = 5 - 3$. ∎

EXAMPLE Evaluate $^-5 + 3$.
Solution

$^-5 + 3 = (^-2 + {}^-3) + 3$	Substituting $^-2 + {}^-3$ for $^-5$
$(^-2 + {}^-3) + 3 = {}^-2 + (^-3 + 3)$	Associative property for addition
$^-5 + 3 = {}^-2 + (^-3 + 3)$	Transitive property for equality
$^-2 + (^-3 + 3) = {}^-2 + 0$	Additive inverse property

$^-5 + 3 = {}^-2 + 0$	Transitive property for equality
$^-2 + 0 = {}^-2$	Additive identity property
$^-5 + 3 = {}^-2$	Transitive property for equality

We now have that $^-5 + 3 = {}^-2$. Also, we know that $5 - 3 = 2$. Hence $^-(5 - 3) = {}^-2$. Since $^-5 + 3 = {}^-2$ and $^-(5 - 3) = {}^-2$, we now have $^-5 + 3 = {}^-(5 - 3)$. ∎

EXAMPLE   Let $a, b \in W$ such that $a > b$. Evaluate $a + {}^-b$.

Solution   We are asked to evaluate the sum of a positive integer with a negative integer. Since $a, b \in W$ and $a > b$, we then have that $a = b + c$, where $c \in N$, by the definition of order on W. Hence

$a + {}^-b = (b + c) + {}^-b$	Substituting $b + c$ for a
$(b + c) + {}^-b = (c + b) + {}^-b$	Commutative property for addition on I
$a + {}^-b = (c + b) + {}^-b$	Transitive property for equality
$(c + b) + {}^-b = c + (b + {}^-b)$	Associative property for addition on I
$a + {}^-b = c + (b + {}^-b)$	Transitive property for equality
$c + (b + {}^-b) = c + 0$	Additive inverse on I
$a + {}^-b = c + 0$	Transitive property for equality
$c + 0 = c$	Additive identity on I
$a + {}^-b = c$	Transitive property for equality
$c = a - b$	Since $a > b$ and $a = b + c$
$a + {}^-b = a - b$	Transitive property for equality

Hence we have that $a + {}^-b = a - b$, if $a > b$. ∎

EXAMPLE   Let $a, b \in W$ such that $a < b$. Evaluate $a + {}^-b$.

Solution   Again, we are asked to evaluate the sum of a positive integer with a negative integer. Since $a, b \in W$ and $a < b$, we then have that $b = a + c$, where $c \in N$, by the definition of order on W. Hence

$a + {}^-b = a + {}^-(a + c)$	Substituting $a + c$ for b
$a + {}^-(a + c) = a + ({}^-a + {}^-c)$	Since $^-(a + c) = {}^-a + {}^-c$ (second example above)
$a + {}^-b = a + ({}^-a + {}^-c)$	Transitive property for equality
$a + ({}^-a + {}^-c) = (a + {}^-a) + {}^-c$	Associative property for addition on I
$a + {}^-b = (a + {}^-a) + {}^-c$	Transitive property for equality
$(a + {}^-a) + {}^-c = 0 + {}^-c$	Additive inverse property on I
$a + {}^-b = 0 + {}^-c$	Transitive property for equality
$0 + {}^-c = {}^-c$	Additive identity on I
$a + {}^-b = {}^-c$	Transitive property for equality
$^-c = {}^-(b - a)$	Since $a < b$ and $b = a + c$
$a + {}^-b = {}^-(b - a)$	Transitive property for equality

Hence we observe that $a + {}^-b = {}^-(b - a)$, if $a < b$. ∎

EXAMPLE   Let $a, b \in W$ such that $a = b$. Evaluate $a + \bar{}b$.
Solution

$a + \bar{}b = a + \bar{}a$	Substituting a for b
$a + \bar{}a = 0$	Additive inverse property on I
$a + \bar{}b = 0$	Transitive property for equality

Hence we observe that $a + \bar{}b = 0$, if $a = b$. ■

If we combine the results of the second through eighth examples above, the additive identity property on I, and the definition for the addition of whole numbers, we now state the following definition for the addition of integers.

DEFINITION 7.3 Let a and b be any two arbitrary whole numbers. Then the *sum of two integers* is defined according to the following cases:
1. $a + b = n(A \cup B)$, where $n(A) = a$, $n(B) = b$, and $A \cap B = \emptyset$.
2. $\bar{}a + \bar{}b = \bar{}(a + b)$.
3. $a + \bar{}b = a - b$, if $a > b$.
4. $a + \bar{}b = \bar{}(b - a)$, if $a < b$.
5. $a + \bar{}b = 0$, if $a = b$.
6. $a + 0 = a$ for all $a \in I$.
7. $\bar{}a + 0 = \bar{}a$ for all $a \in I$.

The above definition covers all possible cases involved in the addition of integers and, in each case, we observe that the sum of the two integers is again an integer. This supports the assumed closure property stated at the beginning of this section.

EXAMPLE

a. $7 + 6 = 13$	By Part 1, Definition 7.3
b. $7 + \bar{}6 = (7 - 6) = 1$	By Part 3, Definition 7.3
c. $\bar{}6 + \bar{}4 = \bar{}(6 + 4) = \bar{}10$	By Part 2, Definition 7.3
d. $\bar{}3 + 6 = (6 - 3) = 3$	By Part 3, Definition 7.3
e. $8 + \bar{}12 = \bar{}(12 - 8) = \bar{}4$	By Part 4, Definition 7.3
f. $9 + \bar{}9 = 0$	By Part 5, Definition 7.3
g. $4 + 0 = 4$	By Part 6, Definition 7.3
h. $\bar{}5 + 0 = \bar{}5$	By Part 7, Definition 7.3

EXERCISES

1. Using the definition for addition of integers, evaluate each of the following expressions.
  a. $3 + 6$. b. $\bar{}2 + \bar{}3$.

  c. $4 + {}^-6$.
  e. $6 + 0$.
  g. ${}^-9 + 9$.
  i. ${}^-3 + (4 + {}^-6)$.
  k. $a + {}^-b$, if $a = 3, b = 2$.
  m. $a + ({}^-b + b)$, if $a = {}^-3, b = 5$.
  n. $(a + b) + ({}^-a + b)$, if $a = 4, b = {}^-3$.
  o. $b + ({}^-a + b) + {}^-(a + {}^-b)$, if $a = 2, b = {}^-1$.

  d. ${}^-5 + 7$.
  f. $0 + 0$.
  h. ${}^-8 + 0$.
  j. $({}^-7 + 6) + ({}^-2 + {}^-3)$.
  l. ${}^-a + {}^-b$, if $a = {}^-2, b = 4$.

2. Identify what property associated with the addition of integers is being illustrated in each of the following.
  a. $4 + {}^-7 = {}^-7 + 4$.
  c. $5 + {}^-5 = {}^-5 + 5 = 0$.
  e. $2 + ({}^-3 + 4) = (2 + {}^-3) + 4$.
  g. ${}^-3 + (6 + 0) = {}^-3 + 6$.

  b. ${}^-6 + 0 = {}^-6$.
  d. $4 + {}^-3 \in I$.
  f. $(5 + {}^-5) + 3 = 0 + 3$.
  h. ${}^-({}^-3 + 4) + ({}^-3 + 4) = 0$.

3. For each of the following fill in the blank with an appropriate integer symbol so that the resulting statement is true.
  a. $3 + $ _____ $= 7$.
  c. _____ $+ 3 = {}^-2$.
  e. $6 + $ _____ $= 0$.
  g. $6 + ({}^-5 + 7) + (2 + {}^-5) = $ _____ .
  h. $(7 + {}^-5) + $ _____ $= 2 + {}^-6$.

  b. ${}^-2 + $ _____ $= 5$.
  d. _____ $+ 4 = 4$.
  f. $(2 + {}^-3) + $ _____ $= {}^-1$.

4. The temperature at 9:00 A.M. on a particular day was reported at 68°F. For the next six hours (on the hour) the changes in temperature were reported as follows: up 2°, up 3°, down 1°, up 2°, down 4°, and down 3°. What was the temperature at 3:00 P.M.?

5. The treasurer of an organization reported that the previous balance in the checking account was $236. Income during the reporting period amounted to $215. Disbursements were as follows: $26, $32, $118, $72, and $51. What should the new balance be?

7.3 MULTIPLICATION OF INTEGERS

In this section we will define the product of two integers as an extension of the multiplication of two whole numbers and we will assume the following properties.

  1. CLOSURE PROPERTY If $a, b \in I$, then $ab \in I$.

  2. COMMUTATIVE PROPERTY FOR MULTIPLICATION If $a, b \in I$, then $ab = ba$.

  3. ASSOCIATIVE PROPERTY FOR MULTIPLICATION If $a, b, c \in I$, then $(ab)c = a(bc)$.

  4. MULTIPLICATIVE IDENTITY There exists $1 \in I$ such that $a \cdot 1 = 1 \cdot a = a$ for *every* $a \in I$.

  5. MULTIPLICATIVE PROPERTY OF 0 If $a \in I$, then $a \cdot 0 = 0 \cdot a = 0$ for *every* $a \in I$.

6. DISTRIBUTIVE PROPERTY OF MULTIPLICATION OVER ADDITION   If a, b, $c \in I$, then $a(b + c) = ab + ac$.

We now will proceed with the definition for the product of two integers. Again, since the integers have been classified as positive, negative, and zero, we must consider the following cases:

1. The product of two positive integers.

2. The product of two negative integers.

3. The product of a positive integer with a negative integer.

The product of two positive integers can be treated exactly the same as the product of two natural numbers. Hence $ab = n(A \times B)$, where $a = n(A)$ and $b = n(B)$.

To motivate the definition for the product of a positive integer with a negative integer, consider the following example.

EXAMPLE   Evaluate: $(^-3 \cdot 2) + (3 \cdot 2)$.
Solution

$(^-3 \cdot 2) + (3 \cdot 2) = (^-3 + 3) \cdot 2$	Distributive property of $\times$ over $+$ on I
$(^-3 + 3) \cdot 2 = 0 \cdot 2$	Additive inverse property on I
$(^-3 \cdot 2) + (3 \cdot 2) = 0 \cdot 2$	Transitive property for equality
$0 \cdot 2 = 0$	Multiplicative property of 0
$(^-3 \cdot 2) + (3 \cdot 2) = 0$	Transitive property for equality ■

From the above example we see that $(^-3 \cdot 2) + (3 \cdot 2) = 0$. But $(^-3 \cdot 2)$, $(3 \cdot 2) \in I$ by the closure property for multiplication of integers. Since we have the sum of two integers equal to zero, each integer, then, must be the additive inverse of the other. In particular $(^-3 \cdot 2)$ is the unique inverse of $(3 \cdot 2)$. Hence we have $(^-3 \cdot 2) = {}^-(3 \cdot 2)$.

EXAMPLE   Evaluate $(^-a \cdot b) + (a \cdot b)$, where $a, b \in W$.
Solution

$(^-a \cdot b) + (a \cdot b) = (^-a + a)b$	Distributive property of $\times$ over $+$ on I
$(^-a + a)b = 0 \cdot b$	Additive inverse property on I
$(^-a \cdot b) + (a \cdot b) = 0 \cdot b$	Transitive property for equality
$0 \cdot b = 0$	Multiplicative property for 0
$(^-a \cdot b) + (a \cdot b) = 0$	Transitive property for equality ■

From the above example we see that $(^-a \cdot b) + (a \cdot b) = 0$. But $(^-a \cdot b)$, $(a \cdot b) \in I$ by the closure property for multiplication of integers. Since we have the sum of two integers equal to zero, each integer, then, must be the unique additive inverse of the other. In particular, $(^-a \cdot b)$ is the unique additive inverse of $(a \cdot b)$. Hence we have $(^-a \cdot b) = {}^-(a \cdot b)$.

EXAMPLE
a. $^-2 \cdot 3 = ^-(2 \cdot 3) = ^-6.$
b. $^-4 \cdot 7 = ^-(4 \cdot 7) = ^-28.$
c. $6 \cdot ^-5 = ^-(6 \cdot 5) = ^-30.$
d. $8 \cdot ^-7 = ^-(8 \cdot 7) = ^-56.$

To motivate the definition for the product of two negative integers, consider the following examples.

EXAMPLE Evaluate $(^-4 \cdot ^-5) + (^-4 \cdot 5).$
Solution

$(^-4 \cdot ^-5) + (^-4 \cdot 5) = ^-4(^-5 + 5)$	Distributive property for $\times$ over $+$ on I
$^-4(^-5 + 5) = ^-4 \cdot 0$	Additive inverse property on I
$(^-4 \cdot ^-5) + (^-4 \cdot 5) = ^-4 \cdot 0$	Transitive property for equality
$^-4 \cdot 0 = 0$	Multiplicative property for 0
$(^-4 \cdot ^-5) + (^-4 \cdot 5) = 0$	Transitive property for equality ∎

From the above example we see that $(^-4 \cdot ^-5) + (^-4 \cdot 5) = 0.$ But $(^-4 \cdot ^-5),$ $(^-4 \cdot 5) \in I$ by the closure property for multiplication on $I.$ Since we have the sum of two integers equal to zero, each integer, then, must be the additive inverse of the other. In particular $(^-4 \cdot ^-5)$ is the unique additive inverse of $(^-4 \cdot 5).$ However, $(^-4 \cdot 5) = ^-(4 \cdot 5) = ^-20.$ Hence $(^-4 \cdot ^-5)$ is the unique additive inverse of $^-20.$ But the unique additive inverse of $^-20$ is 20. Therefore we have that $(^-4 \cdot ^-5) = 20 = (4 \cdot 5).$

EXAMPLE Evaluate $(^-a \cdot ^-b) + (^-a \cdot b),$ where $a, b \in W.$
Solution

$(^-a \cdot ^-b) + (^-a \cdot b) = ^-a(^-b + b)$	Distributive property of $\times$ over $+$ on I
$^-a(^-b + b) = ^-a \cdot 0$	Additive inverse property on I
$(^-a \cdot ^-b) + (^-a \cdot b) = ^-a \cdot 0$	Transitive property for equality
$^-a \cdot 0 = 0$	Multiplicative property for 0
$(^-a \cdot ^-b) + (^-a \cdot b) = 0$	Transitive property for equality ∎

From the above example we see that $(^-a \cdot ^-b) + (^-a \cdot b) = 0.$ But $(^-a \cdot ^-b),$ $(^-a \cdot b) \in I$ by the closure property for multiplication on $I.$ Since we have the sum of two integers equal to zero, each integer, then, must be the additive inverse of the other. In particular $(^-a \cdot ^-b)$ is the unique additive inverse of $(^-a \cdot b).$ But $(^-a \cdot b) = ^-(a \cdot b),$ and the unique additive inverse of $^-(a \cdot b)$ is $ab.$ Hence we have that $(^-a \cdot ^-b) = ab.$

If we combine the results of the second through fifth examples, the multiplicative identity property on $I,$ the multiplicative property for 0, and the definition for the product of two whole numbers, we can state the following definition for the multiplication of integers.

DEFINITION 7.4 Let a and b be any two arbitrary whole numbers. Then the *product of two integers* is defined according to the following cases:

1. $ab = n(A \times B)$, where $a = n(A)$ and $b = n(B)$.
2. $^-a \cdot b = {^-(ab)}$.
3. $^-a \cdot {^-b} = ab$.
4. $a \cdot 1 = 1 \cdot a = a$ for *every* $a \in I$.
5. $a \cdot 0 = 0 \cdot a = 0$ for *every* $a \in I$.

The above definition covers all possible cases involved in the multiplication of integers and, in each case, we observe that the product of the two integers is again an integer. This supports the assumed closure property stated at the beginning of this section.

EXAMPLE

a. $3 \cdot {^-2} = {^-(3 \cdot 2)} = {^-6}$	By Part 2, Definition 7.4
b. $2 \cdot 6 = 12$	By Part 1, Definition 7.4
c. $^-4 \cdot {^-7} = 4 \cdot 7 = 28$	By Part 3, Definition 7.4
d. $^-3 \cdot {^-6} = 3 \cdot 6 = 18$	By Part 3, Definition 7.4
e. $7 \cdot 0 = 0$	By Part 5, Definition 7.4
f. $^-6 \cdot 1 = {^-6}$	By Part 4, Definition 7.4
g. $5 \cdot 1 = 5$	By Part 4, Definition 7.4
h. $^-9 \cdot 0 = 0$	By Part 5, Definition 7.4

EXAMPLE   Verify the distributive property for multiplication over addition for $^-2(5 + {^-3}) = (^-2)(5) + (^-2)(^-3)$.

Solution

$$^-2(5 + {^-3}) = {^-2}(5 - 3) = {^-2}(2) = {^-(2 \cdot 2)} = {^-4}$$

and

$$(^-2)(5) + (^-2)(^-3) = {^-(2 \cdot 5)} + (2 \cdot 3) = {^-10} + 6 = {^-(10 - 6)} = {^-4}$$

Since $^-4 = {^-4}$, we have that $^-2(5 + {^-3}) = (^-2)(5) + (^-2)(^-3)$. ■

EXERCISES

1. Using the definition for multiplication of integers, evaluate each of the following expressions.

   a. $(2)(6)$.
   c. $(^-3)(^-5)$.
   e. $(0)(^-3)$.
   g. $(^-9)(^-8)$.

   b. $(^-2)(4)$.
   d. $(4)(0)$.
   f. $(0)(0)$.
   h. $(^-2)(3)(^-4)$.

    i. $(^-3)(^-2)(^-3)$.   j. $(2)(^-3)(5)(^-2)$.

    k. $(^-2)(^-3)(^-4)(2)$.   !. ab, if $a = 2$, $b = 3$.

    m. abc, if $a = {}^-3$, $b = 2$, $c = {}^-2$.   n. $(^-3)(5)(b)$, if $b = 0$.

    o. $(^-a)(2)(a)$, if $a = 3$.

2. Identify what property associated with the multiplication of integers is being illustrated in each of the following.

    a. $(4)(^-7) = (^-7)(4)$.   b. $(^-6)(0) = 0$.

    c. $[(2)(^-3)](4) = (2)[(^-3)(4)]$.   d. $(3)(4) \in I$.

    e. $(^-5)(0) = (0)(^-5)$.   f. $(^-5)(0) \in I$.

    g. $2(^-3 + 4) = (2)(^-3) + (2)(4)$.

    h. $(7 + {}^-5)(^-3) = (7)(^-3) + (^-5)(^-3)$.

3. For each of the following fill in the blank with an appropriate integer symbol so that the resulting statement is true.

    a. $(2)(\underline{\hspace{1cm}}) = {}^-6$.   b. $(\underline{\hspace{1cm}})(^-3) = 12$.

    c. $(3)(\underline{\hspace{1cm}}) = {}^-(7 + 2)$.   d. $(^-5)(\underline{\hspace{1cm}}) = 0$.

    e. $^-6(2 + 3) = \underline{\hspace{1cm}}$.   f. $(\underline{\hspace{1cm}})(^-3 + {}^-2) = 15$.

    g. $^-3(2 + \underline{\hspace{1cm}}) = 9$.   h. $^-5(4 + {}^-1) = (^-3)(\underline{\hspace{1cm}})$.

4. Evaluate each of the following expressions two different ways.

    a. $^-2(3 + 4)$.   b. $3(^-2 + {}^-3)$.

    c. $^-4(3 + {}^-4)$.   d. $5(^-2 + 0)$.

    e. $0(3 + {}^-2)$.   f. $0(4 + {}^-4)$.

    g. $(^-5 + 2)(^-3)$.   h. $(7 + {}^-4)(^-2)$.

5. Prove each of the following statements by working from left to right one step at a time. Give a reason for each step.

    a. $^-2(3 + 4) = (4)(^-2) + (^-2)(3)$.

    b. $^-3(4 + 2) = (6)(^-3)$.

    c. $(^-2 + {}^-3)(^-1 + 0) = 5$.

    d. $[(2)(^-3)](4) = (^-3)[(4)(2)]$.

    e. $(3 + {}^-4)(^-5) = (^-5)(^-4) + (^-5)(3)$.

    f. $(^-2 + {}^-3)[4 + (^-3 + {}^-1)] = 0$.

6. Let $a, b \in I$ and $ab = 0$. Must a and b both be 0? Explain.

7. Let $a, b \in I$. Is $(^-a)(^-b)$ ever negative? Explain.

7.4 ORDERING OF THE INTEGERS

In this section we will introduce the relation of order on the set of integers as an extension of our definitions of order on the set of whole numbers. Recall that if $a, b \in W$, we defined $a < b$ if there exists a *natural number c* such that $b = a + c$. Since we wish to preserve this definition in formulating a similar definition on the set I and since a natural number corresponds to a positive integer, we readily obtain the following definition.

DEFINITION 7.5 If a and b are arbitrary integers, then a is said to be *less than b*, denoted by $a < b$, if and only if there exists a *positive integer c* such that $b = a + c$.

EXAMPLE

a. The integer 3 is less than the integer 7 since there does exist a positive integer, namely 4, such that $7 = 3 + 4$.

b. $^-3 < 4$ since there exists the positive integer 7 such that $4 = {}^-3 + 7$.

c. $^-6 < {}^-4$ since there exists the positive integer 2 such that $^-4 = {}^-6 + 2$.

d. The integer 7 is not less than the integer 5 since there does not exist a positive integer c such that $5 = 7 + c$.

DEFINITION 7.6 If a and b are arbitrary integers, then a is said to be *greater than b*, denoted by $a > b$, if and only if $b < a$.

EXAMPLE

a. $8 > 3$ since $3 < 8$; $3 < 8$ since there exists a positive integer 5 such that $8 = 3 + 5$.

b. $6 > {}^-3$ since $^-3 < 6$; $^-3 < 6$ since there exists a positive integer 9 such that $6 = {}^-3 + 9$.

c. $^-4 > {}^-7$ since $^-7 < {}^-4$; $^-7 < {}^-4$ since there exists a positive integer 3 such that $^-4 = {}^-7 + 3$.

EXAMPLE

a. $^-7 \not< {}^-7$ since there does not exist a positive integer c such that $^-7 = {}^-7 + c$.

b. $^-7 \not> {}^-7$ since $^-7 \not< {}^-7$.

As with the case of whole numbers we can combine equality and inequality of integers. For instance, if a and b are integers, we may write $a \leq b$ to mean $a < b$ or $a = b$. Similarly, $c \geq d$ means $c > d$ or $c = d$.

7.41 TRICHOTOMY PROPERTY FOR INTEGERS

If a and b are any two integers, then *exactly* one of the following statements is true:

1. $a < b$.

2. $a = b$.

3. $a > b$.

THEOREM 7.1 Transitive Property of Order   The relation "is less than" is a transitive relation on the set I. That is, if a, b, $c \in I$, and if $a < b$ and $b < c$, then $a < c$.

Proof   We are given that a, b, $c \in I$ with $a < b$ and $b < c$. Hence, by Definition 7.5, we have $b = a + p$ and $c = b + q$, where p and q are positive integers. By substituting $a + p$ for b in $c = b + q$, we have $c = (a + p) + q$.

But $(a + p) + q = a + (p + q)$ by the associative property for addition on I. Hence we have $c = a + (p + q)$ by the transitive property for equality. Since $p, q \in I$, then $p + q \in I$ by the closure property for addition on I. Further, since p and q are positive integers, the sum $p + q$ is also a positive integer by Definition 7.3. Therefore, we have the integer c equal to the sum of the integer a and the positive integer $p + q$. Hence, by Definition 7.5, $a < c$.

In a similar manner we can prove that the relation "is greater than" is a transitive relation on the set I. This will be left as an exercise. We will conclude this section with the following useful theorems.

THEOREM 7.2   Let a, b, c, and d be arbitrary integers.
a. If $a = b$, then $a + c = b + c$.
b. If $a + c = b + c$, then $a = b$.
c. If $a = b$, then $ac = bc$.
d. If $a = b$ and $c = d$, then $ac = bd$.
e. If $a = b$ and $c = d$, then $a + c = b + d$.
Proof   We will prove part (b) and leave the other parts as exercises. If $a + c = b + c$, then

$(a + c) + {}^-c = (b + c) + {}^-c$	By Part (a) of this theorem
$a + (c + {}^-c) = b + (c + {}^-c)$	Associative property for addition on I
$a + 0 = b + 0$	Additive inverse property on I
$a = b$	Additive identity on I

THEOREM 7.3   Let a, b, c, and d be arbitrary integers.
a. If $a < b$, then $a + c < b + c$.
b. If $a + c < b + c$, then $a < b$.
c. If $a < b$ and $c > 0$, then $ac < bc$.
d. If $a < b$ and $c < 0$, then $ac > bc$.
Proof   The entire proof is left as an exercise.

EXERCISES

1. Rearrange each of the following sets of integers in order of magnitude using the relation "is less than."
   a. ${}^-2$, 4, 0, ${}^-3$, ${}^-1$, ${}^-5$, 5.
   b. ${}^-5$, 2, ${}^-3$, 4, 1, 0, ${}^-2$.
   c. ${}^-101$, 98, 100, ${}^-103$, 102, ${}^-105$.
   d. 7, ${}^-17$, 23, ${}^-26$, ${}^-27$, ${}^-28$, 13.
   e. 0, 2, ${}^-3$, 5, $({}^-2)^2$, $(3)^2$.
2. Fill in each of the following blanks with the symbol $<$, $>$, or $=$ so that the statement is true.
   a. 2 _____ 3.
   b. ${}^-4$ _____ ${}^-5$.
   c. ${}^-7$ _____ ${}^-5$.
   d. ${}^-3$ _____ 2.
   e. 18 _____ $(16 + 2)$.
   f. $(7 + 2)$ _____ $({}^-3 + 7)$.
   g. $(2 \times {}^-3)$ _____ $({}^-4 \times {}^-1)$.
   h. 8 _____ $({}^-4 \times {}^-2)$.

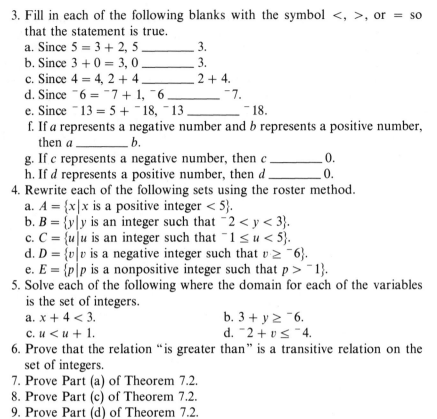

3. Fill in each of the following blanks with the symbol $<$, $>$, or $=$ so that the statement is true.

a. Since $5 = 3 + 2$, 5 _____ 3.

b. Since $3 + 0 = 3$, 0 _____ 3.

c. Since $4 = 4$, $2 + 4$ _____ $2 + 4$.

d. Since $^-6 = {}^-7 + 1$, $^-6$ _____ $^-7$.

e. Since $^-13 = 5 + {}^-18$, $^-13$ _____ $^-18$.

f. If a represents a negative number and b represents a positive number, then a _____ b.

g. If c represents a negative number, then c _____ 0.

h. If d represents a positive number, then d _____ 0.

4. Rewrite each of the following sets using the roster method.

a. $A = \{x \mid x$ is a positive integer $< 5\}$.

b. $B = \{y \mid y$ is an integer such that $^-2 < y < 3\}$.

c. $C = \{u \mid u$ is an integer such that $^-1 \leq u < 5\}$.

d. $D = \{v \mid v$ is a negative integer such that $v \geq {}^-6\}$.

e. $E = \{p \mid p$ is a nonpositive integer such that $p > {}^-1\}$.

5. Solve each of the following where the domain for each of the variables is the set of integers.

a. $x + 4 < 3$.   b. $3 + y \geq {}^-6$.

c. $u < u + 1$.   d. $^-2 + v \leq {}^-4$.

6. Prove that the relation "is greater than" is a transitive relation on the set of integers.

7. Prove Part (a) of Theorem 7.2.

8. Prove Part (c) of Theorem 7.2.

9. Prove Part (d) of Theorem 7.2.

10. Prove Part (e) of Theorem 7.2.

11. Prove Theorem 7.3.

7.5 SUBTRACTION AND DIVISION ON THE SET OF INTEGERS

We defined subtraction on the set of whole numbers and indicated that the difference $a - b$ of the whole numbers a and b was possible if and only if $a \geq b$. In defining subtraction on the set of integers, we will be able to form the difference of any two integers and obtain an integer.

DEFINITION 7.7 If a and b are any two arbitrary integers, then the *difference $a - b$* found by subtracting the integer b from the integer a is equal to the integer c if and only if $a = b + c$.

In Chapter 5 we referred to the operation of subtraction as being the inverse operation of addition on whole number. In this chapter we introduced the concept of an additive inverse for an integer. Specifically, if $a \in I$, then its additive inverse is ^{-}a such that $a + {}^{-}a = 0$. Combining these remarks with Definition 7.7, we can interpret the subtraction of integers as an appropriate addition of integers. That is, instead of subtracting the integer b from the integer a, we could add ^{-}b to a. Hence we have $a - b = a + {}^{-}b$.

EXAMPLE
a. $7 - 3 = 4$ since $4 \in I$ and $7 = 3 + 4$.
b. $8 - 8 = 0$ since $0 \in I$ and $8 = 8 + 0$.
c. $10 - 17 = {}^{-}7$ since $^{-}7 \in I$ and $10 = 17 + {}^{-}7$.
d. $^{-}5 - 8 = {}^{-}13$ since $^{-}13 \in I$ and $^{-}5 = 8 + {}^{-}13$.
e. $^{-}7 - {}^{-}6 = {}^{-}1$ since $^{-}1 \in I$ and $^{-}7 = {}^{-}6 + {}^{-}1$.

It should be clear from Definition 7.7 that the difference can be formed for *any* two integers and that the set of integers, then, is closed under the operation of subtraction. Recall that the set of whole numbers was not. Hence a deficiency on the set of whole numbers has now been corrected.

Just as addition of integers has an inverse operation called subtraction, multiplication of integers has an inverse operation called *division*.

DEFINITION 7.8 If a and b are integers with $b \neq 0$, then a *divided by* b denoted by $a \div b$ or a/b or $\dfrac{a}{b}$ is equal to the integer c if and only if $a = bc$.

It should be clear from Definition 7.8 and the following examples that the set of integers is not closed under the operation of division.

EXAMPLE
a. $18 \div 2 = 9$ since $9 \in I$ and $18 = 2 \times 9$.
b. $27 \div 9 = 3$ since $3 \in I$ and $27 = 9 \times 3$.
c. $^{-}16 \div 4 = {}^{-}4$ since $^{-}4 \in I$ and $^{-}16 = 4 \times {}^{-}4$.
d. $^{-}51 \div {}^{-}3 = 17$ since $17 \in I$ and $^{-}51 = {}^{-}3 \times 17$.
e. $18 \div {}^{-}6 = {}^{-}3$ since $^{-}3 \in I$ and $18 = {}^{-}6 \times {}^{-}3$.

EXAMPLE $17 \div 4 = ?$
Solution If $17 \div 4$ has a solution, then we want an integer c such that $17 = 4 \times c$. But there is no integer c such that the product of 4 and c is 17. Hence $17 \div 4$ does not have a solution on the set of integers. ∎

EXAMPLE
a. $0 \div 7 = 0$ since $0 \in I$ and $0 = 7 \times 0$.
b. $0 \div 13 = 0$ since $0 \in I$ and $0 = 13 \times 0$.

c. $0 \div {}^-4 = 0$ since $0 \in I$ and $0 = {}^-4 \times 0$.
d. $0 \div {}^-9 = 0$ since $0 \in I$ and $0 = {}^-9 \times 0$.

EXAMPLE $5 \div 0 = ?$
Solution   If $5 \div 0$ has a solution on the set of integers, then we want an integer d such that $5 = 0 \times d$. But $0 \times d = 0$ for every integer d by the multiplicative property for 0. Hence no integer d exists and $5 \div 0$ does not have a solution on the set of integers. ■

EXAMPLE $0 \div 0 = ?$
Solution   If $0 \div 0$ has a solution on the set of integers, then we want an integer m such that $0 = 0 \times m$. But $0 \times m = 0$ for *every* integer m by the multiplicative property for 0. Hence $0 \div 0$ has infinitely many solutions on the set of integers. ■

If we consider the last two examples above, the divisor 0 is excluded in Definition 7.8. However, we say that division by 0 on the set of integers is not possible for the following reasons:
1. If $a \neq 0$, then $a \div 0$ is not defined since there is no integer c such that $a = 0 \times c$.
2. $0 \div 0$ is not defined since there is no *unique* solution since $0 = 0 \times d$ for *every* integer d.
We will conclude this section with the following theorems.

THEOREM 7.4   If a, b, and c are integers, then $a(b - c) = ab - ac$.
Proof   Since b, $c \in I$, $b - c = d$, where $d \in I$. Hence $b = c + d$ and $a(b - c) = ad$. Now since $b = c + d$, we have

$ab = a(c + d)$	Theorem 7.2(c)
$a(c + d) = ac + ad$	Distributive property for $\times$ over $+$ on I
$ab = ac + ad$	Transitive property for equality
Hence $ab - ac = ad$	Definition 7.7 since $ad \in I$ (closure property for multiplication on I)
$ad = ab - ac$	Symmetric property for equality

We now have that $a(b - c) = ad$ and $ad = ab - ac$. Therefore, by the transitive property for equality, we conclude that $a(b - c) = ab - ac$.

The above theorem is known as the *distributive property for multiplication over subtraction on the set of integers*.

EXAMPLE   Verify Theorem 7.4 with $a = 2$, $b = {}^-8$, and $c = 5$.
Solution

$$a(b - c) = 2({}^-8 - 5) = 2({}^-8 + {}^-5) = 2({}^-13) = {}^-26$$
$$ab - ac = (2)({}^-8) - (2)(5) = {}^-16 - 10 = {}^-16 + {}^-10 = {}^-26$$

Hence

$$a(b - c) = ab - ac \quad \blacksquare$$

EXAMPLE Verify Theorem 7.4 with $a = 0$, $b = 7$, and $c = {}^-2$.
Solution

$$a(b - c) = 0(7 - {}^-2) = 0(7 + 2) = (0)(9) = 0$$
$$ab - ac = (0)(7) - (0)({}^-2) = 0 - 0 = 0$$

Hence

$$a(b - c) = ab - ac \quad \blacksquare$$

THEOREM 7.5 If a, b, and c are any integers, then $(a + b) - c = a + (b - c)$.
Proof Let $b - c = k$, where $k \in I$. Then

$b = c + k$	Definition 7.7
$a + b = a + (c + k)$	Theorem 7.2(a)
$a + (c + k) = (a + c) + k$	Associative property for addition on I
$a + b = (a + c) + k$	Transitive property for equality
$(a + c) + k = (c + a) + k$	Commutative property for addition on I
$a + b = (c + a) + k$	Transitive property for equality
$(c + a) + k = c + (a + k)$	Associative property for addition on I
$a + b = c + (a + k)$	Transitive property for equality
$(a + b) - c = a + k$	Definition 7.7 since $a + k \in I$ (closure property for addition on I)

Also, since $b - c = k$, we have

$a + (b - c) = a + k$	Theorem 7.2(a)
$a + k = a + (b - c)$	Symmetric property for equality

We now have $(a + b) - c = a + k$ and $a + k = a + (b - c)$. Therefore, by the transitive property of equality, we conclude

$$(a + b) - c = a + (b - c)$$

The above theorem is *not* to be compared with the statement $(a - b) - c \stackrel{?}{=} a - (b - c)$ because, in general, $(a - b) - c \neq a - (b - c)$.

EXAMPLE Using $a = {}^-2$, $b = 6$, and $c = {}^-3$, verify that $(a - b) - c \neq a - (b - c)$.
Solution

$$(a - b) - c = ({}^-2 - 6) - {}^-3 = ({}^-2 + {}^-6) + 3 = {}^-8 + 3 = {}^-5$$
$$a - (b - c) = {}^-2 - (6 - {}^-3) = {}^-2 - (6 + 3) = {}^-2 - 9 = {}^-2 + {}^-9 = {}^-11$$

Since ${}^-5 \neq {}^-11$, $(a - b) - c \neq a - (b - c)$. $\blacksquare$

EXAMPLE Verify Theorem 7.5 with $a = {}^-3$, $b = 9$, and $c = {}^-4$.
Solution

$$(a + b) - c = ({}^-3 + 9) - {}^-4 = 6 - {}^-4 = 6 + 4 = 10$$
$$a + (b - c) = {}^-3 + (9 - {}^-4) = {}^-3 + (9 + 4) = {}^-3 + 13 = 10$$

Hence

$$(a + b) - c = a + (b - c) \quad \blacksquare$$

EXERCISES

1. Using Definition 7.7, express each of the following differences as sums of integers (a and b represent integers).
   - a. $5 - 2$.
   - b. ${}^-9 - 5$.
   - c. ${}^-8 - {}^-5$.
   - d. $7 - {}^-9$.
   - e. $13 - 17$.
   - f. $16 - 11$.
   - g. $b - a$.
   - h. ${}^-a - b$.
   - i. $a - {}^-b$.
   - j. ${}^-a - {}^-b$.

2. Perform the indicated operations and simplify your results.
   - a. ${}^-5 - 2$.
   - b. $4 - {}^-7$.
   - c. $(4 + {}^-3) - 2$.
   - d. $5 - ({}^-6 - 2)$.
   - e. $({}^-5 + {}^-2) - (2 + {}^-3)$.
   - f. ${}^-(2 + {}^-3) - (4 - 5)$.
   - g. $2(3 - 4) + {}^-6$.
   - h. ${}^-3(2 + {}^-6) - 8$.
   - i. $({}^-2)({}^-3)(5 - 7) + 12$.
   - j. $(8 - 2) - (4 - 7) - ({}^-5 - 2)$.

3. Perform the indicated operations and simplify your results.
   - a. $16 \div 4$.
   - b. ${}^-20 \div 5$.
   - c. ${}^-49 \div {}^-7$.
   - d. $0 \div 23$.
   - e. $(13 + {}^-3) \div {}^-5$.
   - f. ${}^-36 \div (54 \div {}^-9)$.
   - g. $(17 - 3) \div ({}^-9 + 2)$.
   - h. $(50 \div 5) \div ({}^-48 \div 24)$.
   - i. ${}^-2(3 - 2) \div ({}^-4 \div 4)$
   - j. ${}^-10 \times 3({}^-4 \times 2) - 50$.

4. Verify Theorem 7.4 with a, b, and c as indicated.
   - a. $a = {}^-2$, $b = 3$, $c = {}^-4$.
   - b. $a = 3$, $b = {}^-5$, $c = {}^-2$.
   - c. $a = 1$, $b = {}^-3$, $c = 4$.
   - d. $a = 0$, $b = 1$, $c = {}^-1$.
   - e. $a = {}^-4$, $b = 0$, $c = 9$.

5. Verify Theorem 7.5 with a, b, and c as indicated.
   - a. $a = 5$, $b = 4$, $c = 3$.
   - b. $a = 4$, $b = {}^-3$, $c = {}^-2$.
   - c. $a = 0$, $b = 2$, $c = {}^-1$.
   - d. $a = {}^-3$, $b = 5$, $c = {}^-7$.
   - e. $a = 9$, $b = {}^-6$, $c = 5$.

6. Verify that $(a - b) - c \neq a - (b - c)$ with a, b, and c as indicated.
   - a. $a = 6$, $b = 4$, $c = 2$.
   - b. $a = {}^-8$, $b = {}^-3$, $c = {}^-2$.

7. For each of the following, if your answer to the question is yes, prove the statement. If your answer is no, give an example to support your answer.
   a. Is subtraction of integers a commutative operation?
   b. Is subtraction of integers an associative operation?
   c. Is division of integers a commutative operation?
   d. Is division of integers an associative operation?
   e. Is the set of all negative integers closed under the operation of subtraction?
   f. Is the set of all negative integers closed under the operation of division?
8. Solve each of the following where the domain for each of the variables is the set of integers.
   a. $x - 3 \leq 4$.   b. $2x + 1 > {}^-1$.
   c. $y|24$.   d. $u|0$.
   e. $2y + 3 = y + 2$.   f. $5u + 2 = 3u$.

7.6 INTEGERS AND THE NUMBER LINE

In Section 5.8 we defined the number line as a geometric representation of the set of whole numbers superimposed upon a line in a particular manner. We located two points, one representing the whole number 0 and the other representing the whole number 1 as illustrated in Figure 7.4. A scale was thus established, and the rest of the whole numbers could be represented, using that scale and the order property of whole numbers as illustrated in Figure 7.5.

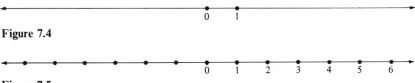

Figure 7.4

Figure 7.5

In this section we will display the integers on the number line as an extension of what was done in Section 5.8. Since we defined the integer ${}^-1$ to be the additive inverse of the integer 1 with the property that ${}^-1 + 1 = 0$, it is not surprising, then, that we would locate the integer ${}^-1$ as far to the left of 0 as 1 is to the right of 0. In a similar manner, the integer ${}^-2$ would be located as far to the left of 0 as the integer 2 is to the right of 0. Continuing in this manner, we can locate all of the integers on the number line as illustrated in Figure 7.6.

Figure 7.6

To represent integers by "arrows" as we did in Section 5.8, the positive integers will be represented in the identical manner as their corresponding natural numbers by arrows to the right. Then negative integers will be illustrated by arrows to the left. In Figure 7.7 we represent the integers 4 and ⁻5.

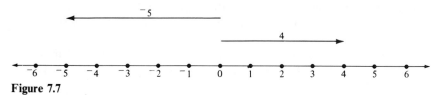

Figure 7.7

Using the number line, we can reaffirm our definition for the addition of two integers. For instance, ⁻3 + ⁻2 = ⁻5 is illustrated as follows. In Figure 7.8 start at the 0-point and move 3 units to the left to represent ⁻3. Then, starting at the terminal point of the ⁻3-vector, construct the ⁻2-vector by moving another 2 units to the left. To determine the sum of these two integers, start over again at the 0-point and proceed to the terminal point of the ⁻2-vector and determine that we moved 5 units to the left, which would illustrate that ⁻3 + ⁻2 = ⁻5.

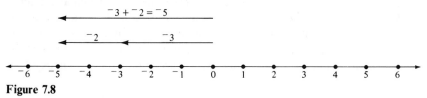

Figure 7.8

In a similar manner we can represent the statements ⁻8 + 5 = ⁻3 and 8 + ⁻5 = 3 as given in Figures 7.9 and 7.10, respectively.

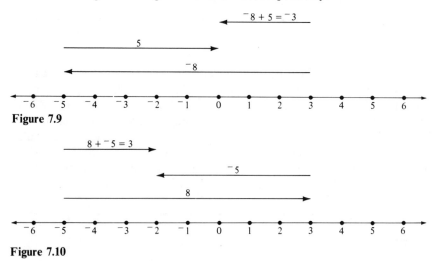

Figure 7.9

Figure 7.10

To represent subtraction of integers using the number line, we recall that if a and b are arbitrary integers, then $a - b$ is another name for $a + {}^-b$. For instance, $8 - 5 = 8 + {}^-5$, and the result, 3, is illustrated in Figure 7.10.

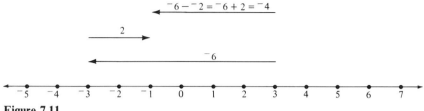

Figure 7.11

Figure 7.11 illustrates the subtraction $^-6 - {}^-2 = {}^-4$. Using the number line, we can also represent the multiplication of two integers provided at least one of the integers is positive. For instance, in Figure 7.12 we illustrate the product $(4)(^-3) = {}^-12$.

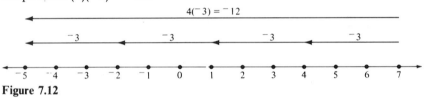

Figure 7.12

The commutative and associative properties of addition and multiplication can be illustrated on the number line as was done in Section 5.8.

EXERCISES

1. Using the number line, evaluate each of the following.
   a. $^-3 + 2$.
   b. $^-4 + {}^-5$.
   c. $9 + {}^-4$.
   d. $(^-3 + 4) + {}^-2$.
   e. $5 - (3 + {}^-4)$.
   f. $^-7 - (5 - 2)$.
   g. $3 \times {}^-2$.
   h. $^-4 \times 2$.
   i. $12 \div 3$.
   j. $^-20 \div 5$.
2. Using the number line, verify each of the following.
   a. $^-3 + 2 = 2 + {}^-3$.
   b. $6 + {}^-4 = {}^-4 + 6$.
   c. $^-3 + (2 + 4) = (^-3 + 2) + 4$.
   d. $^-5 + (^-2 + 3) = (^-2 + {}^-5) + 3$.
   e. $2(^-3 + 4) = (2)(^-3) + (2)(4)$.
   f. $^-3(^-2 + 5) = {}^-9$.
   g. $4(^-1 + 3) = 3 + 5$.
   h. $^-16 \div 4 = {}^-5 + 1$.
   i. $(7 - 4) - 2 \neq 7 - (4 - 2)$.
   j. $^-5 - (^-4 - 6) \neq (^-5 - {}^-4) - 6$.

7.7 A LOOK AT THE SYSTEM OF INTEGERS

Throughout this chapter we have developed the system of integers. We started with the set of integers denoted by I, defined and discussed the operations of addition, multiplication, subtraction, and division of integers with their properties, and introduced the relation of order on the set of integers. In this section we will summarize our results.

If a, b, and c are any integers, then consider the following.

OPERATIONS

ADDITION If $a, b \in W$, then
1. $a + b = n(A \cup B)$, where $n(A) = a$, $n(B) = b$, and $A \cap B = \varnothing$.
2. $^-a + {}^-b = {}^-(a + b)$.
3. $a + {}^-b = a - b$, if $a > b$.
4. $a + {}^-b = {}^-(b - a)$, if $a < b$.
5. $a + {}^-b = 0$, if $a = b$.
6. $a + 0 = a$ for all $a \in I$.
7. $^-a + 0 = a$ for all $a \in I$.
MULTIPLICATION If $a, b \in W$, then
1. $ab = n(A \times B)$, where $n(A) = a$ and $n(B) = b$.
2. $^-a \cdot b = {}^-(ab)$.
3. $^-a \cdot {}^-b = ab$.
4. $a \cdot 1 = 1 \cdot a = a$ for all $a \in I$.
5. $a \cdot 0 = 0 \cdot a = 0$ for all $a \in I$.
SUBTRACTION $a - b = p$ iff $p \in I$ and $a = b + p$.
DIVISION $a \div b = q$ iff $b \neq 0$, $q \in I$ and $a = bq$.

CLOSURE PROPERTIES

1. The set I is closed under addition; that is, if $a, b \in I$, then $a + b \in I$.
2. The set I is closed under multiplication; that is, if $a, b \in I$, then $ab \in I$.
3. The set I is closed under subtraction; that is, if $a, b \in I$, then $a - b \in I$.
4. The set I is *not* closed under the operation of division; that is, if $a, b \in I$, then $a \div b \in I$ iff $b \neq 0$, $q \in I$, and $a = bq$ (e.g., $3 \div {}^-4 \notin I$).

COMMUTATIVE PROPERTIES

FOR ADDITION If $a, b \in I$, then $a + b = b + a$
FOR MULTIPLICATION If $a, b \in I$, then $ab = ba$

ASSOCIATIVE PROPERTIES

FOR ADDITION If $a, b, c \in I$, then $(a + b) + c = a + (b + c)$
FOR MULTIPLICATION If $a, b, c \in I$, then $(ab)c = a(bc)$

DISTRIBUTIVE PROPERTIES

MULTIPLICATION OVER ADDITION If $a, b, c \in I$, then
1. $a(b + c) = ab + ac$.
2. $(a + b)c = ac + bc$.
MULTIPLICATION OVER SUBTRACTION If $a, b, c \in I$, then $a(b - c) = ab - ac$

IDENTITY PROPERTIES

ADDITIVE IDENTITY   If a is an arbitrary integer, then $a + 0 = 0 + a = a$; 0 is the additive identity on the set I

MULTIPLICATIVE IDENTITY   If a is an arbitrary integer, then $a \cdot 1 = 1 \cdot a = a$; 1 is the multiplicative identity on the set I

ADDITIVE INVERSE PROPERTY

If a is an integer, then there exists $^-a \in I$ such that $a + {}^-a = {}^-a + a = 0$

ORDER PROPERTIES

1. If a, $b \in I$, then $a < b$ iff there exists a *positive* integer p such that $b = a + p$.

2. If a, $b \in I$, then $a > b$ iff $b < a$.

3. TRICHOTOMY PROPERTY   If a, $b \in I$, then *exactly* one of the following must hold:

$$\text{Either } a < b \quad \text{or} \quad a = b \quad \text{or} \quad a > b$$

PROPERTIES OF ZERO

1. $0 \cdot a = 0$ for *all* $a \in I$.

2. $0 \div a = 0$, if $a \neq 0$.

3. $0 \div 0$ is not defined since it is *indeterminate* (i.e., $0 \div 0$ does not have a *unique* solution).

4. $a \div 0$, if $a \neq 0$, is not defined since it is *impossible* to divide a nonzero integer by 0.

SUMMARY

In this chapter we developed the system of integers. We defined the set of integers by starting with the natural numbers and, for each natural number, introduced its negative. The set of integers, I, was defined as the union of the set of all these negative numbers with the set of whole numbers. The operations of addition and multiplication, together with their various properties, were defined and discussed. We also defined and discussed the operation of subtraction on the set of integers. With suitable restrictions the operation of division also was defined. The order relation also was introduced and discussed as was the multiplication property of zero. The number line was discussed, and various properties of the integers were illustrated.

During the discussions the following symbols were introduced.

SYMBOL	INTERPRETATION
I	The set of integers
n	A natural number
^-n	The negative of n with the property that $n + \ ^-n = \ ^-n + n = 0$

REVIEW EXERCISES FOR CHAPTER 7

1. Perform the indicated operations, and simplify your results.
   a. $^-3 + 4$.
   b. $^-5 + \ ^-2$.
   c. $7 - 9$.
   d. $^-3 - \ ^-5$.
   e. $(^-5 + 6) - 2$.
   f. $^-5 - (^-7 - 3)$.
   g. $(^-3)(4)$.
   h. $(^-5)(^-4)$.
   i. $(^-2)(^-3)(4)$.
   j. $^-3(^-2 + 4)$.
   k. $^-12 \div 6$.
   l. $^-20 \div \ ^-5$.
   m. $(^-7 + \ ^-3) \div (^-5 + 3)$.
   n. $^-7 - (^-14 \div 2)$.
   o. $^-2(^-18 \div 9) + (0)(^-4)$.
   p. $[^-a(^-b + a)]b$, if $a = 2$, $b = 3$.
2. For each of the following indicate what property of integers is being used.
   a. $2(^-3 + 4) = (2)(^-3) + (2)(4)$.
   b. $^-3 + (4 + \ ^-5) = (^-3 + 4) + \ ^-5$.
   c. $(^-3)(0) = 0$.
   d. $^-2 + 2 = 0$.
   e. $(^-3)(4) = (4)(^-3)$.
   f. $^-4 + 0 = \ ^-4$.
   g. $5 + \ ^-7 = \ ^-7 + 5$.
   h. $[(2)(^-3)](^-4) = (2)[(^-3)(^-4)]$.
   i. $(^-7 + 2)(5) = (^-7)(5) + (2)(5)$.
   j. $^-5 + \ ^-6 = \ ^-(5 + 6)$.
3. In each of the following blanks write the symbol $<$, $>$, or $=$ so that the resulting statement is true.
   a. $^-3 \underline{\quad} \ ^-7$.
   b. Since $^-4 = \ ^-5 + 1$, $^-4 \underline{\quad} \ ^-5$.
   c. $12 \div \ ^-3 \underline{\quad} \ ^-(5 - 1)$.
   d. $5 - 3 \underline{\quad} 3 - 5$.
   e. $(^-2)(3) \underline{\quad} (^-3)(2)$.
   f. Since $^-1 \underline{\quad} \ ^-3$, $^-3 \underline{\quad} \ ^-1$.
   g. $(4 - \ ^-3) \underline{\quad} (7 - 4)$.
   h. $0 \underline{\quad} \ ^-4$.
4. Using the number line, verify each of the following.
   a. $2 + 3 = 5$.
   b. $^-4 + \ ^-3 = \ ^-7$.
   c. $^-3 + 7 = 4$.
   d. $^-7 + 4 = \ ^-3$.
   e. $3 < 5$.
   f. $4 > 2$.
   g. $^-2 > \ ^-4$
   h. $^-2 < 1$.
   i. $^-2 + 3 = 3 + \ ^-2$.
   j. $^-3 + \ ^-4 = \ ^-4 + \ ^-3$.
   k. $(2)(3) = (3)(2)$.
   l. $(^-2)(3) = \ ^-[(2)(3)]$.
   m. $2(^-4 + \ ^-1) = (2)(^-4) + (2)(^-1)$.
   n. $(^-3 + 5)(^-3) = \ ^-4 + \ ^-2$.
5. Prove each of the following by proceeding from the left to the right one step at a time. Give a reason for each step of the proof.
   a. $2(^-3 + \ ^-4) = (^-4)(2) + (^-3)(2)$.
   b. $(^-3 + 4) + \ ^-2 = (^-2 + \ ^-3) + 4$.

c. $(^-2 + 4) \times (^-5 \times 2) = (^-2)(^-5) + (2)(4) + (^-5)(4) + (^-2)(2)$.
d. $[^-(3 + {}^-5) + 3] + {}^-3 = 5 + {}^-3$.
e. $(3 + {}^-2) + (2 + {}^-3) = 0$.

6. Solve each of the following if the domain of each of the variables is the set of integers.

a. $x + 3 = {}^-2$.

b. $y - 7 = 4$.

c. $2(x + 4) = 0$.

d. $3(y - 1) = 6$.

e. $m \div 3 = {}^-4$.

f. $2p \div 5 = {}^-4$.

g. $x + 3 > {}^-2$.

h. $11 - {}^-y < {}^-5$.

i. $2x - 3 < x - 5$.

j. $2x > x$.

7. a. What is the smallest integer?

b. What is the largest nonpositive integer?

c. How many integers exist which are greater than $^-2$ and also less than or equal to 4?

d. What property does the system of integers have that the system of whole numbers does not have?

e. What is the multiplicative identity for the set of integers?

f. What is the trichotomy property for integers?

g. What is $0 \div 2$?

h. What is $2 \div 0$?

i. What is $0 \div 0$?

j. Is the set of integers closed under the operation of division? Why or why not?

CHAPTER 8
AN INTRODUCTION TO NUMBER THEORY

The theory of numbers involves the study of the properties of the integers. The topics involved in the study of number theory have interested prominent mathematicians for many centuries, dating back to the Greeks and even the early Chinese. For instance, during the period about 300 B.C. Euclid proved that there are infinitely many prime numbers and, at least 200 years earlier than that the Chinese knew that if p is prime, then p divides $2^p - 2$.

This chapter will introduce and discuss some of these properties of integers as an extension of what was done in the previous chapter.

8.1 DIVISIBILITY

In the previous chapter we defined division on the set of integers. If a and b are any integers, with $b \neq 0$, we said that a divided by b, denoted by $a \div b$ is the integer c if and only if $a = bc$. If such an integer c exists, we say that a can be divided by b or that b divides a.

DEFINITION 8.1 If a and b are any integers and $a \neq 0$, then a is said to divide b denoted $a|b$ if and only if there exists an integer, c, such that $b = ac$.

DEFINITION 8.2 If a and b are any integers and $a \neq 0$, then if $a|b$ the integer a is called a *divisor* or *factor* of b and the integer b is called a *multiple* of a.

Observe that the expression "*a* divided by *b*" is not the same as the expression "*a* divides *b*." The symbol $a|b$ (i.e., *a* divides *b*) refers to a *relation* between two integers, whereas the symbol $a \div b$ and a/b (i.e., *a* divided by *b*) if meaningful on the set of integers refers to an *integer*.

EXAMPLE $2|8$ since there exists an integer, 4, such that $8 = 2 \times 4$. The integer 2 is a divisor or a factor of 8, whereas 8 is a multiple of 2.

EXAMPLE $4|^-24$ since there exists an integer, $^-6$, such that $^-24 = 4 \times {}^-6$. The integer 4 is a divisor or a factor of $^-24$, whereas $^-24$ is a multiple of 4.

EXAMPLE $^-5|^-20$ since there exists an integer, 4, such that $^-20 = {}^-5 \times 4$. The integer $^-5$ is a divisor or a factor of $^-20$, whereas $^-20$ is a multiple of $^-5$.

If *a* and *b* are any integers with $a \neq 0$ and if *a* does not divide *b*, we symbolize this $a \nmid b$.

EXAMPLE $3 \nmid 7$ since there does not exist an integer, *m*, such that $7 = 3m$.

THEOREM 8.1 If *a* is an arbitrary integer, then
a. $1|a$ for *all* $a \in I$.
b. $a|a$ for *all* $a \in I$ such that $a \neq 0$.
c. $a|0$ for *all* $a \in I$ such that $a \neq 0$.
Proof The proof follows immediately from Definition 8.1 and is left as an exercise.

EXAMPLE
a. $3|9$ since there exists an integer, 3, such that $9 = 3 \times 3$.
b. $9|36$ since there exists an integer, 4, such that $36 = 9 \times 4$.
c. $3|36$ since there exists an integer, 12, such that $36 = 3 \times 12$.

Observe that in the above example we have $3|9$, $9|36$, and $3|36$. In general we have that if $a|b$ and $b|c$, then $a|c$.

THEOREM 8.2 If *a* and *b* are nonzero integers and *c* is any integer and if $a|b$ and $b|c$, then $a|c$.
Proof We are given that $a, b, c \in I$ with $a \neq 0$, $b \neq 0$, $a|b$, and $b|c$. Hence, by Definition 8.1, there exists integers *m* and *n* such that $b = am$ and $c = bn$. Hence

$c = (am)n$	Substituting *am* for *b* in $c = bn$
$(am)n = a(mn)$	Associative property for multiplication on *I*
$c = a(mn)$	Transitive property for equality
But $mn \in I$	Closure property for *I* under multiplication
$\therefore a|c$	Definition 8.1

EXAMPLE
a. $5|15$ since there exists an integer, 3, such that $15 = 5 \times 3$.
b. $5|20$ since there exists an integer, 4, such that $20 = 5 \times 4$.
c. $5|35$ since there exists an integer, 7, such that $35 = 5 \times 7$.

Observe that in the above example we have $5|15$, $5|20$, and $5|(15 + 20)$. In general we have that if $a|b$ and $a|c$, then $a|(b + c)$.

THEOREM 8.3 If $a, b, c \in I$, $a \neq 0$, $a|b$, and $a|c$, then $a|(b + c)$.
Proof We are given that $a, b, c \in I$ with $a \neq 0$, $a|b$, and $a|c$. Hence, by Definition 8.1, there exists $r, s \in I$ such that $b = ar$ and $c = as$. Hence

$b + c = ar + as$	Theorem 7.2(e)
$ar + as = a(r + s)$	Distributive property of $\times$ over $+$ on I
$b + c = a(r + s)$	Transitive property for equality
But $r + s \in I$	Closure property for addition on I
$\therefore a|(b + c)$	Definition 8.1

EXAMPLE
a. $4|12$ since there exists an integer, 3, such that $12 = 4 \times 3$.
b. $4|8$ since there exists an integer, 2, such that $8 = 4 \times 2$.
c. $4|4$ since there exists an integer, 1, such that $4 = 4 \times 1$.

Observe that in the above example, we have $4|12$, $4|8$, and $4|(12 - 8)$. In general we have that if $a|b$ and $a|c$, then $a|(b - c)$.

THEOREM 8.4 If $a, b, c \in I$, $a \neq 0$, $a|b$ and $a|c$, then $a|(b - c)$.
Proof The proof is similar to that for Theorem 8.3 and is left as an exercise.

EXAMPLE
a. $7|14$ since there exists an integer, 2, such that $14 = 7 \times 2$.
b. $7|35$ since there exists an integer, 5, such that $35 = 7 \times 5$.
c. $7|21$ since there exists an integer, 3, such that $21 = 7 \times 3$.

Observe that in the above example we have $7|14$, $7|(14 + 21)$, and $7|21$. In general we have that if $a|b$ and $a|(b + c)$, then $a|c$.

THEOREM 8.5 If $a, b, c \in I$, $a \neq 0$, $a|b$ and $a|(b + c)$, then $a|c$.
Proof The proof is left as an exercise.

EXAMPLE
a. $6|12$ since there exists an integer, 2, such that $12 = 6 \times 2$.
b. $6 \nmid 11$ since there does not exist an integer, q, such that $11 = 6q$.
c. $6|132$ since there exists an integer. 22, such that $132 = 6 \times 22$.

EXAMPLE
a. $8 \nmid 10$ since there does not exist an integer, c, such that $10 = 8c$.
b. $8 \mid 16$ since there does exist an integer, 2, such that $16 = 8 \times 2$.
c. $8 \mid 160$ since there does exist an integer, 20, such that $160 = 8 \times 20$.

EXAMPLE
a. $3 \mid 12$ since there does exist an integer, 4, such that $12 = 3 \times 4$.
b. $3 \mid 15$ since there does exist an integer, 5, such that $15 = 3 \times 5$.
c. $3 \mid 180$ since there does exist an integer, 60, such that $180 = 3 \times 60$.

Observe that in the first example following Theorem 8.5, $6 \mid 12$, $6 \nmid 11$, but $6 \mid (12 \times 11)$. In the next example $8 \nmid 10$, $8 \mid 16$, but $8 \mid (10 \times 16)$. In the last example $3 \mid 12$, $3 \mid 15$, and $3 \mid (12 \times 15)$. In general we have that if $a \mid b$ or $a \mid c$, then $a \mid bc$.

THEOREM 8.6   If a, b, $c \in I$, $a \neq 0$, $a \mid b$ or $a \mid c$, then $a \mid bc$.
Proof   The proof is left as an exercise.

DEFINITION 8.3 If a is an integer, then a is said to be an *even integer* if and only if $2 \mid a$. Every even integer can be expressed as $2n$ where n is an integer.

DEFINITION 8.4 If a is an integer, then a is said to be an *odd integer* if and only if a is not an even integer. Every odd integer can be expressed as $2n + 1$, where n is an integer.

EXAMPLE
a. $4 = 2 \times 2$; hence 4 is an even integer.
b. $^-6 = 2 \times {}^-3$; hence $^-6$ is an even integer.
c. $3 = (2 \times 1) + 1$; hence 3 is an odd integer.
d. $-11 = (2 \times {}^-6) + 1$; hence $^-11$ is an odd integer.
e. $0 = 2 \times 0$; hence 2 is an even integer.
f. $1 = (2 \times 0) + 1$; hence 1 is an odd integer.

EXERCISES

1. Explain the difference between the statements "a divides b" and "a is divided by b."
2. If p divides q, then p is called a _____ or _____ of q and q is called a multiple of _____.
3. a. List all of the positive divisors of 8.
   b. List 3 positive multiples of 8.
   c. List 3 negative multiples of 8.

4. a. List all the negative divisors of 18.
   b. List all the positive factors of 18.
   c. List 3 multiples of 18.
   d. What is the difference between the smallest positive divisor of 18 and the largest negative divisor of 18?
   e. What is the difference between the smallest positive multiple of 18 and its largest negative multiple?

5. a. List 4 distinct positive divisors of 0.
   b. List 4 distinct negative divisors of 0.
   c. Is 0 a divisor of 0? Why or why not?

6. a. Determine a positive integer that is a divisor of 18 and 225.
   b. Determine a positive integer that is a divisor of 144 and also a multiple of 9.

7. For each of the following verify that if $a|b$ and $b|c$, then $a|c$.
   a. $a = 2$, $b = 4$, $c = 24$.
   b. $a = 3$, $b = 9$, $c = 72$.
   c. $a = 5$, $b = 25$, $c = {}^-225$.
   d. $a = {}^-2$, $b = {}^-14$, $c = 98$.
   e. $a = {}^-4$, $b = 20$, $c = {}^-600$.

8. For each of the following verify that if $a|b$ and $a|c$, then $a|(b + c)$.
   a. $a = 2$, $b = 6$, $c = 10$.
   b. $a = {}^-3$, $b = 27$, $c = {}^-9$.
   c. $a = 4$, $b = {}^-64$, $c = {}^-24$.
   d. $a = 5$, $b = {}^-100$, $c = 25$.
   e. $a = {}^-7$, $b = 49$, $c = {}^-147$.

9. For each of the following verify that if $a|b$ and $a|c$, then $a|(b - c)$.
   a. $a = 3$, $b = 12$, $c = 15$.
   b. $a = {}^-2$, $b = 24$, $c = {}^-8$.
   c. $a = 6$, $b = {}^-12$, $c = 102$.
   d. $a = {}^-8$, $b = {}^-24$, $c = {}^-32$.
   e. $a = 4$, $b = 0$, $c = {}^-48$.

10. For each of the following verify that if $a|b$ or $a|c$, then $a|bc$.
    a. $a = 3$, $b = 9$, $c = {}^-6$.
    b. $a = {}^-2$, $b = 8$, $c = {}^-7$.
    c. $a = 4$, $b = 3$, $c = {}^-4$.
    d. $a = {}^-5$, $b = {}^-4$, $c = 20$.
    e. $a = 6$, $b = 6$, $c = {}^-12$.

11. Classify each of the following indicated integers as being odd or even.
    a. ${}^-3 + 2$.
    b. $2({}^-3) + 1$.
    c. $(2)(4) + (2)({}^-5)$.
    d. $(3)(2) + (2)({}^-4) + 1$.
    e. $23 + {}^-47 + {}^-36$.
    f. ${}^-3(2 + {}^-7) - 3$.

12. Prove that the sum of two even integers is always even.

13. Prove that the sum of two odd integers is always even.

14. Prove that the product of two even integers is always even.

15. Prove that the product of two odd integers is always odd.
16. Prove Theorem 8.1.
17. Prove Theorem 8.4.
18. Prove Theorem 8.5.
19. Prove Theorem 8.6.

8.2 TESTS FOR DIVISIBILITY

Using the theorems of the previous section, we can now consider tests of divisibility of integers by certain positive integers. From Theorem 8.1 we know that 1 is a positive integer and 1 divides every integer. Hence every integer is divisible by 1.

Now let's look at divisibility by 2. Does $2|428$? Recall that $428 = (4 \times 100) + (2 \times 10) + 8$. Since $2|100$ and $2|10$, we can use Theorem 8.6 and conclude that $2|(4 \times 100)$ and $2|(2 \times 10)$. Then, by Theorem 8.3, since $2|(4 \times 100)$ and $2|(2 \times 10)$, we conclude that $2|[(4 \times 100) + (2 \times 10)]$ or $2|420$. Finally, since $2|8$ and $2|420$, we conclude that $2|428$ by Theorem 8.3.

In general any positive integer N can be written in the form

$$N = a_n 10^n + a_{n-1} 10^{n-1} + a_{n-2} 10^{n-2} + \cdots + a_2 10^2 + a_1 10 + a_0$$

such that each of the a_is is a single digit and n is a positive integer. Now $2|10$, $2|10^2$, $2|10^3$, and in general $2|10^n$. Hence, by Theorem 8.6 we conclude that $2|(a_1 10)$, $2|(a_2 10^2)$, $2|(a_3 10^3)$, and in general $2|(a_n 10^n)$. Then, by Theorem 8.3 applied repeatedly, we conclude that $2|[a_n 10^n + a_{n-1} 10^{n-1} + \cdots + a_2 10^2 + a_1 10]$ and, therefore, $2|N$ if $2|a_0$. In other words $2|N$ if and only if N is even.

8.21 TEST FOR DIVISIBILITY BY 2

A number, N, is divisible by 2 if and only if the rightmost digit of N, as a base ten numeral, is divisible by 2.

EXAMPLE
a. $2|168$ since 168 is even.
b. $2|{}^-1064$ since ${}^-1064$ is even.
c. $2 \nmid 139$ since 139 is *not* even.
d. $2|0$ since 0 is even.

Now let's look at divisibility by 3. Does $3|768$? Again, we may write $768 = (7 \times 100) + (6 \times 10) + 8$ but $3 \nmid 10$ and $3 \nmid 100$ and, hence, we cannot proceed as we did for the previous test. However, $3|9$ and $3|99$. Since $9 = 10 - 1$ and $99 = 100 - 1$, we may rewrite $768 = [7 \times (100 - 1 + 1)]$

$+ [6 \times (10 - 1 + 1)] + 8$. Now, using the distributive property of multiplication over addition and the commutative and associative properties of addition, we may rewrite $768 = [7 \times (100 - 1)] + [6 \times (10 - 1)] + (7 + 6 + 8)$. Since $3|99$ and $3|9$, we conclude that $3|[7 \times (100 - 1)]$ and $3|[6 \times (10 - 1)]$ by Theorem 8.6. Then, by Theorem 8.3, $3|[7(100 - 1) + 6(10 - 1)]$. Finally, if $3|(7 + 6 + 8)$, then $3|768$. Since each power of 10 may be written as 10^n where n is a nonnegative integer and, since $10^n = 10^n - 1 + 1$ and $3|(10^n - 1)$, we may extend the procedure outlined in this example and conclude that a number is divisible by 3 if the sum of its digits, as a base ten numeral, is divisible by 3.

8.22 TEST FOR DIVISIBILITY BY 3

A number, N, is divisible by 3 if and only if the sum of its digits, as a base ten numeral, is divisible by 3.

EXAMPLE
a. $3|1221$ since $3|(1 + 2 + 2 + 1)$ or $3|6$.
b. $3|2970$ since $3|(2 + 9 + 7 + 0)$ or $3|18$.
c. $3 \nmid 1672$ since $3 \nmid (1 + 6 + 7 + 2)$ or $3 \nmid 16$.
d. $3 \nmid 469$ since $3 \nmid (4 + 6 + 9)$ or $3 \nmid 19$.

Now let's consider a test for divisibility by 4. Although $4 \nmid 10$, $4|100$, and $4|k \cdot 100$, where k is an integer. Now every integer N can be written as $N = 100 \cdot k + p$, where p is a two-digit numeral. Hence $4|(100\,k + p)$ if and only if $4|p$.

8.23 TEST FOR DIVISIBILITY BY 4

A number N is divisible by 4 if and only if the two rightmost digits of its base ten numeral represent a number that is divisible by 4.

EXAMPLE
a. $4|128$ since $4|28$.
b. $4|14,696$ since $4|96$.
c. $4 \nmid 23,035$ since $4 \nmid 35$.
d. $4|198,100$ since $4|00$ or $4|0$.
e. $4 \nmid 2,699,763$ since $4 \nmid 63$.

Now let's consider a test for divisibility by 5. Since every integer N can be written as $N = a_n 10^n + a_{n-1} 10^{n-1} + \cdots + a_2 10^2 + a_1 10 + a_0$ and since 10^n is divisible by 5 for every natural number n, $5|[a_n 10^n + a_{n-1} 10^{n-1} + \cdots + a_2 10^2 + a_1 10]$. Hence we can readily see that $5|N$ if and only if $5|a_0$. But $5|a_0$ if $a_0 = 0$ or 5.

8.24 TEST FOR DIVISIBILITY BY 5

A number, N, is divisible by 5 if and only if the rightmost digit of its base ten numeral is 0 or 5.

EXAMPLE
a. $5|1055$ since $5|5$.
b. $5|2,130$ since $5|0$.
c. $5\nmid119$ since $5\nmid9$.

Now, considering a test for divisibility by 6, we note that $6 = 3 \times 2$ and, hence, a number is divisible by 6 if and only if it is divisible by 2 and also divisible by 3.

8.25 TEST FOR DIVISIBILITY BY 6

A number, N, is divisible by 6 if and only if N is divisible by both 2 and 3. That is, N is divisible by 6 if and only if N is even and divisible by 3.

EXAMPLE
a. $6|234$ since 234 is even (since $234 = 2 \times 117$) and $3|234$ [since $3|(2 + 3 + 4)$ or $3|9$].
b. $6|62,556$ since 62,556 is even (since $62,556 = 2 \times 31,278$) and $3|62,556$ [since $3|(6 + 2 + 5 + 5 + 6)$ or $3|24$].
c. $6\nmid5,961$ since 5961 is not even [since $5,961 = (2 \times 2,980) + 1$].
d. $6\nmid482,144$ since $3\nmid482,144$ [since $3\nmid(4 + 8 + 2 + 1 + 4 + 4)$ or $3\nmid23$].

To determine a test for divisibility by 7 we could write the given number as the sum of a multiple of 7 and another number. For instance, to determine if $7|644$, we could write $644 = 630 + 14$. Now $7|630$ and $7|14$; hence $7|644$. However, there is an alternate test for divisibility by 7 which, although not easily motivated, we will now state.

8.26 TEST FOR DIVISIBILITY BY 7

A number, N, is divisible by 7 if and only if the following procedure is satisfied. Take the rightmost digit, double it, and subtract that quantity from the number represented by the remaining digits. If the difference found is divisible by 7, then $7|N$. The procedure may be repeated until the resulting difference is small enough to determine whether it is divisible by 7. (The number N is written as a base ten numeral.)

EXAMPLE   Does $7|301$?
Solution $30 - 2 = 28$ and $7|28$. Therefore $7|301$. ■

EXAMPLE   Does $7|5964$?
Solution
$596 - 8 = 588$ Does $7|588$?
$58 - 16 = 42$ Since $7|42$, then $7|5964$ ∎

EXAMPLE   Does $7|18,723$?
Solution
$1872 - 6 = 1866$ Does $7|1866$?
$186 - 12 = 174$ Does $7|174$?
$17 - 8 = 9$ Since $7\nmid9$, then $7\nmid18,723$ ∎

A test for divisibility by 8 is similar to the test given for divisibility by 4. Although $8\nmid10$ and $8\nmid100$, $8|1000$ and, hence, $8|k \cdot 1000$, where k is an integer. Now every integer N can be written as $N = 1000 \cdot k + q$, where q is a three-digit numeral. Hence $8|(1000\,k + q)$ if and only if $8|q$.

8.27 TEST FOR DIVISIBILITY BY 8

A number, N, is divisible by 8 if and only if the three rightmost digits of its base ten numeral represents a number that is divisible by 8.

EXAMPLE
a. $8|163,568$ since $8|568$.
b. $8\nmid2,030,070$ since $8\nmid70$.

The test for divisibility by 9 is very similar to the test for divisibility by 3 and is motivated in the same manner.

8.28 TEST FOR DIVISIBILITY BY 9

A number, N, is divisible by 9 if and only if the sum of its digits, as a base ten numeral, is divisible by 9.

EXAMPLE
a. $9|216$ since $9|(2 + 1 + 6)$ or $9|9$.
b. $9|494,874$ since $9|(4 + 9 + 4 + 8 + 7 + 4)$ or $9|36$.
c. $9\nmid364,703$ since $9\nmid(3 + 6 + 4 + 7 + 0 + 3)$ or $9\nmid23$.

A test for divisibility by 10 is obvious.

8.29 TEST FOR DIVISIBILITY BY 10

A number N is divisible by 10 if and only if the rightmost digit of its base ten numeral is 0.

To develop a test for divisibility by 11, we use the facts that $11|11$ or $11|(10 + 1)$, $11|99$ or $11|(100 - 1)$, $11|1001$ or $11|(1000 + 1)$, $11|9,999$ or $11|(10,000 - 1)$, $11|100,001$ or $11|(100,000 + 1)$, and so forth. Does 11 divide 1,254,869? Rewriting, we have

$$1,254,869 = (1 \times 10^6) + (2 \times 10^5) + (5 \times 10^4) + (4 \times 10^3)$$
$$+ (8 \times 10^2) + (6 \times 10) + 9$$

or

$$1,254,869 = 1(10^6 - 1 + 1) + 2(10^5 + 1 - 1) + 5(10^4 - 1 + 1)$$
$$+ 4(10^3 + 1 - 1) + 8(10^2 - 1 + 1) + 6(10 + 1 - 1) + 9$$

or

$$1,254,869 = 1(10^6 - 1) + 2(10^5 + 1) + 5(10^4 - 1) + 4(10^3 + 1)$$
$$+ 8(10^2 - 1) + 6(10 + 1) + (1 - 2 + 5 - 4 + 8 - 6 + 9)$$

Now $11|(10^6 - 1)$, $11|(10^5 + 1)$, ..., $11|(10 + 1)$. Hence $11|1,254,869$ if and only if $11|(1 - 2 + 5 - 4 + 8 - 6 + 9)$, which is true. Hence $11|1,254,869$.

8.210 TEST FOR DIVISIBILITY BY 11

A number, N, is divisible by 11 if and only if the sum of the digits in the odd-numbered columns of its base ten number (counting from right to left) minus the sum of the digits in the even-numbered columns is divisible by 11.

EXAMPLE
a. $11 \nmid 23,169$ since $11 \nmid [(9 + 1 + 2) - (6 + 3)]$ or $11 \nmid (12 - 9)$ or $11 \nmid 3$.
b. $11|4,917$ since $11|[(7 + 9) - (1 + 4)]$ or $11|(16 - 5)$ or $11|11$.

Tests for the other positive divisors also can be developed using the theorems of the previous section.

EXERCISES

1. Without actually dividing, test each of the following for divisibility by 2.

a. 21.	b. 46.	c. 69.
d. 128.	e. 237.	f. 462.
g. 701.	h. 800.	i. 916.
j. 1121.	k. 1326.	l. 4669.
m. 6372.	n. 9764.	o. 11,217.
p. 15,609	q. 17,876.	r. 20,692.
s. 86,769.	t. 123,468.	u. 376,871.
v. 999,862.	w. 1,605,987.	x. 716,890,239.

2. Without actually dividing, test each of the numbers indicated in Exercise 1 for divisibility by 3.
3. Repeat Exercise 1 for divisibility by 4.
4. Repeat Exercise 1 for divisibility by 5.
5. Repeat Exercise 1 for divisibility by 6.
6. Repeat Exercise 1 for divisibility by 7.
7. Repeat Exercise 1 for divisibility by 8.
8. Repeat Exercise 1 for divisibility by 9.
9. Repeat Exercise 1 for divisibility by 10.
10. Repeat Exercise 1 for divisibility by 11.
11. If a number is divisible by 3, is it also divisible by 9? Explain.
12. If a number is divisible by 6, is it also divisible by 3? Explain.
13. If a number is divisible by 4, must it also be divisible by 8? Explain.
14. Determine a test for divisibility by 15.
15. Determine the missing digit(s) in order for the divisibility to hold. If more than one answer is possible, list all possible answers. Where two blanks are included in a numeral, the same digit must be used.
   a. $2|12\_$.　　　　b. $3|\_45$.
   c. $4|3\_5$.　　　　d. $5|4\_40$.
   e. $6|5\_48$.　　　　f. $7|\_876$.
   g. $8|92\_4$.　　　　h. $9|8\_76\_23$,
   i. $10|231\_$.　　　j. $11|65\_234\_$.

8.3 PRIMES AND COMPOSITES

In Section 8.1 we discussed division of integers, and it may have been noted that some integers have many divisors while others do not. For instance, the integer 8 has for its divisors ⁻8, ⁻4, ⁻2, ⁻1, 1, 2, 4, and 8; the integer ⁻4 has for its divisors ⁻4, ⁻2, ⁻1, 1, 2, and 4; the integer 5 has for its divisors ⁻5, ⁻1, 1, and 5; and so forth. From the definition that a divides b if a and b are integers and $a \neq 0$, it follows that the integer b would have for its divisors ⁻b, ⁻1, 1, and b and possibly others unless, of course, $b = 1$ or ⁻1. If b is 1 or ⁻1, the only divisors are 1 and ⁻1. For most of our work in this section we will be concerned with positive divisors of an integer. If a positive integer has for its positive divisors only the integers 1 and itself, we say that the integer is prime.

DEFINITION 8.5 If x is a positive integer greater than 1, then x is said to be a *prime number* if and only if the only positive integer divisors of x are 1 and x.

DEFINITION 8.6 A positive integer is said to be *composite* if and only if it has positive integer divisors other than 1 and itself.

It should be noted from the two definitions above that the positive integer 1 is neither prime nor composite.

EXAMPLE
a. The positive integer 5 is a prime since the only positive divisors of 5 are 1 and 5.

b. The positive integer 7 is prime since the only positive divisors of 7 are 1 and 7.

c. The positive integer 79 is prime since the only positive divisors of 79 are 1 and 79 (Do you agree?)

d. The positive integer 36 is composite since the positive integers 2, 3, 4, 6, 9, 12, and 18 are divisors of 36 in addition to the positive divisors 1 and 36.

e. The positive integer 25 is composite since the positive integer divisors of 25 are 1, 5, and 25.

It is easy to determine the first few primes by applying Definition 8.5. There is a procedure, however, for finding all of the prime numbers less than or equal to any given positive integer. This procedure involves "the sieve of Eratosthenes" and is named after the Greek mathematician who formulated the procedure about 200 B.C. We will apply the procedure for finding all of the primes less than 100.

We will arrange the first 100 positive integers in ten rows with each row containing ten integers as indicated in Table 8.1. Then, since 1 is not a prime, we will cross it out. Since 2 is a prime, we will not cross it out but will cross out all multiples of 2 except 2 itself. Since 3 is a prime, we will not cross it out but will cross out all multiples of 3 except 3 itself. (Notice that 6 is a multiple of 3, but it was already crossed out since 6 is also a multiple of 2.) 4 has already been crossed out since it is a multiple of 2, and every multiple of 4 also has been crossed out for the same reason. Since 5 is a prime we will not cross it out but will cross out all multiples of 5 except 5 itself. We will continue in this manner, and in so doing we will cross out all of the composite integers from the table. The result will appear as shown in Table 8.1. Those numerals which have not been crossed out represent prime numbers less than or equal to 100.

Table 8.1

1	2	3	4	5	6	7	8	9	10
11	12	13	14	15	16	17	18	19	20
21	22	23	24	25	26	27	28	29	30
31	32	33	34	35	36	37	38	39	40
41	42	43	44	45	46	47	48	49	50
51	52	53	54	55	56	57	58	59	60
61	62	63	64	65	66	67	68	69	70
71	72	73	74	75	76	77	78	79	80
81	82	83	84	85	86	87	88	89	90
91	92	93	94	95	96	97	98	99	100

An examination of Table 8.1 will reveal that primes appear less and less frequently as we continue from 1 to 100. One may be tempted to conclude, then, that if the procedure outlined above were continued indefinitely, we may run out of primes. However, that conjecture would be false.

THEOREM 8.7 There are infinitely many primes.

Proof Assume that there are only a finite number of primes, say n, where n is a natural number. We could list these as $p_1, p_2, p_3, \ldots, p_n$. We next form an integer T by multiplying all these primes together; that is, $T = p_1 \cdot p_2 \cdot p_3 \cdots p_n$. Now consider $T + 1$. Either $T + 1$ is a prime or $T + 1$ is a composite number, since it is greater than 1. If $T + 1$ is a prime, then we have a contradiction to our original assumption that there are only n primes. Let us assume, then, that $T + 1$ is composite. If $T + 1$ is composite, then some p_i is a divisor of $T + 1$. Now since $p_i | T$ and $p_i | (T + 1)$. $p_i | 1$ by Theorem 8.5. But no prime number divides 1 and, hence, our original assumption is false and there are infinitely many prime numbers.

Since there are infinitely many primes, we also can conclude that there are infinitely many composite numbers since the product of any two primes is a composite. It follows, then, that each composite number can be expressed as the product of $^-1$ or 1 and a finite number of prime factors. We will state this more precisely but in a modified form in the following theorem.

THEOREM 8.8 The Fundamental Theorem of Arithmetic Every composite positive integer can be expressed as the product of a finite number of prime numbers in a unique way, except for the order in which the factors appear.

The proof of this theorem is not given since it is beyond the level of this course.

EXAMPLE Find a complete factorization of (a) 205, (b) 160, and (c) 1368.

Solution

a. $205 = 5 \times 41$

b. $160 = 2 \times 2 \times 2 \times 2 \times 2 \times 5$

c. $1368 = 2 \times 2 \times 2 \times 3 \times 3 \times 19$ ∎

Returning to Table 8.1 we observe that the first two primes are 2 and 3, which are consecutive positive integers. The primes 3 and 5 are consecutive primes but differ by two; likewise with the primes 5 and 7, 17 and 19, and 29 and 31.

DEFINITION 8.7 Pairs of prime numbers that differ by two are called *twin primes*. Three consecutive prime numbers with a common difference of two are called *prime triplets*.

EXAMPLE
a. 17 and 19 are twin primes.
b. 3, 5, and 7 are prime triplets.

Many interesting problems in number theory involve divisors of numbers and concern such things as perfect numbers and amicable numbers.

DEFINITION 8.8 A *proper divisor* of a number is a divisor of a number other than the number itself.

DEFINITION 8.9 A positive integer is said to be *perfect* if and only if it is equal to the sum of all its proper divisors.

EXAMPLE
a. 12 is not a perfect number since the proper divisors of 12 are 1, 2, 3, 4, and 6; $1 + 2 + 3 + 4 + 6 = 16 \neq 12$.
b. 28 is a perfect number since the proper divisors of 28 are 1, 2, 4, 7, and 14; $1 + 2 + 4 + 7 + 14 = 28$.

The answer to the question "How many perfect numbers are there?" is not known. No one knows. Further, all the perfect numbers that have been determined to date are even. That is, there are no known odd perfect numbers.

DEFINITION 8.10 Two positive integers are said to be *amicable* if and only if each is equal to the sum of the proper divisors of the other.

EXAMPLE
a. 18 and 27 are not amicable numbers. The proper divisors of 18 are 1, 2, 3, 6, and 9; $1 + 2 + 3 + 6 + 9 = 21 \neq 27$. The proper divisors of 27 are 1, 3, and 9; $1 + 3 + 9 = 13 \neq 18$.
b. 220 and 284 are amicable numbers. The proper divisors of 220 are 1, 2, 4, 5, 10, 11, 20, 22, 44, 55, and 110; $1 + 2 + 4 + 5 + 10 + 11 + 20 + 22 + 44 + 55 + 110 = 284$. The proper divisors of 284 are 1, 2, 4, 71, and 142; $1 + 2 + 4 + 71 + 142 = 220$.

Again, the answer to the question "How many amicable numbers are there?" is not known at this time.

EXERCISES

1. Classify each of the numbers indicated as being prime, composite, or neither.

   a. 16.   b. 23.   c. 35.
   d. 47.   e. 17.   f. 13.
   g. 51.   h. 1.   i. $2 + 3$.
   j. $3 + 5$.   k. $16 \div 2$.   l. 3×5.
   m. $(2 \times 8) + 1$.   n. $(15 \div 3) \div 5$.   o. 231.
   p. 401.   q. 987.   r. 1001.

2. List all of the even prime numbers.

3. List all of the odd prime numbers less than 30.

4. For each of the following list all the prime divisors.

   a. 24.   b. 35.   c. 54.
   d. 81.   e. 123.   f. 83.
   g. 59.   h. 987.   i. 231.
   j. 761.   k. 236.   l. 144.
   m. 862.   n. 101.   o. 319.

5. Using the sieve of Eratosthenes, determine all of the primes less than 200.

6. For each of the following determine a complete prime factorization.

   a. 68.   b. 124.   c. 210.
   d. 305.   e. 320.   f. 1462.
   g. 2360.   h. 4090.   i. 5633.
   j. 6194.   k. 7235.   l. 8649.

7. List all of the twin primes less than 100.

8. List all of the prime triplets less than 150.

9. Determine all of the proper divisors for the numbers indicated in Exercise 6.

10. Determine which of the following represent perfect numbers.

    a. 6.   b. 8.   c. 10.
    d. 18.   e. 24.   f. 28.
    g. 296.   h. 496.   i. 8128.

11. Show that 1184 and 1210 are amicable numbers.

12. Write each of the following as the sum of two odd primes.

    a. 68.   b. 192.   c. 106.
    d. 144.   e. 200.   f. 216.

8.4 GREATEST COMMON DIVISOR

Let a and b be positive integers. Let A be the set of all the divisors of a, and let B be the set of all the divisors of b. Clearly $A \cap B$ will contain the positive integer 1 since 1 divides every integer. It is possible, however, that $A \cap B$ contains integers other than 1. For instance, if $a = 12$, then $A = \{1, 2, 3, 4, 6, 12\}$, and if $b = 18$, then $B = \{1, 2, 3, 6, 9, 18\}$. Hence $A \cap B = \{1, 2, 3, 6\}$. All of those elements in $A \cap B$ are common divisors of 12 and 18, and 6 is the greatest of these common divisors.

DEFINITION 8.11 The *greatest common divisor* of two nonzero integers a and b is the largest positive integer d such that $d|a$ and $d|b$. The greatest common divisor of a and b is denoted by g.c.d.(a, b).

To determine the greatest common divisor of two or more integers, we could find the set of all common divisors of the integers and then determine the greatest element of the set. However, this procedure is very time consuming and involves finding all of the divisors of each integer. A more direct approach would be to utilize the fundamental theorem of arithmetic and write each integer as a product of prime factors. Next determine all of the prime factors that are common to the integers involved, and determine the smallest power of each prime factor that occurs in any of these prime factorizations. The g.c.d., then, will be the product of all these smallest powers of primes.

To determine g.c.d. (24, 36) we proceed as follows:

1. Write the prime factorization of 24 and 36: $24 = 2 \times 2 \times 2 \times 3$ and $36 = 2 \times 2 \times 3 \times 3$.

2. Write as a product each of the prime factors occurring in these factorizations: 2×3.

3. Raise each of the factors in 2 above to the smallest power that it occurs in either of the factorizations. Note that 2 occurs 3 times in the factorization of 24 and 2 times in the factorization of 36; hence we want 2^2. Likewise, 3 occurs once in the factorization of 24 and twice in the factorization of 36; hence we want 3^1. The g.c.d. $(24, 36) = 2^2 \cdot 3^1 = 12$.

EXAMPLE   Determine the g.c.d. of 88, 128, and 180.
Solution   Factoring, we have

$$88 = 2 \times 2 \times 2 \times 11$$

$$128 = 2 \times 2 \times 2 \times 2 \times 2 \times 2 \times 2$$

$$180 = 2 \times 2 \times 3 \times 3 \times 5$$

g.c.d.$(88, 128, 180) = 2^2 \cdot 3^0 \cdot 5^0 \cdot 11^0 = 4 \cdot 1 \cdot 1 \cdot 1 = 4.$ ∎

EXAMPLE Determine the g.c.d. of 60, 70, 80, and 90.
Solution Factoring, we have

$$60 = 2 \times 2 \times 3 \times 5$$

$$70 = 2 \times 5 \times 7$$

$$80 = 2 \times 2 \times 2 \times 2 \times 5$$

$$90 = 2 \times 3 \times 3 \times 5$$

g.c.d.(60, 70, 80, 90) $= 2^1 \cdot 3^0 \cdot 5^1 \cdot 7^0 = 2 \cdot 1 \cdot 5 \cdot 1 = 10$. ∎

EXAMPLE Determine the g.c.d. of 27 and 8.
Solution Factoring, we have
$$27 = 3 \times 3 \times 3$$

$$8 = 2 \times 2 \times 2$$

g.c.d.(27, 8) $= 2^0 \cdot 3^0 = 1 \times 1 = 1$. ∎

In the example above we note that the positive integer 1 is the greatest common divisor of 27 and 8. Neither 27 nor 8 is prime but each is said to be prime relative to the other.

DEFINITION 8.12 Two integers, a and b, are said to be *relatively prime* if and only if their greatest common divisor is 1.

So far we have concerned ourselves with determining the greatest common divisor of positive integers. However, we recall that if a is an integer, then a and ^-a have the same divisors. For instance, the divisors of 6 are $^-6$, $^-3$, $^-2$, $^-1$, 1, 2, 3, and 6, and these integers are also the divisors of $^-6$. Therefore the g.c.d.($^-12$, 27) would be equal to the g.c.d.(12, 27). For the remainder of our discussion we will be concerned with the greatest common divisors of positive integers.

EXAMPLE Determine the g.c.d.(12, 30, 57).
Solution Factoring, we have

$$12 = 2 \times 2 \times 3$$

$$30 = 2 \times 3 \times 5$$

$$57 = 3 \times 19$$

Hence, g.c.d.(12, 30, 57) $= 2^0 \cdot 3^1 \cdot 5^0 \cdot 19^0 = 1 \times 3 \times 1 \times 1 = 3$. ∎

EXAMPLE Consider the integers 12, 30, and 57.
a. Determine g.c.d.(12, 30).
b. Determine g.c.d.(12, 57).
c. Determine g.c.d.(30, 57).

Solution

a. Factoring, we have

$$12 = 2 \times 2 \times 3$$

$$30 = 2 \times 3 \times 5$$

$$\text{g.c.d.}(12, 30) = 2^1 \times 3^1 \times 5^0 = 2 \times 3 \times 1 = 6$$

b. Factoring, we have

$$12 = 2 \times 2 \times 3$$

$$57 = 3 \times 19$$

$$\text{g.c.d.}(12, 57) = 2^0 \times 3^1 \times 19^0 = 1 \times 3 \times 1 = 3$$

c. Factoring, we have

$$30 = 2 \times 3 \times 5$$

$$57 = 3 \times 19$$

$$\text{g.c.d.}(30, 57) = 2^0 \times 3^1 \times 5^0 \times 19^0 = 1 \times 3 \times 1 \times 1 = 3. \quad \blacksquare$$

Observe that in the above example g.c.d.$(12, 30) = 6$, g.c.d.$(12, 57) = 3$, and g.c.d.$(30, 57) = 3$. Now considering the g.c.d.$(3, 6)$, we would get 3, which is the g.c.d.$(12, 30, 57)$ as was determined in the example following Definition 8.12. In general, if g.c.d.$(a, b) = p$ and g.c.d.$(b, c) = q$, then g.c.d.$(a, b, c) = $ g.c.d.(p, q).

To determine the g.c.d. of large positive integers, the procedure outlined above is cumbersome. A more practical method involves the use of the division algorithm, which we will now state without proof.

THEOREM 8.9 The Division Algorithm If a and b are any two positive integers, then there exist nonnegative integers q and r such that $a = b \cdot q + r$, where $0 \leq r < b$. The integer q is called the *quotient*, and the integer r is called the *remainder*.

The procedure used to find the greatest common divisor of two positive integers, which we will now discuss, includes the repeated use of the division algorithm and is known as *Euclid's algorithm*.

Suppose that we wish to find the g.c.d. of 2244 and 418 using Euclid's algorithm. By the use of the division algorithm, we have $2244 = 5(418) + 154$, where $q = 5$, $r = 154$, and $0 \leq 154 < 418$. Since the g.c.d.$(2244, 418)$ divides both 2244 and 418, then it must also divide 154 by Theorem 8.8. Hence g.c.d.$(2244, 418) = $ g.c.d.$(418, 154)$. Repeating the division algorithm, we have $418 = 2(154) + 110$, where $q = 2$, $r = 110$, and $0 \leq 110 < 154$. Again, since g.c.d.$(418, 154)$ divides both 418 and 154 it must also divide 110. Hence g.c.d.$(418, 154) = $ g.c.d.$(154, 110)$. Repeating the division algorithm again, we have $154 = 1(110) + 44$ with $q = 1$, $r = 44$, and $0 \leq 44 < 110$. Again, g.c.d.$(154, 110)$ also will divide 44 and g.c.d.$(154, 110) =$ g.c.d.$(110, 44)$. Repeating the division algorithm again, we have $110 = 2(44) + 22$, where $q = 2$, $r = 22$, and $0 \leq 22 < 44$. Again, g.c.d.$(110, 44) =$ g.c.d.$(44, 22)$ and, by the division algorithm, we have $44 = 2(22) + 0$ with $q = 2$, $r = 0$, and $0 \leq 0 < 22$. Now g.c.d.$(44, 22) =$ g.c.d.$(22, 0)$. But g.c.d.$(22, 0) = 22$. Hence g.c.d.$(44, 22) =$ g.c.d.$(110, 44) =$ g.c.d.$(154, 110) =$ g.c.d.$(418, 154) =$ g.c.d. $(2244, 418) = 22$.

EXAMPLE Using Euclid's algorithm, determine g.c.d.$(7286, 416)$.
Solution

$$7286 = 17(416) + 214$$
$$416 = 1(214) + 202$$
$$214 = 1(202) + 12$$
$$202 = 16(12) + 10$$
$$12 = 1(10) + 2$$
$$10 = 5(2) + 0$$

Hence g.c.d.$(7286, 416) = 2$ since g.c.d.$(2, 0) = 2$. ■

EXERCISES

1. Determine the greatest common divisor for each of the following pairs of indicated numbers.
   a. 14 and 26.   b. 13 and 52.
   c. 102 and 210.   d. 105 and 315.
   e. 81 and 729.   f. 245 and 360.
   g. 110 and 132.   h. 88 and 96.
   i. 7 and 10.   j. 17 and 53.
2. Determine which of the following pairs of indicated numbers are relatively prime.
   a. 7 and 26.   b. 38 and 51.
   c. 49 and 196.   d. 101 and 113.
   e. 47 and 74.   f. 25 and 36.
   g. 81 and 216.   h. 95 and 106.
   i. 216 and 801.   j. 341 and 424.
3. Determine the greatest common divisor for each of the following sets of indicated numbers.
   a. 14, 26, and 38.   b. 25, 50, and 72.
   c. 24, 48, and 96.   d. 13, 52, and 69.
   e. 48, 243, and 316.   f. 125, 625, and 1275.
   g. 44, 72, 186, and 490.   h. 23, 48, 92, and 210.
   i. 34, 51, 136, and 270.   j. 81, 135, 260, and 372.

4. Using Euclid's algorithm, determine the greatest common divisor for each of the following pairs of indicated numbers.
   a. 105 and 1236.   b. 217 and 3962.
   c. 1126 and 5972.   d. 1024 and 6973.
   e. 3235 and 5345.   f. 486 and 9276.
   g. 895 and 6340.   h. 916 and 2315.
   i. 524 and 6396.   j. 2365 and 21,625.
5. a. Determine all of the divisors of 5.
   b. Determine all the divisors of 15.
   c. Verify that the set of all divisors of 5 is a proper subset of the set of all the divisors of 15.
   d. What is the g.c.d.(5, 15)?
6. a. Determine all of the divisors of 6.
   b. Determine all of the divisors of 24.
   c. Verify that the set of all the divisors of 6 is a proper subset of the set of all the divisors of 24.
   d. What is the g.c.d.(6, 24)?
7. Let A be the set of all the divisors of a, and let B be the set of all the divisors of b. Further, suppose that $a|b$.
   a. Prove that the g.c.d.$(a, b) = a$.
   b. Prove that $A \subset B$.
8. Suppose that a and b are relatively prime.
   a. Determine the g.c.d.(a, b).
   b. Must either a or b be prime? Explain.
9. Suppose that c and d are two distinct prime numbers. Determine the g.c.d.(c, d).
10. Suppose that $a|b$, $b|c$, and $c|d$.
    a. Determine the g.c.d.(a, b).
    b. Determine the g.c.d.(a, c).
    c. Determine the g.c.d.(a, b, c, d).

8.5 LEAST COMMON MULTIPLE

If b is an integer and a is a nonzero integer, then we defined $a|b$ if and only if there exists an integer c such that $b = ac$. Recall that a is a factor or divisor of b whereas b is a multiple of a. In the previous section we considered the greatest common divisor of two positive integers. In this section we will consider the least common multiple of two positive integers.

Let a and b be positive integers. Let A be the set of all the multiples of a, and let B be the set of all the multiples of b. Clearly, $A \cap B$ will contain the integer ab that is a multiple of both a and b. It is possible, however, that $A \cap B$ will contain integers other than ab. For instance, if

$a = 6$, then $A = \{6, 12, 18, 24, 30, 36, 42, 48, 54, 60, 66, 72, 78, 84, 90, 96, \ldots\}$ and if $b = 8$, then $B = \{8, 16, 24, 32, 40, 48, 56, 64, 72, 80, 88, 96, \ldots\}$. Hence $A \cap B = \{24, 48, 72, 96, \ldots\}$. All of those elements in $A \cap B$ are common multiples of 6 and 8, and 24 is the least common multiple.

DEFINITION 8.13 The *least common multiple* of two positive integers a and b is the positive integer m if and only if $a\,|\,m$, $b\,|\,m$, and m is the least of the positive integers divisible by both a and b. The least common multiple of a and b is denoted by l.c.m.(a, b).

To determine the least common multiple of two or more positive integers, we could find the set of all the common multiples of the integers and then determine the least of all these common multiples. A more direct approach involves the use of the fundamental theorem of arithmetic in a manner similar to that used for determining the greatest common divisor. The procedure involves writing the prime factorization of each integer and then determining all those prime factors common to the integers involved. Next determine the largest power of each prime factor that occurs in any of these prime factorizations. The l.c.m., then will be the product of all those largest powers of primes.

To determine l.c.m.$(24, 36)$, we proceed as follows:

1. Write the prime factorization of 24 and 36: $24 = 2 \times 2 \times 2 \times 3$ and $36 = 2 \times 2 \times 3 \times 3$.

2. Write as a product each of the prime factors occurring in these factorizations: 2×3.

3. Raise each of the factors in 2 above to the largest power that it occurs in either of the factorizations. Note that 2 occurs 3 times in the factorization of 24 and 2 times in the factorization of 36; hence we want 2^3. Likewise, 3 occurs once in the factorization of 24 and twice in the factorization of 36; hence we want 3^2. The l.c.m.$(24, 36)$ $= 2^3 \times 3^2 = 72$.

EXAMPLE Determine the l.c.m. of 88, 128, and 180.
Solution Factoring, we have

$$88 = 2 \times 2 \times 2 \times 11$$

$$128 = 2 \times 2 \times 2 \times 2 \times 2 \times 2 \times 2$$

$$180 = 2 \times 2 \times 3 \times 3 \times 5$$

l.c.m.$(88, 128, 180) = 2^7 \times 3^2 \times 5^1 \times 11^1 = 128 \times 9 \times 5 \times 11 = 63{,}360$ ∎

EXAMPLE   Determine the l.c.m. of 60, 70, 80, and 90.
Solution   Factoring, we have

$$60 = 2 \times 2 \times 3 \times 5$$

$$70 = 2 \times 5 \times 7$$

$$80 = 2 \times 2 \times 2 \times 2 \times 5$$

$$90 = 2 \times 3 \times 3 \times 5$$

l.c.m.(60, 70, 80, 90) = $2^4 \times 3^2 \times 5^1 \times 7^1 = 16 \times 9 \times 5 \times 7 = 5{,}040$ ∎

EXAMPLE   Determine the l.c.m.(27, 8).
Solution   Factoring, we have

$$27 = 3 \times 3 \times 3$$

$$8 = 2 \times 2 \times 2$$

l.c.m.(27, 8) = $2^3 \times 3^3 = 8 \times 27 = 216$ ∎

In the example above we note that the positive integer 216 is the least common multiple of 27 and 8. That is, the l.c.m.(27, 8) is the product of 27 and 8. In the third example in Section 8.4 we observed that g.c.d.(27, 8) = 1. In general, if a and b are two positive integers and g.c.d.$(a, b) = 1$, then l.c.m.$(a, b) = ab$.

If we are given three positive integers, a, b, and c and wish to determine l.c.m.(a, b, c), we would first determine l.c.m.$(a, b) = p$ and then determine l.c.m.$(p, c) = q$. Finally, l.c.m.$(a, b, c) = q$.

EXAMPLE   Determine l.c.m.(21, 35, 42).
Solution   Factoring, we have

$$21 = 3 \times 7$$

$$35 = 5 \times 7$$

$$42 = 2 \times 3 \times 7$$

l.c.m.(21, 35) = $3^1 \times 5^1 \times 7^1 = 3 \times 5 \times 7 = 105$

Factoring, we have

$$105 = 3 \times 5 \times 7$$

l.c.m.(105, 42) = $2^1 \times 3^1 \times 5^1 \times 7^1 = 210$

Hence, l.c.m.(21, 35, 42) = 210   ∎

To determine l.c.m.(a, b, c, d), we could find l.c.m.$(a, b) = p$ and l.c.m.$(c, d) = q$; then, l.c.m.$(a, b, c, d) =$ l.c.m.(p, q).

EXAMPLE Determine l.c.m.(8, 12, 15, 21).
Solution Factoring, we have

$$8 = 2 \times 2 \times 2$$

$$12 = 2 \times 2 \times 3$$

$$15 = 3 \times 5$$

$$21 = 3 \times 7$$

$$\text{l.c.m.}(8, 12) = 2^3 \times 3^1 = 24$$

$$\text{l.c.m.}(15, 21) = 3^1 \times 5^1 \times 7^1 = 105$$

Factoring, we have

$$24 = 2 \times 2 \times 2 \times 3$$

$$105 = 3 \times 5 \times 7$$

$$\text{l.c.m.}(24, 105) = 2^3 \times 3^1 \times 5^1 \times 7^1 = 840 \quad \blacksquare$$

We will now state two theorems without proof that are also useful in determining the least common multiple of two positive integers.

THEOREM 8.10 If a and b are any two positive integers such that g.c.d.$(a, b) = d$ and l.c.m.$(a, b) = m$, then $ab = dm$.

THEOREM 8.11 If a and b are any two positive integers, then

$$\text{l.c.m.}(a, b) = \frac{ab}{\text{g.c.d.}(a, b)}$$

We will conclude this section with another procedure for finding the least common multiple of two or more positive integers. Consider finding the l.c.m.(12, 20, 24, 32). We will list the integers in the order written and divide each by 2, obtaining

$$\begin{array}{c|cccc} 2 & 12 & 20 & 24 & 32 \\ \hline & 6 & 10 & 12 & 16 \end{array}$$

Divide each of the quotients obtained by 2, obtaining

$$\begin{array}{c|cccc} 2 & 6 & 10 & 12 & 16 \\ \hline & 3 & 5 & 6 & 8 \end{array}$$

Divide each of the quotients obtained by 2 again. If the quotient is not divisible by 2, just "bring down" the integer, obtaining

$$\begin{array}{c|cccc} 2 & 3 & 5 & 6 & 8 \\ \hline & 3 & 5 & 3 & 4 \end{array}$$

Divide each of the quotients obtained by 2 again, obtaining

$$
\begin{array}{r|cccc}
2 & 3 & 5 & 3 & 4 \\
\hline
 & 3 & 5 & 3 & 2
\end{array}
$$

Dividing by 2 again, we obtain

$$
\begin{array}{r|cccc}
2 & 3 & 5 & 3 & 2 \\
\hline
 & 3 & 5 & 3 & 1
\end{array}
$$

Since all quantities are odd, we will now divide by 3, obtaining

$$
\begin{array}{r|cccc}
3 & 3 & 5 & 3 & 1 \\
\hline
 & 1 & 5 & 1 & 1
\end{array}
$$

Finally, dividing by 5, we have

$$
\begin{array}{r|cccc}
5 & 1 & 5 & 1 & 1 \\
\hline
 & 1 & 1 & 1 & 1
\end{array}
$$

Since the last line of quotients consists of 1's only, we determine that the l.c.m.$(12, 20, 24, 32) = 2 \times 2 \times 2 \times 2 \times 2 \times 3 \times 5 = 480$.

EXAMPLE   Determine l.c.m.$(8, 18, 20, 35, 42)$.
Solution

$$
\begin{array}{r|ccccc}
2 & 8 & 18 & 20 & 35 & 42 \\
\hline
2 & 4 & 9 & 10 & 35 & 21 \\
\hline
2 & 2 & 9 & 5 & 35 & 21 \\
\hline
3 & 1 & 9 & 5 & 35 & 21 \\
\hline
3 & 1 & 3 & 5 & 35 & 7 \\
\hline
5 & 1 & 1 & 5 & 35 & 7 \\
\hline
7 & 1 & 1 & 1 & 7 & 7 \\
\hline
 & 1 & 1 & 1 & 1 & 1
\end{array}
$$

Hence, l.c.m.$(8, 18, 20, 35, 42) = 2 \times 2 \times 2 \times 3 \times 3 \times 5 \times 7 = 2520$. ■

EXERCISES

1. Determine the least common multiple for each of the following pairs of indicated numbers.

a. 12 and 26.   b. 13 and 42.
c. 102 and 210.   d. 100 and 305.
e. 81 and 126.   f. 88 and 96.
g. 7 and 10.   h. 17 and 102.
i. 245 and 360.   j. 112 and 309.

2. Determine the least common multiple for each of the following sets of indicated numbers.
   a. 44, 64, and 90.   b. 30, 35, 40, and 45.
   c. 8, 27, 64, and 81.   d. 42, 70, and 84.
   e. 16, 24, 30, and 42.   f. 36, 42, 54, and 70.
   g. 312, 524, and 628.   h. 210, 250, 305, and 410.
   i. 512, 680, 1024, and 2000.   j. 35, 210, 820, and 1320.

3. Determine the least common multiple for each of the pairs of indicated numbers given in Exercise 1 using Theorem 8.11.

4. Determine the least common multiple for each of the sets of indicated numbers given in Exercise 2, using the process outlined in the last example above.

5. a. Determine the 15 smallest positive multiples of 5. Denote this set of multiples by A.
   b. Determine the 5 smallest positive multiples of 15. Denote this set of multiples by B.
   c. Verify that $B \subset A$.
   d. What is the l.c.m.$(5, 15)$?
   e. What is the g.c.d.$(5, 15)$?
   f. Verify that l.c.m.$(5, 15) \times$ g.c.d.$(5, 15) = 5 \times 15$.

6. a. Determine the 20 smallest positive multiples of 6. Denote this set of multiples by C.
   b. Determine the 5 smallest positive multiples of 24. Denote this set of multiples by D.
   c. Verify that $D \subset C$.
   d. What is the l.c.m.$(6, 24)$?
   e. What is the g.c.d.$(6, 24)$?
   f. Verify that l.c.m.$(6, 24) \times$ g.c.d.$(6, 24) = 6 \times 24$.

7. Let P be the set of all the multiples of a, and let Q be the set of all the multiples of b. Further, suppose that $a|b$.
   a. Prove that l.c.m.$(a, b) = b$.
   b. Prove that $Q \subset P$.
   c. Verify that g.c.d.$(a, b) = ab/$l.c.m.(a, b).

8. Suppose that $a|b$, $b|c$, and $c|d$.
   a. Determine the l.c.m.(a, b).
   b. Determine the l.c.m.(b, c).
   c. Determine the l.c.m.(c, d).
   d. Let $p = $ l.c.m.(a, b). Determine the l.c.m.(p, c).
   e. Let $q = $ l.c.m.(b, c). Determine the l.c.m.(q, d).
   f. Let $r = $ l.c.m.(p, c) and $s = $ l.c.m.(q, d). Verify that the l.c.m.$(r, s) = d$.

8.6 CONGRUENCES

Using the division algorithm, we can readily determine that two integers can be divided by the same positive integer and the remainders obtained may or may not be the same. In this section we will discuss a relation on the set of integers that involves the remainders when integers are divided by the same positive integers.

DEFINITION 8.14 Two integers a and b are said to be *congruent modulo m* denoted $a \equiv b$ (mod m) if and only if m is a positive integer and $m|(a - b)$.

In the above definition we see that two integers are congruent modulo m if and only if m divides their difference. Some authors define the congruence of integers modulo m if and only if the remainders obtained when two integers are divided by m are the same. The two definitions are equivalent, but throughout this section we will use Definition 8.14.

EXAMPLE
a. $12 \equiv 2$ (mod 5) since $5|(12 - 2)$ (that is, $5|10$).
b. $17 \equiv 3$ (mod 7) since $7|(17 - 3)$ (that is, $7|14$).
c. $9 \equiv 0$ (mod 9) since $9|(9 - 0)$ (that is, $9|9$).
d. $1 \equiv 9$ (mod 8) since $8|(1 - 9)$ (that is, $8|(^-8)$).
e. $2 \not\equiv 7$ (mod 6) since $6\nmid(2 - 7)$ (that is, $6\nmid(^-5)$).
f. $3 \not\equiv 8$ (mod 2) since $2\nmid(3 - 8)$ (that is, $2\nmid(^-5)$).

EXAMPLE
a. $15 \equiv 23$ (mod 4) since the remainder obtained when 15 is divided by 4 is 3 and 3 is also the remainder obtained when 23 is divided by 4. Also note that $4|(15 - 23)$ since $4|(^-8)$.
b. $232 \equiv 124$ (mod 6) since the remainder obtained when 232 is divided by 6 is 4 and 4 is also the remainder obtained when 124 is divided by 6. Also note that $6|(232 - 124)$ since $6|108$.
c. $625 \not\equiv 209$ (mod 5) since the remainder obtained when 625 is divided by 5 is 0 but the remainder obtained when 209 is divided by 5 is 4. Also note that $5\nmid(625 - 209)$ since $5\nmid416$.

EXAMPLE
a. If $a \equiv 0$ (mod m), then $m|(a - 0)$ or $m|a$. Hence $a \equiv 0$ (mod m) implies that $m|a$.
b. If $m|a$, then $a \equiv 0$ (mod m) since m also divides 0. Hence $m|a$ implies that $a \equiv 0$ (mod m).

EXAMPLE

a. $3 \equiv 3 \pmod 4$ since $4|(3 - 3)$ or $4|0$.

b. $6 \equiv 9 \pmod 3$ since $3|(6 - 9)$ and $9 \equiv 6 \pmod 3$ since $3|(9 - 6)$.

c. $7 \equiv 12 \pmod 5$ since $5|(7 - 12)$, $12 \equiv 27 \pmod 5$ since $5|(12 - 27)$, and $7 \equiv 27 \pmod 5$ since $5|(7 - 27)$.

Part (a) of the above example may suggest that the relation "congruence, modulo m" may be a reflexive relation on the set I. Part (b) may suggest that the relation is symmetric on I, and part (c) may suggest that the relation is a transitive relation. It is indeed easy to prove that this relation is an equivalence relation on I.

THEOREM 8.12   The relation "congruence, modulo m" on the set I is an equivalence relation.

Proof   To prove this theorem we must establish that the relation is reflexive, symmetric, and transitive.

a. **Reflexive**   If a is an arbitrary integer, is $a \equiv a \pmod m$? Since $a - a = 0$ and $m|0$, then $m|(a - a)$ and $a \equiv a \pmod m$.

b. **Symmetric**   If a and b are arbitrary integers with $a \equiv b \pmod m$, is $b \equiv a \pmod m$? Since $a \equiv b \pmod m$, then $m|(a - b)$. But if $m|(a - b)$, then $m|(b - a)$. If $m|(b - a)$, then $b \equiv a \pmod m$.

c. **Transitive**   If a, b, and c are arbitrary integers with $a \equiv b \pmod m$ and $b \equiv c \pmod m$, is $a \equiv c \pmod m$? If $a \equiv b \pmod m$ and $b \equiv c \pmod m$, then $m|(a - b)$ and $m|(b - c)$. If $m|(a - b)$ and $m|(b - c)$, then there exist integers p and q such that $a - b = mp$ and $b - c = mq$. By Theorem 7.2(e) we have $a - c = m(p + q)$. Since $p, q \in I$, $p + q \in I$ and we have that $m|(a - c)$. Since $m|(a - c)$, then $a \equiv c \pmod m$.

It should be clear from the examples of this section that we are really only concerned with the remainders obtained when integers are divided by a positive integer m. Hence in a system of integers modulo m, we could perform the operation of addition provided that if the sum of integers is greater than or equal to m, we will divide the sum by m and use the remainder in place of the sum. For instance, in a mod 5 system the only possible answers for the sum of two or more integers would be 0, 1, 2, 3, or 4.

Table 8.2 is a mod 4 addition table. Tables 8.3 and 8.4 are mod 6 and mod 9 addition tables, respectively.

Table 8.2 Addition table, mod 4

+	0	1	2	3
0	0	1	2	3
1	1	2	3	0
2	2	3	0	1
3	3	0	1	2

Table 8.3 Addition table, mod 6

+	0	1	2	3	4	5
0	0	1	2	3	4	5
1	1	2	3	4	5	0
2	2	3	4	5	0	1
3	3	4	5	0	1	2
4	4	5	0	1	2	3
5	5	0	1	2	3	4

Table 8.4 Addition table, mod 9

+	0	1	2	3	4	5	6	7	8
0	0	1	2	3	4	5	6	7	8
1	1	2	3	4	5	6	7	8	0
2	2	3	4	5	6	7	8	0	1
3	3	4	5	6	7	8	0	1	2
4	4	5	6	7	8	0	1	2	3
5	5	6	7	8	0	1	2	3	4
6	6	7	8	0	1	2	3	4	5
7	7	8	0	1	2	3	4	5	6
8	8	0	1	2	3	4	5	6	7

EXAMPLE

a. $3 + 2 \equiv 5 \pmod 6$.
b. $4 + 5 \equiv 3 \pmod 6$.
c. $2 + 3 \equiv 1 \pmod 4$.
d. $3 + 3 \equiv 2 \pmod 4$.
e. $4 + 4 \equiv 8 \pmod 9$.
f. $7 + 6 \equiv 4 \pmod 9$.

Multiplication of integers, modulo m, also can be performed provided that if the product of integers is greater than or equal to m, we will divide the product by m and use the remainder in place of of the product.

Table 8.5 is a mod 4 multiplication table. Tables 8.6 and 8.7 are mod 6 and mod 9 multiplication tables, respectively.

Table 8.5 Multiplication table, mod 4

×	0	1	2	3
0	0	0	0	0
1	0	1	2	3
2	0	2	0	2
3	0	3	2	1

Table 8.6 Multiplication table, mod 6

×	0	1	2	3	4	5
0	0	0	0	0	0	0
1	0	1	2	3	4	5
2	0	2	4	0	2	4
3	0	3	0	3	0	3
4	0	4	2	0	4	2
5	0	5	4	3	2	1

Table 8.7 Multiplication table, mod 9

×	0	1	2	3	4	5	6	7	8
0	0	0	0	0	0	0	0	0	0
1	0	1	2	3	4	5	6	7	8
2	0	2	4	6	8	1	3	5	7
3	0	3	6	0	3	6	0	3	6
4	0	4	8	3	7	2	6	1	5
5	0	5	1	6	2	7	3	8	4
6	0	6	3	0	6	3	0	6	3
7	0	7	5	3	1	8	6	4	2
8	0	8	7	6	5	4	3	2	1

EXAMPLE
a. $3 \times 3 \equiv 1$ (mod 4).
b. $2 \times 2 \equiv 0$ (mod 4).
c. $4 \times 4 \equiv 4$ (mod 6).
d. $5 \times 3 \equiv 3$ (mod 6).
e. $3 \times 6 \equiv 0$ (mod 9).
f. $5 \times 7 \equiv 8$ (mod 9).

Just as we discussed the properties of ordinary addition and ordinary multiplication of integers, we also can discuss properties of addition (mod m) and multiplication (mod m). Observe in Tables 8.2, 8.3, and 8.4 that if we interchange the first row of the table with the first column, the resulting table is identical to the original table. In a similar manner, interchanging the second row with the second column produces a table identical to the original table. In fact, for each of these tables, if we interchange a particular row with the corresponding column, the table is left unchanged. Whenever this can be done for all rows and columns, we conclude that the operation defined by the table is a commutative operation. Hence addition (mod 4), addition (mod 6), and addition (mod 9) are commutative operations. In a similar manner an examination of Tables 8.5, 8.6, and 8.7 reveals that multiplication (mod 4), multiplication (mod 6), and multiplication (mod 9) are commutative operations. In general addition (mod m) and multiplication (mod m) are commutative operations.

The associative property for addition (mod m) and multiplication (mod m) also are valid, but these cannot be established readily by an examination of the tables. We also have the distributive property of multiplication (mod m) over addition (mod m) for the same value of m. By examining the tables it is easy to verify that 0 is the additive identity (mod m) and 1 is the multiplicative identity (mod m).

Also, we can readily verify that each entry 0, 1, 2, ..., $(m-1)$ has an additive inverse (mod m). For instance, for $m = 4$ we conclude that the additive inverses of 0, 1, 2, 3 are, respectively, 0, 3, 2, 1. For $m = 6$, the additive inverses of 0, 1, 2, 3, 4, 5 are, respectively, 0, 5, 4, 3, 2, 1.

Finally, we note that in the system (mod 4) there exists a multiplicative identity 1, but only 1 and 3 have multiplicative inverses which are, respectively, 1 and 3. For $m = 6$, only 1 and 5 have multiplicative inverses which are, respectively, 1 and 5. For $m = 9$, only 1, 2, 4, 5, 7, 8 have multiplicative inverses which are, respectively, 1, 5, 7, 2, 4, 8.

Other properties for the relation congruence mod m will be established in a later chapter of this text.

EXERCISES

1. Determine which of the following statements are true and which are false.
   a. $2 \equiv 4$ (mod 3).   b. $3 \equiv 7$ (mod 4).

c. $4 \equiv 4 \pmod 5$.

d. $6 \equiv 7 \pmod 5$.

e. $^-7 \equiv 6 \pmod{13}$.

f. $^-8 \equiv {}^-5 \pmod 4$.

g. $4 + 3 \equiv 5 - 2 \pmod 2$.

h. $^-5 - 6 \equiv 3 \pmod 7$.

i. $(^-2)(3) \equiv (4)(^-5) \pmod{14}$.

j. $(16 \div {}^-4) \equiv (^-8 \div 2) \pmod 9$.

2. Perform the indicated operations, and express each result as a whole number less than the modulus.

a. $5 + 2 \pmod 6$.

b. $4 \times 3 \pmod 5$.

c. $2 - 5 \pmod 3$.

d. $4 + (^-2 + 5) \pmod 4$.

e. $1 - 12 \pmod 7$.

f. $2(3 + 4) \pmod 6$.

g. $(^-2)(3)(^-4) \pmod{13}$.

h. $26 \div {}^-2 \pmod 8$.

i. $(^-2 + {}^-3)(^-4) \pmod 9$.

j. $(7 - 6) \times (^-2 - {}^-6) \pmod{11}$.

3. For each of the following determine the smallest whole number value for x such that the statement is true.

a. $x + 3 \equiv 2 \pmod 4$.

b. $x - 4 \equiv 5 \pmod 6$.

c. $2x + 1 \equiv 3 \pmod 5$.

d. $x + 2 \equiv 10 \pmod{11}$.

e. $4x \equiv 13 \pmod 7$.

f. $3x - 8 \equiv 3 \pmod 4$.

g. $14 \equiv x \pmod{13}$.

h. $16 \equiv 13 \pmod x$.

i. $2(17 - 2) \equiv 15 \pmod x$.

j. $4[^-3 + (^-2 + 5)] \equiv {}^-2 \pmod x$.

SUMMARY

In this chapter we discussed some elements from number theory. We discussed some applications of divisibility and considered tests for divisibility by certain positive integers. Prime and composite numbers were introduced and discussed, and the fundamental theorem of arithmetic was stated. Terms such as twin primes and prime triplets were defined as were proper divisor, perfect number, and amicable numbers. The greatest common divisor, least common multiple, relatively prime numbers, and the division algorithm also were introduced and discussed. Congruences, modulo m, were introduced, and properties associated with this relation were discussed. Addition (mod m) and multiplication (mod m) also were discussed.

During our discussions the following symbols were introduced.

SYMBOL	INTERPRETATION
$a|b$	The positive integer a divides the integer b.
g.c.d.(a, b)	The greatest common divisor of the integers a and b.
l.c.m.(a, b)	The least common multiple of the integers a and b.
$a \equiv b \pmod m$	The integers a and b are congruent, modulo m $(m \in N)$.

REVIEW EXERCISES FOR CHAPTER 8

1. a. List all the positive divisors of 36.

   b. List all the negative divisors of 45.

2. Verify that if $a|b$ and $b|c$, then $a|c$, given
   a. $a = 3, b = 9, c = {}^-72$.
   b. $a = 5, b = {}^-25, c = 625$.
   c. $a = {}^-4, b = 12, c = {}^-96$.

3. Classify each of the following indicated integers as being odd or even.
   a. $4 + {}^-3$.   b. $7(2 - {}^-1)$.
   c. $(2)({}^-3) + (4)({}^-3)$.   d. ${}^-2(3 + 4) - 7$.
   e. $(25 \div {}^-5)(2)$.   f. $5 + [4 + ({}^-3 + 2)]$.
   g. $2a + 1$, if $a = {}^-3$.   h. $3n + 1$, if $n = 2$.

4. Without actually dividing, determine which of the following statements are true and which are false.
   a. $2|31692$.   b. $3|40563$.
   c. $4|9762$.   d. $5|43,209$.
   e. $6|7692$.   f. $7|88207$.
   g. $8|23124$.   h. $9|630297$.
   i. $10|6230$.   j. $11|897545$.

5. Classify each of the numbers indicated as being prime, composite, or neither.
   a. 23.   b. 39.
   c. $13 - 5$.   d. $3 + 5$.
   e. 5×7.   f. 361.
   g. $(3 \times 8) + 5$.   h. $63 - 36$.
   i. $(16 \div 4) \div 4$.   j. $(2 + 1) + {}^-2$.
   k. $({}^-2)(4) \div {}^-4$. l. ${}^-(8 \times {}^-2) - {}^-13$.

6. For each of the following determine a complete factorization.
   a. 70.   b. 122.   c. 208.
   d. 315.   e. 1581.   f. 6432.

7. a. Give an example of twin primes.
   b. Give an example of prime triplets.
   c. What is the smallest whole number that is a perfect number?

8. For each of the following pairs of indicated numbers determine the g.c.d.
   a. 12 and 26.   b. 13 and 17.
   c. 100 and 208.   d. 86 and 94.
   e. 17 and 49.   f. 130 and 215.

9. For each of the pairs of indicated numbers given in Exercise 8 determine the l.c.m..

10. For each of the following sets of indicated numbers determine the g.c.d..
    a. 12, 28, and 32.   b. 15, 35, and 50.
    c. 27, 36, and 81.   d. 75, 125, and 250.
    e. 33, 55, and 132.   f. 18, 42, 54, and 70.

11. For each of the sets of indicated numbers given in Exercise 10 determine the l.c.m.

12. Using Euclid's algorithm, determine the g.c.d. for each of the following pairs of indicated numbers.
    a. 812 and 3624. b. 968 and 4168.
    c. 1,265 and 23,902. d. 1,372 and 43,691.
13. Determine which of the following statements are true and which are false.
    a. $3 \equiv 5 \pmod 6$. b. $7 - 2 \equiv 10 \pmod 5$.
    c. $(2)(^-3) \equiv 7 \pmod{13}$. d. $(12 \div {}^-3) \equiv 9 \pmod 7$.
    e. $^-5 + {}^-6 \equiv {}^-3 \pmod 4$. f. $4 - 3 \equiv 3 - 4 \pmod 3$.
14. Perform the indicated operations, and express each result as a whole number less than the modulus.
    a. $3 + 4 \pmod 5$. b. $7 - 6 \pmod 8$.
    c. $4 \times 5 \pmod 9$. d. $36 \div {}^-9 \pmod 6$.
    e. $^-2(3 + {}^-9) \pmod 7$. f. $7 \times {}^-4 \pmod{15}$.

CHAPTER 9
THE SYSTEM OF RATIONAL NUMBERS

Chapter 5 developed the system of whole numbers and defined the operations of addition, multiplication, subtraction, and division on W. We concluded that the set W was closed under the operations of addition and multiplication but was not closed under the operations of subtraction and division. In fact, if a, $b \in W$, $a - b$ was meaningful only if $a \geq b$. Hence a solution for the equation $a + x = b$ did not always exist on W. In a similar manner there were restrictions for division.

In Chapter 7 we developed the system of integers as an extension of the system of whole numbers. In defining the basic operations with integers we basically extended the corresponding definitions for whole numbers. By introducing negative integers we defined subtraction on I such that it was always possible to solve the equation $a + x = b$ on I, thus eliminating a deficiency that existed on W; the property of the additive inverse was introduced. However, the equation $ax = b$ did not always have a solution on the set I. In fact, the equation had a solution only if $a|b$. In this chapter we will introduce and develop a new system of numbers such that the equation $ax = b$ will always have a solution, excepting when $a = 0$ and $b \neq 0$.

9.1 FRACTIONS

For most students the word "fraction" probably means "part of." For instance, consider the diagram of Figure 9.1 as representing a unit of

Figure 9.1

something such as a piece of paper, a candy bar, a stick, or a bar of soap. Now suppose that the unit is divided exactly into two parts as indicated in Figure 9.2. Then each part could be considered as 1 of the 2 equal parts

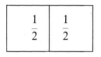

Figure 9.2

which we could represent by the symbol 1/2. Similarly, the unit in Figure 9.1 could be divided exactly into 3 parts as indicated in Figure 9.3. Then each part could be considered as 1 of the 3 equal parts which we could represent by the symbol 1/3.

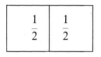

Figure 9.3

The symbols 1/2 and 1/3 are called *fractions*, and each fraction consists of two parts. In the fraction 1/2, 1 is called the *numerator* and 2 is called the *denominator*. The denominator indicates the number of parts into which the unit is divided and the numerator indicates the number of those equal parts which were taken. For instance, the fraction 3/7 has a numerator of 3 and a denominator of 7, indicating that the unit was divided into 7 equal parts and that 3 of those equal parts are being taken or considered.

DEFINITION 9.1 A *fraction* is a numeral consisting of two integers a and b such that $b \neq 0$ and written in the form $\dfrac{a}{b}$ or a/b or $a \div b$. The integer a is called the *numerator* of the fraction, and the nonzero integer b is called the *denominator* of the fraction.

From the above definition we readily determine that a fraction is a quotient of two integers provided that the denominator is not equal to zero. Also, a fraction, being a numeral, will represent a number which will be defined in the next section. The numerator of a fraction may be less than, equal to, or greater than its denominator.

Consider now the fraction 6/2. From Chapter 7 we know that 6/2 represents the number 3. In a similar manner the fraction $^-27/^-9$ also represents the number 3. We now have two fractions representing the same number. What if any is the relationship between these two fractions?

DEFINITION 9.2 Two fractions a/b and c/d represent the same number or are said to be *equivalent*, denoted $a/b \simeq c/d$, if and only if $ad = bc$.

EXAMPLE

a. $\dfrac{2}{3} \simeq \dfrac{4}{6}$ since $(2)(6) = (3)(4) = 12$.

b. $\dfrac{^-2}{7} \simeq \dfrac{10}{^-35}$ since $(^-2)(^-35) = (7)(10) = 70$.

c. $\dfrac{0}{3} \simeq \dfrac{0}{^-2}$ since $(0)(^-2) = (3)(0) = 0$.

It should be noted that Definition 9.2 gives a relationship between two fractions. We will be able to prove that this relation is an equivalence relation on the set of fractions.

THEOREM 9.1 The relation "is equivalent to" on the set of fractions is an equivalence relation.

Proof To prove this theorem, we must establish that the given relation is reflexive, symmetric, and transitive on the set of fractions.

a. **Reflexive** For any fraction a/b, $a/b \simeq a/b$ since $ab = ba$ by the commutative property of multiplication of integers. Therefore the relation "is equivalent to" is a reflexive relation on the set of integers.

b. **Symmetric** If a/b and c/d are any two fractions with $a/b \simeq c/d$, then $ad = bc$ by Definition 9.2. Further, $da = cb$ by the commutative property of multiplication of integers and $cb = da$ by the symmetric property of equality. Therefore, by Definition 9.2, since $cb = da$, $c/d \simeq a/b$ and we have established that the relation "is equivalent to" on the set of fractions is a symmetric relation.

c. **Transitive** If a/b, c/d, and e/f are fractions with $a/b \simeq c/d$ and $c/d \simeq e/f$, we must prove that $a/b \simeq e/f$. If $a/b \simeq c/d$ and $c/d \simeq e/f$, then $ad = bc$ and $cf = de$ by Definition 9.2. Since $ad = bc$ and $cf = de$, we have, by

Theorem 7.2(c), $(ad)f = (bc)f$. Using the commutative and associative properties for multiplication of integers, we may write $(ad)f = (bc)f$ as $(af)d = b(cf)$. Since $cf = de$, we may rewrite $(af)d = b(cf)$ as $(af)d = b(de)$. Again using the commutative and associative properties of multiplication of integers, $(af)d = b(de)$ becomes $(af)d = (be)d$. Since $d \neq 0$ (d is a denominator), we may use the cancellation property for multiplication and rewrite $(af)d = (be)d$ as $af = be$. Hence by Definition 9.2 we have $a/b \simeq e/f$, and the transitive relation has been established.

EXERCISES

1. Determine which of the following pairs of fractions are equivalent.

a. $\dfrac{1}{2}$ and $\dfrac{2}{3}$.

b. $\dfrac{2}{3}$ and $\dfrac{4}{6}$.

c. $\dfrac{^-2}{5}$ and $\dfrac{10}{^-4}$.

d. $\dfrac{2}{7}$ and $\dfrac{^-4}{^-14}$.

e. $\dfrac{^-5}{9}$ and $\dfrac{5}{^-9}$.

f. $\dfrac{2+3}{5}$ and $\dfrac{6-2}{4}$.

g. $\dfrac{^-2 \times 3}{5}$ and $\dfrac{12}{^-10}$.

h. $\dfrac{a+b}{2}$ and $\dfrac{^-2(a+b)}{4}$.

i. $\dfrac{4-4}{5}$ and $\dfrac{^-5+5}{6}$.

j. $\dfrac{m}{n+p}$ and $\dfrac{2m}{2n+2p}$ if $n + p \neq 0$.

2. Determine the value of x such that each of the following pairs of fractions will be equivalent.

a. $\dfrac{2}{3}$ and $\dfrac{x}{^-6}$.

b. $\dfrac{3}{x}$ and $\dfrac{6}{8}$.

c. $\dfrac{8}{5}$ and $\dfrac{^-16}{x}$.

d. $\dfrac{x}{3}$ and $\dfrac{0}{5}$.

e. $\dfrac{2}{5}$ and $\dfrac{2x}{5}$.

f. $\dfrac{3x}{4}$ and $\dfrac{^-9}{^-12}$.

g. $\dfrac{^-4x}{7}$ and $\dfrac{16}{^-14}$.

h. $\dfrac{2x}{3}$ and $\dfrac{3x}{6}$.

i. $\dfrac{2x+1}{^-6}$ and $\dfrac{^-10}{12}$.

j. $\dfrac{3x-1}{5}$ and $\dfrac{4}{4}$.

3. Write five distinct fractions that are equivalent to each of the following.

a. $\dfrac{2}{3}$.

b. $\dfrac{^-3}{4}$.

c. $\dfrac{0}{5}$.

d. $\dfrac{^-2}{13}$.

e. $\dfrac{4}{5}$.

f. $\dfrac{^-7}{7}$.

g. $\dfrac{13}{^-7}$.

h. $\dfrac{^-4}{9}$.

i. $\dfrac{10}{^-11}$.

j. $\dfrac{1-3}{4}$.

k. $\dfrac{^-2+5}{3}$.

l. $\dfrac{2+3}{4-5}$.

m. $\dfrac{3 \times 2}{5}$.

n. $\dfrac{^-2 \times 5}{2}$.

o. $\dfrac{^-3 \times 4}{2-3}$.

9.2 RATIONAL NUMBERS

In the previous section we defined a fraction as a quotient of two integers provided the denominator is not equal to zero. The fraction 1/3 is a symbol that represents one part of three equal parts as indicated in Figure 9.4. Now

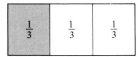

Figure 9.4

if the unit in Figure 9.4 is divided into six equal parts, we could consider two of those equal parts, as in Figure 9.5, and observe that 2/6 would represent the same part of the unit as 1/3 does. Of course this is not surprising since 2/6 ≃ 1/3. Either fraction may be considered as a ratio,

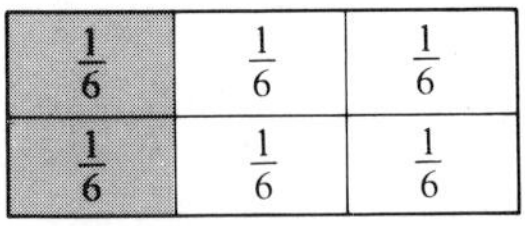

Figure 9.5

then, and the ratio 1/3 would be equivalent to the ratio 2/6. In a similar manner the fractions 6/8 and 3/4 represent the same ratio or part of the unit as indicated in Figure 9.6.

The fractions 2/3 and 4/6 are distinct but equivalent fractions which represent the same ratio or number. The number, called a *rational number* (or ratio number), can be represented by any fraction which is equivalent

Figure 9.6

to 2/3. Since the relation "is equivalent to" on the set of all fractions is an equivalence relation, the relation divides or partitions the set of fractions into disjoint equivalence classes. Hence the fraction 2/3 belongs to such an equivalence class and the rational number being discussed can be represented by any fraction from that equivalence class. The rational number, then, is the equivalence class containing the fraction 2/3. Therefore, the set $\{\ldots, {}^-4/{}^-6, {}^-2/{}^-3, 2/3, 4/6, 6/9, \ldots\}$ is the rational number that represents the ratio of 2 to 3 or 2/3. To distinguish between the symbol used to denote a rational number and the fraction which represents it, we will enclose the fraction between braces to denote the rational number. Hence 2/3 will represent a fraction but $\{2/3\}$ will represent the rational number as the equivalence class of all fractions which are equivalent to 2/3.

DEFINITION 9.3 A *rational number* denoted by $\left\{\dfrac{a}{b}\right\}$ or $\{a/b\}$ or $\{a \div b\}$, with $b \neq 0$, is the set of all fractions p/q such that $p/q \simeq a/b$. The set of all rational numbers will be denoted by Q.

EXAMPLE

a. $\left\{\dfrac{1}{3}\right\} = \left\{\ldots, \dfrac{{}^-3}{{}^-9}, \dfrac{{}^-2}{{}^-6}, \dfrac{{}^-1}{{}^-3}, \dfrac{1}{3}, \dfrac{2}{6}, \dfrac{3}{9}, \dfrac{4}{12}, \ldots\right\}$

b. $\left\{\dfrac{{}^-4}{7}\right\} = \left\{\ldots, \dfrac{{}^-12}{21}, \dfrac{{}^-8}{14}, \dfrac{{}^-4}{7}, \dfrac{4}{{}^-7}, \dfrac{8}{{}^-14}, \dfrac{12}{{}^-21}, \dfrac{16}{{}^-28}, \ldots\right\}$

c. $\left\{\dfrac{5}{2}\right\} = \left\{\ldots, \dfrac{{}^-15}{{}^-6}, \dfrac{{}^-10}{{}^-4}, \dfrac{{}^-5}{{}^-2}, \dfrac{5}{2}, \dfrac{10}{4}, \dfrac{15}{6}, \ldots\right\}$

d. $\left\{\dfrac{0}{1}\right\} = \left\{\ldots, \dfrac{0}{{}^-3}, \dfrac{0}{{}^-2}, \dfrac{0}{{}^-1}, \dfrac{0}{1}, \dfrac{0}{2}, \dfrac{0}{3}, \ldots\right\}$

The fractions 2/3 and 6/9 are distinct fractions. However, they are equivalent fractions, and we may write $2/3 \simeq 6/9$. However, the rational numbers $\{2/3\}$ and $\{6/9\}$ are not distinct since the set of all fractions equivalent to 2/3 is exactly the same as the set of all fractions equivalent to 6/9. Hence we say that the rational numbers $\{2/3\}$ and $\{6/9\}$ are equal and write $\{2/3\} = \{6/9\}$.

DEFINITION 9.4 If $\{a/b\}$ and $\{c/d\}$ are rational numbers, then $\{a/b\}$ is *equal* to $\{c/d\}$, denoted by $\{a/b\} = \{c/d\}$, if and only if $a/b \simeq c/d$.

EXAMPLE

a. $\left\{\dfrac{2}{7}\right\} = \left\{\dfrac{-6}{-21}\right\}$ since $\dfrac{2}{7} \simeq \dfrac{-6}{-21}$ since $(2)(-21) = (7)(-6)$

b. $\left\{\dfrac{3}{4}\right\} = \left\{\dfrac{12}{16}\right\}$ since $\dfrac{3}{4} \simeq \dfrac{12}{16}$ since $(3)(16) = (4)(12)$

c. $\left\{\dfrac{3}{5}\right\} \neq \left\{\dfrac{2}{4}\right\}$ since $\dfrac{3}{5} \neq \dfrac{2}{4}$ since $(3)(4) \neq (5)(2)$

Since a rational number is an equivalence class of fractions, it may be represented by any one of the fractions in the set. However, it is sometimes convenient to have one of the fractions in the equivalence set denoted as the "simplest" fraction to represent the rational number. This "simplest" fraction also is called the "reduced representative."

DEFINITION 9.5 If $\{a/b\}$ is a rational number and p/q is a fraction representing $\{a/b\}$, then p/q is called the reduced representative of $\{a/b\}$ if and only if $q > 0$ and g.c.d.$(p, q) = 1$.

EXAMPLE

a. The reduced representative of $\left\{\dfrac{10}{15}\right\}$ is $\dfrac{2}{3}$ since $\dfrac{10}{15} \simeq \dfrac{2}{3}$, $3 > 0$, and g.c.d.$(2, 3) = 1$.

b. The reduced representative of $\left\{\dfrac{-8}{-6}\right\}$ is $\dfrac{4}{3}$ since $\dfrac{-8}{-6} \simeq \dfrac{4}{3}$, $3 > 0$, and g.c.d.$(4, 3) = 1$.

c. The reduced representative of $\left\{\dfrac{4}{-5}\right\}$ is $\dfrac{-4}{5}$ since $\dfrac{4}{-5} \simeq \dfrac{-4}{5}$, $5 > 0$, and g.c.d.$(-4, 5) = 1$.

d. The reduced representative of $\left\{\dfrac{0}{6}\right\}$ is $\dfrac{0}{1}$ since $\dfrac{0}{6} \simeq \dfrac{0}{1}$, $1 > 0$, and g.c.d.$(0, 1) = 1$.

To determine the reduced representative or the simplest form of a rational number we use the prime factorization of both the numerator and denominator and the following theorem.

THEOREM 9.2 Fundamental Law of Fractions If a, b, and c are integers with $b \neq 0$ and $c \neq 0$, then $a/b \simeq ac/bc$.

Proof The proof of this theorem follows almost immediately from Definition 9.2 and is left as an exercise.

THEOREM 9.3 If $\{a/b\}$ is any rational number and $c \neq 0$ is an integer, then $\{a/b\} = \{ac/bc\}$.

Proof The proof of this theorem follows immediately from Definition 9.4 and Theorem 9.2 and is left as an exercise.

EXAMPLE Express $\{34/85\}$ in simplest form.

Solution Since $34 = 2 \times 17$ and $85 = 5 \times 17$, we have

$$\left\{\frac{34}{85}\right\} = \left\{\frac{(2)(17)}{(5)(17)}\right\}$$

$$= \left\{\frac{2}{5}\right\} \quad \text{by Theorem 9.3} \quad \blacksquare$$

EXAMPLE Determine the reduced representative for $\{224/{}^-368\}$.

Solution The required reduced representative will be the fraction a/b such that $b > 0$, g.c.d.$(a, b) = 1$ and $a/b \simeq 224/{}^-368$. Now $224/{}^-368 \simeq {}^-224/368$. Also, ${}^-224 = ({}^-1)(2)(2)(2)(2)(2)(7)$ and $368 = (2)(2)(2)(2)(23)$. Hence

$$\left\{\frac{{}^-224}{368}\right\} = \left\{\frac{({}^-1)(2)(2)(2)(2)(2)(7)}{(2)(2)(2)(2)(23)}\right\}$$

$$= \left\{\frac{({}^-1)(2)(7)(2 \times 2 \times 2 \times 2)}{(23)(2 \times 2 \times 2 \times 2)}\right\}$$

$$= \left\{\frac{({}^-1)(2)(7)}{23}\right\} \quad \text{by Theorem 9.2 with } c = 2 \times 2 \times 2 \times 2$$

$$= \left\{\frac{{}^-14}{23}\right\}$$

Hence the reduced representative is ${}^-14/23$. $\blacksquare$

EXERCISES

1. What is the relationship between a fraction and a rational number?
2. For each of the following identify whether the indicated expression is a fraction, a rational number, or neither.

a. $\dfrac{2}{3}$.

b. $\left\{\dfrac{{}^-2}{5}\right\}$.

c. $\left\{\dfrac{5}{0}\right\}$.

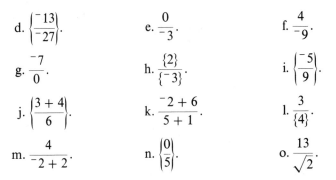

d. $\left|\dfrac{^-13}{^-27}\right|$. e. $\dfrac{0}{^-3}$. f. $\dfrac{4}{^-9}$.

g. $\dfrac{^-7}{0}$. h. $\dfrac{\{2\}}{\{^-3\}}$. i. $\left|\dfrac{^-5}{9}\right|$.

j. $\left|\dfrac{3+4}{6}\right|$. k. $\dfrac{^-2+6}{5+1}$. l. $\dfrac{3}{\{4\}}$.

m. $\dfrac{4}{^-2+2}$. n. $\left|\dfrac{0}{5}\right|$. o. $\dfrac{13}{\sqrt{2}}$.

3. Determine which of the following pairs of rational numbers are equal.

a. $\left|\dfrac{2}{^-3}\right|$ and $\left|\dfrac{^-4}{6}\right|$. b. $\left|\dfrac{^-1}{7}\right|$ and $\left|\dfrac{^-2}{9}\right|$.

c. $\left|\dfrac{0}{4}\right|$ and $\left|\dfrac{0}{5}\right|$, d. $\left|\dfrac{^-13}{2}\right|$ and $\left|\dfrac{39}{^-6}\right|$.

e. $\left|\dfrac{108}{26}\right|$ and $\left|\dfrac{^-54}{^-13}\right|$. f. $\left|\dfrac{11}{20}\right|$ and $\left|\dfrac{9}{17}\right|$.

g. $\left|\dfrac{^-2+3}{4}\right|$ and $\left|\dfrac{2}{5+3}\right|$. h. $\left|\dfrac{4}{^-5}\right|$ and $\left|\dfrac{^-14+2}{15}\right|$.

i. $\left|\dfrac{5\times5}{4}\right|$ and $\left|\dfrac{4\times4}{5}\right|$. j. $\left|\dfrac{^-2\times3}{5}\right|$ and $\left|\dfrac{2-26}{^-20}\right|$.

4. Determine which of the following fractions are reduced representatives for rational numbers.

a. $\dfrac{2}{5}$. b. $\dfrac{3}{^-4}$. c. $\dfrac{0}{2}$.

d. $\dfrac{2}{^-3}$. e. $\dfrac{14}{26}$. f. $\dfrac{^-15}{20}$.

g. $\dfrac{^-13}{17}$. h. $\dfrac{23}{^-7}$. i. $\dfrac{6}{6}$.

j. $\dfrac{^-15}{8}$. k. $\dfrac{^-51}{4}$. l. $\dfrac{105}{^-201}$.

m. $\dfrac{391}{209}$. n. $\dfrac{^-369}{100}$. o. $\dfrac{3125}{100}$.

5. If N represents the set of natural numbers, W the set of whole numbers, I the set of integers, and Q the set of rational numbers, identify which of the following statements are true and which are false.

a. If $a \in I$, then $\dfrac{a}{1}$ represents an element in Q.

b. If $b \in N$, then $b \in W$ and $b \in I$.

c. If $\left\{\dfrac{a}{b}\right\} \in Q$, then $a \in I$ and $b \in N$.

d. If $\left\{\dfrac{c}{d}\right\} \in Q$, then $c \in I$ and $d \in I$ but $d \notin (W - N)$.

6. Prove Theorem 9.2.
7. Prove Theorem 9.3.

9.3 GRAPHICAL REPRESENTATION OF RATIONAL NUMBERS

In Chapter 5 we represented the whole numbers on the so-called number line and extended this representation to the integers in Chapter 7. In this section we will discuss the representation of the rational numbers on the number line as an extension of what has already been done.

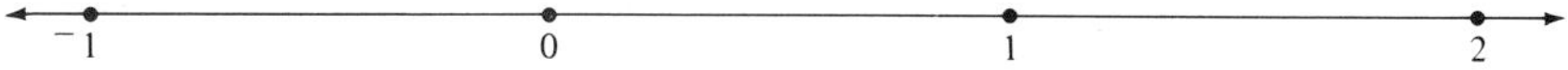

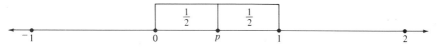

Figure 9.7

Consider the number line with the integers indicated such as in Figure 9.7. On the portion of the line between 0 and 1, construct a "rectangle" with a length of one unit as indicated in Figure 9.8. Then, as in Section 9.1, divide the unit into two equals parts. Now since the length of the "rectangle" is one unit long, the distance along the line from the whole number 0-point to the whole number 1-point, is also one unit. It should be somewhat obvious, then, that the distance from the whole number 0-point to the p-point is 1/2. Hence we could locate the fraction 1/2 at p. But $2/4 \simeq 1/2$ and, hence, 2/4 would also be located at p. Similarly, any fraction equivalent to 1/2 also would be located at p. Hence the point p would correspond to the rational number $\{1/2\}$.

$\dfrac{1}{2}$	$\dfrac{1}{2}$	

Figure 9.8

In a similar manner we could locate the rational numbers $\{1/4\}$, $\{2/4\}$, and $\{3/4\}$ on the number line as illustrated in Figure 9.9.

Extending this procedure, we could locate additional rational numbers on the number line as indicated in Figure 9.10. Observe that the rational

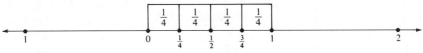

Figure 9.9

number $\{2/2\}$ corresponds to the integer 1 and the rational number $\{0/1\}$ corresponds to the integer 0. In general the rational number $\{n/1\}$ corresponds to the integer n.

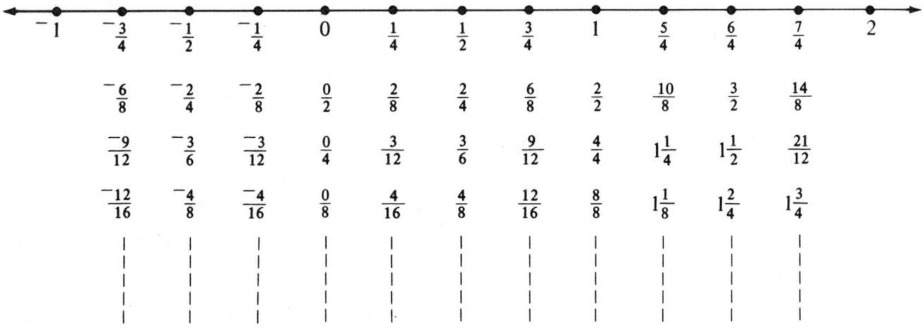

Figure 9.10

EXERCISES

Represent each of the following rational numbers on the number line.

1. $\left\{\dfrac{2}{3}\right\}$,

2. $\left\{\dfrac{^-3}{4}\right\}$.

3. $\left\{\dfrac{0}{6}\right\}$.

4. $\left\{\dfrac{^-4}{8}\right\}$.

5. $\left\{\dfrac{3}{^-12}\right\}$.

6. $\left\{\dfrac{7}{7}\right\}$.

7. $\left\{\dfrac{13}{^-13}\right\}$.

8. $\left\{\dfrac{50}{^-25}\right\}$.

9. $\left\{\dfrac{^-125}{250}\right\}$.

10. $\left\{\dfrac{486}{^-162}\right\}$.

9.4 ADDITION OF RATIONAL NUMBERS

In this section we will define the addition of rational numbers as an extension of the definition of addition of integers. In formulating the definition, we will attempt to retain all of the properties of addition on the set I.

On the number line the rational number $\{2/1\}$ corresponds to the integer 2 and the rational number $\{3/1\}$ corresponds to the integer 3. Since we know that $2 + 3 = 5$, we should attempt to define addition of rational numbers such that $\{2/1\} + \{3/1\} = \{5/1\}$, as illustrated in Figure 9.11.

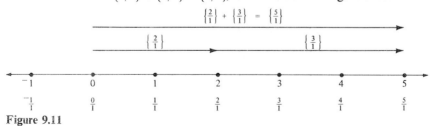

Figure 9.11

In computations involving rational numbers we generally use the fractions $2/3$ and $4/5$ instead of the rational numbers $\{2/3\}$ and $\{4/5\}$ and also write $2/3 = 4/6$ instead of $2/3 \simeq 4/6$ to mean $\{2/3\} = \{4/6\}$. Throughout the balance of this text we will not continue to make the distinction between a rational number such as $\{a/b\}$ and its fraction numeral a/b. We will use the symbol a/b to denote either the fraction a/b or the rational number $\{a/b\}$; the context in which the symbol is used will indicate what is being represented.

Returning now to the addition of rational numbers, we see, then, that $2/1 + 3/1 = (2 + 3)/1 = 5/1$. In general we will define the addition of rational numbers such that $a/1 + b/1 = (a + b)/1$.

What happens when we attempt to add two rational numbers represented by fractions with the same denominators but different from 1? For instance, how can we define $2/3 + 5/3$? On the number line in Figure 9.12 we illustrate $2/3 + 5/3$ and observe that the sum is $(2 + 5)/3$ or $7/3$.

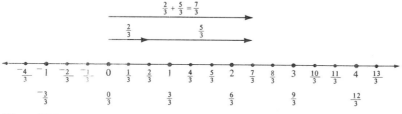

Figure 9.12

In general, then, we will define the addition of rational numbers such that $\{a/c\} + \{b/c\} = \{(a + b)/c\}$, or simply $a/c + b/c = (a + b)/c$.

We next consider what happens when we attempt to add two rational numbers represented by fractions with different denominators. For instance, how can we define $2/3 + 1/2$? By Theorem 9.2 we have $2/3 \simeq (2 \times 2)/(3 \times 2)$ or $2/3 \simeq 4/6$ and $1/2 \simeq (3 \times 1)/(3 \times 2)$ or $1/2 \simeq 3/6$. Hence

$$2/3 + 1/2 = 4/6 + 3/6$$

Now the denominators of the fractions representing the two rational numbers are the same. Thus

$$\frac{4}{6} + \frac{3}{6} = \frac{4+3}{6} = \frac{7}{6}$$

Combining these results, we have

$$\frac{2}{3} + \frac{1}{2} = \frac{2 \times 2}{3 \times 2} + \frac{3 \times 1}{3 \times 2} = \frac{(2 \times 2) + (3 \times 1)}{3 \times 2}$$

In general, if $\{a/b\}$ and $\{c/d\}$ are rational numbers, we have

$$\left\{\frac{a}{b}\right\} + \left\{\frac{c}{d}\right\} = \left\{\frac{a \cdot d}{b \cdot d}\right\} + \left\{\frac{b \cdot c}{b \cdot d}\right\} = \left\{\frac{(a \cdot d) + (b \cdot c)}{b \cdot d}\right\}$$

or simply

$$\frac{a}{b} + \frac{c}{d} = \frac{ad}{bd} + \frac{bc}{bd} = \frac{ad + bc}{bd}$$

DEFINITION 9.6 If $\{a/b\}$ and $\{c/d\}$ are any two rational numbers, then $\{a/b\} + \{c/d\} = \{(ad + bc)/bd\}$.

EXAMPLE

a. $\dfrac{1}{2} + \dfrac{2}{5} = \dfrac{(1)(5) + (2)(2)}{(2)(5)} = \dfrac{5 + 4}{10} = \dfrac{9}{10}$

b. $\dfrac{3}{7} + \dfrac{5}{4} = \dfrac{(3)(4) + (7)(5)}{(7)(4)} = \dfrac{12 + 35}{28} = \dfrac{47}{28}$

c. $\dfrac{^-2}{3} = \dfrac{4}{^-5} = \dfrac{(^-2)(^-5) + (3)(4)}{(3)(^-5)} = \dfrac{10 + 12}{^-15} = \dfrac{22}{^-15} = \dfrac{^-22}{15}$

d. $\dfrac{4}{7} + \dfrac{0}{^-1} = \dfrac{(4)(^-1) + (7)(0)}{(7)(^-1)} = \dfrac{^-4 + 0}{^-7} = \dfrac{^-4}{^-7} = \dfrac{4}{7}$

It is easy to show that the properties for addition that were assumed on the set I are also valid on the set of rational numbers.

9.41 CLOSURE PROPERTY

We wish to show that if a/b, $c/d \in Q$, then $a/b + c/d \in Q$. From the definition of addition of rational numbers we have $a/b + c/d = (ad + bc)/bd$. Now $a/b + c/d \in Q$ if and only if $(ad + bc)/bd$ is a fraction. Since a/b, $c/d \in Q$, then a/b and c/d are fractions. Hence a, b, c, $d \in I$ with $b \neq 0$ and $d \neq 0$ by the definition of a fraction. Therefore ad, bc, $bd \in I$ by the closure property of I under the operation of multiplication. Also, $ad + bc \in I$

by the closure property of I under addition. Hence $(ad + bc)/bd$ is a quotient of integers. Also, $bd \neq 0$ by the multiplication property of zero on I since $b \neq 0$ and $d \neq 0$. Therefore, $(ad + bc)/bd$ is a fraction and $(ad + bc)/bd \in Q$.

EXAMPLE

a. $\dfrac{2}{3}, \dfrac{4}{5} \in Q$ and $\dfrac{2}{3} + \dfrac{4}{5} \in Q$

since

$$\frac{2}{3} + \frac{4}{5} = \frac{(2)(5) + (3)(4)}{(3)(5)} = \frac{10 + 12}{15} = \frac{22}{15} \in Q$$

b. $\dfrac{^-3}{7}, \dfrac{4}{^-9} \in Q$ and $\dfrac{^-3}{7} + \dfrac{4}{^-9} \in Q$

since

$$\frac{^-3}{7} + \frac{4}{^-9} = \frac{(^-3)(^-9) + (7)(4)}{(7)(^-9)} = \frac{27 + 28}{^-63} = \frac{55}{^-63} \in Q$$

9.42 COMMUTATIVE PROPERTY

We wish to show that if a/b, $c/d \in Q$, then $a/b + c/d = c/d + a/b$. This property follows immediately from the definition of addition of rational numbers and properties of integers and is left as an exercise.

EXAMPLE   Verify that $\dfrac{3}{7} + \dfrac{^-4}{5} = \dfrac{^-4}{5} + \dfrac{3}{7}$.

Solution

$$\frac{3}{7} + \frac{^-4}{5} = \frac{(3)(5) + (7)(^-4)}{(7)(5)} = \frac{15 + ^-28}{35} = \frac{^-13}{35}$$

$$\frac{^-4}{5} + \frac{3}{7} = \frac{(^-4)(7) + (5)(3)}{(5)(7)} = \frac{^-28 + 15}{35} = \frac{^-13}{35}$$

Hence

$$\frac{3}{7} + \frac{^-4}{5} = \frac{^-4}{5} + \frac{3}{7} \quad \blacksquare$$

9.43 ASSOCIATIVE PROPERTY

We wish to show that if a/b, c/d, $e/f \in Q$, then $(a/b + c/d) + e/f = a/b + (c/d + e/f)$. This is also left as an exercise.

EXAMPLE   Verify that $(1/3 + 2/5) + {}^-3/4 = 1/3 + (2/5 + {}^-3/4)$.
Solution

$$\left(\frac{1}{3} + \frac{2}{5}\right) + \frac{{}^-3}{4} = \frac{(1)(5) + (3)(2)}{(3)(5)} + \frac{{}^-3}{4} = \frac{5 + 6}{15} + \frac{{}^-3}{4}$$

$$= \frac{11}{15} + \frac{{}^-3}{4} = \frac{(11)(4) + (15)({}^-3)}{(15)(4)} = \frac{44 + {}^-45}{60} = \frac{{}^-1}{60}$$

$$\frac{1}{3} + \left(\frac{2}{5} + \frac{{}^-3}{4}\right) = \frac{1}{3} + \frac{(2)(4) + (5)({}^-3)}{(5)(4)} = \frac{1}{3} + \frac{8 + {}^-15}{20}$$

$$= \frac{1}{3} + \frac{{}^-7}{20} = \frac{(1)(20) + (3)({}^-7)}{(3)(20)} = \frac{20 + {}^-21}{60} = \frac{{}^-1}{60}$$

Hence

$$\left(\frac{1}{3} + \frac{2}{5}\right) + \frac{{}^-3}{4} = \frac{1}{3} + \left(\frac{2}{5} + \frac{{}^-3}{4}\right) \quad\blacksquare$$

9.44 ADDITIVE IDENTITY PROPERTY

This property states that there exists a unique rational number, $0/1$, such that if $a/b \in Q$, then $a/b + 0/1 = 0/1 + a/b = a/b$. To prove that $a/b + 0/1 = a/b$ involves the use of the definition for the addition of rational numbers. Hence $a/b + 0/1 = [(a)(1) + (b)(0)]/(b)(1) = (a + 0)/b = a/b$. The proof that $0/1 + a/b = a/b$ is very similar. To prove that the additive identity $0/1$ is unique, we would assume the existence of two distinct additive identities and arrive at a contradiction. We leave this part of the proof as an exercise.

EXAMPLE

a. $\dfrac{2}{3} + \dfrac{0}{1} = \dfrac{(2)(1) + (3)(0)}{(3)(1)} = \dfrac{2 + 0}{3} = \dfrac{2}{3}$

b. $\dfrac{{}^-3}{5} + \dfrac{0}{1} = \dfrac{({}^-3)(1) + (5)(0)}{(5)(1)} = \dfrac{{}^-3 + 0}{5} = \dfrac{{}^-3}{5}$

c. $\dfrac{0}{1} + \dfrac{4}{{}^-9} = \dfrac{(0)({}^-9) + (1)(4)}{(1)({}^-9)} = \dfrac{0 + 4}{{}^-9} = \dfrac{4}{{}^-9}$

d. $\dfrac{0}{1} + \dfrac{0}{1} = \dfrac{(0)(1) + (1)(0)}{(1)(1)} = \dfrac{0 + 0}{1} = \dfrac{0}{1}$

9.45 ADDITIVE INVERSE PROPERTY

For each rational number a/b we will define the rational number ${}^-a/b$ as its unique *additive inverse* such that $a/b + {}^-a/b = {}^-a/b + a/b = 0/1$, where $0/1$ is the additive identity in Q. We leave the proof of this property as an exercise.

EXAMPLE

a. $\dfrac{^-4}{7}$ is the additive inverse of $\dfrac{4}{7}$

since

$$\frac{^-4}{7} + \frac{4}{7} = \frac{(^-4)(7) + (7)(4)}{(7)(7)} = \frac{^-28 + 28}{49} = \frac{0}{49} = \frac{0}{1}$$

b. $\dfrac{5}{9}$ is the additive inverse of $\dfrac{5}{^-9}$

since

$$\frac{5}{9} + \frac{5}{^-9} = \frac{(5)(^-9) + (9)(5)}{(9)(^-9)} = \frac{^-45 + 45}{^-81} = \frac{0}{^-81} = \frac{0}{1}$$

c. $\dfrac{0}{1}$ is the additive inverse of $\dfrac{0}{1}$

since

$$\frac{0}{1} + \frac{0}{1} = \frac{(0)(1) + (1)(0)}{(1)(1)} = \frac{0 + 0}{1} = \frac{0}{1}.$$

The numeral that represents a rational number is a fraction, and the numerator of the fraction may be less than, equal to, or larger than the denominator of the fraction. For instance, the fractions 2/3, 4/4, and 17/6 all represent rational numbers. Sometimes the fraction 17/6 is written as $2\frac{5}{6}$ and is referred to as representing a *mixed number* insofar as $2\frac{5}{6}$ means $2 + \frac{5}{6}$ where 2 represents an integer and $\frac{5}{6}$ represents a rational number. To change the mixed number $2\frac{5}{6}$ back to a fraction we simply write $2\frac{5}{6} = 2 + \frac{5}{6}$ and rewrite 2 as $\frac{2}{1}$. Then

$$2\frac{5}{6} = \frac{2}{1} + \frac{5}{6} = \frac{(2)(6)}{(1)(6)} + \frac{5}{6} = \frac{12}{6} + \frac{5}{6} = \frac{17}{6}$$

EXAMPLE Rewrite the fraction $\frac{63}{29}$ as a mixed number.
Solution

$$\frac{63}{29} = \frac{58 + 5}{29} = \frac{58}{29} + \frac{5}{29} = 2 + \frac{5}{29} = 2\frac{5}{29} \quad \blacksquare$$

EXAMPLE Rewrite the mixed number $3\frac{4}{17}$ as a fraction.
Solution

$$3\frac{4}{17} = 3 + \frac{4}{17} = \frac{3}{1} + \frac{4}{17} = \frac{(3)(17)}{(1)(17)} + \frac{4}{17} = \frac{51}{17} + \frac{4}{17} = \frac{55}{17} \quad \blacksquare$$

Alternate solution: A method the student may be familiar with to rewrite the mixed number $3\frac{4}{17}$ as a fraction follows. Multiply the integer part, 3, by the denominator of the fraction, 17, to obtain 51, and add to this the numerator of the fraction, 4, to obtain 55. Write this sum, 55, over the denominator, 17, as $\frac{55}{17}$.

EXAMPLE   To add two mixed numbers, we rewrite them as equivalent fractions and add the fractions. For instance

$$3\frac{1}{2} + 4\frac{2}{7} = \frac{7}{2} + \frac{30}{7} = \frac{(7)(7)}{(2)(7)} + \frac{(2)(30)}{(2)(7)} = \frac{49}{14} + \frac{60}{14} = \frac{109}{14}$$

which may now be written as the mixed number $7\frac{11}{14}$. Observe that the same result would have been obtained by adding the integer portion of the two mixed numbers, obtaining $3 + 4 = 7$, adding the fraction portions, obtaining $\frac{1}{2} + \frac{2}{7} = \frac{11}{14}$, and combining the results.

We will conclude this section with some useful theorems.

THEOREM 9.4   If a/b, c/d, and e/f are any rational numbers with $a/b = c/d$, then $a/b + e/f = c/d + e/f$.
Proof   The proof is left as an exercise.

THEOREM 9.5   Cancellation Property for Addition   If a/b, c/d, and e/f are any rational numbers with $a/b + e/f = c/d + e/f$, then $a/b = c/d$.
Proof   The proof is left as an exercise.

THEOREM 9.6   If a/b, c/d, e/f, and g/h are any rational numbers with $a/b = c/d$ and $e/f = g/h$, then $a/b + e/f = c/d + g/h$.
Proof   The proof is left as an exercise.

EXERCISES

1. Perform the following additions using Definition 9.6, and write each result in its simplest form with a positive denominator.

   a. $\dfrac{2}{5} + \dfrac{3}{4}$.

   b. $\dfrac{^-1}{5} + \dfrac{2}{^-7}$.

   c. $\dfrac{4}{13} + \dfrac{^-5}{9}$.

   d. $\dfrac{5}{^-11} + \dfrac{^-2}{7}$.

   e. $\dfrac{6}{15} + \dfrac{0}{^-3}$.

   f. $\dfrac{10}{19} + \dfrac{^-1}{3}$.

   g. $\left(\dfrac{2}{3} + \dfrac{^-1}{4}\right) + \dfrac{2}{5}$.

   h. $\dfrac{^-2}{7} + \left(\dfrac{4}{5} + \dfrac{^-3}{7}\right)$.

i. $\left(\dfrac{2}{5} + \dfrac{^-1}{7}\right) + \left(\dfrac{1}{3} + \dfrac{^-2}{5}\right).$ j. $\left(\dfrac{2}{9} + \dfrac{0}{^-3}\right) + \left(\dfrac{6}{^-8} + \dfrac{0}{4}\right).$

2. Verify each of the following to illustrate that the addition of rational numbers is a commutative operation. Use Definition 9.6 to perform the additions.

a. $\dfrac{2}{3} + \dfrac{^-3}{7} = \dfrac{^-3}{7} + \dfrac{2}{3}.$ b. $\dfrac{4}{9} + \dfrac{0}{^-3} = \dfrac{0}{^-3} + \dfrac{4}{9}.$

c. $\dfrac{7}{11} + \dfrac{^-1}{2} = \dfrac{^-1}{2} + \dfrac{7}{11}.$ d. $\dfrac{^-2}{5} + \dfrac{^-3}{7} = \dfrac{^-3}{7} + \dfrac{^-2}{5}.$

3. Verify each of the following to illustrate that the addition of rational numbers is an associative operation. Use Definition 9.6 to perform the additions.

a. $\left(\dfrac{1}{2} + \dfrac{1}{3}\right) + \dfrac{1}{4} = \dfrac{1}{2} + \left(\dfrac{1}{3} + \dfrac{1}{4}\right).$

b. $\left(\dfrac{^-2}{5} + \dfrac{1}{4}\right) + \dfrac{^-2}{7} = \dfrac{^-2}{5} + \left(\dfrac{1}{4} + \dfrac{^-2}{7}\right).$

c. $\left(\dfrac{3}{10} + \dfrac{0}{4}\right) + \dfrac{^-1}{2} = \dfrac{3}{10} + \left(\dfrac{0}{4} + \dfrac{^-1}{2}\right).$

4. Determine the additive inverse (the negative or the opposite) of each of the following.

a. $\dfrac{2}{3}.$ b. $\dfrac{^-1}{4}.$ c. $\dfrac{0}{3}.$

d. $\dfrac{^-4}{5}.$ e. $\dfrac{3}{^-7}.$ f. $\dfrac{^-2}{^-9}.$

g. $\dfrac{0}{^-11}.$ h. $\dfrac{^-13}{^-21}.$ i. $\dfrac{4}{8}.$

j. $\dfrac{4}{^-13}.$ k. $\dfrac{^-26}{1}.$ l. $\dfrac{37}{^-53}.$

5. Rewrite each of the following fractions as a mixed number.

a. $\dfrac{23}{12}.$ b. $\dfrac{37}{12}.$ c. $\dfrac{53}{17}.$

d. $\dfrac{119}{17}.$ e. $\dfrac{201}{17}.$ f. $\dfrac{692}{25}.$

g. $\dfrac{^-27}{11}.$ h. $\dfrac{^-89}{26}.$ i. $\dfrac{^-196}{57}.$

6. For each of the following determine the least common denominator for the fractions given. Then rewrite each fraction in equivalent form with the least common denominator. Finally, determine the indicated sum and express your results in simplest form.

a. $\dfrac{2}{3} + \dfrac{3}{6}$.

b. $\dfrac{^-3}{5} + \dfrac{4}{15}$.

c. $\dfrac{5}{9} + \dfrac{7}{36}$.

d. $\dfrac{^-5}{11} + \dfrac{2}{3}$.

e. $\dfrac{11}{36} + \dfrac{0}{54}$.

f. $\dfrac{2}{17} + \dfrac{3}{7}$.

g. $\dfrac{2}{3} + \dfrac{3}{4} + \dfrac{4}{5}$.

h. $\dfrac{1}{7} + \dfrac{^-2}{4} + \dfrac{3}{14}$.

i. $\dfrac{^-7}{51} + \dfrac{2}{3} + \dfrac{^-5}{17}$.

j. $\dfrac{^-1}{16} + \dfrac{3}{48} + \dfrac{^-5}{24}$.

7. Prove that for each rational number $\dfrac{a}{b}$ there exists a rational number $\dfrac{^-a}{b}$ such that $\dfrac{a}{b} + \dfrac{^-a}{b} = \dfrac{^-a}{b} + \dfrac{a}{b} = \dfrac{0}{1}$, where $\dfrac{0}{1}$ is the additive identity in Q.

8. Prove Theorem 9.4.
9. Prove Theorem 9.5.
10. Prove Theorem 9.6.

9.5 MULTIPLICATION OF RATIONAL NUMBERS

In the previous section we defined the operation of addition of rational numbers in such a manner that we retained the properties for addition of the integers. In this section we continue the development of the system of rational numbers by defining the operation of multiplication of rational numbers by retaining the definition and properties of multiplication of integers.

Since the rational numbers $\frac{3}{1}$, $\frac{4}{1}$, and $\frac{12}{1}$ correspond, respectively, to the integers 3, 4, and 12, and since $3 \times 4 = 12$, we could define the product of the rational numbers $\frac{3}{1}$ and $\frac{4}{1}$ to be the rational number $\frac{12}{1}$. In a similar manner, since the rational numbers $\frac{7}{1}$, $\frac{5}{1}$, and $\frac{35}{1}$ correspond, respectively, to the integers 7, 5, and 35, and since $7 \times 5 = 35$, we could define the product of the rational numbers $\frac{7}{1}$ and $\frac{5}{1}$ to be the rational number $\frac{35}{1}$. In general we could define the product of the rational numbers $a/1$ and $b/1$ to be the rational number $ab/1$ where the fraction is formed by multiplying the numerators of the fractions $a/1$ and $b/1$ together, obtaining ab, and multiplying their denominators, obtaining 1.

What would happen, however, if we tried to multiply the rational numbers $\frac{2}{3}$ and $\frac{3}{4}$ in a similar manner, obtaining the rational number $\frac{6}{12}$ or $\frac{1}{2}$?

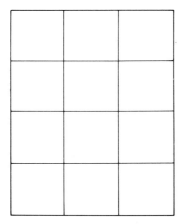

Figure 9.13

Would this be feasible? To answer the question let's consider the following. The fraction $\frac{2}{3}$ means 2 of 3 equal parts, and the fraction $\frac{3}{4}$ means 3 of 4 equal parts. Now consider the rectangular region in Figure 9.13. Divide the region vertically into 3 equal parts and horizontally into 4 equal parts as indicated. Next shade any two of the 3 equal vertical parts as in Figure 9.14. This corresponds to the fraction $\frac{2}{3}$. Also shade any 3 of the 4 equal horizontal parts as indicated in Figure 9.15. This corresponds to the fraction $\frac{3}{4}$. Finally, observe that there are 12 equal parts comprising the rectangular region and that 6 of them are shaded twice. Hence we have 6 of 12 equal parts or the fraction $\frac{6}{12}$.

Observe that if we multiply the numerators of the fractions $\frac{2}{3}$ and $\frac{3}{4}$, we do get 6 and the product of their denominators is 12. This would suggest that the product of the rational numbers $\frac{2}{3}$ and $\frac{3}{4}$ is the rational number $\frac{6}{12}$.

Figure 9.14

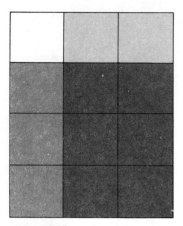

Figure 9.15

Another procedure that could be used to motivate the definition for multiplication of rational numbers is the following. Again, consider the rational numbers represented by the fractions $\frac{2}{3}$ and $\frac{3}{4}$. Since $\frac{2}{3} + \frac{2}{3} + \frac{2}{3} = \frac{6}{3} = \frac{2}{1}$ and $\frac{3}{4} + \frac{3}{4} + \frac{3}{4} + \frac{3}{4} = \frac{12}{4} = \frac{3}{1}$, consider the rectangular region in Figure 9.16, which is 2 units long and 3 units wide.

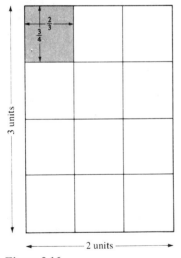

Figure 9.16

The rectangular region in the upper left-hand corner in Figure 9.16 represents 1 of 12 equal parts in the total region. Now since $2 \times 3 = 6$, the 12 equal parts would represent 6 units and, therefore, any 2 of the 12 equal parts would represent 1 unit. The shaded portion, then, would represent 1 of 2 equal units or the fraction $\frac{1}{2}$. Now the fraction $\frac{1}{2}$ is equivalent to the fraction $\frac{6}{12}$, again suggesting that the product of the two rational numbers $\frac{2}{3}$ and $\frac{3}{4}$ is the rational number $\frac{6}{12}$.

DEFINITION 9.7 If a/b and c/d are any two rational numbers, then
$$a/b \times c/d = ac/bd.$$

EXAMPLE

a. $\dfrac{1}{2} \times \dfrac{1}{2} = \dfrac{(1)(1)}{(2)(2)} = \dfrac{1}{4}$

b. $\dfrac{^-2}{3} \times \dfrac{4}{5} = \dfrac{(^-2)(4)}{(3)(5)} = \dfrac{^-8}{15}$

c. $\dfrac{3}{^-7} \times \dfrac{^-2}{11} = \dfrac{(3)(^-2)}{(^-7)(11)} = \dfrac{^-6}{^-77} = \dfrac{6}{77}$

d. $\dfrac{2}{9} \times \dfrac{0}{1} = \dfrac{(2)(0)}{(9)(1)} = \dfrac{0}{9} = \dfrac{0}{1}$

e. $\dfrac{0}{1} \times \dfrac{0}{1} = \dfrac{(0)(0)}{(1)(1)} = \dfrac{0}{1}$

In the previous section we introduced mixed numbers. Multiplication of mixed numbers can be accomplished by treating the mixed numbers as rational numbers since a mixed number can be written as a fraction.

EXAMPLE

a. $\left(5\dfrac{2}{3}\right) \cdot \left(4\dfrac{1}{2}\right) = \dfrac{17}{3} \times \dfrac{9}{2} = \dfrac{(17)(9)}{(3)(2)} = \dfrac{153}{6} = 25\dfrac{1}{2}$

b. $\left(3\dfrac{1}{2}\right) \cdot \left(5\dfrac{4}{7}\right) = \dfrac{7}{2} \times \dfrac{39}{7} = \dfrac{(7)(39)}{(2)(7)} = \dfrac{273}{14} = 19\dfrac{1}{2}$

c. $\left(^-2\dfrac{4}{9}\right) \cdot \left(7\dfrac{1}{3}\right) = \dfrac{^-22}{9} \times \dfrac{22}{3} = \dfrac{(^-22)(22)}{(9)(3)} = \dfrac{^-484}{27} = {}^-17\dfrac{25}{27}$

Observe that in part (a) of the above example the product of $5\frac{2}{3}$ and $4\frac{1}{2}$ is *not* obtained by multiplying the integer parts, multiplying the fractional parts, and adding the two products. Actually, the product is obtained by finding the sum of $(5)(4) + (5)(\frac{1}{2}) + (\frac{2}{3})(4) + (\frac{2}{3})(\frac{1}{2})$.

With the definition for the multiplication of rational numbers as given in Definition 9.7 and using the properties of integers, it is easy to prove that the properties assumed for multiplication of integers also are valid for rational numbers. We will state these properties and leave the proofs as exercises.

9.51 CLOSURE PROPERTY

If a/b, $c/d \in Q$, then $a/b \times c/d \in Q$.

EXAMPLE

a. $\dfrac{2}{3}, \dfrac{4}{7} \in Q$ and $\dfrac{2}{3} \times \dfrac{4}{7} \in Q$ since $\dfrac{2}{3} \times \dfrac{4}{7} = \dfrac{(2)(4)}{(3)(7)} = \dfrac{8}{21} \in Q.$

b. $\dfrac{^-3}{4}, \dfrac{^-5}{9} \in Q$ and $\dfrac{^-3}{4} \times \dfrac{^-5}{9} \in Q$ since $\dfrac{^-3}{4} \times \dfrac{^-5}{9} = \dfrac{(^-3)(^-5)}{(4)(9)} = \dfrac{15}{36} \in Q.$

9.52 COMMUTATIVE PROPERTY

If a/b and c/d are any two rational numbers, then $a/b \times c/d = c/d \times a/b$.

EXAMPLE Verify that

$$\frac{4}{7} \times \frac{^-3}{8} = \frac{^-3}{8} \times \frac{4}{7}$$

Solution

$$\frac{4}{7} \times \frac{^-3}{8} = \frac{(4)(^-3)}{(7)(8)} = \frac{^-12}{56}$$

$$\frac{^-3}{8} \times \frac{4}{7} = \frac{(^-3)(4)}{(8)(7)} = \frac{^-12}{56}$$

Hence

$$\frac{4}{7} \times \frac{^-3}{8} = \frac{^-3}{8} \times \frac{4}{7} \quad \blacksquare$$

9.53 ASSOCIATIVE PROPERTY

If a/b, c/d, $e/f \in Q$, then $(a/b \times c/d) \times e/f = a/b \times (c/d \times e/f)$.

EXAMPLE Verify that

$$\left(\frac{^-1}{3} \times \frac{2}{5}\right) \times \frac{^-7}{9} = \frac{^-1}{3} \times \left(\frac{2}{5} \times \frac{^-7}{9}\right)$$

Solution

$$\left(\frac{^-1}{3} \times \frac{2}{5}\right) \times \frac{^-7}{9} = \frac{(^-1)(2)}{(3)(5)} \times \frac{^-7}{9} = \frac{^-2}{15} \times \frac{^-7}{9} = \frac{(^-2)(^-7)}{(15)(9)} = \frac{14}{135}$$

$$\frac{^-1}{3} \times \left(\frac{2}{5} \times \frac{^-7}{9}\right) = \frac{^-1}{3} \times \frac{(2)(^-7)}{(5)(9)} = \frac{^-1}{3} \times \frac{^-14}{45} = \frac{(^-1)(^-14)}{(3)(45)} = \frac{14}{135}$$

Hence

$$\left(\frac{^-1}{3} \times \frac{2}{5}\right) \times \frac{^-7}{9} = \frac{^-1}{3} \times \left(\frac{2}{5} \times \frac{^-7}{9}\right) \quad \blacksquare$$

9.54 DISTRIBUTIVE PROPERTY OF MULTIPLICATION OVER ADDITION

If a/b, c/d, $e/f \in Q$, then $a/b \times (c/d + e/f) = (a/b \times c/d) + (a/b \times e/f)$.

EXAMPLE Verify that

$$\frac{^-1}{3} \times \left(\frac{2}{5} + \frac{^-2}{7}\right) = \left(\frac{^-1}{3} \times \frac{2}{5}\right) + \left(\frac{^-1}{3} \times \frac{^-2}{7}\right)$$

Solution

$$\frac{^-1}{3} \times \left(\frac{2}{5} + \frac{^-2}{7}\right) = \frac{^-1}{3} \times \frac{(2)(7) + (5)(^-2)}{(5)(7)} = \frac{^-1}{3} \times \frac{14 + {}^-10}{35}$$

$$= \frac{^-1}{3} \times \frac{4}{35} = \frac{(^-1)(4)}{(3)(35)} = \frac{^-4}{105}$$

$$\left(\frac{^-1}{3} \times \frac{2}{5}\right) + \left(\frac{^-1}{3} \times \frac{^-2}{7}\right) = \frac{(^-1)(2)}{(3)(5)} + \frac{(^-1)(^-2)}{(3)(7)}$$

$$= \frac{^-2}{15} + \frac{2}{21} = \frac{(^-2)(21) + (15)(2)}{(15)(21)} = \frac{^-42 + 30}{315}$$

$$= \frac{^-12}{315} = \frac{(^-4)(3)}{(105)(3)} = \frac{^-4}{105}$$

Hence

$$\frac{^-1}{3} \times \left(\frac{2}{5} + \frac{^-2}{7}\right) = \left(\frac{^-1}{3} \times \frac{2}{5}\right) + \left(\frac{^-1}{3} \times \frac{^-2}{7}\right) \quad \blacksquare$$

9.55 MULTIPLICATIVE IDENTITY

There exists a unique rational number $1/1$ such that $a/b \times 1/1 = 1/1 \times a/b = a/b$ for *every* $a/b \in Q$.

EXAMPLE

a. $\dfrac{2}{3} \times \dfrac{1}{1} = \dfrac{(2)(1)}{(3)(1)} = \dfrac{2}{3}$

b. $\dfrac{^-3}{4} \times \dfrac{1}{1} = \dfrac{(^-3)(1)}{(4)(1)} = \dfrac{^-3}{4}$

c. $\dfrac{1}{1} \times \dfrac{^-2}{9} = \dfrac{(1)(^-2)}{(1)(9)} = \dfrac{^-2}{9}$

d. $\dfrac{0}{1} \times \dfrac{1}{1} = \dfrac{(0)(1)}{(1)(1)} = \dfrac{0}{1}$

9.56 MULTIPLICATIVE INVERSE

For each rational number $a/b \neq 0/1$ we define the rational number b/a as the *unique multiplicative inverse* such that $a/b \times b/a = b/a \times a/b = 1/1$.

EXAMPLE

a. $\dfrac{^-2}{7}$ is the multiplicative inverse of $\dfrac{7}{^-2}$

since $\dfrac{^-2}{7} \times \dfrac{7}{^-2} = \dfrac{(^-2)(7)}{(7)(^-2)} = \dfrac{^-14}{^-14} = \dfrac{1}{1}$.

b. $\dfrac{4}{5}$ is the multiplicative inverse of $\dfrac{5}{4}$ since $\dfrac{4}{5} \times \dfrac{5}{4} = \dfrac{(4)(5)}{(5)(4)} = \dfrac{20}{20} = \dfrac{1}{1}$.

c. $\dfrac{^-3}{^-2}$ is the multiplicative inverse of $\dfrac{2}{3}$ since $\dfrac{^-3}{^-2} \times \dfrac{2}{3} = \dfrac{(^-3)(2)}{(^-2)(3)} = \dfrac{^-6}{^-6} = \dfrac{1}{1}$.

Observe that the rational number $0/1$ *does not* have a multiplicative inverse.

9.57 MULTIPLICATION PROPERTY OF $0/1$

For every rational number a/b, $a/b \times 0/1 = 0/1 \times a/b = 0/1$.

EXAMPLE

a. $\dfrac{2}{3} \times \dfrac{0}{1} = \dfrac{(2)(0)}{(3)(1)} = \dfrac{0}{3} = \dfrac{0}{1}$

b. $\dfrac{^-3}{8} \times \dfrac{0}{1} = \dfrac{(^-3)(0)}{(8)(1)} = \dfrac{0}{8} = \dfrac{0}{1}$

c. $\dfrac{0}{1} \times \dfrac{0}{1} = \dfrac{(0)(0)}{(1)(1)} = \dfrac{0}{1}$

We conclude this section with some useful theorems.

THEOREM 9.7 If a/b, c/d, and e/f are any rational numbers with $a/b = c/d$, then $a/b \times e/f = c/d \times e/f$.
Proof The proof is left as an exercise.

THEOREM 9.8 Cancellation Property for Multiplication If a/b, c/d, and e/f are any rational numbers with $a/b \times e/f = c/d \times e/f$, then $a/b = c/d$, provided $e/f \neq 0/1$.
  Proof The proof is left as an exercise.

THEOREM 9.9 If a/b, c/d, e/f, and g/h are any rational umbers with $a/b = c/d$ and $e/f = g/h$, then $a/b \times e/f = c/d \times g/h$.
  Proof The proof is left as an exercise.

EXERCISES

1. Perform the following multiplications using Definition 9.7. Express your results in simplest form.

a. $\dfrac{2}{3} \times \dfrac{4}{7}$.

b. $\dfrac{^-1}{5} \times \dfrac{^-2}{7}$.

c. $\dfrac{4}{13} \times \dfrac{^-5}{9}$.

d. $\dfrac{3}{7} \times \dfrac{^-2}{11}$.

e. $\dfrac{^-3}{5} \times \dfrac{^-4}{9}$.

f. $\dfrac{0}{^-7} \times \dfrac{^-11}{18}$.

g. $\left(\dfrac{2}{3} \times \dfrac{1}{4}\right) \times \dfrac{^-2}{5}$.

h. $\dfrac{^-3}{7} \times \left(\dfrac{4}{5} \times \dfrac{^-3}{4}\right)$.

i. $\left(\dfrac{2}{5} + \dfrac{^-1}{7}\right) \times \left(\dfrac{1}{3} + \dfrac{^-2}{5}\right)$.

j. $\left(\dfrac{2}{9} + \dfrac{0}{^-3}\right) \times \left(\dfrac{6}{^-8} + \dfrac{0}{4}\right)$.

2. Perform the indicated operations by first rewriting the mixed numbers as rational numbers.

a. $4\dfrac{1}{2} \times 5\dfrac{1}{3}$.

b. $^-3\dfrac{2}{7} \times 6\dfrac{1}{9}$.

c. $7\dfrac{1}{2} \times {}^-3\dfrac{1}{2}$.

d. $^-2\dfrac{4}{5} \times {}^-3\dfrac{1}{4}$.

e. $\left(6\dfrac{1}{2} + 2\dfrac{2}{3}\right) \times 5\dfrac{1}{4}$.

f. $7\dfrac{1}{3} + \left(^-2\dfrac{3}{4} \times 5\dfrac{1}{3}\right)$.

g. $\left(2\dfrac{1}{3} \times 3\dfrac{2}{5}\right) \times {}^-4\dfrac{1}{2}$.

h. $5\dfrac{1}{6} \times \left(^-3\dfrac{2}{9} \times 2\dfrac{1}{6}\right)$.

3. Verify each of the following to illustrate that multiplication of rational numbers is a commutative operation. Use Definition 9.7 to perform the multiplications.

a. $\dfrac{2}{3} \times \dfrac{4}{5} = \dfrac{4}{5} \times \dfrac{2}{3}$.

b. $\dfrac{6}{7} \times \dfrac{0}{3} = \dfrac{0}{3} \times \dfrac{6}{7}$.

c. $\dfrac{7}{11} \times \dfrac{^-1}{2} = \dfrac{^-1}{2} \times \dfrac{7}{11}$.

d. $\dfrac{^-2}{5} \times \dfrac{^-3}{4} = \dfrac{^-3}{4} \times \dfrac{^-2}{5}$.

4. Verify each of the following to illustrate that multiplication of rational numbers is an associative operation. Use Definition 9.7 to perform the multiplications.

a. $\left(\dfrac{1}{2} \times \dfrac{1}{3}\right) \times \dfrac{1}{4} = \dfrac{1}{2} \times \left(\dfrac{1}{3} \times \dfrac{1}{4}\right)$.

b. $\left(\dfrac{^-2}{5} \times \dfrac{1}{7}\right) \times \dfrac{2}{9} = \dfrac{^-2}{5} \times \left(\dfrac{1}{7} \times \dfrac{2}{9}\right)$.

c. $\left(\dfrac{^-3}{5} \times \dfrac{^-2}{3}\right) \times \dfrac{^-3}{11} = \dfrac{^-3}{5} \times \left(\dfrac{^-2}{3} \times \dfrac{^-3}{11}\right)$.

5. Verify each of the following to illustrate the distributive property of multiplication over addition on the set Q.

a. $\dfrac{2}{3} \times \left(\dfrac{1}{6} + \dfrac{2}{5}\right) = \left(\dfrac{2}{3} \times \dfrac{1}{6}\right) + \left(\dfrac{2}{3} \times \dfrac{2}{5}\right)$.

b. $\dfrac{^-1}{6} \times \left(\dfrac{2}{3} + \dfrac{^-3}{4}\right) = \left(\dfrac{^-1}{6} \times \dfrac{2}{3}\right) + \left(\dfrac{^-1}{6} \times \dfrac{^-3}{4}\right)$.

c. $\dfrac{^-3}{4} \times \left(\dfrac{^-1}{2} + \dfrac{^-2}{3}\right) = \left(\dfrac{^-3}{4} \times \dfrac{^-1}{2}\right) + \left(\dfrac{^-3}{4} \times \dfrac{^-2}{3}\right)$.

6. Determine the multiplicative inverse for each of the following.

a. $\dfrac{1}{2}$.

b. $\dfrac{^-2}{3}$.

c. $\dfrac{4}{7}$.

d. $\dfrac{^-5}{6}$.

e. $\dfrac{^-2}{^-5}$.

f. $\dfrac{13}{21}$.

g. $\dfrac{17}{^-23}$.

h. $\dfrac{^-35}{7}$.

i. $\dfrac{^-4}{4}$.

j. $\dfrac{^-5}{^-5}$.

k. $\dfrac{1}{26}$.

l. $\dfrac{^-5}{4}$.

7. a. Which rational numbers if any do not have a multiplicative inverse in Q?
   b. Which integers if any do not have a multiplicative inverse in I?
   c. Which whole numbers if any do not have a multiplicative inverse in W?
   d. Which natural numbers if any do not have a multiplicative inverse in N?

8. Prove that the set Q is closed under the operation of multiplication.
9. Prove that the operation of multiplication is a commutative operation on the set Q.

10. Prove that the operation of multiplication is an associative operation on the set Q.
11. Prove that the operation of multiplication can be distributed over the operation of addition on the set Q.
12. Prove that the rational number $1/1$ is the multiplicative identity in Q.
13. Prove that each rational number $a/b \neq 0/1$ has a unique multiplicative inverse b/a in Q.
14. Prove that for every rational number a/b that $a/b \times 0/1 = 0/1$.
15. Prove Theorem 9.7.
16. Prove Theorem 9.8.
17. Prove Theorem 9.9.

9.6 SUBTRACTION AND DIVISION OF RATIONAL NUMBERS

In this section we will continue the development of the system of rational numbers by introducing the operations of subtraction and division. Since subtraction and division of integers were defined in terms of addition and multiplication, respectively, and since addition and multiplication of rational numbers have been defined in such a manner as to retain the respective definitions for integers, it should not be surprising that the new operations being introduced here will be analogous to those for integers.

DEFINITION 9.8 If a/b and c/d are any two rational numbers, then the *difference* $a/b - c/d$ formed by subtracting the rational number c/d from the rational number a/b is equal to the rational number e/f if and only if $a/b = c/d + e/f$.

In Chapter 7 we referred to the operation of subtraction as being the inverse operation of addition of integers. In this chapter we introduced the concept of an additive inverse for a rational number. Combining these remarks with Definition 9.8, we can interpret the subtraction of rational numbers as an appropriate addition of rational numbers. That is, instead of subtracting the rational number c/d from the rational number a/b, we could add the rational number $^-c/d$ to the rational number a/b. Hence we have $a/b - c/d = a/b + {}^-c/d$.

EXAMPLE

a. $\dfrac{6}{1} - \dfrac{2}{1} = \dfrac{4}{1}$ since $\dfrac{4}{1} \in Q$ and $\dfrac{6}{1} = \dfrac{2}{1} + \dfrac{4}{1}$.

b. $\dfrac{2}{3} - \dfrac{2}{3} = \dfrac{0}{1}$ since $\dfrac{0}{1} \in Q$ and $\dfrac{2}{3} = \dfrac{2}{3} + \dfrac{0}{1}$.

c. $\dfrac{4}{7} - \dfrac{^-2}{3} = \dfrac{26}{21}$ since $\dfrac{26}{21} \in Q$ and $\dfrac{4}{7} = \dfrac{^-2}{3} + \dfrac{26}{21}$.

d. $\dfrac{^-3}{5} - \dfrac{^-4}{9} = \dfrac{^-7}{45}$ since $\dfrac{^-7}{45} \in Q$ and $\dfrac{^-3}{5} = \dfrac{^-4}{9} + \dfrac{^-7}{45}$

EXAMPLE

a. $\dfrac{6}{1} - \dfrac{2}{1} = \dfrac{6}{1} + \dfrac{^-2}{1} = \dfrac{(6)(1) + (1)(^-2)}{(1)(1)} = \dfrac{6 + \,^-2}{1} = \dfrac{4}{1}$

b. $\dfrac{2}{3} - \dfrac{2}{3} = \dfrac{2}{3} + \dfrac{^-2}{3} = \dfrac{(2)(3) + (3)(^-2)}{(3)(3)} = \dfrac{6 + \,^-6}{9} = \dfrac{0}{9} = \dfrac{0}{1}$

c. $\dfrac{4}{7} - \dfrac{^-2}{3} = \dfrac{4}{7} + \dfrac{2}{3} = \dfrac{(4)(3) + (7)(2)}{(7)(3)} = \dfrac{12 + 14}{21} = \dfrac{26}{21}$

d. $\dfrac{^-3}{5} - \dfrac{^-4}{9} = \dfrac{^-3}{5} + \dfrac{4}{9} = \dfrac{(^-3)(9) + (5)(4)}{(5)(9)} = \dfrac{^-27 + 20}{45} = \dfrac{^-7}{45}$.

The example above illustrates that the subtraction of rational numbers can be changed to an appropriate addition of rational numbers.

It should be clear from Definition 9.8 that the difference can be formed for *any* two rational numbers and that the set of rational numbers, then, is closed under the operation of subtraction. Recall that the set of integers also was closed under the operation of subtraction but the set of whole numbers was not.

Just as addition of rational numbers has an inverse operation called subtraction, multiplication of rational numbers has an inverse operation called *division*.

DEFINITION 9.9 If a/b and c/d are integers with $c/d \neq 0/1$, then a/b divided by c/d, denoted by $a/b \div c/d$, is equal to the rational number e/f if and only if $a/b = c/d \times e/f$.

EXAMPLE

a. $\dfrac{2}{3} \div \dfrac{1}{2} = \dfrac{4}{3}$ since $\dfrac{4}{3} \in Q$ and $\dfrac{2}{3} = \dfrac{1}{2} \times \dfrac{4}{3}$.

b. $\dfrac{^-3}{7} \div \dfrac{^-4}{5} = \dfrac{15}{28}$ since $\dfrac{15}{28} \in Q$ and $\dfrac{^-3}{7} = \dfrac{^-4}{5} \times \dfrac{15}{28}$.

c. $\dfrac{^-3}{8} \div \dfrac{3}{1} = \dfrac{^-1}{8}$ since $\dfrac{^-1}{8} \in Q$ and $\dfrac{^-3}{8} = \dfrac{3}{1} \times \dfrac{^-1}{8}$.

d. $\dfrac{0}{1} \div \dfrac{3}{11} = \dfrac{0}{1}$ since $\dfrac{0}{1} \in Q$ and $\dfrac{0}{1} = \dfrac{3}{11} \times \dfrac{0}{1}$.

EXAMPLE   Evaluate (a) $2/3 \div 4/5$ and (b) $2/3 \times 5/4$.
Solution

a. $\dfrac{2}{3} \div \dfrac{4}{5} = \dfrac{5}{6}$ since $\dfrac{5}{6} \in Q$ and $\dfrac{2}{3} = \dfrac{4}{5} \times \dfrac{5}{6}$.

b. $\dfrac{2}{3} \times \dfrac{5}{4} = \dfrac{(2)(5)}{(3)(4)} = \dfrac{10}{12} = \dfrac{5}{6}$. ■

EXAMPLE   Evaluate (a) $\dfrac{^-3}{7} \div \dfrac{1}{2}$ and (b) $\dfrac{^-3}{7} \times \dfrac{2}{1}$.

Solution

a. $\dfrac{^-3}{7} \div \dfrac{1}{2} = \dfrac{^-6}{7}$ since $\dfrac{^-6}{7} \in Q$ and $\dfrac{^-3}{7} = \dfrac{1}{2} \times \dfrac{^-6}{7}$.

b. $\dfrac{^-3}{7} \times \dfrac{2}{1} = \dfrac{(^-3)(2)}{(7)(1)} = \dfrac{^-6}{7}$. ■

The two examples above seem to suggest that if a/b and c/d are rational numbers with $c/d \neq 0/1$, then $a/b \div c/d = a/b \times d/c$. We will consider this statement to be our *operational definition* for the division of rational numbers.

DEFINITION 9.10 If a/b and c/d are rational numbers with $c/d \neq 0/1$, then
$$a/b \div c/d = a/b \times d/c.$$

Every rational number except the rational number $0/1$ has a multiplicative inverse. Hence to divide a rational number by a nonzero rational number, we simply multiply the first rational number by the multiplicative inverse of the second. It should be obvious then that division on the set of rational numbers can always be performed if the divisor is not zero. Hence the set of all *nonzero* rational numbers is closed under the operation of division.

From the remarks in the preceding paragraph we see that the solution of the equation $ax = b$ such that a, $b \in I$ and $a \neq 0$ always has a solution in the set Q. This was not true for the set I and, hence, another deficiency has been corrected by introducing the system of rational numbers.

EXAMPLE   The equation $2x = 1$ has no solution in the set I since there is not an integer, x, such that $2x = 1$. However, the same equation has the solution $x = 1/2$ in the set of rational numbers since $2/1 \times 1/2 = (2)(1)/(1)(2) = (1)(2)/(1)(2) = 1/1$.

EXAMPLE   The equation $3y = 9$ has a solution $y = 3$ in the set of integers since $3 \times 3 = 9$. The same equation also has the solution $y = 3/1$. in the set of rational numbers since $3/1 \times 3/1 = (3)(3)/(1)(1) = 9/1$.

EXAMPLE $5/1 \div 0/1 = ?$

Solution If $5/1 \div 0/1$ has a solution on the set of rationals, then we want a rational number a/b such that $5/1 = a/b \times 0/1$. But $a/b \times 0/1 = 0/1$ by the multiplicative property for $0/1$. Hence no rational number a/b exists and $5/1 \div 0/1$ does not have a solution on the set Q. ∎

EXAMPLE $0/1 \div 0/1 = ?$

Solution If $0/1 \div 0/1$ has a solution on the set Q, then we want a rational number c/d such that $0/1 = 0/1 \times c/d$. But $0/1 \times c/d = 0/1$ for *every* rational number c/d by the multiplicative property of $0/1$. Hence $0/1 \div 0/1$ has infinitely many solutions in the set Q. ∎

Considering the latter two examples above, we say that division by $0/1$ on the set Q is not possible for the following reasons:

1. If $a/b \neq 0/1$, then $a/b \div 0/1$ is not defined since there is no rational number c/d such that $a/b = 0/1 \times c/d$.

2. $0/1 \div 0/1$ is not defined since there is no *unique* solution such that $0/1 = 0/1 \times c/d$ for *every* rational number c/d.

We will conclude this section with the following theorems.

THEOREM 9.10 If a/b, c/d, and e/f are rational numbers, then $a/b \times (c/d - e/f) = (a/b \times c/d) - (a/b \times e/f)$.

Proof The proof of this theorem is analogous to that for Theorem 7.4 and is left as an exercise.

The above theorem is known as the *distributive property for multiplication over subtraction on the set of rational numbers.*

EXAMPLE Verify Theorem 9.10 with $a/b = 2/3, c/d = 1/7$ and $e/f = 3/4$.

Solution

$$\frac{a}{b} \times \left(\frac{c}{d} - \frac{e}{f}\right) = \frac{2}{3} \times \left(\frac{1}{7} - \frac{3}{4}\right) = \frac{2}{3} \times \frac{(1)(4) - (7)(3)}{(7)(4)} = \frac{2}{3} \times \frac{4 - 21}{28}$$

$$= \frac{2}{3} \times \frac{^-17}{28} = \frac{(2)(^-17)}{(3)(28)} = \frac{^-34}{84} = \frac{^-17}{42}$$

$$\left(\frac{a}{b} \times \frac{c}{d}\right) - \left(\frac{a}{b} \times \frac{e}{f}\right) = \left(\frac{2}{3} \times \frac{1}{7}\right) - \left(\frac{2}{3} \times \frac{3}{4}\right) = \frac{(2)(1)}{(3)(7)} - \frac{(2)(3)}{(3)(4)} = \frac{2}{21} - \frac{6}{12}$$

$$= \frac{(2)(12) - (21)(6)}{(21)(12)} = \frac{24 - 126}{252} = \frac{^-102}{252} = \frac{^-17}{42}$$

Hence

$$\frac{a}{b} \times \left(\frac{c}{d} - \frac{e}{f}\right) = \left(\frac{a}{b} \times \frac{c}{d}\right) - \left(\frac{a}{b} \times \frac{e}{f}\right) \qquad ∎$$

THEOREM 9.11 If a/b, c/d, and e/f are any rational numbers, then $(a/b + c/d) - e/f = a/b + (c/d - e/f)$.

Proof The proof of this theorem is analogous to that for Theorem 7.5 and is left as an exercise.

The above theorem is *not* to be compared with the statement

$$\left(\frac{a}{b} - \frac{c}{d}\right) - \frac{e}{f} \overset{?}{=} \frac{a}{b} - \left(\frac{c}{d} - \frac{e}{f}\right)$$

since in general

$$\left(\frac{a}{b} - \frac{c}{d}\right) - \frac{e}{f} \neq \frac{a}{b} - \left(\frac{c}{d} - \frac{e}{f}\right)$$

EXAMPLE Using $a/b = {}^-2/3$, $c/d = 1/2$, and $e/f = {}^-3/4$, verify that $(a/b - c/d) - e/f \neq a/b - (c/d - e/f)$.

Solution

$$\left(\frac{a}{b} - \frac{c}{d}\right) - \frac{e}{f} = \left(\frac{{}^-2}{3} - \frac{1}{2}\right) - \frac{{}^-3}{4} = \frac{({}^-2)(2) - (3)(1)}{(3)(2)} - \frac{{}^-3}{4} = \frac{{}^-4 - 3}{6} - \frac{{}^-3}{4}$$

$$= \frac{{}^-7}{6} - \frac{{}^-3}{4} = \frac{({}^-7)(4) - (6)({}^-3)}{(6)(4)} = \frac{{}^-28 - {}^-18}{24} = \frac{{}^-10}{24} = \frac{{}^-5}{12}$$

$$\frac{a}{b} - \left(\frac{c}{d} - \frac{e}{f}\right) = \frac{{}^-2}{3} - \left(\frac{1}{2} - \frac{{}^-3}{4}\right) = \frac{{}^-2}{3} - \frac{(1)(4) - (2)({}^-3)}{(2)(4)} = \frac{{}^-2}{3} - \frac{4 - {}^-6}{8}$$

$$= \frac{{}^-2}{3} - \frac{10}{8} = \frac{({}^-2)(8) - (3)(10)}{(3)(8)} = \frac{{}^-16 - 30}{24} = \frac{{}^-46}{24} = \frac{{}^-23}{12}$$

Since

$$\frac{{}^-5}{12} \neq \frac{{}^-23}{12}, \quad \left(\frac{a}{b} - \frac{c}{d}\right) - \frac{e}{f} \neq \frac{a}{b} - \left(\frac{c}{d} - \frac{e}{f}\right) \quad ■$$

EXAMPLE Verify Theorem 9.11 with $a/b = 2/7$, $c/d = {}^-1/3$, and $e/f = 2/5$.

Solution

$$\left(\frac{a}{b} + \frac{c}{d}\right) - \frac{e}{f} = \left(\frac{2}{7} + \frac{{}^-1}{3}\right) - \frac{2}{5} = \frac{(2)(3) + (7)({}^-1)}{(7)(3)} - \frac{2}{5} = \frac{6 + {}^-7}{21} - \frac{2}{5}$$

$$= \frac{{}^-1}{21} - \frac{2}{5} = \frac{({}^-1)(5) - (21)(2)}{(21)(5)} = \frac{{}^-5 - 42}{105} = \frac{{}^-47}{105}$$

$$\frac{a}{b} + \left(\frac{c}{d} - \frac{e}{f}\right) = \frac{2}{7} + \left(\frac{{}^-1}{3} - \frac{2}{5}\right) = \frac{2}{7} + \frac{({}^-1)(5) - (3)(2)}{(3)(5)} = \frac{2}{7} + \frac{{}^-5 - 6}{15}$$

$$= \frac{2}{7} + \frac{{}^-11}{15} = \frac{(2)(15) + (7)({}^-11)}{(7)(15)} = \frac{30 + {}^-77}{105} = \frac{{}^-47}{105}$$

Hence

$$\left(\frac{a}{b}+\frac{c}{d}\right)-\frac{e}{f}=\frac{a}{b}+\left(\frac{c}{d}-\frac{e}{f}\right) \quad ■$$

EXERCISES

1. Perform the indicated subtractions using Definition 9.8. Express your results in simplest form.

a. $\dfrac{2}{3}-\dfrac{1}{4}$.

b. $\dfrac{3}{7}-\dfrac{1}{2}$.

c. $\dfrac{^-2}{5}-\dfrac{4}{3}$.

d. $\dfrac{0}{1}-\dfrac{^-2}{9}$.

e. $\dfrac{^-5}{7}-\dfrac{0}{^-2}$.

f. $\dfrac{2}{3}-\dfrac{2}{3}$.

g. $\dfrac{3}{5}-\dfrac{^-3}{5}$.

h. $\dfrac{1}{9}-\dfrac{1}{5}$.

i. $\dfrac{^-13}{26}-\dfrac{2}{^-4}$.

j. $\dfrac{2}{^-7}-\dfrac{1}{3}$.

2. Perform the indicated subtractions by considering that $a/b - c/d = a/b + {}^-c/d$. Simply your results.

a. $\dfrac{2}{5}-\dfrac{1}{3}$.

b. $-3\dfrac{1}{7}-2\dfrac{1}{4}$.

c. $\dfrac{3}{9}-\dfrac{^-2}{5}$.

d. $-3\dfrac{4}{5}-\dfrac{6}{7}$.

e. $\dfrac{0}{1}-\dfrac{2}{9}$.

f. $\dfrac{11}{^-13}-\dfrac{0}{3}$.

g. $\dfrac{^-21}{17}-\dfrac{13}{5}$.

h. $\dfrac{^-4}{11}-\dfrac{^-3}{9}$.

i. $\left(\dfrac{11}{5}-\dfrac{16}{3}\right)-\dfrac{2}{7}$.

j. $4\dfrac{1}{3}-\left(3\dfrac{1}{2}-2\dfrac{1}{4}\right)$.

3. Perform the indicated divisions using Definition 9.9. Express your results in simplest form.

a. $\dfrac{3}{5}\div\dfrac{1}{4}$.

b. $\dfrac{2}{3}\div\dfrac{3}{7}$.

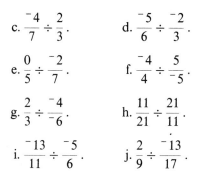

c. $\dfrac{^-4}{7} \div \dfrac{2}{3}$.

d. $\dfrac{^-5}{6} \div \dfrac{^-2}{3}$.

e. $\dfrac{0}{5} \div \dfrac{^-2}{7}$.

f. $\dfrac{^-4}{4} \div \dfrac{5}{^-5}$.

g. $\dfrac{2}{3} \div \dfrac{^-4}{^-6}$.

h. $\dfrac{11}{21} \div \dfrac{21}{11}$.

i. $\dfrac{^-13}{11} \div \dfrac{^-5}{6}$.

j. $\dfrac{2}{9} \div \dfrac{^-13}{17}$.

4. Perform the indicated divisions using Definition 9.10. Express your results in simplest form.

a. $\dfrac{^-3}{7} \div \dfrac{1}{2}$.

b. $3\dfrac{4}{9} \div -2\dfrac{2}{3}$.

c. $\dfrac{0}{6} \div \dfrac{^-2}{5}$.

d. $\dfrac{5}{^-5} \div \dfrac{^-6}{6}$.

e. $\dfrac{23}{^-11} \div \dfrac{^-11}{23}$.

f. $\dfrac{21}{13} \div \dfrac{^-5}{7}$.

g. $\dfrac{63}{^-21} \div \dfrac{16}{^-32}$.

h. $\dfrac{49}{^-98} \div \dfrac{^-6}{3}$.

i. $\dfrac{^-2}{1} \div \dfrac{^-6}{1}$.

j. $^-5\dfrac{1}{3} \div 6\dfrac{1}{4}$.

5. Solve the following equations on the set of whole numbers. If no solution exists, indicate this.

a. $2x = 1$.

b. $3y = 6$.

c. $3u + 1 = 2u - 2$.

d. $4x = 3x - 1$.

e. $\dfrac{y - 3}{2} = 1$.

f. $\dfrac{2u - 3}{2} = 1$.

g. $\dfrac{1 - 2x}{3} = {}^-2$.

h. $\dfrac{1 - 2y}{3} = {}^-2y$.

i. $\dfrac{2}{3 - x} = 4$.

j. $\dfrac{u}{u + 1} = \dfrac{2}{3}$.

6. Solve each of the equations in Exercise 5 on the set of integers. If no solution exists, indicate this.

7. Solve each of the equations in Exercise 5 on the set of rational numbers. If no solution exists, indicate this.

8. With each of the following, verify Theorem 9.10.

a. $\dfrac{a}{b} = \dfrac{2}{3}, \dfrac{c}{d} = \dfrac{1}{2},$ and $\dfrac{e}{f} = \dfrac{^-3}{5}$.

b. $\dfrac{a}{b} = \dfrac{1}{7}, \dfrac{c}{d} = \dfrac{^-2}{3},$ and $\dfrac{e}{f} = \dfrac{^-5}{9}$.

c. $\dfrac{a}{b} = \dfrac{^-3}{4}, \dfrac{c}{d} = \dfrac{^-5}{9},$ and $\dfrac{e}{f} = \dfrac{^-3}{5}$.

d. $\dfrac{a}{b} = \dfrac{1}{9}, \dfrac{c}{d} = \dfrac{1}{3},$ and $\dfrac{e}{f} = \dfrac{^-1}{5}$.

9. With each of the following verify Theorem 9.11.

a. $\dfrac{a}{b} = \dfrac{1}{2}, \dfrac{c}{d} = \dfrac{1}{3},$ and $\dfrac{e}{f} = \dfrac{1}{4}$.

b. $\dfrac{a}{b} = \dfrac{^-2}{5}, \dfrac{c}{d} = \dfrac{^-3}{4},$ and $\dfrac{e}{f} = \dfrac{^-1}{7}$.

c. $\dfrac{a}{b} = \dfrac{2}{11}, \dfrac{c}{d} = \dfrac{^-1}{9},$ and $\dfrac{e}{f} = \dfrac{^-3}{7}$.

d. $\dfrac{a}{b} = \dfrac{4}{7}, \dfrac{c}{d} = \dfrac{0}{2},$ and $\dfrac{e}{f} = \dfrac{0}{^-3}$,

10. Identify each of the following as being true or false.
    a. The sets of natural numbers, whole numbers, integers, and rational numbers are all closed under the operation of subtraction.
    b. The sets of natural numbers, whole numbers, integers, and rational numbers are all closed under the operation of division.
    c. The set of rational numbers is closed under the operation of division.
    d. The set of nonzero rational numbers is closed under the operation of division.
11. Prove Theorem 9.10.
12. Prove Theorem 9.11.

9.7 ORDERING OF RATIONAL NUMBERS

In this section we will introduce the relation of order on the set of rational numbers as an extension of our definition of order on the set of integers. Recall that if $a, b \in I$, we defined $a < b$ if there exists a *positive* integer, c, such that $b = a + c$. Since we wish to preserve this definition in formulating a similar definition on the set Q, we could define the rational number a/b to be less than the rational number c/d if and only if there exists a positive rational number e/f such that $c/d = a/b + e/f$. However, we will state a definition differently but in such a manner as to facilitate working with the order relation.

DEFINITION 9.11 Let a/b and c/d be rational numbers such that $b > 0$ and $d > 0$. Then the rational number a/b is said to be *less than* the rational number c/d denoted by $a/b < c/d$ if and only if $ad < bc$.

EXAMPLE

a. The rational number $\frac{2}{3} < \frac{4}{5}$ since $(2)(5) < (3)(4)$ since $10 < 12$.

b. The rational number $\frac{2}{^-3} < \frac{^-3}{5}$ since $\frac{^-2}{3} < \frac{^-3}{5}$ since $(^-2)(5) < (3)(^-3)$

since $^-10 < ^-9$.

c. The rational number $\frac{0}{1} < \frac{3}{11}$ since $(0)(11) < (1)(3)$ since $0 < 3$.

d. The rational number $\frac{^-2}{3} < \frac{0}{1}$ since $(^-2)(1) < (3)(0)$ since $^-2 < 0$.

DEFINITION 9.12 Let a/b and c/d be rational numbers such that $b > 0$ and $d > 0$. Then the rational number a/b is said to be *greater than* the rational number c/d, denoted by $a/b > c/d$, if and only if $c/d < a/b$.

EXAMPLE

a. $\frac{4}{5} > \frac{1}{2}$ since $\frac{1}{2} < \frac{4}{5}$.

$\frac{1}{2} < \frac{4}{5}$ since $(1)(5) < (2)(4)$ since $5 < 8$.

b. $\frac{^-3}{5} > \frac{^-2}{3}$ since $\frac{^-2}{3} < \frac{^-3}{5}$.

$\frac{^-2}{3} < \frac{^-3}{5}$ since $(^-2)(5) < (3)(^-3)$ since $^-10 < ^-9$.

c. $\frac{0}{1} > \frac{^-3}{11}$ since $\frac{^-3}{11} < \frac{0}{1}$.

$\frac{^-3}{11} < \frac{0}{1}$ since $(^-3)(1) < (11)(0)$ since $^-3 < 0$.

d. $\dfrac{5}{9} > \dfrac{0}{1}$ since $\dfrac{0}{1} < \dfrac{5}{9}$.

$\dfrac{0}{1} < \dfrac{5}{9}$ since $(0)(9) < (1)(5)$ since $0 < 5$.

EXAMPLE

a. $\dfrac{2}{3} \nless \dfrac{2}{3}$ since $(2)(3) \nless (3)(2)$ since $6 \nless 6$.

b. $\dfrac{2}{3} \ngtr \dfrac{2}{3}$ since $\dfrac{2}{3} \nless \dfrac{2}{3}$.

As with the case of integers, we can combine equality and inequality of rational numbers. For instance, if a/b and c/d are rational numbers, we may write $a/b \le c/d$ to mean $a/b < c/d$ or $a/b = c/d$. Similarly, $e/f \ge g/h$ means $e/f > g/h$ or $e/f = g/h$.

9.71 TRICHOTOMY PROPERTY FOR RATIONAL NUMBERS

If a/b and c/d are any two rational numbers, then *exactly* one of the following statements is true:
1. $a/b < c/d$.
2. $a/b = c/d$.
3. $a/b > c/d$.

THEOREM 9.12 Transitive Property of Order   The relation "is less than" is a transitive relation on the set Q. That is, is a/b, c/d, $e/f \in Q$, and if $a/b < c/d$ and $c/d < e/f$, then $a/b < e/f$.
Proof   The proof of this theorem is analogous to that for Theorem 7.1 and is left as an exercise.

Similarly, the relation "is greater than" is a transitive relation on the set Q.

THEOREM 9.13   Let a/b, c/d, e/f, and g/h be arbitrary rational numbers.

a. If $\dfrac{a}{b} < \dfrac{c}{d}$, then $\left(\dfrac{a}{b} + \dfrac{e}{f}\right) < \left(\dfrac{c}{d} + \dfrac{e}{f}\right)$

b. If $\left(\dfrac{a}{b} + \dfrac{e}{f}\right) < \left(\dfrac{c}{d} + \dfrac{e}{f}\right)$, then $\dfrac{a}{b} < \dfrac{c}{d}$.

c. If $\dfrac{a}{b} < \dfrac{c}{d}$ and $\dfrac{e}{f} > 0$, then $\left(\dfrac{a}{b} \times \dfrac{e}{f}\right) < \left(\dfrac{c}{d} \times \dfrac{e}{f}\right)$.

d. If $\dfrac{a}{b} < \dfrac{c}{d}$ and $\dfrac{e}{f} < 0$, then $\left(\dfrac{a}{b} \times \dfrac{e}{f}\right) > \left(\dfrac{c}{d} \times \dfrac{e}{f}\right)$.

Proof The proof of this theorem is analogous to that for Theorem 7.3 and is left as an exercise.

EXAMPLE Arrange the rational numbers $^-23/^-6$, $2/^-3$, $^-13/17$, and $0/1$ in increasing order.

Solution We will first write each of the rational numbers in equivalent form with positive denominators as such: $23/6$, $^-2/3$, $^-13/17$, and $0/1$. Next we will determine the l.c.m.$(17, 3, 1, 6) = 102$ and write each of the rational numbers in equivalent form with a denominator of 102 as follows:

$$\frac{391}{102}, \frac{^-68}{102}, \frac{^-78}{102}, \text{ and } \frac{0}{102}$$

Finally, we will arrange the rational numbers according to increasing size of their numerators such as

$$\frac{^-78}{102}, \frac{^-68}{102}, \frac{0}{102}, \frac{391}{102}$$

or

$$\frac{^-13}{17}, \frac{2}{^-3}, \frac{0}{1}, \frac{^-23}{^-6} \quad \blacksquare$$

We will conclude this section with a remark characterizing a property for the set Q that is not true for N, W, or I. Observe that the rational numbers seem to be much closer to each other than the integers were. For instance, between the integers $^-3$ and 2 there exists the integers $^-2$, $^-1$, 0, and 1 and no others. However, between the rational numbers $\frac{1}{3}$ and $\frac{1}{2}$ there are infinitely many rational numbers. For instance, the average of $\frac{1}{3}$ and $\frac{1}{2}$ found by taking one half their sum, or $(\frac{1}{3} + \frac{1}{2}) \times \frac{1}{2} = \frac{5}{12}$ is such that $\frac{1}{3} < \frac{5}{12} < \frac{1}{2}$. The average of $\frac{1}{3}$ and $\frac{5}{12}$ is $\frac{3}{8}$, and the average of $\frac{5}{12}$ and $\frac{1}{2}$ is $\frac{11}{24}$. Further, $\frac{1}{3} < \frac{3}{8} < \frac{5}{12} < \frac{11}{24} < \frac{1}{2}$. Continuing in this manner, we could include infinitely many rational numbers between $\frac{1}{3}$ and $\frac{1}{2}$. This property of the rationals is called *denseness*.

9.72 DENSENESS PROPERTY OF RATIONAL NUMBERS

The set Q is said to be *dense*. That is, between any two *distinct* rational numbers, there always exists another rational number.

EXAMPLE Insert three rational numbers between the rational numbers $^-2/5$ and $3/4$.

Solution Rewriting each of the given rational numbers in equivalent form with denominators $d = $ l.c.m.$(5, 4) = 20$, we have $^-8/20$ and $15/20$. It is easy to see, then, that $^-8/20 < ^-6/20 < ^-2/20 < 0/20 < 15/20$. Hence $^-3/10$, $^-1/10$, and $0/1$ can be inserted between $^-2/5$ and $3/4$. $\blacksquare$

EXERCISES

1. Using Definitions 9.11 and 9.12, determine which of the following are true and which are false.

a. $\dfrac{2}{3} < \dfrac{3}{4}$.

b. $\dfrac{^-1}{2} > \dfrac{1}{3}$.

c. $\dfrac{0}{2} < \dfrac{^-2}{^-3}$.

d. $\dfrac{13}{^-17} < \dfrac{23}{^-29}$.

e. $\dfrac{4}{^-9} > \dfrac{0}{^-2}$.

f. $\dfrac{13}{17} < \dfrac{^-23}{^-26}$.

g. $\dfrac{15}{^-23} > \dfrac{^-19}{21}$.

h. $\dfrac{^-5}{13} > \dfrac{1}{^-27}$.

i. $\dfrac{4}{^-4} > \dfrac{^-16}{8}$.

j. $\dfrac{56}{^-23} < \dfrac{47}{^-24}$.

2. Determine an integer value of x such that each of the following statements will be true.

a. $\dfrac{2}{3} < \dfrac{x}{7}$.

b. $\dfrac{^-1}{2} < \dfrac{2}{x}$.

c. $\dfrac{0}{x} > \dfrac{^-4}{5}$.

d. $\dfrac{x}{^-3} > \dfrac{^-2}{5}$.

e. $\dfrac{4}{^-7} < \dfrac{x}{8}$.

f. $\dfrac{^-5}{9} \le \dfrac{6}{x}$.

g. $\dfrac{3}{x} \ge \dfrac{2}{7}$.

h. $\dfrac{x}{^-4} \ge \dfrac{0}{2}$.

3. For each of the blanks between the two rational numbers given, write the symbol <, >, or = so that the resulting statement is true.

a. $\dfrac{1}{2} \underline{\hspace{1cm}} \dfrac{^-2}{3}$.

b. $\dfrac{^-3}{4} \underline{\hspace{1cm}} \dfrac{5}{^-9}$.

c. $\dfrac{0}{2} \underline{\hspace{1cm}} \dfrac{0}{^-3}$.

d. $\dfrac{4}{^-5} \underline{\hspace{1cm}} \dfrac{18}{10}$.

e. $\dfrac{9}{^-13} \underline{\hspace{1cm}} \dfrac{^-11}{17}$.

f. $\dfrac{3}{^-7} \underline{\hspace{1cm}} \dfrac{^-6}{14}$.

g. $\dfrac{2}{3} + \dfrac{^-1}{2} \underline{\hspace{1cm}} \dfrac{4}{^-7}$.

h. $\dfrac{^-3}{5} - \dfrac{2}{3} \underline{\hspace{1cm}} \dfrac{^-1}{2}$.

i. $\dfrac{4}{9} \times \dfrac{^-2}{3} \underline{\hspace{1cm}} \dfrac{^-3}{5}$. j. $\dfrac{4}{7} \underline{\hspace{1cm}} \dfrac{^-2}{3} \times \dfrac{4}{^-9}$.

4. Arrange each of the following sets of rational numbers in increasing order.

a. $\dfrac{^-1}{2}, \dfrac{2}{3}, \dfrac{^-4}{7}, \dfrac{0}{^-6}$. b. $\dfrac{13}{5}, \dfrac{51}{^-95}, \dfrac{17}{^-19}, \dfrac{6}{1}$.

c. $\dfrac{^-1}{3}, \dfrac{2}{5}, \dfrac{^-4}{15}, \dfrac{6}{^-90}, \dfrac{23}{45}$. d. $\dfrac{^-3}{14}, \dfrac{1}{7}, \dfrac{^-41}{^-98}, \dfrac{^-8}{28}, \dfrac{9}{^-49}$.

5. Between each pair of rational numbers given, insert three distinct rational numbers. (Hint: Use the average of the two rational numbers.)

a. $\dfrac{2}{3}$ and $\dfrac{4}{9}$. b. $\dfrac{^-3}{7}$ and $\dfrac{5}{14}$.

c. $\dfrac{6}{13}$ and $\dfrac{^-5}{9}$. d. $\dfrac{^-1}{4}$ and $\dfrac{0}{2}$.

e. $\dfrac{^-13}{16}$ and $\dfrac{^-5}{64}$. f. $\dfrac{13}{25}$ and $\dfrac{3}{20}$.

6. a. How many distinct integers are there between $^-4$ and 2?
   b. How many distinct rational numbers are there between $^-4/1$ and $2/1$?
   c. What is the smallest positive integer?
   d. What is the smallest positive rational number?
   e. What is the largest negative integer?
   f. What is the largest negative rational number?
   g. What is the largest nonpositive rational number?
   h. What is the smallest nonnegative rational number?
   i. Is the average of two integers always an integer? Explain.
   j. Is the average of two rational numbers always a rational number? Explain.

7. Which of the following statements are true and which are false. If a statement is false, give an example to substantiate your answer. All denominators are nonzero.

a. If $\dfrac{a}{b} < \dfrac{c}{d}$, then $\dfrac{a}{b} + \dfrac{e}{f} < \dfrac{c}{d} + \dfrac{e}{f}$.

b. If $\dfrac{a}{b} + \dfrac{c}{d} \geq \dfrac{e}{f} + \dfrac{c}{d}$, then $\dfrac{a}{b} > \dfrac{e}{f}$.

c. If $\dfrac{a}{b} < \dfrac{c}{d}$, then $\dfrac{a}{b} \times \dfrac{p}{q} < \dfrac{c}{d} \times \dfrac{p}{q}$.

d. If $\dfrac{a}{b} > \dfrac{c}{d}$, then $\dfrac{a}{b} \times \dfrac{p}{q} = \dfrac{c}{d} \times \dfrac{p}{q}$, if $p = 0$.

e. If $\dfrac{a}{b} \times \dfrac{p}{q} = \dfrac{c}{d} \times \dfrac{p}{q}$ and $\dfrac{a}{b} > \dfrac{c}{d}$, then $p = q$.

8. Prove Theorem 9.12.

9. Prove Theorem 9.13.

9.8 A LOOK AT THE SYSTEM OF RATIONAL NUMBERS

Throughout this chapter we have developed the system of rational numbers. We started with the set of rational numbers denoted by Q, defined and discussed the operations of addition, multiplication, subtraction, and division of rational numbers together with their properties, and introduced the relation of order on the set of rational numbers. In this section we will summarize our results.

9.81 OPERATIONS

1. ADDITION   If $\dfrac{a}{b}, \dfrac{c}{d} \in Q$, then $\dfrac{a}{b} + \dfrac{c}{d} = \dfrac{ad + bc}{bd}$.

2. MULTIPLICATION   If $\dfrac{a}{b}, \dfrac{c}{d} \in Q$, then $\dfrac{a}{b} \times \dfrac{c}{d} = \dfrac{ac}{bd}$.

3. SUBTRACTION   If $\dfrac{a}{b}, \dfrac{c}{d} \in Q$, then $\dfrac{a}{b} - \dfrac{c}{d} = \dfrac{e}{f}$ if and only if $\dfrac{e}{f} \in Q$ and

$\dfrac{a}{b} = \dfrac{c}{d} + \dfrac{e}{f}; \dfrac{e}{f} = \dfrac{ad - bc}{bd}$.

4. DIVISION   If $\dfrac{a}{b}, \dfrac{c}{d} \in Q$ with $\dfrac{c}{d} \neq \dfrac{0}{1}$, then $\dfrac{a}{b} \div \dfrac{c}{d} = \dfrac{e}{f}$ if and only if

$\dfrac{e}{f} \in Q$ and $\dfrac{a}{b} = \dfrac{c}{d} \times \dfrac{e}{f}$; operationally, we have $\dfrac{a}{b} \div \dfrac{c}{d} = \dfrac{a}{b} \times \dfrac{d}{c}$, provided

$\dfrac{c}{d} \neq \dfrac{0}{1}$.

9.82 CLOSURE PROPERTIES

1. The set Q is closed under the operation of addition; that is, if

$\dfrac{a}{b}, \dfrac{c}{d} \in Q$, then $\dfrac{a}{b} + \dfrac{c}{d} \in Q$.

2. The set Q is closed under the operation of multiplication; that is, if $\frac{a}{b}, \frac{c}{d} \in Q$, then $\frac{a}{b} \times \frac{c}{d} \in Q$.

3. The set Q is closed under the operation of subtraction; that is, if $\frac{a}{b}, \frac{c}{d} \in Q$, then $\frac{a}{b} - \frac{c}{d} \in Q$.

4. The set Q is *not* closed under the operation of division; for example, $\frac{a}{b} \div \frac{c}{d} \notin Q$ if $\frac{c}{d} = \frac{0}{1}$.

5. The set of all *nonzero* rational numbers is closed under the operation of division; that is, if $\frac{a}{b}, \frac{c}{d} \in Q$ with $\frac{c}{d} \neq \frac{0}{1}$, then $\frac{a}{b} \div \frac{c}{d} \in Q$.

9.83 COMMUTATIVE PROPERTIES

1. FOR ADDITION   If $\frac{a}{b}, \frac{c}{d} \in Q$, then $\frac{a}{b} + \frac{c}{d} = \frac{c}{d} + \frac{a}{b}$.

2. FOR MULTIPLICATION   If $\frac{a}{b}, \frac{c}{d} \in Q$, then $\frac{a}{b} \times \frac{c}{d} = \frac{c}{d} \times \frac{a}{b}$.

9.84 ASSOCIATIVE PROPERTIES

1. FOR ADDITION   If $\frac{a}{b}, \frac{c}{d}, \frac{e}{f} \in Q$, then $\left(\frac{a}{b} + \frac{c}{d}\right) + \frac{e}{f} = \frac{a}{b} + \left(\frac{c}{d} + \frac{e}{f}\right)$.

2. FOR MULTIPLICATION   If $\frac{a}{b}, \frac{c}{d}, \frac{e}{f} \in Q$, then $\left(\frac{a}{b} \times \frac{c}{d}\right) \times \frac{e}{f} = \frac{a}{b} \times \left(\frac{c}{d} \times \frac{e}{f}\right)$

9.85 DISTRIBUTIVE PROPERTIES

1. MULTIPLICATION OVER ADDITION   If $\frac{a}{b}, \frac{c}{d}, \frac{e}{f} \in Q$, then

$$\frac{a}{b} \times \left(\frac{c}{d} + \frac{e}{f}\right) = \left(\frac{a}{b} \times \frac{c}{d}\right) + \left(\frac{a}{b} \times \frac{e}{f}\right)$$

and

$$\left(\frac{a}{b} + \frac{c}{d}\right) \times \frac{e}{f} = \left(\frac{a}{b} \times \frac{e}{f}\right) + \left(\frac{c}{d} \times \frac{e}{f}\right).$$

2. MULTIPLICATION OVER SUBTRACTION   If $\dfrac{a}{b}$, $\dfrac{c}{d}$, $\dfrac{e}{f} \in Q$,   then

$$\frac{a}{b} \times \left(\frac{c}{d} - \frac{e}{f} \right) = \left(\frac{a}{b} \times \frac{c}{d} \right) - \left(\frac{a}{b} \times \frac{e}{f} \right).$$

9.86 IDENTITY PROPERTIES

1. ADDITIVE IDENTITY   If $\dfrac{a}{b}$ is an arbitrary rational number, then

$\dfrac{a}{b} + \dfrac{0}{1} = \dfrac{0}{1} + \dfrac{a}{b} = \dfrac{a}{b}$; $\dfrac{0}{1}$ is the additive identity in the set Q.

2. MULTIPLICATIVE IDENTITY   If $\dfrac{a}{b}$ is an arbitrary rational number, then

$\dfrac{a}{b} \times \dfrac{1}{1} = \dfrac{1}{1} \times \dfrac{a}{b} = \dfrac{a}{b}$; $\dfrac{1}{1}$ is the multiplicative identity in the set Q.

9.87 INVERSE PROPERTIES

1. ADDITIVE INVERSE   If $\dfrac{a}{b}$ is a rational number, then there exists

$\dfrac{^-a}{b} \in Q$ such that $\dfrac{a}{b} + \dfrac{^-a}{b} = \dfrac{^-a}{b} + \dfrac{a}{b} = \dfrac{0}{1}$; $\dfrac{^-a}{b}$ is called the additive in-

verse of $\dfrac{a}{b}$ and $\dfrac{a}{b}$ is called the additive inverse of $\dfrac{^-a}{b}$.

2. MULTIPLICATIVE INVERSE   If $\dfrac{a}{b}$ is a rational number and $\dfrac{a}{b} \neq \dfrac{0}{1}$, then

there exists $\dfrac{b}{a} \in Q$ such that $\dfrac{a}{b} \times \dfrac{b}{a} = \dfrac{b}{a} \times \dfrac{a}{b} = \dfrac{1}{1}$; $\dfrac{b}{a}$ is called the multi-

plicative inverse of $\dfrac{a}{b}$ and $\dfrac{a}{b}$ is called the multiplicative inverse of $\dfrac{b}{a}$;

the rational number $\dfrac{0}{1}$ does *not* have a multiplicative inverse.

9.88 ORDER PROPERTIES

1. If $\dfrac{a}{b}$, $\dfrac{c}{d} \in Q$ with $b > 0$ and $d > 0$, then $\dfrac{a}{b} < \dfrac{c}{d}$ if and only if $ad < bc$.

2. If $\frac{a}{b}, \frac{c}{d} \in Q$, then $\frac{a}{b} > \frac{c}{d}$ if and only if $\frac{c}{d} < \frac{a}{b}$.

3. TRICHOTOMY PROPERTY   If $\frac{a}{b}, \frac{c}{d} \in Q$, then *exactly* one of the following must hold:

$$\text{Either } \frac{a}{b} < \frac{c}{d} \quad \text{or} \quad \frac{a}{b} = \frac{c}{d} \quad \text{or} \quad \frac{a}{b} > \frac{c}{d}$$

9.89 PROPERTIES OF ZERO

1. $\frac{0}{1} \times \frac{a}{b} = \frac{0}{1}$ for *all* $\frac{a}{b} \in Q$.

2. $\frac{0}{1} \div \frac{a}{b} = \frac{0}{1}$, if $\frac{a}{b} \neq \frac{0}{1}$.

3. $\frac{0}{1} \div \frac{0}{1}$ is not defined since it is *indeterminate* (that is, $\frac{0}{1} \div \frac{0}{1}$ does not have a *unique* solution).

4. $\frac{a}{b} \div \frac{0}{1}$, if $\frac{a}{b} \neq \frac{0}{1}$ is *not* defined since it is *impossible* to divide a nonzero rational number by $\frac{0}{1}$.

9.810 DENSENESS PROPERTY

The set Q is *dense*. That is, between any two *distinct* rational numbers there always exists another rational number.

SUMMARY

In this chapter we developed the system of rational numbers. We defined the set of rational numbers as equivalence classes represented by fractions. A fraction was defined as a quotient of an integer by a nonzero integer. The operations of addition, multiplication, subtraction, and division, together with their various properties, were defined and discussed. The order relation was also introduced and discussed as was the multiplication property of zero. The property of denseness also was discussed as it relates to the rational numbers.

During our discussions the following symbols were introduced.

SYMBOL	INTERPRETATION
$\dfrac{a}{b}$	A fraction where $a, b \in I$ and $b \neq 0$
$\dfrac{a}{b} \simeq \dfrac{c}{d}$	Equivalence of fractions $\left(\dfrac{a}{b} \simeq \dfrac{c}{d} \text{ iff } ad = bc\right)$
$\left|\dfrac{a}{b}\right|$	A rational number
Q	The set of rational numbers
$\left|\dfrac{a}{b}\right| = \left|\dfrac{c}{d}\right|$	Equality of rational numbers $\left(\left|\dfrac{a}{b}\right| = \left|\dfrac{c}{d}\right| \text{ iff } \dfrac{a}{b} \simeq \dfrac{c}{d}\right)$
$\dfrac{a}{b} + \dfrac{c}{d}$	The sum of two rational numbers
$\dfrac{a}{b} \times \dfrac{c}{d}$	The product of two rational numbers
$\dfrac{a}{b} - \dfrac{c}{d}$	The difference of two rational numbers
$\dfrac{a}{b} \div \dfrac{c}{d}, \text{ iff } \dfrac{c}{d} \neq \dfrac{0}{1}$	The quotient of two rational numbers
$\dfrac{a}{b} < \dfrac{c}{d}$	The rational number $\dfrac{a}{b}$ is less than the rational number $\dfrac{c}{d}$
$\dfrac{a}{b} > \dfrac{c}{d}$	The rational number $\dfrac{a}{b}$ is greater than the rational number $\dfrac{c}{d}$

REVIEW EXERCISES FOR CHAPTER 9

1. Perform the indicated operations on the rational numbers indicated and write your results in simplest form.

a. $\dfrac{^-2}{3} + \dfrac{4}{7}$.

b. $\dfrac{3}{4} - \dfrac{^-2}{7}$.

c. $\dfrac{^-4}{9} \times \dfrac{5}{^-11}$.

d. $\dfrac{3}{^-7} \div \dfrac{^-4}{13}$.

e. $\dfrac{-3}{7} \times \left(\dfrac{11}{13} + \dfrac{-9}{5} \right).$ f. $\left(\dfrac{-6}{7} - \dfrac{1}{2} \right) \div \dfrac{-2}{3}.$

g. $\left(\dfrac{-5}{11} \times \dfrac{2}{-3} \right) + \dfrac{4}{7}.$ h. $\left(\dfrac{6}{11} \div \dfrac{-3}{8} \right) - \dfrac{5}{9}.$

i. $\dfrac{4}{5} - \left(\dfrac{-1}{2} - \dfrac{1}{3} \right).$ j. $\left(\dfrac{2}{5} \div \dfrac{-1}{7} \right) \div \dfrac{-2}{9}.$

2. Rewrite each of the following mixed numbers as rational numbers.

a. $4\dfrac{2}{3}.$ b. $^-5\dfrac{1}{2}.$ c. $7\dfrac{13}{15}.$

d. $^-3\dfrac{1}{2} + 5\dfrac{1}{3}.$ e. $2\dfrac{1}{3} \times 5\dfrac{1}{4}.$ f. $^-7\dfrac{1}{4} \div 2\dfrac{3}{4}.$

3. Determine an integer value for x such that each of the following statements is true.

a. $\dfrac{x}{3} < \dfrac{-1}{9}.$ b. $\dfrac{-2}{x} > \dfrac{0}{3}.$

c. $\dfrac{2}{3} + \dfrac{4}{7} \leq \dfrac{x}{5}.$ d. $\dfrac{2x+1}{3} > \dfrac{-4}{-5}.$

e. $\dfrac{-4}{5} \times \dfrac{x}{3} \geq \dfrac{1}{-7}.$ f. $\dfrac{0}{2} \div \dfrac{-2}{3} < \dfrac{x}{-4}.$

4. Determine the additive inverse for each of the following rational numbers.

a. $\dfrac{2}{3}.$ b. $\dfrac{1}{-4}.$ c. $\dfrac{-3}{-5}.$

d. $\dfrac{2-3}{4}.$ e. $\dfrac{-5 \times 6}{-7}.$ f. $\dfrac{-3+3}{9}.$

g. $\dfrac{-a}{b}, b \neq 0.$ h. $\dfrac{a+b}{c}, c \neq 0.$ i. $\dfrac{a}{b+c}, b+c \neq 0.$

5. For parts (a) through (f) of Exercise 4 determine the multiplicative inverse for each of the indicated rational numbers. If the multiplicative inverse does not exist, indicate this.

6. For parts (g) through (i) of Exercise 4 indicate under what conditions if any the multiplicative inverse will exist for each of the indicated rational numbers.

7. Rewrite each of the following sets of rational numbers in decreasing order.

a. $\dfrac{2}{3}, \dfrac{1}{4}, \dfrac{2}{5}, \dfrac{1}{10}.$

b. $\dfrac{^-2}{7}, \dfrac{^-1}{9}, \dfrac{3}{^-4}, \dfrac{^-1}{5}$.

c. $\dfrac{4}{^-5}, \dfrac{^-2}{3}, \dfrac{0}{^-4}, \dfrac{^-11}{9}$.

d. $\dfrac{^-1}{5}, \dfrac{^-12}{^-25}, \dfrac{^-6}{125}, \dfrac{^-201}{625}$.

8. Let N represent the set of natural numbers, W the set of whole numbers, I the set of integers, and Q the set of rational numbers.
   a. Which of the above sets, if any, are closed under the operation of addition?
   b. Which of the above sets, if any, are closed under the operation of subtraction?
   c. Which of the above sets, if any, are closed under the operation of multiplication?
   d. Which of the above sets, if any, are closed under the operation of division?
   e. Which of the above sets, if any, are closed under the operation of division if division by zero is excluded?
   f. Which of the above sets, if any, are dense?
   g. Which of the above sets, if any, contain an additive inverse for each of its elements?
   h. Which of the above sets, if any, contain a multiplicative inverse for each of its elements?
   i. Which of the above sets, if any, are infinite sets?
   j. Which of the above sets, if any, can be considered as the union of two or more of the remaining sets? Explain.
9. Which of the following statements are true and which are false?
   a. There are as many rational numbers between $a/1$ and $b/1$ as there are integers between a and b, if a and b are integers.
   b. The smallest positive integer is 1 but the smallest positive rational number is $0/1$.
   c. The reduced representative for the rational number $\left(\dfrac{^-6}{8}\right)$ is $\dfrac{3}{^-4}$.
   d. If p and q represent rational numbers and $pq = 0$, then the numerator of p and the numerator of q must both be 0.
   e. The operation of subtraction on the set of rational numbers is an associative operation.
   f. Addition and multiplication were defined on the set of rational numbers in such a manner so as to preserve the definitions for the respective operations on the set of integers.
   g. Every rational number has a multiplicative inverse.

h. $\{x \mid x \in I,\ 7 < x < 9\} = \{8\}$.

i. $\left\{ y \mid y \in Q,\ \dfrac{7}{1} < y < \dfrac{9}{1} \right\} = \left\{ \dfrac{8}{1} \right\}$.

j. Every integer a can be written as the rational number $a/1$ even if $a = 0$.

CHAPTER 10
DECIMALS AND THE SYSTEM OF REAL NUMBERS

As we extended the system of whole numbers to the system of integers and the system of rational numbers, we obtained a more complete system of numbers with additional properties satisfied for the operations on these numbers. As was noted earlier in this text, when we extended the system of whole numbers to the system of integers, we gained closure for our new set I under the operation of subtraction which we did not have on the set W. Also, each integer now had an additive inverse and the solution to the equation $a + x = b$ $(a, b \in I)$ was always possible. Further extension to the system of rational numbers produced a multiplicative inverse for every nonzero rational number, and the solution to the equation $ax = b$ $(a, b \in I,$ $a \neq 0)$ was always possible on the set Q. It would seem that by the time we had completed the development of the system of rational numbers our job was done.

On the set Q we could always add, subtract, and multiply with rational numbers. Also, excepting for a divisor of zero, division was possible. Finally, we noted that the set Q was dense; that is, Q was nonempty and between any two distinct rational numbers we could always find another rational number. Relative to the number line it appeared that each point of the line would correspond to a rational number because of this denseness property.

10.1 THE NEED FOR NEW NUMBERS

If we concern ourselves only with the operations of addition, subtraction, multiplication, and division there would be no need to extend our system beyond the system of rational numbers. However, relative to the operation of multiplication, we could multiply a rational number by itself; this process is known as *squaring*. Further, if $x \in Q$, then $x^2 \in Q$ since x^2 simply means x times x and the set Q is closed under the operation of multiplication.

Now suppose that we start with a value for $x^2 \in Q$. Can we always find $x \in Q$? For instance, if $x^2 = 4$, we could take $x = 2 \in Q$. But if $x^2 = 2$, is $x \in Q$? That is, does there exist a rational number which when squared will yield the rational number 2? Stated equivalently, we could ask, "Does the equation $x^2 = 2$ have a solution in Q?" We will return to this question later.

First consider another problem. If we start with a number and multiply it by itself, the process is known as squaring. If we start with a number and try to determine another number which, when squared, will yield the first number, the process is known as *extracting the square root*. Now we may ask, "Do the negative rational numbers have square roots?" Clearly, the square of a positive rational number is positive and the square of a negative number is also positive. (Do you agree?) Finally, the square of zero is zero. Hence the square of every rational number is a nonnegative rational number and no negative rational number has a square root which is a rational number.

We have cited two reasons for extending the system of rational numbers. The first of these will involve the introduction of *irrational numbers*, which will be done in the next section. The second problem involves the introduction of *complex numbers*. We will have no need for complex numbers for the remainder of this text; hence complex numbers will not be introduced.

10.2 IRRATIONAL NUMBERS

Irrational numbers are much used in mathematics today; one encounters them frequently in algebra, trigonometry, and calculus. Although fractions (and rational numbers) were used by the ancient Egyptians, it was not until about 530 B.C. that a need for irrational numbers existed. It was about that time that Pythagoras showed that the ratio, x, of the length of the hypothenuse of an isosceles right triangle to the length of a side satisfies the equation $x^2 = 2$.

Consider the "square" in Figure 10.1 with each side having a length of 1 unit. Then, by the theorem of Pythagoras, which states the "the square of the length of the hypothenuse of a right triangle is equal to the sum of the

Figure 10.1

squares of the lengths of the other two sides," we readily determine that the diagonal of the square (corresponding to the hypotenuse of one of the right triangles in Figure 10.1) is equal to $\sqrt{2}$.

THEOREM 10.1 $\sqrt{2}$ is not a rational number.

Proof The proof will be by contradiction. That is, we will assume that $\sqrt{2}$ is a rational number and show that this leads to a contradiction.

Assume that $\sqrt{2}$ is a rational number. This implies that $\sqrt{2}$ is an equivalence class of fractions that can be represented by the fraction p/q such that p and q are integers, $q \neq 0$, and p and q are relatively prime. Hence we have $\sqrt{2} = p/q$. Squaring both sides of this equation, we have $2 = p^2/q^2$ and $2 \cdot q^2 = p^2$. Since $2 \cdot q^2 = p^2$, p^2 is even since p^2 is equal to the product of 2 and an integer (since $p \in I$, $p^2 \in I$). Now if p^2 is even, p must be even since if p is odd, we may write $p = 2u + 1$ where $u \in I$ and $p^2 = (2u + 1)(2u + 1) = 4u^2 + 4u + 1 = 2(2u^2 + 2u) + 1$, which is odd. If p is even, let $p = 2k$ where $k \in I$. Hence $2 \cdot q^2 = p^2 = (2k)^2 = 4k^2$. From the equation $2 \cdot q^2 = 4 \cdot k^2$, we obtain $q^2 = 2 \cdot k^2$ and conclude that q^2 is even, since q^2 is equal to the product of 2 and an integer (since $k \in I$, $k^2 \in I$). If q^2 is even, then q must be even. We now have that p is even and q is even and, hence, 2 must divide both p and q. But p and q were chosen to be relatively prime and, therefore, our assumption that $\sqrt{2}$ is a rational number is false.

We have proved that $\sqrt{2}$ is not a rational number. However, this does not mean that we proved that $\sqrt{2}$ is an irrational number. There are numbers such as $\sqrt{-2}$, that are neither rational nor irrational. However, for the time being we will assume that all numbers are either rational or irrational. With that assumption we will give the following tentative definition.

DEFINITION 10.1 (Tentative) An *irrational number* is a number that is not a rational number.

According to this tentative definition, then, $\sqrt{2}$ is an irrational number. Also, $\sqrt{2}$ added to any integer will yield another irrational number. Hence $1 + \sqrt{2}$, $5 + \sqrt{2}$, and $^-7 + \sqrt{2}$ are all examples of irrational numbers.

Another example of an irrational number is π, which is used to denote the ratio of the circumference of a circle to the length of its diameter. Additional irrational numbers can be "constructed" as suggested in Figure 10.2, where the successive triangles are right triangles. Of course some of the numbers so constructed (such as $\sqrt{4}$, $\sqrt{9}$, etc.) are not irrational numbers.

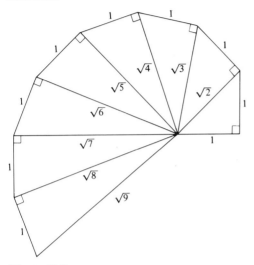

Figure 10.2

Throughout the balance of this chapter we shall strive to develop a system known as the real number system which, basically, will be the union of the rationals together with the irrationals. Although different approaches may be used for this development, most are beyond the level of this text. To develop this new system, we will introduce the concept of decimal numbers.

EXERCISES

1. Classify each of the following as representing rational or irrational numbers.
   a. $\sqrt{3}$.
   b. $\sqrt{4}$.
   c. $\sqrt{5}$.
   d. $\sqrt{6}$.
   e. $\sqrt{7}$.
   f. $\sqrt{8}$.
   g. $\sqrt{9}$.
   h. $\sqrt{10}$.
   i. $\sqrt{25}$.
   j. $2 + \sqrt{3}$.
   k. $1 - \sqrt{2}$.
   l. $(\sqrt{2} - 1) - \sqrt{2}$.
   m. $(3 - \sqrt{2}) - (\sqrt{2} - 3)$.
   n. $\sqrt{3}/2\sqrt{3}$.
   o. $\sqrt{68} \div 17$.
2. Verify that the set of irrational numbers is not closed under the operation of addition by forming the indicated sums.
   a. $-\sqrt{2} + \sqrt{2}$.
   b. $-\sqrt{3} + \sqrt{3}$.
   c. $-\sqrt{7} + \sqrt{7}$.
   d. $-\sqrt{11} + \sqrt{11}$.

3. If $a > 0$ and $(\sqrt{a})^2 = \sqrt{a} \cdot \sqrt{a} = a$, verify that the set of irrational numbers is not closed under the operation of multiplication by forming the indicated products.

   a. $(\sqrt{2})^2$.   b. $(\sqrt{3})^2$.

   c. $(\sqrt{11})^2$.   d. $(\sqrt{19})^2$.

4. a. Is the sum of a rational number and an irrational number a rational number or an irrational number?

   b. Is the product of a rational number and an irrational number a rational number or an irrational number?

   c. Is the sum of two irrational numbers always an irrational number? Explain.

   d. Is the product of two irrational numbers always an irrational number? Explain.

   e. Is 0 an irrational number? Why or why not?

10.3 TERMINATING DECIMALS AND RATIONAL NUMBERS

Performing operations on or comparing rational numbers is not always a simple task, as was noted in Chapter 9. For instance, to add the rational numbers 2/3 and 4/7, we first obtained a common denominator, 21, and considered adding the rational numbers 14/21 and 12/21, which simplified matters somewhat. Similarly, to determine the lesser of the two rational numbers $^-2/5$ and $4/^-17$, we considered the rational numbers in the equivalent forms of $^-34/85$ and $^-20/85$ and then compared the numerators of the representative fractions. Further ease of computation involving rational numbers results when the denominators of the fractions representing the rational numbers can be written as a power of ten. Using place value and a dot called a *decimal point*, we can eliminate the need for denominators. The decimal point is placed after the units digit, and we refer to the first digit to the right of the dot as tenths, the second digit as hundredths, the third digit as thousandths, and so forth.

EXAMPLE

a. $\frac{7}{10} = 0.7$ (read seven tenths)

b. $\frac{17}{10} = 1.7$ (read one and seven tenths)

c. $\frac{175}{10} = 17.5$ (read seventeen and five tenths).

d. $\frac{1752}{10} = 175.2$ (read one hundred seventy-five and two tenths).

EXAMPLE

a. $\frac{7}{100} = 0.07$ (read seven hundredths).

b. $\frac{17}{100} = 0.17$ (read seventeen hundredths).

c. $\frac{175}{100} = 1.75$ (read one and seventy-five hundredths).

d. $\frac{1752}{100} = 17.52$ (read seventeen and fifty-two hundredths).

EXAMPLE

a. $\frac{7}{1000} = 0.007$ (read seven thousandths).

b. $\frac{17}{1000} = 0.017$ (read seventeen thousandths).

c. $\frac{175}{1000} = 0.175$ (read one hundred seventy-five thousandths).

d. $\frac{1752}{1000} = 1.752$ (read one and seven hundred fifty-two thousandths).

EXAMPLE

a. $23.12 = 23 + \frac{1}{10} + \frac{2}{100}$

$\qquad = \frac{2300}{100} + \frac{10}{100} + \frac{2}{100}$

$\qquad = \frac{2312}{100}$

b. $137.207 = 137 + \frac{2}{10} + \frac{0}{100} + \frac{7}{1000}$

$\qquad = \frac{137000}{1000} + \frac{200}{1000} + \frac{0}{1000} + \frac{7}{1000}$

$\qquad = \frac{137207}{1000}$

EXAMPLE

a. $47.16 = 47 + \frac{1}{10} + \frac{6}{100}$

$\qquad = (4 \times 10^1) + (7 \times 10^0) + (1 \times 10^{-1}) + (6 \times 10^{-2})$

b. $8.0169 = 8 + \frac{0}{10} + \frac{1}{100} + \frac{6}{1000} + \frac{9}{10000}$

$\qquad = (8 \times 10^0) + (0 \times 10^{-1}) + (1 \times 10^{-2}) + (6 \times 10^{-3})$

$\qquad\quad + (9 \times 10^{-4})$

c. $0.023 = 0 + \frac{0}{10} + \frac{2}{100} + \frac{3}{1000}$

$\qquad = (0 \times 10^0) + (0 \times 10^{-1}) + (2 \times 10^{-2}) + (3 \times 10^{-3})$

Observe in the above examples that the dot or decimal point separates the whole number part from the fractional part of the number. Also notice that the fractions in the above examples also are represented by decimals that terminate or have a last digit.

DEFINITION 10.2 (Terminating Decimal) A decimal that can be represented by a fraction whose denominator is an integral power of ten is said to be a *terminating decimal*.

All of the decimals used so far in this section are examples of terminating decimals. In the following example we will note that the rational numbers represented by the fractions given also can be represented by terminating decimals. The use of the fundamental theorem of fractions is involved.

EXAMPLE

a. $\dfrac{1}{8} = \dfrac{1(125)}{8(125)} = \dfrac{125}{1000}$

$\quad$ Hence $\quad \dfrac{1}{8} = 0.125$

b. $\dfrac{3}{20} = \dfrac{3(5)}{20(5)} = \dfrac{15}{100}$

Hence $\dfrac{3}{20} = 0.15$

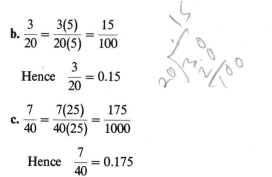

c. $\dfrac{7}{40} = \dfrac{7(25)}{40(25)} = \dfrac{175}{1000}$

Hence $\dfrac{7}{40} = 0.175$

A close examination of the above example may suggest an algorithm for rewriting the fractions given in equivalent decimal form. This algorithm involves dividing the numerator of the fraction by its denominator, placing a decimal point after the numerator, and annexing as many zeros as needed for the division. Hence $\frac{1}{8}$ becomes:

$$
\begin{array}{r}
0.125 \\
8{\overline{)1.000}} \\
8 \\
\hline
20 \\
16 \\
\hline
40 \\
40 \\
\hline
0
\end{array}
$$

EXAMPLE By using division, find a decimal representation for $\frac{7}{40}$.
Solution

$$
\begin{array}{r}
0.175 \\
40{\overline{)7.000}} \\
40 \\
\hline
300 \\
280 \\
\hline
200 \\
200 \\
\hline
0
\end{array}
$$

Since the division is complete (that is, the remainder is 0) we have $\frac{7}{40} = 0.175$. ∎

It should be noted that not all rational numbers can be expressed as terminating decimals. For instance, the rational number represented by the fraction $\frac{1}{3}$ cannot be written as a terminating decimal. There is no integer whose product with 3 yields an integral power of ten. We have a useful theorem that enables us to determine whether a fraction can be expressed as a terminating decimal. We will now state the theorem without proof.

THEOREM 10.2 Let the fraction a/b be the reduced representative for some rational number. Then a/b can be written as a terminating decimal if and only if b has no prime factors other than 2 or 5.

EXAMPLE
a. $\frac{5}{8}$ can be expressed as a terminating decimal since $\frac{5}{8}$ is in reduced representative form and the only prime factor of 8 is 2.
b. $\frac{17}{24}$ cannot be expressed as a terminating decimal since $\frac{17}{24}$ is in reduced representative form but the prime factors of 24 are 2 and 3.

EXAMPLE The decimals 1.7 and 1.70 are equivalent decimals and represent the same rational number. This follows since the decimal 1.7 can be written as the fraction $\frac{17}{10}$ and the decimal 1.70 can be written as the fraction $\frac{170}{100}$. But $\frac{17}{10} \simeq \frac{170}{100}$ since $(17)(100) = (10)(170)$. Since the two fractions represent the same rational number, the decimals 1.7 and 1.70 also represent the same rational number and thus are equivalent.

The last example above suggests if we have a terminating decimal, then we may annex a finite number of zeros after the last digit to the right of the decimal point. Thus the decimals 2.7, 2.700, 2.70000, and 2.7000000000000000 all represent the same rational number.

EXERCISES

1. Rewrite each of the following fractions as equivalent decimals.
   a. $\frac{2}{10}$, b. $\frac{13}{10}$. c. $\frac{23}{10}$.
   d. $\frac{34}{10}$. e. $\frac{41}{10}$. f. $\frac{53}{10}$.
   g. $\frac{72}{10}$. h. $\frac{89}{10}$. i. $\frac{101}{10}$.
   j. $\frac{231}{10}$. k. $\frac{469}{10}$. l. $\frac{613}{10}$.
   m. $\frac{829}{10}$. n. $\frac{999}{10}$. o. $\frac{1023}{10}$.

2. Rewrite each of the following fractions as equivalent decimals.
   a. $\frac{2}{100}$. b. $\frac{13}{100}$. c. $\frac{23}{100}$.
   d. $\frac{34}{100}$. e. $\frac{41}{100}$. f. $\frac{53}{100}$.
   g. $\frac{72}{100}$. h. $\frac{89}{100}$. i. $\frac{101}{100}$.
   j. $\frac{231}{100}$. k. $\frac{469}{100}$. l. $\frac{613}{100}$.
   m. $\frac{829}{100}$. n. $\frac{999}{100}$. o. $\frac{1023}{100}$.

3. Rewrite each of the following decimal numerals in equivalent fraction form.

a. 1.3.	b. 11.2.	c. 23.9.
d. 2.13.	e. 46.19.	f. 57.27.
g. 14.123.	h. 23.296.	i. 37.307.
j. 40.029.	k. 57.167.	l. 70.002.

4. Using division, determine a decimal representation for each of the following.

a. $\frac{3}{40}$.	b. $\frac{7}{20}$.	c. $\frac{9}{80}$.
d. $\frac{13}{20}$.	e. $\frac{17}{200}$.	f. $\frac{23}{40}$.
g. $\frac{37}{20}$.	h. $\frac{169}{50}$.	i. $\frac{96}{80}$.
j. $\frac{137}{200}$.	k. $\frac{897}{400}$.	l. $\frac{769}{2000}$.

5. Use Theorem 10.2 to determine which of the following fractions may be written as terminating decimals.

a. $\frac{6}{35}$.	b. $\frac{9}{80}$.	c. $\frac{7}{60}$.
d. $\frac{14}{40}$.	e. $^-17/50$.	f. $^-83/90$.
g. $^-75/10$.	h. $\frac{69}{200}$.	i. $^-97/400$.
j. $\frac{7}{81}$.	k. $^-19/57$.	l. $\frac{169}{300}$.
m. $\frac{23}{36}$.	n. $\frac{123}{150}$.	o. $\frac{231}{360}$.

10.4 OPERATIONS WITH TERMINATING DECIMALS

Performing arithmetic operations with terminating decimals is relatively easy if we remember the place value associated with each digit in the decimal. For instance, if we wish to add 2.31 to 1.46, we could first rewrite the decimals in equivalent fraction form. Hence $2.31 = \frac{231}{100}$ and $1.46 = \frac{146}{100}$. We now form the sum $\frac{231}{100} + \frac{146}{100} = \frac{377}{100}$ and rewrite $\frac{377}{100}$ in equivalent decimal form as 3.77.

Next consider the sum of 0.169 and 0.46. Rewriting the decimals, we have $0.169 = \frac{169}{1000}$ and $0.46 = \frac{46}{100} \simeq \frac{460}{1000}$. Now $\frac{169}{1000} + \frac{460}{1000} = \frac{629}{1000}$, which may be written as the decimal 0.629.

A close examination of these two addition problems may reveal an algorithm for the addition of decimal numbers. Basically, arrange the decimals one under the other, line up the decimal points, and add the corresponding digits. Reconsidering the two preceding problems, we would have

$$\begin{array}{r} 2.31 \\ +1.46 \\ \hline 3.77 \end{array} \quad \text{and} \quad \begin{array}{r} 0.169 \\ +0.46 \\ \hline 0.629 \end{array}$$

EXAMPLE   Add 3.462, 12.03, 7.0192, and 0.069.
Solution

$$
\begin{array}{r}
3.462 \\
12.03 \\
7.0192 \\
0.069 \\
\hline
22.5802
\end{array}
$$
■

Subtraction would be performed in a similar manner.

EXAMPLE   Subtract 3.72 from 12.069.
Solution

$$
\begin{array}{r}
12.069 \\
-\ 3.720 \\
\hline
8.349
\end{array}
$$
■

EXAMPLE   From the sum of 10.03 and 2.691, subtract 4.1234.
Solution

$$
\begin{array}{r}
10.03 \\
+\ 2.691 \\
\hline
12.721 \\
-\ 4.1234 \\
\hline
\end{array}
$$
which we may rewrite as
$$
\begin{array}{r}
12.7210 \\
-\ 4.1234 \\
\hline
8.5976
\end{array}
$$
■

To motivate the algorithm for multiplying decimals, consider multiplying 0.32 by 0.17. Since $0.32 = 32/100$ and $0.17 = 17/100$, we have $(0.32)(0.17) = (32/100)(17/100) = (32)(17)/10000 = 544/10000 = 0.0544$. A close examination of this multiplication may suggest that we could have multiplied 0.32 by 0.17 as though they were whole numbers, obtained 544, and then "marked off" four decimal places to obtain 0.0544. The four decimal places would correspond to the two decimal places in 0.32 plus the two decimal places in 0.17. To mark off four decimal places in the answer, it was necessary to insert a 0 between the decimal point and 5.

EXAMPLE   Multiply 1.23 by 0.694.
Solution

$$
\begin{array}{r}
1.23 \\
\times 0.694 \\
\hline
492 \\
1107 \\
738 \\
\hline
0.85362
\end{array}
$$

1.23   A 2 decimal place numeral
× 0.694   A 3 decimal place numeral

492

1107   Multiply as though the decimal numbers were whole numbers
738

0.85362   A $2 + 3 = 5$ decimal place numeral   ■

EXAMPLE Multiply 23.192 by 3.72.
Solution

$$
\begin{array}{r}
23.192 \\
\times\ \ \ \ 3.72 \\
\hline
46384 \\
162344 \\
69576 \\
\hline
86.27424
\end{array}\ \ \blacksquare
$$

To divide the decimal 0.786 by the decimal 2.62, we consider multiplying both the dividend and the divisor by a power of ten, which will make the divisor a whole number. Hence,

$$0.786 \div 2.62 = \frac{0.786}{2.62} = \frac{0.786 \times 100}{2.62 \times 100} = \frac{78.6}{262}$$

or

$$
\begin{array}{r}
0.3\ \ \ \\
262\overline{)78.6} \\
786 \\
\hline
0
\end{array}
$$

This algorithm suggests that to divide a decimal by a decimal, we must move the decimal point in both the dividend and the divisor the same number of places to the right so that the divisor becomes a whole number. Then we proceed as in long division, lining up the decimal point in the quotient.

EXAMPLE Divide 133.66 by 8.2.
Solution

$$
8.2\overline{)133.66} \quad \text{or} \quad
\begin{array}{r}
16.3\ \ \ \\
82\overline{)1336.6} \\
82\ \ \ \ \\
\hline
516\ \ \\
492\ \ \\
\hline
246 \\
246 \\
\hline
0
\end{array}
$$

Hence $133.66 \div 8.2 = 16.3$. $\blacksquare$

EXAMPLE Divide 15.903 by 0.93.
Solution

$$0.93\overline{)15.903} \quad \text{or} \quad 93\overline{)1590.3}$$

with quotient 17.1:

$$
\begin{array}{r}
17.1 \\
93\overline{)1590.3} \\
93 \\
\hline
660 \\
651 \\
\hline
93 \\
93 \\
\hline
0
\end{array}
$$

Hence $15.903 \div 0.93 = 17.1$. ■

Now consider the decimal 0.23, which we may write as the fraction $\frac{23}{100}$. The fraction $\frac{23}{100}$ may be considered as 23 parts per 100 parts or as 23 *percent*, which may be denoted by the symbol 23%. A percent, then, represents a fraction whose denominator is 100. Hence certain fractions may be written as equivalent decimals or as equivalent percents.

EXAMPLE Convert each of the following decimals to percents: (a) 0.47, (b) 0.53, (c) 0.81, (d) 1.09, and (e) 0.238.
Solution
a. $0.47 = \frac{47}{100} = 47\%$

b. $0.53 = \frac{53}{100} = 53\%$

c. $0.81 = \frac{81}{100} = 81\%$

d. $1.09 = \frac{109}{100} = 109\%$

e. $0.238 = \frac{238}{100} = 23.8\%$ ■

EXAMPLE Convert each of the following percents to decimals: (a) 57%, (b) 63%, (c) 14.4%, (d) 123%, and (e) 0.2%.
Solution
a. $57\% = \frac{57}{100} = 0.57.$

b. $63\% = \frac{63}{100} = 0.63$

c. $14.4\% = \frac{14.4}{100} \simeq \frac{144}{1000} = 0.144.$

d. $123\% = \frac{123}{100} = 1.23.$

e. $0.2\% = \frac{0.2}{100} \simeq \frac{2}{1000} = 0.002.$ ■

From an examination of the two examples above, it should be clear that to change a decimal to a percent, we move the decimal point *two* places to the *right* and annex the % symbol. Hence $0.36 = 36\%$ and $0.147 = 14.7\%$. To change a percent to a decimal, we drop the % symbol and move the decimal point *two* places to the *left*. Hence $59\% = 0.59$, $143\% = 1.43$, and $5.9\% = 0.059$.

EXAMPLE 32% of the students in Professor Cube's mathematics class are veterans. If there are 75 students in the class, how many students are not veterans?

Solution Since the total class would be considered as 100% of the students, the percent of non-veterans in Professor Cube's class would be 100% − 32% = 68%. Hence our desired result is 68% of 75 or (0.68)(75) = 51. ∎

EXAMPLE If a student earns $86.40 per week and has $12.96 deducted for withholding taxes, what percent of his earnings is being deducted for withholding taxes?

Solution $12.96/$86.40 represents the ratio of deducted withholding taxes to earnings. This may also be represented as 1296/8640 (in dollars and cents). But by division 1296/8640 = 0.15 = 15%. Hence $12.96 is 15% of $86.40. ∎

EXERCISES

1. Perform the indicated additions and subtractions.
   a. 2.136 + 23.27.
   b. 39.69 − 27.1
   c. 36.024 − 19.697.
   d. 56.2 + 69.701.
   e. 96.23 − ⁻69.17.
   f. ⁻396.23 + 127.697.
   g. 63.2 + 79.89 + 6.039.
   h. (709.26 + 209.5) − 469.7.
   i. (89.69 − 23.027) − 96.2.
   j. 607.8 − (49.08 + 209.201).

2. Perform the indicated multiplications.
   a. 23.2 × 36.4.
   b. 69.17 × 21.02.
   c. 96.023 × 16.3.
   d. ⁻23.12 × 87.9.
   e. (8.09)(10.024).
   f. (⁻1.23)(⁻3.046).
   g. 2(69.17)(24.2).
   h. 0.04(6.09)(17.2).
   i. (0.02)(⁻0.03)(0.004).
   j. (6.209)(13.01)(⁻13).

3. Perform the indicated divisions.
   a. 29.92 ÷ 2.2.
   b. 38.391 ÷ 19.1.
   c. 24.824 ÷ 1.07.
   d. 41.580 ÷ 0.9.
   e. 46.224 ÷ 43.2.
   f. 31.68 ÷ 2.2.
   g. 197.442 ÷ 47.01.
   h. 218.691 ÷ 51.7.
   i. 315.1 ÷ 0.023.
   j. 0.812 ÷ 1.16.

4. Convert each of the following decimals to percents.
   a. 0.46.
   b. 0.59.
   c. 0.63.
   d. 0.109.
   e. 0.213.
   f. 0.369.
   g. 0.639.
   h. 0.832.
   i. 0.976.
   j. 2.13.
   k. 13.16.
   l. 11.231.
   m. 2.312.
   n. 0.0371.
   o. 0.0023.

5. Convert each of the following percents to decimals.
   a. 49%.
   b. 67%.
   c. 19.1%.
   d. 87.68%.
   e. 99.9%.
   f. 121%.
   g. 132.4%.
   h. 0.3%.
   i. 0.19%.

j. 0.019%.   k. 0.001%.   l. 1247%.
m. 2317.8%.   n. 0.999%.   o. 0.0101%.

6. a. What is 6% of 231?
   b. 12 is what percent of 48?
   c. A person's salary is doubled. What is the percent increase?
   d. A person received an increase that was double his salary. What was the percent increase?

7. An employee received a salary of $10,000. Due to poor market conditions, he agreed to take a 10% reduction in salary. Six months later, due to improved market conditions, his employer gave him a 10% increase in salary. What is the employee's new salary? (Assume that no other changes in salary were made.)

8. Employee A works at Three Dimensional Industries and receives a salary of $14,000 per year. Fringe benefits paid by the employer on behalf of the employee amount to 23.6% of the employee's salary. How much is paid by the employer to and on behalf of Employee A?

9. A contract was renegotiated at the Three Dimensional Industries that called for a 6.8% increase in salary plus additional fringe benefits, bringing the total fringe benefit package to 27.2% of the salary. How much is now paid to and on behalf of employee A by the employer? (See Exercise 8.)

10. An individual's income tax amounts to 23.8% of salary, or $5000, whichever is larger. If the individual's income is $22,000, how much is the income tax due?

10.5 REPEATING DECIMALS AND RATIONAL NUMBERS

In Section 10.3 we considered some rational numbers being expressed as terminating decimals. That is, if we considered the reduced representative of the rational number and divided its numerator by its denominator, we eventually obtained a remainder of 0; the division terminated. Now consider the rational number represented by the fraction $\frac{1}{3}$. If we divide the numerator, 1, by the denominator, 3, the division does not terminate:

$$
\begin{array}{r}
0.333333 \\
3\overline{)1.000000} \\
9 \\
\overline{10} \\
9 \\
\overline{10} \\
9 \\
\overline{10} \\
9 \\
\overline{10} \\
9 \\
\overline{1}
\end{array}
$$

For each step in the division process we obtained a remainder of 1 and the division repeated. We may represent $\frac{1}{3} = 0.333333\ldots$ with the symbol "..." indicating the repeating nature of the decimal.

Next consider the rational number represented by the fraction $\frac{1}{7}$. Again, dividing we obtain

$$
\begin{array}{r}
0.142857 \\
7\overline{)1.000000} \\
\underline{7} \\
30 \\
\underline{28} \\
20 \\
\underline{14} \\
60 \\
\underline{56} \\
40 \\
\underline{35} \\
50 \\
\underline{49} \\
1
\end{array}
$$

In this division the last remainder obtained is 1 and the digits in the quotient will now repeat. We may represent $\frac{1}{7} = 0.142857142857\ldots$ (The block of digits that repeats is 142857.)

In general, if a rational number cannot be represented as a terminating decimal, it must be represented as a repeating decimal as we will now discuss. Let a/b be a fraction representing a rational number. Then by the division algorithm, $a = bq + r$ with $0 \leq r < b$. The remainder r must be one of the elements in the set $\{0, 1, 2, 3, 4, \ldots, b - 1\}$. Hence if we have repeated applications of the division algorithm, not more than b divisions can be made before a remainder will appear that appeared previously. When this happens, the digits in the quotient will repeat themselves. Therefore we conclude that every rational number must be represented either by a terminating decimal or a repeating decimal. If the decimal is repeating, we will indicate which digits repeat by placing a bar across them. Hence $\frac{1}{7} = 0.\overline{142857}$ and $\frac{1}{3} = 0.3\overline{3}$.

EXAMPLE

a. $\frac{2}{3} = 0.66\overline{6}$

b. $\frac{3}{11} = 0.27\overline{27}$

c. $\frac{2}{9} = 0.22\overline{2}$

d. $\frac{5}{9} = 0.5\overline{5}$

e. $\frac{1}{2} = 0.50\overline{0}$.

f. $\frac{5}{8} = 0.6250\overline{0}$

The last two parts of the preceding example suggest that terminating decimals also may be written as repeating decimals with the digit 0 repeating.

We now have established that every rational number may be represented either as a terminating decimal or a repeating decimal. We also know that a number represented by a terminating decimal is always a rational number. However, if a number is represented by a repeating decimal, is it always a rational number? For instance, consider the repeating decimal $2.43\overline{3}$. We could let $N = 2.4333\ldots$. Since only the digit 3 repeats, we could multiply N by 10, obtaining

$$10N = 24.3333\ldots$$

and

$$N = 2.4333\ldots$$

Subtracting, we obtain

$$9N = 21.9$$

or

$$N = \frac{21.9}{9} = \frac{219}{90} \simeq \frac{73}{30}$$

Since N is written as a fraction, $2.4333\ldots$ represents a rational number.

EXAMPLE Determine a fraction which represents $0.147147147\ldots$.
Solution Let $N = 0.147147147\ldots$ with the digits 147 repeating. Since three digits repeat, we form

$$1000N = 147.147147147\ldots$$

and

$$N = 0.147147147\ldots$$

Subtracting, we obtain

$$999N = 147$$

or

$$N = \frac{147}{999} \simeq \frac{49}{333} \quad \blacksquare$$

We conclude this section with the following theorem, which has been verified by the examples given in this section and in Section 10.3.

THEOREM 10.3 Every rational number can be represented either by a terminating decimal or a repeating decimal. Conversely, every decimal that either terminates or repeats represents a rational number.

EXERCISES

1. Rewrite each of the following rational numbers as a terminating or repeating decimal.

   a. $\frac{2}{3}$.
   b. $\frac{1}{4}$.
   c. $\frac{2}{7}$.

   d. $\frac{3}{11}$.
   e. $\frac{4}{17}$.
   f. $\frac{1}{8}$.

   g. $\frac{2}{9}$.
   h. $\frac{3}{13}$.
   i. $\frac{5}{11}$.

   j. $\frac{4}{7}$.
   k. $\frac{2}{19}$.
   l. $\frac{5}{6}$.

   m. $\frac{3}{4}$.
   n. $\frac{11}{16}$.
   o. $\frac{9}{11}$.

   p. $\frac{23}{46}$.
   q. $\frac{38}{86}$.
   r. $\frac{49}{96}$.

2. Rewrite each of the following terminating decimals as a fraction in simplest form.

   a. 0.31.
   b. 0.29.
   c. 0.85.

   d. 0.609.
   e. 0.213.
   f. 0.896.

   g. 1.24.
   h. 3.19.
   i. 5.607.

   j. 7.08.
   k. 23.96.
   l. 37.002.

3. Rewrite each of the following repeating decimals as a fraction in simplest form.

   a. $0.111\overline{1}$.
   b. $0.2\overline{323}$.
   c. $0.145\overline{145}$.

   d. $0.312\overline{312}$.
   e. $0.1596\overline{1596}$.
   f. $0.2021\overline{2021}$.

   g. $0.03\overline{97}$.
   h. $0.002\overline{34}$.
   i. $0.000069^{\cdot}$.

   j. $0.123\overline{456}$.
   k. $2.398\overline{767}$.
   l. $14.2\overline{467}$.

4. Rewrite each of the following mixed numbers as a decimal.

   a. $4\frac{1}{2}$.
   b. $5\frac{2}{3}$.
   c. $7\frac{1}{9}$.

   d. $8\frac{3}{4}$.
   e. $2\frac{7}{11}$.
   f. $3\frac{6}{7}$.

   g. $9\frac{3}{8}$.
   h. $10\frac{1}{5}$.
   i. $11\frac{2}{13}$.

   j. $16\frac{7}{8}$.
   k. $23\frac{2}{21}$.
   l. $37\frac{4}{15}$.

5. Let $P = 1.212112111211112\ldots$
   a. Does P represent a terminating decimal?
   b. Does P represent a repeating decimal?
   c. Does P represent a rational number? Explain.

6. a. Determine a decimal expression for $\frac{1}{4}$.
   b. Determine a decimal expression for $\frac{2}{8}$.
   c. Determine a decimal expression for $\frac{3}{5}$.
   d. Determine a decimal expression for $\frac{6}{10}$.
   e. Determine a decimal expression for $\frac{1}{3}$.
   f. Determine a decimal expression for $\frac{2}{6}$.
   g. Do equivalent fractions have the same decimal representation? Explain.

10.6 DECIMALS AND BASES OTHER THAN TEN

So far in this chapter we have been discussing decimals relative to base ten numerals. Of course the word decimal only pertains to base ten since it means "a tenth part of." However, in earlier chapters we have worked with numerals represented in bases other than ten, and the idea of decimal representation of rational numbers expressed in other bases is meaningful. Of course the dot used in the numeral would not be called a decimal point. For instance, the numeral 12.43_{five} represents 1 fives, 2 ones, 4 fifths, and 3 twenty-fifths. Similarly, the numeral 0.346_{seven} represents 3 sevenths, 4 forty-ninths, and 6 three hundred forty-thirds.

EXAMPLE Add 513.6_{seven}, 1.34_{seven}, and 23.654_{seven}.
Solution As with base ten, we would line up the dots and add, in base seven, the numerals as though they represented whole numbers as follows:

$$
\begin{array}{r}
513.6_{\text{seven}} \\
1.34_{\text{seven}} \\
23.654_{\text{seven}} \\
\hline
542.224_{\text{seven}} \quad \blacksquare
\end{array}
$$

EXAMPLE Multiply 23.4_{five} by 3.42_{five}.
Solution

$$
\begin{array}{r}
23.4_{\text{five}} \\
3.42_{\text{five}} \\
\hline
1023 \\
2101 \\
1312 \\
\hline
203.233_{\text{five}} \quad \blacksquare
\end{array}
$$

EXAMPLE Express 45.32_{eight} as a base ten fraction.
Solution

$$
\begin{aligned}
45.23_{\text{eight}} &= [(4 \times 10^1) + (5 \times 10^0) + (2 \times 10^{-1}) + (3 \times 10^{-2})]_{\text{eight}} \\
&= [(4 \times 8^1) + (5 \times 8^0) + (2 \times 8^{-1}) + (3 \times 8^{-2})]_{\text{ten}} \\
&= \left(32 + 5 + \frac{2}{8} + \frac{3}{64}\right)_{\text{ten}} \\
&= \left(37 + \frac{19}{64}\right)_{\text{ten}} \\
&= \frac{2387}{64}_{\text{ten}} \quad \blacksquare
\end{aligned}
$$

EXERCISES

1. Perform the indicated operations in the given base.

   a. $21.364_{seven} + 3.05_{seven} + 143.3_{seven}$.

   b. $35.62_{eight} - 23.712_{eight}$.

   c. $93T.E_{twelve} \times 89.2T_{twelve}$.

   d. $(34.12_{five} + 43.2_{five}) - 102.234_{five}$.

   e. $10110.1_{two} + 110.0011_{two} + 1.011_{two}$.

2. Express each of the following as a base ten fraction.

   a. 23.65_{nine}.   b. 6.234_{eight}.

   c. $T.E3_{twelve}$.   d. 101.011_{two}.

   e. 201.102_{three}.   f. 23.45_{seven}.

   g. 3.122_{four}.   h. 31.424_{five}.

   i. 44.44_{six}.   j. 40.23_{eleven}.

10.7 IRRATIONAL NUMBERS REVISITED

In Section 10.5 we established that every rational number can be represented as a terminating or a repeating decimal. We also established that every terminating or repeating decimal represents a rational number. So far in our discussion we have not considered the possibility of a nonterminating, nonrepeating decimal. Do such decimals exist and, if so, what do they represent?

Consider the decimal 0.7272727222722227222227222227 .... This decimal is neither terminating nor repeating since each 7 in the decimal has one more 2 preceding it than the preceding 7. Hence we have demonstrated the existence of a nonterminating, nonrepeating decimal. Other such decimals could easily be "constructed." What do these decimals represent?

DEFINITION 10.3 A decimal that is neither repeating nor terminating represents an *irrational number*.

Recall in Definition 10.1 that we gave a tentative definition for an irrational number as a number that is not a rational number. This definition was made with the assumption that all numbers are either rational or irrational, which is not true. However, *as decimals, all numbers are either rational or irrational.*

Also in Section 10.2 we proved that $\sqrt{2}$ is not rational and accepted it as being irrational. Now what is the decimal representation for $\sqrt{2}$? To evaluate $\sqrt{2}$, we wish to determine a value n such that $n^2 = 2$. Since $(1.4)^2 = 1.96$ and $(1.5)^2 = 2.25$, we have $1.4 < \sqrt{2} < 1.5$. Hence we conclude that the value of $\sqrt{2}$ is between 1.4 and 1.5. To obtain a better approximation, we note that $(1.41)^2 = 1.9881$ and $(1.42)^2 = 2.0164$ and, hence, $1.41 <$

$\sqrt{2} < 1.42$. Also, $(1.414)^2 = 1.999396$ and $(1.415)^2 = 2.002225$. Therefore $1.414 < \sqrt{2} < 1.415$. We could continue in this manner and obtain a value of $\sqrt{2}$ to as many decimal places as desired. However, no matter how long we continue in this manner, we will never obtain a repeating block of digits nor will the process terminate. The reason for this, of course, is that $\sqrt{2}$ is irrational.

Another example of an irrational number is $\sqrt{3}$. To obtain a decimal representation of $\sqrt{3}$, we may use what is known as the method of averaging. First we determine an approximation for $\sqrt{3}$. This first approximation may be too large or too small, and we next try to improve upon it. Suppose we approximate $\sqrt{3}$ by 2. Using this approximation, we will divide 3 by 2, obtaining 1.5. Since $2 \neq 1.5$, we will average the quotient, 1.5, and the approximated value, 2. To average we add the two and divide by 2. Hence $(2 + 1.5)/2 = 1.75$, which value we now use as our new approximation. We now divide 3 by 1.75, obtaining 1.714 to three decimal places. Since $1.75 \neq 1.714$, we average the two, obtaining 1.7321. Dividing 3 by 1.7321, we now obtain 1.7320 to four decimal places. We would now conclude that $\sqrt{3}$ is approximately equal to 1.73 (if we desired a two-place decimal approximation) or 1.732 (for a three-place decimal approximation). If a better approximation were desired, we would continue the process of averaging. It should be emphasized, however, that any approximation, regardless of the number of decimal places involved, will be a rational approximation for this irrational number.

Other examples of irrational numbers are $\sqrt{5}, \sqrt{7}, \sqrt{13}, 1 + \sqrt{2}, 2 - \sqrt{3}$, π, and e. The last two examples are called *transcendental numbers*, whereas the others are called *algebraic numbers* since they are solutions to equations of the form $x^2 = b$ or $x^3 + ax^2 + bx + c = 0$. The irrational number π may be written as $\pi = 3.14159$ to five decimal places, and $e = 2.718281828459045$ to fifteen decimal places. There are infinitely many irrational numbers and, in fact, there are more irrational numbers than rational numbers. Considering the denseness property for rational numbers, this may surprise the reader.

EXERCISES

1. Classify each of the following as being a rational number or an irrational number.

   a. $2\sqrt{3}$.

   b. $2 - \sqrt{3}$.

   c. $\sqrt{2} \cdot \sqrt{3}$.

   d. $\sqrt{3} \cdot \sqrt{3}$.

   e. $4\sqrt{3} \div 2\sqrt{3}$.

   f. $(2 + \sqrt{2}) \div 2$.

   g. 3π.

   h. $\pi - \sqrt{4}$.

   i. $^-4 + \sqrt{4}$.

   j. $3\sqrt{121}$.

   k. $\sqrt{5}\sqrt{125}$.

   l. $\sqrt{9 + 16}$.

   m. $\sqrt{25 + 36}$.

   n. $\sqrt{50 \div 2}$.

   o. $\sqrt{\pi}$.

2. Classify each of the following decimals as representing a rational number or an irrational number.

   a. $23.16\overline{9}$.

   b. 4.0.

   c. $1.3333 \ldots$.

d. 57.1666 ....   e. 1.8181181118 ....   f. 0.01010101.

g. $\frac{1}{3}(0.5)$.   h. $\frac{1}{2}(0.625)$.   i. $\frac{1}{10}(0.3333\ldots)$.

j. $639.72196\overline{4}$.   k. 4.000923.   l. 0.697654321.

3. Determine a rational approximation to two decimal places for each of the following.

     a. $\sqrt{2}$.   b. $\sqrt{3}$.   c. $\sqrt{5}$.

     d. $\sqrt{27}$.   e. $\sqrt{31}$.   f. $\sqrt{43}$.

     g. $\sqrt{101}$.   h. $\sqrt{205}$.   i. $\sqrt{300}$.

4. Determine a rational approximation to three decimal places for each of the following.

     a. $\sqrt{7}$.   b. $\sqrt{11}$.   c. $\sqrt{17}$.

     d. $\sqrt{39}$.   e. $\sqrt{57}$.   f. $\sqrt{82}$.

     g. $\sqrt{110}$.   h. $\sqrt{201}$.   i. $\sqrt{307}$.

10.8 REAL NUMBERS

In Section 10.1 we indicated a need for a new type of number beyond the rational numbers. The approach used to introduce these new numbers was the decimal number approach. We classified the decimals as being terminating, repeating, or nonterminating, nonrepeating. Further, we classified rational numbers as those which could be represented by terminating or repeating decimals, and classified irrational numbers as those which could be represented by the nonterminating, nonrepeating decimals. The totality of all these numbers constitutes what we call the real numbers.

DEFINITION 10.4 The set of *real numbers*, denoted by *Re*, is defined to be the union of the set of all rational numbers with the set of all irrational numbers. Or, equivalently, the set of all real numbers is the set of all decimal numbers.

The relationship between the set of real numbers and the proper subsets of *Re* that we have discussed so far in this text is given by the diagram in Figure 10.3. Relative to the so-called number line, every point on the line will now correspond to a unique real number and every real number will correspond to a unique point on the line. There are no "holes" in the line.

The set *Re*, together with the operations of addition and multiplication, form a system satisfying all the properties satisfied by the system of rational numbers. Division by zero is still not possible on *Re*, and order between real numbers is defined as an extension of order between rational numbers. An additional property of the set of real numbers is known as the *completeness property*. Before we state this property, we need the following definitions.

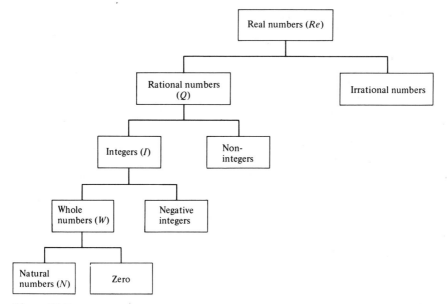

Figure 10.3

DEFINITION 10.5 An *upper bound* of a set of real numbers is a number that is greater than or equal to every element of the set. A *lower bound* of a set of real numbers is a number that is less than or equal to every element of the set.

DEFINITION 10.6 A *least upper bound* of a set of real numbers is an upper bound of the set that is less than or equal to all the upper bounds of the set. A *greatest lower bound* of a set of real numbers is a lower bound of the set that is greater than or equal to all the lower bounds of the set.

COMPLETENESS PROPERTY This property states that every nonempty subset of the set of real numbers that has an upper bound has a least upper bound and every nonempty subset of the set of real numbers that has a lower bound has a greatest lower bound.

EXERCISES

1. Identify each of the following as being true or false.
   a. Every integer can be expressed as a rational number.
   b. The set of all real numbers, *Re*, is the union of the set of integers with the set of irrational numbers.

c. The intersection of the set of rational numbers with the set of irrational numbers is 0.

d. No whole number can be expressed as an irrational number.

e. An irrational number can be approximated as a rational number to any desired number of decimal places.

f. The set of irrational numbers is closed under the operation of addition.

g. The set of rational numbers is closed under the operation of multiplication.

h. Every natural number can be expressed as an integer, and every rational number can be expressed as a non-integer.

i. Zero may be classified as a whole number, as an integer, as a rational number, and also as a real number.

j. The set of irrational numbers is a proper subset of the set Re.

2. For each of the following sets, determine an upper bound if it exists.

a. $\{^-2, 0, 8, 1, ^-3, 4\}$.

b. $\{^-2\frac{1}{2}, 3\frac{1}{4}, 0.17, 4\frac{2}{3}\}$.

c. $\{x \mid x \in Re, ^-2 \le x \le 7\}$.

d. $\{y \mid y \in Re, 0 \le y < 5\}$.

e. $\{u \mid u \in Re, ^-5 < u \le 9\}$.

f. $\left\{ \left| \frac{1}{n} \right| n \in N \right\}$.

g. $\left\{ \left| \frac{n+1}{n} \right| n \in N \right\}$.

h. $\{a \mid a = b + 3\}$, where $b \in \{^-2, 1, \sqrt{2}, 5\}$.

i. $\{^-2, 4, 7, 9\} \cap \{^-3, 4, 6, 8\}$.

j. $\{^-2, 4, 7, 9\} - \{^-3, 4, 6, 8\}$.

3. For each of the sets given in Exercise 2, determine a lower bound if it exists.

10.9 A LOOK AT THE SYSTEM OF REAL NUMBERS

Using the decimal representation for numbers, we have developed the system of real numbers as an extension of the system of rational numbers. The system of real numbers satisfies the following properties.

10.91 CLOSURE PROPERTIES

1. The set Re is closed under the operation of addition; that is, if $a, b \in Re$, then $a + b \in Re$.

2. The set Re is closed under the operation of multiplication; that is, if $a, b \in Re$, then $a \cdot b \in Re$.

3. The set *Re* is closed under the operation of subtraction; that is, if $a, b \in Re$, then $a - b \in Re$.
4. The set *Re* is *not* closed under the operation of division; that is, if $a, b \in Re$, then $a \div b \notin Re$ if $b = 0$.
5. The set of all nonzero real numbers is closed under the operation of division; that is, if $a, b \in Re$ with $b \neq 0$, then $a \div b \in Re$.

10.92 COMMUTATIVE PROPERTIES

1. FOR ADDITION   If $a, b \in Re$, then $a + b = b + a$.
2. FOR MULTIPLICATION   If $a, b \in Re$, then $ab = ba$.

10.93 ASSOCIATIVE PROPERTIES

1. FOR ADDITION   If $a, b, c \in Re$, then $(a + b) + c = a + (b + c)$.
2. FOR MULTIPLICATION   If $a, b, c \in Re$, then $(ab)c = a(bc)$.

10.94 DISTRIBUTIVE PROPERTIES

1. MULTIPLICATION OVER ADDITION   If $a, b, c \in Re$, then
   a. $a(b + c) = ab + ac$.
   b. $(a + b)c = ac + bc$.
2. MULTIPLICATION OVER SUBTRACTION   If $a, b, c \in Re$, then $a(b - c) = ab - ac$.

10.95 IDENTITY PROPERTIES

1. ADDITIVE IDENTITY   There exists $0 \in Re$ such that $a + 0 = 0 + a = a$ for *every* $a \in Re$; 0 is the additive identity in the set *Re*.
2. MULTIPLICATIVE IDENTITY   There exists $1 \in Re$ such that $a \cdot 1 = 1 \cdot a = a$ for *every* $a \in Re$; 1 is the multiplicative identity in the set *Re*.

10.96 INVERSE PROPERTIES

1. ADDITIVE IDENTITY   There exists $0 \in Re$ such that $a + 0 = 0 + a = a$ for *every* $a \in Re$; 0 is the additive ientity in the set *Re*.
   of *b*.
2. MULTIPLICATIVE INVERSE   If $a \in Re$ and $a \neq 0$, then there exists $b \in Re$ such that $ab = ba = 1$; b is the multiplicative inverse of a and $b = 1/a$; the real number 0 *does not* have a multiplicative inverse.

10.97 ORDER PROPERTIES

1. If $a, b \in Re$, then $a < b$ if and only if there exists $c \in Re$ such that $c > 0$ and $b = a + c$.

2. If $a, b \in Re$, then $a > b$ if and only if $b < a$.

3. TRICHOTOMY PROPERTY   If $a, b \in Re$, then *exactly* one of the following must hold:

$$\text{Either } a < b \quad \text{or} \quad a = b \quad \text{or} \quad a > b$$

10.98 PROPERTIES OF ZERO

1. $0 \cdot a = 0$ for *all* $a \in Re$.

2. $0 \div a = 0$, if $a \neq 0$.

3. $0 \div 0$ is not defined since it is *indeterminate*; that is, $0 \div 0$ does not have a *unique* solution.

4. $a \div 0$, if $a \neq 0$, is not defined since it is *impossible* to divide a nonzero real number by 0.

10.99 DENSENESS PROPERTY

The set *Re* is dense. That is, between any two *distinct* real numbers there always exists another real number.

10.910 COMPLETENESS PROPERTY

Every nonempty subset of the set of real numbers that has an upper bound has a least upper bound, and every nonempty subset of the real numbers that has a lower bound has a greatest lower bound.

SUMMARY

In this chapter we developed the system of real numbers. The set of real numbers, *Re*, was introduced using decimal number representation. The set *Re* was defined as the union of the set of all rational numbers with the set of all the irrational numbers. The rational numbers were represented as either terminating or repeating decimals, whereas the irrational numbers were represented as nonterminating, nonrepeating decimals. The system of real numbers was developed as an extension of the system of rational numbers, preserving all the properties already obtained. Upper and lower bounds and least upper and greatest lower bounds for sets of real numbers were defined, and the completeness property for *Re* was stated. Percents also were discussed, and we converted from decimals to percents and from percents to decimals.

During the discussions the following symbols were introduced.

SYMBOL	INTERPRETATION
%	Percent symbol
Re	The set of real numbers

REVIEW EXERCISES FOR CHAPTER 10

1. Under which of the four basic arithmetic operations if any is the set *Re* closed?
2. Which of the four basic arithmetic operations if any are commutative operations on *Re*?
3. Which of the four basic arithmetic operations if any are associative operations on *Re*?
4. What real number if any is the additive identity in *Re*?
5. What real number if any is the multiplicative identity in *Re*?
6. What real numbers if any do not have an additive inverse in *Re*?
7. What real numbers if any do not have a multiplicative inverse in *Re*?
8. What is meant by the denseness property for *Re*?
9. How may we classify a real number as being either rational or irrational according to its decimal numeral?
10. Is it possible for a proper subset of *Re* to be dense? Explain.
11. Is is possible for a finite subset of *Re* to be dense? Explain.
12. a. Give an example of a subset of *Re* that has a lower bound.
    b. Give an example of a subset of *Re* that does not have a lower bound.
13. a. Give an example of a subset of *Re* that has an upper bound.
    b. Give an example of a subset of *Re* that does not have an upper bound.
14. Is the set of irrational numbers closed under the operation of addition? Explain.
15. Is the set of irrational numbers closed under the operation of multiplication? Explain.
16. If p represents a prime number, classify $\sqrt{p}$ as being rational or irrational.
17. If 8.6 percent of the total workforce of 18,263 in a particular community are unemployed, how many people in that community are unemployed?
18. 81 is what percent of 540?
19. Without actually dividing, indicate whether $\frac{7}{23}$ can be represented by (a) a repeating or terminating decimal or (b) a nonrepeating, nonterminating decimal. Explain.
20. An unsuspecting hourly employee approached his employer and asked for a 5 percent raise in pay. Without much hesitation, the employer proposed to increase the employee's hours by 25% and his gross pay by 30%. Did the employee in fact receive a net 5% raise? Explain.

CHAPTER II
THE SYSTEM OF MATRICES

This chapter will introduce the student to a new mathematical system—the system of matrices. That is, we will develop a system consisting of a nonempty set whose elements will be called matrices (plural of matrix) and on which set will be defined various operations and relations.

The use of matrices is becoming increasingly important in just about every discipline. A knowledge of matrices and properties thereof should enable the student to read the literature that uses matrices and matrix algebra.

11.1 A MATRIX

Frequently data are presented in the form of rectangular arrays of elements, such as the statistics for a particular ball team. Consider, for instance, the rectangular array given in the table. The first column from the left contains the names of five baseball players. The second column lists the number of hits during the particular game being summarized; the third column, the number of runs; the fourth column, the number of errors; and the fifth column, the batting average to date. Such a rectangular array is an example of a matrix.

	Hits	Runs	Errors	Average
Craig Davis	2	1	1	0.331
Dennis Fusco	1	0	0	0.315
John Rea	1	1	0	0.295
Dan Wilson	3	2	1	0.293
Mark Wuest	2	1	0	0.288

DEFINITION 11.1 A *matrix A* is a rectangular array of entries denoted by

$$A = \begin{bmatrix} a_{11} & a_{12} & a_{13} & \cdots & a_{1n} \\ a_{21} & a_{22} & a_{23} & \cdots & a_{2n} \\ \vdots & & & & \\ a_{m1} & a_{m2} & a_{m3} & \cdots & a_{mn} \end{bmatrix}$$

The entries $a_{11}, a_{12}, \ldots, a_{mn}$ are called the *elements* of the matrix.

In the above definition the entry a_{11} refers to the element in the first row and first column, the entry a_{22} refers to the element in the second row and second column, and the entry a_{mn} refers to the element in the mth row and nth column. In general the entry a_{ij} will refer to the element in the ith row and jth column.

In this chapter, unless otherwise indicated, the entries of our matrices will be *scalars*.

DEFINITION 11.2 A *scalar* is a real number, or a function of a real number.

EXAMPLE

a. $\begin{bmatrix} 2 & 0 \\ -1 & 1 \end{bmatrix}$ is a matrix with integer entries.

b. $\begin{bmatrix} \frac{3}{4} & \pi & 0 \\ 3 & -2 & \frac{1}{2} \end{bmatrix}$ is a matrix with real number entries.

c. $\begin{bmatrix} \sqrt{2} & -\sqrt{3} \\ \sqrt{34} & \sqrt{19} \\ -\sqrt{5} & \sqrt{17} \end{bmatrix}$ is a matrix with irrational number entires.

d. $\begin{bmatrix} 3u & u+2 & u^2 \\ u-1 & -2u & 1-u \\ u^3 & 1-u^2 & -3u^2 \end{bmatrix}$ is a matrix whose entries are functions of u where u is real.

DEFINITION 11.3 The horizontal arrays of a matrix are called its *rows*, and the vertical arrays are called its *columns*.

EXAMPLE

a. The matrix in part (a) of the example above has 2 rows and 2 columns.
b. The matrix in part (b) of the example above has 2 rows and 3 columns.
c. The matrix in part (c) of the example above has 3 rows and 2 columns.
d. The matrix in part (d) of the example above has 3 rows and 3 columns.

DEFINITION 11.4 A matrix A with m rows and n columns is called a matrix of *order* (m, n) or $m \times n$ (read "m by n"). When $m = n$ the matrix is called a *square matrix* and is said to be of order n. If A is a square matrix of order n, then the elements $a_{11}, a_{22}, \ldots, a_{nn}$ are said to form the *main diagonal* of A and the elements $a_{1n}, a_{2(n-1)}, \ldots, a_{mn}$ are said to form the *secondary diagonal*.

EXAMPLE

a. The matrix $A = \begin{bmatrix} 3 & {}^-1 & 0 \\ 1 & 4 & 5 \end{bmatrix}$ is a matrix of order $(2, 3)$ or 2×3.

b. The matrix $B = \begin{bmatrix} 0 & 1 \\ 1 & 2 \end{bmatrix}$ is a square matrix of order 2 with the main diagonal consisting of entries 0 and 2 and the secondary diagonal consisting of entries 1 and 1.

DEFINITION 11.5 If all the entries of a matrix are zero, then the matrix is called the *zero matrix* or the *null matrix*.

The zero matrix is not to be confused with the scalar zero. The scalar zero has a value associated with it, whereas the zero matrix is just a rectangular array of zeros. We will denote the zero matrix using bold type as $\mathbf{0}$.

Generally there are several ways by which we can denote a matrix. Let

$$A = \begin{bmatrix} a_{11} & a_{12} & a_{13} & \cdots & a_{1n} \\ a_{21} & a_{22} & a_{23} & \cdots & a_{2n} \\ \vdots & & & & \\ a_{m1} & a_{m2} & a_{m3} & \cdots & a_{mn} \end{bmatrix} \tag{11.1}$$

Then the matrix may be denoted by the symbol on the right (11.1) or simply by the letter A. A third way for convenience when operating with matrices would be to denote the matrix given in (11.1) by the symbol $[a_{ij}]_{m \times n}$ representing the matrix with elements of the form a_{ij} which is of the order m by n. The zero or null matrix which contains 3 rows and 4 columns, then, would be denoted by $\mathbf{0}_{3 \times 4}$. The square zero matrix of order 5 would be denoted by $\mathbf{0}_5$. If B is a matrix of order 2×3, we may also denote this as $B_{2 \times 3}$ or $B_{(2, 3)}$. If C is a square matrix of order 4, we may denote this by C_4.

EXERCISES

1. Two companies A and B manufacture the same three items R, S, and T.
   Company A produces per day 6 of item R, 4 of item S, and 8 of item T.
   Company B produces, per day, 4 of item R, 7 of item S, and 6 of item T.
   a. Represent the number of items R, S, and T produced by companies
      A and B in matrix form with three rows and two columns.
   b. Represent the number of items R, S, and T produced by companies
      A and B in matrix form with two rows and three columns.
2. Consider a matrix of order 4×3.
   a. How many rows are in the matrix?
   b. How many columns are in the matrix?
   c. How many entries are in the matrix?
3. a. Write the zero matrix of order 3×2.
   b. Write the matrix $\mathbf{0}_3$.
4. Display the matrix $[a_{ij}]_{2 \times 4}$ with entries $a_{11} = 2$, $a_{21} = 0$, $a_{13} = {}^-1$,
   $a_{24} = 5$, $a_{12} = 3$, $a_{22} = 7$, $a_{14} = 1$, and $a_{23} = {}^-2$.
5. a. Can a matrix have only one row? If so, what is its order if the row
      has 4 entries?
   b. Can a matrix have only one column? If so, what is its order if the
      column has 3 entries?
6. Is it possible for a matrix to have only 5 entries? Explain.
7. Consider the matrix $A = \begin{bmatrix} 1 & 2 & {}^-3 \\ 4 & 6 & 0 \end{bmatrix}$.

   a. What is a_{12}?
   b. What is a_{23}?
   c. What is a_{31}?

8. Consider the matrix $B = \begin{bmatrix} \begin{bmatrix} 1 & 0 \\ 2 & 1 \end{bmatrix} & \begin{bmatrix} 0 & 1 \\ 1 & 0 \end{bmatrix} \\ [2 \ \ 1] & \begin{bmatrix} 1 \\ 2 \end{bmatrix} \end{bmatrix}$.

   a. How many entries are there in B?
   b. What is the order of B?
   c. What is b_{11}?
   d. What is b_{21}?

9. Consider the matrix $C = \begin{bmatrix} 3 & 0 & 1 \\ {}^-1 & 2 & 0 \\ 0 & 4 & 1 \end{bmatrix}$.

   a. What are the main diagonal elements of C?
   b. What are the secondary diagonal elements of C?
10. If a matrix has more than one row and more than one column, is it
    possible for the matrix to have only 7 entries? Explain.

11. An automobile dealer has six salespersons and sells four different models of a particular automobile. During a 20-day period the record of sales for each salesperson, per model of automobile, is given in the matrix below.

Salespersons

	1	2	3	4	5	6
Model A	2	3	2	0	4	1
Model B	0	2	4	3	0	5
Model C	6	2	1	5	0	2
Model D	5	3	0	2	4	1

a. Which salesperson sold the greatest number of automobiles?
b. Which model was the best seller?
c. How many automobiles were sold during the period?

11.2 EQUALITY OF MATRICES

DEFINITION 11.6 Two matrices A and B are said to be *equal*, denoted by $A = B$, if and only if A and B are of the same order and have equal corresponding elements; that is, $a_{ij} = b_{ij}$ for all i and j.

EXAMPLE

a. The matrix $A = \begin{bmatrix} 1 & 2 \\ -2 & 0 \end{bmatrix}$ is equal to the matrix $B = \begin{bmatrix} 1 & 2 \\ -2 & 0 \end{bmatrix}$ since both A and B are of order 2 and their corresponding elements are equal.

b. The matrix $C = \begin{bmatrix} 2 & 1 & 3 \\ 0 & 4 & -1 \end{bmatrix}$ is *not* equal to the matrix $D = \begin{bmatrix} 2 & 1 & -3 \\ 0 & 4 & -1 \end{bmatrix}$ since $c_{13} = 3 \neq -3 = d_{13}$; note, however, the order of C is equal to the order of D.

c. The matrix $E = \begin{bmatrix} 1 & 2 \\ 3 & 4 \end{bmatrix}$ is *not* equal to the matrix $F = \begin{bmatrix} 1 & 2 & 0 \\ 3 & 4 & 0 \end{bmatrix}$ since E is of order 2×2 whereas F is of order 2×3.

d. The matrix $G = \begin{bmatrix} 2 & x \\ -3 & 1 \end{bmatrix}$ is equal to the matrix $H = \begin{bmatrix} y & 2 \\ -3 & 1 \end{bmatrix}$ if and only if $x = 2$ and $y = 2$.

THEOREM 11.1 The relation "is equal to" on the set of matrices is an equivalence relation.

Proof The proof follows immediately from the definition of an equivalence relation and the definition of equality of matrices and is left as an exercise.

One important difference between matrices and real numbers is relative to order. Recall that on the set *Re* we had the trichotomy property. That is, if a, $b \in Re$, then either $a < b$ or $a = b$ or $a > b$ and exactly one of these statements holds true for every pair of real numbers. This is not true for matrices.

DEFINITION 11.7 A matrix A with real number entries is said to be *greater than* a matrix B with real number entries, denoted by $A > B$, if and only if the two matrices are of the same order and each of the entries of A is greater than the corresponding entries of B. The matrix C with real number entries is said to be *less than* the matrix D with real number entries, denoted by $C < D$, if and only if $D > C$.

EXAMPLE

a. The matrix $A = \begin{bmatrix} 2 & 0 \\ 1 & 3 \end{bmatrix}$ is *greater than* the matrix $B = \begin{bmatrix} 1 & {}^-3 \\ 0 & 2 \end{bmatrix}$ since A and B are of the same order and every entry of A is greater than the corresponding entry of B.

b. The matrix $C = \begin{bmatrix} 2 & {}^-1 & 4 \\ 0 & 3 & {}^-2 \end{bmatrix}$ is *less than* the matrix $D = \begin{bmatrix} 4 & 0 & 6 \\ 2 & 5 & 1 \end{bmatrix}$ since C and D are of the same order and every entry of C is less than the corresponding entry of D.

EXAMPLE   Consider the two matrices $A = \begin{bmatrix} 2 & {}^-4 \\ 0 & 1 \end{bmatrix}$ and $B = \begin{bmatrix} 1 & {}^-5 \\ 2 & 0 \end{bmatrix}$ which are both of order 2.

a. A is not greater than B since $a_{21} = 0$ is not greater than $b_{21} = 2$.

b. A is not less than B since B is not greater than A since $b_{11} = 1$ is not greater than $a_{11} = 2$.

c. A is not equal to B since $a_{11} = 2$ is not equal to $b_{11} = 1$.

EXERCISES

1. Is the matrix $A = \begin{bmatrix} 1 & 2 \\ 3 & 0 \\ 0 & 4 \end{bmatrix}$ equal to the matrix $B = \begin{bmatrix} 1 & 3 & 0 \\ 2 & 0 & 4 \end{bmatrix}$? Why or why not?

2. Consider the matrices $C = \begin{bmatrix} 3 & 4 & 0 \\ {}^-1 & 0 & 5 \end{bmatrix}$ and $D = \begin{bmatrix} 2 & 3 & 0 \\ 1 & {}^-2 & 5 \end{bmatrix}$.

   a. Is C greater than D? Why or why not?

   b. Is C less than D? Why or why not?

   c. Is $C = D$? Why or why not?

3. Consider the matrix $E = \begin{bmatrix} 2 & 0 & ^-1 \\ 4 & ^-5 & 1 \end{bmatrix}$.

   a. Write two distinct matrices that are greater than E.

   b. Write two distinct matrices that are less than E.

4. For each of the following determine a value for each unknown, if possible, such that the following will be true.

   a. $\begin{bmatrix} 1 & 2 \\ x & 4 \end{bmatrix} = \begin{bmatrix} 1 & 2 \\ 3 & 4 \end{bmatrix}$.

   b. $\begin{bmatrix} 1 & 4 \\ x & 5 \end{bmatrix} = \begin{bmatrix} 1 & y \\ 2 & 5 \end{bmatrix}$.

   c. $\begin{bmatrix} 2 & 3 \\ 5 & x \end{bmatrix} = \begin{bmatrix} 3 & 2 \\ 5 & 4 \end{bmatrix}$.

   d. $\begin{bmatrix} ^-1 & 0 \\ 1 & 6 \end{bmatrix} > \begin{bmatrix} x & ^-1 \\ 0 & y \end{bmatrix}$.

   e. $\begin{bmatrix} 4 & 1 & 0 \\ x & 2 & 0 \end{bmatrix} = \begin{bmatrix} 4 & 1 \\ 0 & 2 \end{bmatrix}$.

   f. $\begin{bmatrix} x & 2 & 1 \\ 3 & y & 0 \end{bmatrix} < \begin{bmatrix} 0 & 3 & 2 \\ 4 & 1 & 1 \end{bmatrix}$.

   g. $\begin{bmatrix} 1 & 4 & 5 \\ x & y & 0 \\ 2 & 3 & 4 \end{bmatrix} = \begin{bmatrix} 1 & 4 & 5 \\ 2x+1 & 2y & 0 \\ 2 & 3 & 4 \end{bmatrix}$.

   h. $\begin{bmatrix} x & 0 & 4 \\ ^-1 & y & 0 \end{bmatrix} \geq \begin{bmatrix} 1 & 0 & 3 \\ ^-2 & 1 & 0 \end{bmatrix}$.

   i. $\begin{bmatrix} 3 & 0 \\ x & 4 \end{bmatrix} \leq \begin{bmatrix} 4 & 0 \\ 2x & 3 \end{bmatrix}$.

   j. $\begin{bmatrix} x+1 \\ y \\ 2u-3 \end{bmatrix} = \begin{bmatrix} 1-x \\ 2y \\ u+4 \end{bmatrix}$.

5. Is $\mathbf{0}_5 = \mathbf{0}_4$? Why or why not.

6. Prove that the relation "is equal to" on the set of matrices is an equivalence relation.

11.3 ADDITION OF MATRICES

Matrices can be added only if they are of the same order. Matrices that are of the same order are said to be *conformable for addition.*

DEFINITION 11.8 If $A = [a_{ij}]_{m \times n}$ and $B = [b_{ij}]_{m \times n}$ are two matrices of order $m \times n$, then the *sum* of A and B, denoted by $A + B$, is the matrix $C = [c_{ij}]_{m \times n}$ of order $m \times n$ such that $c_{ij} = (a_{ij} + b_{ij})$ for all i and j.

From the above definition we observe that the sum of two conformable for addition matrices is obtained by adding the corresponding entries of the matrices.

EXAMPLE Find the sum of $A = \begin{bmatrix} 2 & ^-3 \\ 1 & 0 \end{bmatrix}$ and $B = \begin{bmatrix} ^-4 & 2 \\ 0 & ^-3 \end{bmatrix}$.

Solution $A + B = \begin{bmatrix} 2 & ^-3 \\ 1 & 0 \end{bmatrix} + \begin{bmatrix} ^-4 & 2 \\ 0 & ^-3 \end{bmatrix} = \begin{bmatrix} (2-4) & (^-3+2) \\ (1+0) & (0-3) \end{bmatrix}$

$$= \begin{bmatrix} ^-2 & ^-1 \\ 1 & ^-3 \end{bmatrix} \quad \blacksquare$$

EXAMPLE Find the sum of $C = \begin{bmatrix} 1 & 0 \\ 0 & 2 \end{bmatrix}$ and $D = \begin{bmatrix} ^-2 & 0 \\ 1 & 4 \\ 0 & 1 \end{bmatrix}$.

Solution Matrix C is of order 2×2 and matrix D is of order 3×2. Since the orders are not equal, the matrices are not conformable for addition and $C + D$ does not exist. $\blacksquare$

It is easy to show that the operation of matrix addition is both a commutative and an associative operation. If A and B are matrices that are conformable for addition, then

$$\begin{aligned} A + B &= [a_{ij}] + [b_{ij}] && \text{Renaming} \\ &= [a_{ij} + b_{ij}] && \text{Definition 11.8} \\ &= [b_{ij} + a_{ij}] && \text{Commutative property of ordinary } + \\ &= [b_{ij}] + [a_{ij}] && \text{Definition 11.8} \\ &= B + A && \text{Renaming} \end{aligned}$$

Since A and B are arbitrary matrices of the same order and $A + B = B + A$, we have established the commutative property for matrix addition.

Now let A, B, and C be three arbitrary matrices all of the same order. Then

$$\begin{aligned} (A + B) + C &= ([a_{ij}] + [b_{ij}]) + [c_{ij}] && \text{Renaming} \\ &= [a_{ij} + b_{ij}] + [c_{ij}] && \text{Definition 11.8} \\ &= [(a_{ij} + b_{ij}) + c_{ij}] && \text{Definition 11.8} \\ &= [a_{ij} + (b_{ij} + c_{ij})] && \text{Associative property for} \\ &&& \text{ordinary } + \\ &= [a_{ij}] + [b_{ij} + c_{ij}] && \text{Definition 11.8} \\ &= [a_{ij}] + ([b_{ij}] + [c_{ij}]) && \text{Definition 11.8} \\ &= A + (B + C) && \text{Renaming} \end{aligned}$$

Since A, B, and C are arbitrary matrices conformable for matrix addition and since $(A + B) + C = A + (B + C)$, we have established the associative property for matrix addition.

We also will observe that if A is an arbitrary matrix of order $m \times n$ and $\mathbf{0}$ is the zero or null matrix of order $m \times n$, then $A + \mathbf{0} = \mathbf{0} + A = A$ for every $m \times n$ matrix A. Hence $\mathbf{0}_{m \times n}$ is the identity matrix for matrix addition on the subset of matrices of order $m \times n$.

EXAMPLE

a. $\begin{bmatrix} 2 & 3 \\ 1 & {}^-1 \end{bmatrix} + \begin{bmatrix} 0 & 0 \\ 0 & 0 \end{bmatrix} = \begin{bmatrix} (2+0) & (3+0) \\ (1+0) & ({}^-1+0) \end{bmatrix} = \begin{bmatrix} 2 & 3 \\ 1 & {}^-1 \end{bmatrix}$

b. $\begin{bmatrix} 1 & {}^-2 \\ 4 & 0 \\ {}^-1 & 1 \end{bmatrix} + \begin{bmatrix} 0 & 0 \\ 0 & 0 \\ 0 & 0 \end{bmatrix} = \begin{bmatrix} (1+0) & ({}^-2+0) \\ (4+0) & (0+0) \\ ({}^-1+0) & (1+0) \end{bmatrix} = \begin{bmatrix} 1 & {}^-2 \\ 4 & 0 \\ {}^-1 & 1 \end{bmatrix}$

EXAMPLE Verify that $A + B = B + A$ given $A = \begin{bmatrix} 2 & 1 & 3 \\ 0 & {}^-1 & {}^-2 \end{bmatrix}$ and

$B = \begin{bmatrix} 3 & 5 & 0 \\ {}^-3 & 2 & {}^-1 \end{bmatrix}$.

Solution

$$A + B = \begin{bmatrix} 2 & 1 & 3 \\ 0 & {}^-1 & {}^-2 \end{bmatrix} + \begin{bmatrix} 3 & 5 & 0 \\ {}^-3 & 2 & {}^-1 \end{bmatrix}$$

$$= \begin{bmatrix} 2+3 & 1+5 & 3+0 \\ 0+{}^-3 & {}^-1+2 & {}^-2+{}^-1 \end{bmatrix}$$

$$= \begin{bmatrix} 5 & 6 & 3 \\ {}^-3 & 1 & {}^-3 \end{bmatrix}$$

$$B + A = \begin{bmatrix} 3 & 5 & 0 \\ {}^-3 & 2 & {}^-1 \end{bmatrix} + \begin{bmatrix} 2 & 1 & 3 \\ 0 & {}^-1 & {}^-2 \end{bmatrix}$$

$$= \begin{bmatrix} 3+2 & 5+1 & 0+3 \\ {}^-3+0 & 2+{}^-1 & {}^-1+{}^-2 \end{bmatrix}$$

$$= \begin{bmatrix} 5 & 6 & 3 \\ {}^-3 & 1 & {}^-3 \end{bmatrix}$$

Hence $A + B = B + A$. ■

EXAMPLE Verify that $(A + B) + C = A + (B + C)$
given $A = \begin{bmatrix} {}^-2 & 0 & {}^-1 \\ 1 & 4 & 3 \end{bmatrix}$, $B = \begin{bmatrix} 5 & {}^-2 & 0 \\ 1 & {}^-1 & 1 \end{bmatrix}$ and $C = \begin{bmatrix} 0 & {}^-3 & 4 \\ 2 & {}^-1 & 3 \end{bmatrix}$.

Solution

$$(A + B) + C = \left(\begin{bmatrix} {}^-2 & 0 & {}^-1 \\ 1 & 4 & 3 \end{bmatrix} + \begin{bmatrix} 5 & {}^-2 & 0 \\ 1 & {}^-1 & 1 \end{bmatrix} \right) + \begin{bmatrix} 0 & {}^-3 & 4 \\ 2 & {}^-1 & 3 \end{bmatrix}$$

$$= \begin{bmatrix} ({}^-2+5) & (0-2) & ({}^-1+0) \\ (1+1) & (4-1) & (3+1) \end{bmatrix} + \begin{bmatrix} 0 & {}^-3 & 4 \\ 2 & {}^-1 & 3 \end{bmatrix}$$

$$= \begin{bmatrix} 3 & {}^-2 & {}^-1 \\ 2 & 3 & 4 \end{bmatrix} + \begin{bmatrix} 0 & {}^-3 & 4 \\ 2 & {}^-1 & 3 \end{bmatrix}$$

$$= \begin{bmatrix} (3+0) & ({}^-2-3) & ({}^-1+4) \\ (2+2) & (3-1) & (4+3) \end{bmatrix} = \begin{bmatrix} 3 & {}^-5 & 3 \\ 4 & 2 & 7 \end{bmatrix}$$

$$A + (B + C) = \begin{bmatrix} ^-2 & 0 & ^-1 \\ 1 & 4 & 3 \end{bmatrix} + \left(\begin{bmatrix} 5 & ^-2 & 0 \\ 1 & ^-1 & 1 \end{bmatrix} + \begin{bmatrix} 0 & ^-3 & 4 \\ 2 & ^-1 & 3 \end{bmatrix} \right)$$

$$= \begin{bmatrix} ^-2 & 0 & ^-1 \\ 1 & 4 & 3 \end{bmatrix} + \begin{bmatrix} (5+0) & (^-2-3) & (0+4) \\ (1+2) & (^-1-1) & (1+3) \end{bmatrix}$$

$$= \begin{bmatrix} ^-2 & 0 & ^-1 \\ 1 & 4 & 3 \end{bmatrix} + \begin{bmatrix} 5 & ^-5 & 4 \\ 3 & ^-2 & 4 \end{bmatrix}$$

$$= \begin{bmatrix} (^-2+5) & (0-5) & (^-1+4) \\ (1+3) & (4-2) & (3+4) \end{bmatrix} = \begin{bmatrix} 3 & ^-5 & 3 \\ 4 & 2 & 7 \end{bmatrix}$$

Hence $(A + B) + C = A + (B + C)$. ∎

EXERCISES

1. Let $A = \begin{bmatrix} 2 & 0 \\ ^-3 & 1 \end{bmatrix}$, $B = \begin{bmatrix} ^-1 & 2 \\ 0 & 1 \end{bmatrix}$, $C = \begin{bmatrix} 3 & ^-1 \\ 2 & 0 \end{bmatrix}$, $D = \begin{bmatrix} ^-1 & 1 \\ 1 & ^-1 \end{bmatrix}$,

   $E = \begin{bmatrix} 1 & 0 \\ 0 & 1 \end{bmatrix}$, and $F = \begin{bmatrix} 0 & 0 \\ 0 & 0 \end{bmatrix}$.

   a. Form $A + B$.
   b. Form $C + D$.
   c. Form $A + E$.
   d. Form $C + F$.
   e. Form $(A + C) + E$.
   f. Form $B + (D + F)$.
   g. Form $(A + E) + (B + D)$.

2. Using the matrices of Exercises 1, verify the commutative property for addition of matrices for each of the following.

   a. $A + B = B + A$.
   b. $C + D = D + C$.
   c. $B + E = E + B$.
   d. $B + C = C + B$.
   e. $A + F = F + A$.

3. Using the matrices of Exercise 1, verify the associative property for addition of matrices for each of the following.

   a. $(A + B) + C = A + (B + C)$.
   b. $(B + D) + E = B + (D + E)$.
   c. $(C + E) + F = C + (E + F)$.
   d. $(A + C) + E = A + (C + E)$.

4. If G is a matrix of order 2×3 and H is a matrix of order 3×2, are the two matrices conformable for matrix addition? Explain.

5. If T is an arbitrary matrix of order $m \times n$, what is $T + 0_{m \times n}$?

6. For each of the following form the indicated sum or indicate why the sum cannot be formed.

a. $\begin{bmatrix} 2 & x \\ 3 & 0 \end{bmatrix} + \begin{bmatrix} y & ^-1 \\ 0 & 6 \end{bmatrix}$.

b. $\begin{bmatrix} 1 \\ 2 \end{bmatrix} + \begin{bmatrix} 2 & 3 \end{bmatrix}$.

c. $\begin{bmatrix} \sqrt{2} & 0 & ^-\sqrt{3} \\ 1 & ^-1 & 2 \end{bmatrix} + \begin{bmatrix} 0 & 1 & 0 \\ 0 & 1 & 2 \end{bmatrix}$.

d. $\begin{bmatrix} 1 & 2 & 3 \\ ^-4 & 0 & 5 \\ ^-1 & ^-2 & 1 \end{bmatrix} + \begin{bmatrix} ^-1 & ^-2 & ^-3 \\ 4 & 0 & ^-5 \\ 1 & 2 & ^-1 \end{bmatrix}$.

e. $\begin{bmatrix} \begin{bmatrix} 1 & 0 \\ 0 & 1 \end{bmatrix} \begin{bmatrix} y & 2 \\ 3 & 4 \end{bmatrix} \\ \begin{bmatrix} 0 & x \\ 1 & 0 \end{bmatrix} \begin{bmatrix} ^-3 & 1 \\ 0 & 2 \end{bmatrix} \end{bmatrix} + \begin{bmatrix} \begin{bmatrix} 0 & x \\ 1 & 0 \end{bmatrix} \begin{bmatrix} ^-1 & 2 \\ 3 & ^-4 \end{bmatrix} \\ \begin{bmatrix} y & 0 \\ 0 & 1 \end{bmatrix} \begin{bmatrix} 1 & ^-3 \\ 2 & 0 \end{bmatrix} \end{bmatrix}$.

f. $\left(\begin{bmatrix} 2 & ^-3 \\ 4 & 1 \end{bmatrix} + \begin{bmatrix} 0 & 6 \\ ^-5 & 4 \end{bmatrix} \right) + \begin{bmatrix} 1 & 2 & 3 \\ 0 & ^-3 & 4 \end{bmatrix}$.

7. A manufacturer produces three different items A, B, and C in two different locations R and S. For each item produced, skilled, semiskilled, and unskilled labor are required as follows:

	A	B	C	
	169	194	210	Skilled labor (in hours)
Location R	234	207	305	Semi-skilled labor (in hours)
	226	276	410	Unskilled labor (in hours)

	A	B	C	
	262	156	126	Skilled labor (in hours)
Location S	301	173	184	Semi-skilled labor (in hours)
	297	198	203	Unskilled labor (in hours)

Write the matrix which displays the total labor required, per type of labor, to produce the items A, B, and C.

11.4 SCALAR MULTIPLES OF A MATRIX

In the next section we will consider the product of matrices. First, however, we will consider the product of a matrix with a scalar, or what is known as a scalar multiple of a matrix.

DEFINITION 11.9 Let $A = [a_{ij}]_{m \times n}$ be an arbitrary $m \times n$ matrix and c be an arbitrary scalar. Then $cA = [ca_{ij}]_{m \times n}$ is an $m \times n$ matrix called a *scalar multiple* of A.

From the above definition we observe that to multiply a matrix by a scalar means to multiply each entry in the matrix by the same scalar.

EXAMPLE

a. $2\begin{bmatrix} 1 & 2 \\ 0 & -1 \end{bmatrix} = \begin{bmatrix} 2 & 4 \\ 0 & -2 \end{bmatrix}.$

b. $-3\begin{bmatrix} a & b & c \\ d & e & f \end{bmatrix} = \begin{bmatrix} -3a & -3b & -3c \\ -3d & -3e & -3f \end{bmatrix}.$

c. $-1\begin{bmatrix} a & b \\ c & d \end{bmatrix} = \begin{bmatrix} -a & -b \\ -c & -d \end{bmatrix}.$

d. $-1\begin{bmatrix} 2 & -1 & 3 \\ 4 & 0 & -2 \end{bmatrix} = \begin{bmatrix} -2 & 1 & -3 \\ -4 & 0 & 2 \end{bmatrix}.$

THEOREM 12.2 If A and B are matrices conformable for matrix addition and r and s are arbitrary scalars, then
a. $(r + s)A = rA + sA.$
b. $r(A + B) = rA + rB.$
c. $r(sA) = (rs)A.$
Proof

a. $(r + s)A = (r + s)[a_{ij}]$ Renaming

$= [(r + s)a_{ij}]$ Definition 11.9 since $r + s$ is a scalar

$= [ra_{ij} + sa_{ij}]$ Distributive property for ordinary $\times$ over ordinary $+$

$= [ra_{ij}] + [sa_{ij}]$ Definition 11.8

$= r[a_{ij}] + s[a_{ij}]$ Definition 11.9

$= rA + sA$ Renaming

The proofs of (b) and (c) are similar and are left as exercises.

With the introduction of the scalar multiple of a matrix we can now define subtraction of matrices in a similar manner as subtraction of scalars.

DEFINITION 11.10 If A and B are arbitrary matrices of the same order, then the *difference* $A - B = A + (-B)$, where $-B$ means $(^-1)B$.

EXAMPLE

a.
$$\begin{bmatrix} 2 & 3 \\ 1 & 0 \end{bmatrix} - \begin{bmatrix} ^-3 & 1 \\ 0 & 4 \end{bmatrix} = \begin{bmatrix} 2 & 3 \\ 1 & 0 \end{bmatrix} + \left(-\begin{bmatrix} ^-3 & 1 \\ 0 & 4 \end{bmatrix} \right)$$

$$= \begin{bmatrix} 2 & 3 \\ 1 & 0 \end{bmatrix} + \begin{bmatrix} 3 & ^-1 \\ 0 & ^-4 \end{bmatrix} = \begin{bmatrix} 5 & 2 \\ 1 & ^-4 \end{bmatrix}.$$

b.
$$\begin{bmatrix} a & b & c \\ d & e & f \end{bmatrix} - \begin{bmatrix} a & b & c \\ d & e & f \end{bmatrix} = \begin{bmatrix} a & b & c \\ d & e & f \end{bmatrix} + \left(-\begin{bmatrix} a & b & c \\ d & e & f \end{bmatrix} \right)$$

$$= \begin{bmatrix} a & b & c \\ d & e & f \end{bmatrix} + \begin{bmatrix} ^-a & ^-b & ^-c \\ ^-d & ^-e & ^-f \end{bmatrix}$$

$$= \begin{bmatrix} 0 & 0 & 0 \\ 0 & 0 & 0 \end{bmatrix}.$$

Part (b) above suggests that if $A = [a_{ij}]_{m \times n}$ is an $m \times n$ matrix, then its *additive inverse* is the matrix $-A = [^-a_{ij}]_{m \times n}$ such that $A + (-A) = (-A) + A = \mathbf{0}_{m \times n}$. This is similar to the additive inverse of a scalar.

EXERCISES

1. Consider the matrices $A = \begin{bmatrix} 2 & 2 & 1 \\ ^-1 & 0 & 2 \end{bmatrix}$, $B = \begin{bmatrix} 1 & 0 & 4 \\ 0 & ^-2 & 3 \end{bmatrix}$,

$C = \begin{bmatrix} 0 & 2 & ^-1 \\ 3 & 1 & 2 \end{bmatrix}$, and $D = \begin{bmatrix} 3 & ^-1 & 1 \\ 1 & 0 & 2 \end{bmatrix}$.

Form
a. $2A$.
b. $(^-3)B$.
c. $\frac{1}{2}C$.
d. ^-D.
e. $2A + 3B$.
f. $3B - 4C$.
g. $2C + 5D$.
h. $4D - 2A$.
i. $2A - 3B - D$.
j. $^-3B + 4C - 2D$.

2. Let A, B, C, and D be the matrices of Exercise 1.
   a. Determine a matrix E such that $A + E = C$.
   b. Determine a matrix F such that $2B + F = C$.
   c. Determine a matrix G such that $C - 2G = A$.
   d. Determine a matrix H such that $2A + H = B - 2D$.
3. For each of the following determine a value for each unknown, if possible, such that the following will be true.

a. $2\begin{bmatrix} 3 & x \\ ^-1 & 0 \end{bmatrix} - 3\begin{bmatrix} 1 & 2 \\ y & 0 \end{bmatrix} = \begin{bmatrix} u & 2 \\ ^-5 & v \end{bmatrix}.$

b. $3\begin{bmatrix} 1 & x & 0 \\ -1 & 0 & 2 \end{bmatrix} - 2\begin{bmatrix} y & 1 & -1 \\ 0 & 2 & u \end{bmatrix} > \begin{bmatrix} 1 & 4 & -1 \\ -6 & -10 & 0 \end{bmatrix}.$

4. Let $A = \begin{bmatrix} 2 & -1 & 3 \\ 0 & 2 & 1 \end{bmatrix}$, $B = \begin{bmatrix} -3 & 2 & 1 \\ 1 & -3 & -2 \end{bmatrix}$, $r = 2$, and $s = 3$.

a. Verify that $(r + s)A = rA + sA$.
b. Verify that $(r + s)B = rB + sB$.
c. Verify that $r(A + B) = rA + rB$.
d. Verify that $s(A + B) = sA + sB$.
e. Verify that $r(sA) = (rs)A$.
f. Verify that $r(sB) = (rs)B$.

5. Prove Part (b) of Theorem 11.2.
6. Prove Part (c) of Theorem 11.2.

11.5 MULTIPLICATION OF MATRICES

In this section we will consider the product of matrices. Just as the matrices had to be conformable for matrix addition, two matrices also must be conformable for matrix multiplication.

DEFINITION 11.11 A matrix that contains only one row is called a *row matrix*. A matrix that contains only one column is called a *column matrix*.

We will now define the product obtained by multiplying a row matrix by a column matrix *provided the number of rows in the column matrix is equal to the number of columns in the row matrix.*

DEFINITION 12.12 Let $A = [a_{ij}]$ be a $1 \times m$ row matrix and $B = [b_{ij}]$ be an $m \times 1$ column matrix. Then the *product* $C = AB$ is the 1×1 matrix given by

$$C = AB = [a_{11} + a_{12} + \cdots + a_{1n}]\begin{bmatrix} b_{11} \\ b_{21} \\ \vdots \\ b_{m1} \end{bmatrix}$$

$$= [a_{11}b_{11} + a_{12}b_{21} + \cdots + a_{1m}b_{m1}]$$

Observe that in forming the product AB of the matrices A and B that the number of rows in B is equal to the number of columns in A. The row matrix preceded the column matrix in the product.

EXAMPLE From the product AB if $A = [2 \quad 0 \quad 3]$ and $B = \begin{bmatrix} 1 \\ 4 \\ -2 \end{bmatrix}$.

Solution

$$AB = [2 \quad 0 \quad 3] \begin{bmatrix} 1 \\ 4 \\ -2 \end{bmatrix} = [(2)(1) + (0)(4) + (3)(-2)]$$

$$= [2 + 0 - 6] = [-4] \quad \blacksquare$$

EXAMPLE If $C = [-1 \quad 4 \quad -2 \quad 1]$ and $D = \begin{bmatrix} 3 \\ -4 \end{bmatrix}$, then the product CD does *not* exist since the number of rows in D is 2, which is not equal to the number of columns in C, which is 4.

We will now consider the product of two arbitrary matrices using Definition 11.12 successively. In general, in order to multiply the matrix A by the matrix B, the number of rows in the matrix B must be equal to the number of columns in the matrix A. Such matrices are said to be *conformable for matrix multiplication*.

DEFINITION 11.13 Let $A = [a_{ij}]$ be an $m \times p$ matrix and let $B = [b_{ij}]$ be a $p \times n$ matrix. Then the *product* $C = AB$ is the $m \times n$ matrix such that each entry c_{ij} of the matrix $C = [c_{ij}]$ is obtained by multiplying the entries in the ith row of A by the corresponding entries in the jth column of B and then adding the results. The element $c_{ij} = a_{i1}b_{1j} + a_{i2}b_{2j} + a_{i3}b_{3j} + \cdots + a_{ip}b_{pj}$ for every c_{ij} in C.

To illustrate the above definition consider the matrices $A = [a_{ij}]_{2 \times 3}$ and $B = [b_{ij}]_{3 \times 2}$. Then

$$AB = \begin{bmatrix} a_{11} & a_{12} & a_{13} \\ a_{21} & a_{22} & a_{23} \end{bmatrix} \begin{bmatrix} b_{11} & b_{12} \\ b_{21} & b_{22} \\ b_{31} & b_{32} \end{bmatrix} = \begin{bmatrix} c_{11} & c_{12} \\ c_{21} & c_{22} \end{bmatrix}$$

where

$$c_{11} = a_{11}b_{11} + a_{12}b_{21} + a_{13}b_{31}$$

$$c_{12} = a_{11}b_{12} + a_{12}b_{22} + a_{13}b_{32}$$

$$c_{21} = a_{21}b_{11} + a_{22}b_{21} + a_{23}b_{31}$$

$$c_{22} = a_{21}b_{12} + a_{22}b_{22} + a_{23}b_{32}$$

EXAMPLE Form the product AB given $A = \begin{bmatrix} 2 & 3 \\ -1 & 0 \end{bmatrix}$ and $B = \begin{bmatrix} 1 & -1 & 3 \\ 4 & 0 & 2 \end{bmatrix}$.

Solution

$$AB = \begin{bmatrix} 2 & 3 \\ -1 & 0 \end{bmatrix}\begin{bmatrix} 1 & -1 & 3 \\ 4 & 0 & 2 \end{bmatrix}$$

$$= \begin{bmatrix} (2)(1) + (3)(4) & (2)(-1) + (3)(0) & (2)(3) + (3)(2) \\ (-1)(1) + (0)(4) & (-1)(-1) + (0)(0) & (-1)(3) + (0)(2) \end{bmatrix}$$

$$= \begin{bmatrix} 2 + 12 & -2 + 0 & 6 + 6 \\ -1 + 0 & 1 + 0 & -3 + 0 \end{bmatrix} = \begin{bmatrix} 14 & -2 & 12 \\ -1 & 1 & -3 \end{bmatrix} \quad \blacksquare$$

Now consider the following matrix equation:

$$\begin{bmatrix} 2 & 3 \\ 1 & -5 \end{bmatrix}\begin{bmatrix} x \\ y \end{bmatrix} = \begin{bmatrix} 4 \\ 1 \end{bmatrix}$$

Performing the indicated matrix multiplication on the left-hand side, we obtain

$$\begin{bmatrix} 2x + 3y \\ x - 5y \end{bmatrix} = \begin{bmatrix} 4 \\ 1 \end{bmatrix}$$

Using the definition for equality of matrices, we obtain

$$2x + 3y = 4$$

$$x - 5y = 1$$

Hence we have a system of linear equations that can be expressed as a matrix equation. Later in this chapter we will consider the use of matrices to solve such systems of equations.

If A and B are two matrices that are conformable for matrix multiplication such that the product AB exists, we say that the matrix B is *premultiplied* by the matrix A and that the matrix A is *postmultiplied* by the matrix B. Further, if A is a matrix of order $m \times p$ and B is a matrix of order $p \times n$, then the product AB can be formed and AB will be a matrix of order $m \times n$. The product BA *cannot* be formed unless $m = n$.

EXAMPLE Consider $A = \begin{bmatrix} 2 & 3 \\ 1 & 0 \end{bmatrix}$ and $B = \begin{bmatrix} -2 & 1 \\ 0 & -1 \end{bmatrix}$. Form (a) AB and (b) BA.

Solution

a. $AB = \begin{bmatrix} 2 & 3 \\ 1 & 0 \end{bmatrix}\begin{bmatrix} -2 & 1 \\ 0 & -1 \end{bmatrix} = \begin{bmatrix} (2)(-2) + (3)(0) & (2)(1) + (3)(-1) \\ (1)(-2) + (0)(0) & (1)(1) + (0)(-1) \end{bmatrix}$

$$= \begin{bmatrix} -4 + 0 & 2 - 3 \\ -2 + 0 & 1 + 0 \end{bmatrix} = \begin{bmatrix} -4 & -1 \\ -2 & 1 \end{bmatrix}$$

b. $BA = \begin{bmatrix} -2 & 1 \\ 0 & -1 \end{bmatrix}\begin{bmatrix} 2 & 3 \\ 1 & 0 \end{bmatrix} = \begin{bmatrix} (-2)(2) + (1)(1) & (-2)(3) + (1)(0) \\ (0)(2) + (-1)(1) & (0)(3) + (-1)(0) \end{bmatrix}$

$$= \begin{bmatrix} -4 + 1 & -6 + 0 \\ 0 - 1 & 0 + 0 \end{bmatrix} = \begin{bmatrix} -3 & -6 \\ -1 & 0 \end{bmatrix} \quad \blacksquare$$

In this last example the matrices A and B are conformable for multiplication such that both AB and BA can be formed. However, $AB \neq BA$. In general, if A is conformable to B for matrix multiplication, this does not imply that B is conformable to A for matrix multiplication. Even if each matrix is conformable to the other for matrix multiplication, the two products obtained in general are not equal. Hence we see that matrix multiplication is *not* a commutative operation.

EXAMPLE   Let $C = \begin{bmatrix} 2 & 3 \\ 0 & 0 \end{bmatrix}$ and $D = \begin{bmatrix} {}^-3 & 0 \\ 2 & 0 \end{bmatrix}$. Form CD.

Solution

$$
\begin{aligned}
CD &= \begin{bmatrix} 2 & 3 \\ 0 & 0 \end{bmatrix} \begin{bmatrix} {}^-3 & 0 \\ 2 & 0 \end{bmatrix} \\
&= \begin{bmatrix} (2)({}^-3) + (3)(2) & (2)(0) + (3)(0) \\ (0)({}^-3) + (0)(2) & (0)(0) + (0)(0) \end{bmatrix} \\
&= \begin{bmatrix} {}^-6 + 6 & 0 + 0 \\ 0 + 0 & 0 + 0 \end{bmatrix} = \begin{bmatrix} 0 & 0 \\ 0 & 0 \end{bmatrix} \quad \blacksquare
\end{aligned}
$$

Observe that in the example above we have the product of two nonzero matrices which is the zero matrix. Hence we see that the multiplication property for the scalar zero does not hold for matrix multiplication. Hence, from the matrix equation $AB = 0$, we *cannot* conclude that at least one of the matrices is the zero matrix.

EXAMPLE   Given the matrices $A = \begin{bmatrix} 2 & 1 \end{bmatrix}$, $B = \begin{bmatrix} 3 \\ -2 \end{bmatrix}$, and $C = \begin{bmatrix} 2 \\ 0 \end{bmatrix}$, form (a) AB and (b) AC.

Solution

a. $AB = \begin{bmatrix} 2 & 1 \end{bmatrix} \begin{bmatrix} 3 \\ -2 \end{bmatrix} = [4]$

b. $AC = \begin{bmatrix} 2 & 1 \end{bmatrix} \begin{bmatrix} 2 \\ 0 \end{bmatrix} = [4]$ $\blacksquare$

Observe that in the example above the matrices A, B, and C are all nonzero matrices, the matrix $AB = AC$, but the matrix $B \neq C$. Hence we see that the cancellation property does not hold for matrix multiplication. That is, if $AB = AC$, it does *not* follow that $B = C$ even if $A \neq 0$.

We have now observed that the set of matrices is not closed under matrix multiplication, the operation of matrix multiplication is not commutative, the multiplication property for zero does not apply to matrices, and the cancellation property does not hold for matrix multiplication. It is interesting to note, however, that matrix multiplication is an associative operation and that the distributive property of matrix multiplication over matrix addition does hold true. The proof of these two statements is somewhat cumbersome because of the need for what is known as the sigma notation and, hence, we will state these properties as theorems without proofs.

THEOREM 11.3 Associative Property for Matrix Multiplication If A, B, and C are matrices, then $(AB)C = A(BC)$, provided the matrices are conformable for the respective matrix multiplications involved.

THEOREM 11.4 Distributive Property for Matrices If A, B, and C are matrices, then $A(B + C) = AB + AC$, provided the matrices are conformable for the respective matrix additions and matrix multiplications involved.

EXAMPLE Verify Theorem 11.3 with $A = \begin{bmatrix} 2 & -1 & 0 \\ 1 & 0 & -2 \end{bmatrix}$,

$B = \begin{bmatrix} 1 & 0 & 1 & 0 \\ -1 & 2 & 0 & 3 \\ 2 & -1 & 1 & 0 \end{bmatrix}$, and $C = \begin{bmatrix} 3 & 2 \\ 1 & -3 \\ 0 & 1 \\ -1 & 0 \end{bmatrix}$.

Solution

$$(AB)C = \left(\begin{bmatrix} 2 & -1 & 0 \\ 1 & 0 & -2 \end{bmatrix} \begin{bmatrix} 1 & 0 & 1 & 0 \\ -1 & 2 & 0 & 3 \\ 2 & -1 & 1 & 0 \end{bmatrix} \right) \begin{bmatrix} 3 & 2 \\ 1 & -3 \\ 0 & 1 \\ -1 & 0 \end{bmatrix}$$

$$= \begin{bmatrix} 3 & -2 & 2 & -3 \\ -3 & 2 & -1 & 0 \end{bmatrix} \begin{bmatrix} 3 & 2 \\ 1 & -3 \\ 0 & 1 \\ -1 & 0 \end{bmatrix} = \begin{bmatrix} 10 & 14 \\ -7 & -13 \end{bmatrix}$$

$$A(BC) = \begin{bmatrix} 2 & -1 & 0 \\ 1 & 0 & -2 \end{bmatrix} \left(\begin{bmatrix} 1 & 0 & 1 & 0 \\ -1 & 2 & 0 & 3 \\ 2 & -1 & 1 & 0 \end{bmatrix} \begin{bmatrix} 3 & 2 \\ 1 & -3 \\ 0 & 1 \\ -1 & 0 \end{bmatrix} \right)$$

$$= \begin{bmatrix} 2 & -1 & 0 \\ 1 & 0 & -2 \end{bmatrix} \begin{bmatrix} 3 & 3 \\ -4 & -8 \\ 5 & 8 \end{bmatrix} = \begin{bmatrix} 10 & 14 \\ -7 & -13 \end{bmatrix}$$

Hence $(AB)C = A(BC)$. ∎

EXAMPLE Verify Theorem 11.4 with $A = \begin{bmatrix} 4 & -1 & 0 \\ 1 & 2 & -1 \end{bmatrix}$,

$B = \begin{bmatrix} 1 & 0 & 1 \\ -1 & 1 & 2 \\ 2 & 0 & 1 \end{bmatrix}$, and $C = \begin{bmatrix} 2 & -1 & 0 \\ 3 & 0 & -1 \\ 1 & 1 & 2 \end{bmatrix}$.

Solution

$$A(B + C) = \begin{bmatrix} 4 & -1 & 0 \\ 1 & 2 & -1 \end{bmatrix} \left(\begin{bmatrix} 1 & 0 & 1 \\ -1 & 1 & 2 \\ 2 & 0 & 1 \end{bmatrix} + \begin{bmatrix} 2 & -1 & 0 \\ 3 & 0 & -1 \\ 1 & 1 & 2 \end{bmatrix} \right)$$

$$= \begin{bmatrix} 4 & -1 & 0 \\ 1 & 2 & -1 \end{bmatrix} \begin{bmatrix} 3 & -1 & 1 \\ 2 & 1 & 1 \\ 3 & 1 & 3 \end{bmatrix} = \begin{bmatrix} 10 & -5 & 3 \\ 4 & 0 & 0 \end{bmatrix}$$

$$AB + AC = \begin{bmatrix} 4 & -1 & 0 \\ 1 & 2 & -1 \end{bmatrix} \begin{bmatrix} 1 & 0 & 1 \\ -1 & 1 & 2 \\ 2 & 0 & 1 \end{bmatrix} + \begin{bmatrix} 4 & -1 & 0 \\ 1 & 2 & -1 \end{bmatrix} \begin{bmatrix} 2 & -1 & 0 \\ 3 & 0 & -1 \\ 1 & 1 & 2 \end{bmatrix}$$

$$= \begin{bmatrix} 5 & -1 & 2 \\ -3 & 2 & 4 \end{bmatrix} + \begin{bmatrix} 5 & -4 & 1 \\ 7 & -2 & -4 \end{bmatrix} = \begin{bmatrix} 10 & -5 & 3 \\ 4 & 0 & 0 \end{bmatrix}$$

Hence $A(B + C) = AB + AC$. ∎

We have introduced the matrix 0_n which is the square matrix of order n all of whose elements are equal to zero. This matrix was called the identity matrix for matrix addition of square matrices of order n. Also of interest is the matrix I_n which is a square matrix of order n such that every element on the main diagonal is 1 and all other elements are 0. I_n is called the *identity* matrix for matrix multiplication of square matrices of order n. We observe that

$$I_n = \begin{bmatrix} 1 & 0 & 0 & 0 & \cdots & 0 \\ 0 & 1 & 0 & 0 & \cdots & 0 \\ 0 & 0 & 1 & 0 & \cdots & 0 \\ \vdots & & & & & \\ 0 & 0 & 0 & 0 & \cdots & 1 \end{bmatrix}$$

and as a special case where $n = 3$ we have

$$I_3 = \begin{bmatrix} 1 & 0 & 0 \\ 0 & 1 & 0 \\ 0 & 0 & 1 \end{bmatrix}$$

EXAMPLE

a. $\begin{bmatrix} 2 & 1 \\ 0 & -1 \end{bmatrix} \begin{bmatrix} 1 & 0 \\ 0 & 1 \end{bmatrix} = \begin{bmatrix} 2 & 1 \\ 0 & -1 \end{bmatrix}$.

b. $\begin{bmatrix} 3 & 0 & 1 \\ -1 & 2 & 0 \\ 4 & 6 & -2 \end{bmatrix} \begin{bmatrix} 1 & 0 & 0 \\ 0 & 1 & 0 \\ 0 & 0 & 1 \end{bmatrix} = \begin{bmatrix} 3 & 0 & 1 \\ -1 & 2 & 0 \\ 4 & 6 & -2 \end{bmatrix}$.

EXERCISES

1. Let A and B be 3×4 matrices, let C be a 4×1 matrix, D be a 3×1 matrix, and E be a 4×3 matrix. Determine which of the following matrix operations can be performed. For those which can be performed state the order of the resulting matrix. For those which cannot be performed give an appropriate reason for your answer.

a. AB.
b. AC.
c. EA.
d. $A + B$.
e. $E + B$.
f. $2C$.
g. $BC + D$.
h. $AE + 2B$.
i. $E(A + 2B)$.
j. $(EA)C$.
k. $BE - 2I_3$.
l. $AC - 4D$.

2. Let $F = \begin{bmatrix} 1 & 2 \\ -1 & 0 \\ 3 & 1 \end{bmatrix}$, $G = \begin{bmatrix} 4 & 1 \\ 0 & 3 \end{bmatrix}$, $H = \begin{bmatrix} 2 & 4 & -1 \\ 1 & 0 & 5 \end{bmatrix}$, $R = \begin{bmatrix} 1 & 2 & 3 \\ 0 & -1 & 3 \\ -2 & 0 & 1 \end{bmatrix}$,

and $S = \begin{bmatrix} 2 & -1 & 1 \\ 1 & 0 & 2 \\ 4 & -1 & 3 \end{bmatrix}$.

Perform the indicated operations if possible. If not possible, give an appropriate reason for your answer.

a. FG.
b. $R - 2S$.
c. $2G - HF$.
d. $(SR)F$.
e. G^2 (where $G^2 = GG$).
f. R^3 (where $R^3 = R^2 R$).
g. $(FG)H$.
h. $(2G)H - 3H$.
i. $R^2 - FH$.
j. $G^2 - 2F$.
k. $(GH)(3S)$.
l. H^2.

3. Express the system of equations

$$2x - 3y = 1$$
$$x + 2y = {}^-3$$

in matrix form.

4. Express the system of equations

$$a_{11}x_1 + a_{12}x_2 + a_{13}x_3 = b_1$$
$$a_{21}x_1 + a_{22}x_2 + a_{23}x_3 = b_2$$
$$a_{31}x_1 + a_{32}x_2 + a_{33}x_3 = b_3$$

in matrix form.

5. Let $A = \begin{bmatrix} 1 & 0 \\ 2 & -1 \end{bmatrix}$ and $B = \begin{bmatrix} -1 & 1 \\ 1 & 2 \end{bmatrix}$.

Show that $(A + B)^2 \neq A^2 + 2AB + B^2$.

6. Verify that matrix multiplication is an associative operation by showing that $(AB)C = A(BC)$ for each of the following.

a. $A = \begin{bmatrix} 2 & 3 \\ -1 & 0 \end{bmatrix}$, $B = \begin{bmatrix} 2 & 1 & 3 \\ 1 & 4 & -1 \end{bmatrix}$, and $C = \begin{bmatrix} 1 & 0 \\ 2 & -1 \\ -2 & 1 \end{bmatrix}$.

b. $A = \begin{bmatrix} 1 & 2 & 3 \\ 0 & -1 & 2 \\ -2 & 0 & 1 \end{bmatrix}$, $B = \begin{bmatrix} 2 & -1 & 3 \\ 1 & 0 & 1 \\ 2 & 1 & 0 \end{bmatrix}$, and $C = \begin{bmatrix} 1 & -1 & 0 \\ 0 & 2 & 1 \\ 3 & 1 & 2 \end{bmatrix}$.

7. Verify that matrix multiplication can be distributed over matrix addition by showing that $A(B + C) = AB + AC$ for each of the following.

a. $A = \begin{bmatrix} 1 & 2 & -1 \\ 0 & 1 & 2 \\ 3 & 2 & 1 \end{bmatrix}$, $B = \begin{bmatrix} 1 & 0 \\ -1 & 2 \\ 2 & 1 \end{bmatrix}$, and $C = \begin{bmatrix} 2 & -1 \\ 0 & 1 \\ 1 & 3 \end{bmatrix}$.

b. $A = \begin{bmatrix} 1 & 2 \\ 3 & 4 \end{bmatrix}$, $B = \begin{bmatrix} 1 & 2 & 3 \\ 3 & 2 & -1 \end{bmatrix}$, and $C = \begin{bmatrix} 4 & 0 & 2 \\ -1 & 2 & 3 \end{bmatrix}$.

11.6 EQUIVALENT MATRICES

In this section we will consider equivalent matrices and an application. Two matrices are equivalent if we can transform one matrix into the other by what are called elementary operations. These operations will be defined for rows but also are valid for columns. Throughout the balance of this chapter we will be primarily concerned with elementary row operations.

DEFINITION 11.14 An *elementary row operation* on a matrix consists of the following operations:
1. Interchanging any two rows.
2. Multiplying any row by a nonzero scalar.
3. Adding to any row a nonzero scalar multiple of another row.

EXAMPLE The matrix $\begin{bmatrix} 2 & 1 \\ 3 & -4 \end{bmatrix}$ can be transformed to the matrix $\begin{bmatrix} 3 & -4 \\ 2 & 1 \end{bmatrix}$ by the elementary row operation of interchanging the two rows.

EXAMPLE The matrix $\begin{bmatrix} 1 & 2 & 3 \\ 0 & 4 & -2 \end{bmatrix}$ can be transformed to the matrix $\begin{bmatrix} 1 & 2 & 3 \\ -1 & 2 & -5 \end{bmatrix}$ by the elementary row operation of multiplying row one by -1 and adding it to row two. Note that row one was left unchanged.

EXAMPLE The matrix $\begin{bmatrix} 4 & 3 & 1 \\ -6 & 3 & 0 \\ 1 & 0 & 2 \end{bmatrix}$ can be transformed to the matrix

$\begin{bmatrix} 4 & 3 & 1 \\ -2 & 1 & 0 \\ 1 & 0 & 2 \end{bmatrix}$ by multiplying row two by the nonzero scalar 1/3.

DEFINITION 11.15 Two matrices A and B are said to be *equivalent*, denoted by $A \sim B$, if and only if the matrix A can be transformed to the matrix B using only elementary row (column) operations.

There are several applications for elementary row operations. In this section we will use them in the solution of systems of linear equations. In the next section we will use them to find the inverse of a matrix.

Now consider the system of two linear equations in the two variables x and y:

$$x - y = 2$$

$$2x + 3y = 1$$

To solve this system we may multiply equation one by ($^-2$) and add the resulting equation to equation two, obtaining the equivalent systems

$$\begin{cases} x - y = 2 \\ 2x + 3y = 1 \end{cases} \rightarrow \begin{cases} -2x + 2y = -4 \\ 2x + 3y = 1 \end{cases} \rightarrow \begin{cases} -2x + 2y = -4 \\ 5y = {}^-3 \end{cases} \rightarrow \begin{cases} x - y = 2 \\ 5y = {}^-3 \end{cases}$$

We may now multiply the second equation of the last system by 1/5, obtaining the equivalent system

$$\begin{cases} x - y = 2 \\ y = \dfrac{{}^-3}{5} \end{cases}$$

If we now add the second equation to the first we obtain the equivalent system

$$\begin{cases} x \quad\;\; = \dfrac{7}{5} \\ y = \dfrac{{}^-3}{5} \end{cases}$$

which is the required solution. We can check this by substituting 7/5 for x and $^-3/5$ for y in the two equations of the original system.

Observe that the systems

$$\begin{cases} x - y = 2 \\ 2x + 3y = 1 \end{cases}, \begin{cases} x - y = 2 \\ 5y = {}^-3 \end{cases}, \begin{cases} x - y = 2 \\ y = \dfrac{{}^-3}{5} \end{cases}, \text{ and } \begin{cases} x = \dfrac{7}{5} \\ y = \dfrac{{}^-3}{5} \end{cases}$$

are all equivalent since they have the same solutions.

Now return to the original system of equations

$$\begin{cases} x - y = 2 \\ 2x + 3y = 1 \end{cases}$$

When rewriting this system as an equivalent system, the important parts were the numerical values in the system. The variables x and y just serve to line things up, so to speak. In fact, if we change the variables x and y to u and v, for instance, then the resulting system would have the same solutions.

Now consider the matrix

$$A = \begin{bmatrix} 1 & {}^-1 \\ 2 & 3 \end{bmatrix}$$

which we will call the *matrix of coefficients* of the system of equations being discussed. Also consider the matrices

$$X = \begin{bmatrix} x \\ y \end{bmatrix} \quad \text{and} \quad B = \begin{bmatrix} 2 \\ 1 \end{bmatrix}$$

Clearly, the system of equations we are working with could then be represented by the matrix equation

$$AX = B$$

We also will introduce a new matrix, A^*, formed by adding another column to the matrix A. This column will be the column of matrix B. However, we will separate the entries of matrix A from those of matrix B by a broken line segment as indicated by

$$A^* = \begin{bmatrix} 1 & {}^-1 & \vdots & 2 \\ 2 & 3 & \vdots & 1 \end{bmatrix}$$

Matrix A^* is called the *augmented matrix* of the system of equations.

Now let's solve the system of equations as before but this time do to the corresponding augmented matrix representing the system of equations whatever we do to the system itself. Starting, then, we have

$$\begin{cases} x - y = 2 \\ 2x + 3y = 1 \end{cases} \qquad A^* = \begin{bmatrix} 1 & {}^-1 & \vdots & 2 \\ 2 & 3 & \vdots & 1 \end{bmatrix}$$

Multiplying the first equation by ($^-2$) and adding this equation to the second equation, we obtain

$$\begin{cases} x - y = 2 \\ \quad\ 5y = {}^-3 \end{cases} \qquad A^* \sim \begin{bmatrix} 1 & {}^-1 & \vdots & 2 \\ 0 & 5 & \vdots & {}^-3 \end{bmatrix}$$

Multiplying the second equation of the new system by the nonzero scalar 1/5, we obtain

$$\begin{cases} x - y = 2 \\ \quad\ y = \dfrac{{}^-3}{5} \end{cases} \qquad A^* \sim \begin{bmatrix} 1 & {}^-1 & \vdots & 2 \\ 0 & 1 & \vdots & \dfrac{{}^-3}{5} \end{bmatrix}$$

Adding the second equation of the new system to the first equation, we obtain

$$\begin{cases} x \quad\ = \dfrac{7}{5} \\ \quad\ y = \dfrac{{}^-3}{5} \end{cases} \qquad A^* \sim \begin{bmatrix} 1 & 0 & \vdots & \dfrac{7}{5} \\ 0 & 1 & \vdots & \dfrac{{}^-3}{5} \end{bmatrix}$$

Finally, observe that the last equivalent matrix for A^* corresponds to the augmented matrix for the last equivalent system of equations. Replacing the variables x and y, we can readily obtain

$$\begin{cases} 1x + 0y = \dfrac{7}{5} \\ 0x + 1y = \dfrac{{}^-3}{5} \end{cases}$$

or, simply, $x = 7/5$ and $y = {}^-3/5$ for the solutions.

EXAMPLE Solve the system of equations

$$\begin{cases} x + 2y + 3z = 4 \\ 2x + \ y - \ z = 0 \\ x - 2y + \ z = 1 \end{cases}$$

using matrix methods.

Solution The augmented matrix for the system is

$$A^* = \begin{bmatrix} 1 & 2 & 3 & \vdots & 4 \\ 2 & 1 & {}^-1 & \vdots & 0 \\ 1 & {}^-2 & 1 & \vdots & 1 \end{bmatrix}$$

By performing elementary row operations on A^* we will obtain equivalent matrices as follows:

$$A^* \sim \begin{bmatrix} 1 & 2 & 3 & \vdots & 4 \\ 0 & ^-3 & ^-7 & \vdots & ^-8 \\ 1 & ^-2 & 1 & \vdots & 1 \end{bmatrix}$$

Multiplying row one by $^-2$ and adding to row two

$$A^* \sim \begin{bmatrix} 1 & 2 & 3 & \vdots & 4 \\ 0 & ^-3 & ^-7 & \vdots & ^-8 \\ 0 & ^-4 & ^-2 & \vdots & ^-3 \end{bmatrix}$$

Multiplying row one by $^-1$ and adding to row three

$$A^* \sim \begin{bmatrix} 1 & 2 & 3 & \vdots & 4 \\ 0 & 1 & \dfrac{7}{3} & \vdots & \dfrac{8}{3} \\ 0 & ^-4 & ^-2 & \vdots & ^-3 \end{bmatrix}$$

Multiplying row two by the nonzero scalar $\dfrac{^-1}{3}$

$$A^* \sim \begin{bmatrix} 1 & 0 & \dfrac{^-5}{3} & \vdots & \dfrac{^-4}{3} \\ 0 & 1 & \dfrac{7}{3} & \vdots & \dfrac{8}{3} \\ 0 & ^-4 & ^-2 & \vdots & ^-3 \end{bmatrix}$$

Multiplying row two by $(^-2)$ and adding to row one

$$A^* \sim \begin{bmatrix} 1 & 0 & \dfrac{^-5}{3} & \vdots & \dfrac{^-4}{3} \\ 0 & 1 & \dfrac{7}{3} & \vdots & \dfrac{8}{3} \\ 0 & 0 & \dfrac{22}{3} & \vdots & \dfrac{23}{3} \end{bmatrix}$$

Multiplying row two by 4 and adding to row three

$$A^* \sim \begin{bmatrix} 1 & 0 & \dfrac{^-5}{3} & \vdots & \dfrac{^-4}{3} \\ 0 & 1 & \dfrac{7}{3} & \vdots & \dfrac{8}{3} \\ 0 & 0 & 1 & \vdots & \dfrac{23}{22} \end{bmatrix}$$

Multiplying row three by the nonzero scalar $\dfrac{3}{22}$

$$A^* \sim \begin{bmatrix} 1 & 0 & 0 & \vdots & \dfrac{9}{22} \\ 0 & 1 & \dfrac{7}{3} & \vdots & \dfrac{8}{3} \\ 0 & 0 & 1 & \vdots & \dfrac{23}{22} \end{bmatrix}$$

Multiplying row three by $\dfrac{5}{3}$ and adding to row one

$$A^* \sim \begin{bmatrix} 1 & 0 & 0 & \vdots & \dfrac{9}{22} \\ 0 & 1 & 0 & \vdots & \dfrac{5}{22} \\ 0 & 0 & 1 & \vdots & \dfrac{23}{22} \end{bmatrix}$$

Multiplying row three by $\dfrac{-7}{3}$ and adding to row two

This last matrix is the augmented matrix for the equivalent system of equations

$$\begin{cases} 1x + 0y + 0z = \dfrac{9}{22} \\ 0x + 1y + 0z = \dfrac{5}{22} \\ 0x + 0y + 1z = \dfrac{23}{22} \end{cases}$$

and hence our solutions are

$$\begin{cases} x = \dfrac{9}{22} \\ y = \dfrac{5}{22} \\ z = \dfrac{23}{22} \end{cases}$$

which can be verified by substituting these values for x, y, and z in *each* of the three original equations. ∎

For each of the two systems of equations considered there was a unique solution. However, for a particular system of linear equations there may be no solutions or there may be more than one solution. We will not discuss these cases in this text since our intent is simply to provide an introduction to the system of matrices. The interested reader may refer to any text on matrix algebra or elementary linear algebra for a more complete discussion of the solution of such systems of equations.

EXERCISES

1. Determine the matrix of coefficients for each of the following systems of equations.

a. $x + 2y = 3$
   $2x - y = 1$

b. $4x = 3 - y$
   $y + 2x = 1$

c. $x - y + 2z = 5$
$2x + y - 2z = 0$
$3x + 2y - 4z = 2$

d. $2x + y - 3z - 4 = 0$
$x - 3y + z + 5 = 0$
$2x + y + 2z = 1$
$3x - 2y - z = 7$

2. Determine the augmented matrix for each of the systems of equations given in Exercise 1.

3. Determine the appropriate elementary row operation(s) used to transform each matrix A to the corresponding matrix B.

a. $A = \begin{bmatrix} 2 & 1 \\ 0 & -3 \end{bmatrix}$, $B = \begin{bmatrix} 0 & -3 \\ 2 & 1 \end{bmatrix}$.

b. $A = \begin{bmatrix} 3 & 1 & 2 \\ -1 & 0 & 1 \end{bmatrix}$, $B = \begin{bmatrix} 3 & 1 & 2 \\ -4 & -1 & -1 \end{bmatrix}$.

c. $A = \begin{bmatrix} 2 & 0 & -4 \\ -1 & 1 & 0 \\ 1 & 0 & 0 \end{bmatrix}$, $B = \begin{bmatrix} 1 & 0 & -2 \\ 2 & -2 & 0 \\ 1 & 0 & 0 \end{bmatrix}$.

d. $A = \begin{bmatrix} 2 & 0 & 3 \\ 1 & -1 & 2 \\ 5 & 0 & -3 \end{bmatrix}$, $B = \begin{bmatrix} 5 & 0 & -3 \\ -3 & 3 & -6 \\ 2 & 0 & 3 \end{bmatrix}$.

4. Solve the following systems of equations using matrix methods.

a. $x + 2y = 7$
$3x + 5y = 11$

b. $2x - y = 5$
$x + 3y = 4$

c. $x - z = 2$
$2x + y + z = 3$
$-y + 2z = 4$

d. $2x + 4y - 3z = 1$
$x + y + 2z = 9$
$3x + 6y - 5z = 0$

11.7 INVERSE OF A MATRIX

DEFINITION 11.16 If A is a square matrix of order n, then the *inverse matrix* of A, denoted by A^{-1}, if it exists is a square matrix of order n such that $AA^{-1} = A^{-1}A = I_n$ where I_n is the identity matrix of order n.

For the inverse of a square matrix to exist certain conditions must be satisfied. Different methods can be used to find the inverse of a matrix, and some of these methods would involve introducing new terms such as the transpose of a matrix, the cofactor matrix, and the adjoint matrix. The method we will use, however, is the result of a theorem we will state without proof.

THEOREM 11.5 Let A be a square matrix of order n, and let I be the identity matrix of the same order as A. If there exists a sequence of elementary row operations that can be used to transform the matrix A into the matrix I, then the same sequence of elementary row operations will transform the augmented matrix $[A \vdots I]$ into the augmented matrix $[I \vdots B]$. The matrix B will be the inverse matrix of matrix A.

Theorem 11.5 does not guarantee the existence of the inverse for every square matrix of order n. It simply suggests a procedure for finding the inverse of a matrix if it exists. We will illustrate the procedure suggested by Theorem 11.5 by the following example.

EXAMPLE Find the inverse matrix if it exists of the matrix

$$A = \begin{bmatrix} 1 & 1 & 2 \\ {}^-1 & 0 & 3 \\ 4 & 1 & 2 \end{bmatrix}.$$

Solution We will form the augmented matrix of the form $[A \vdots I]$:

$$[A \vdots I] = \begin{bmatrix} 1 & 1 & 2 & \vdots & 1 & 0 & 0 \\ {}^-1 & 0 & 3 & \vdots & 0 & 1 & 0 \\ 4 & 1 & 2 & \vdots & 0 & 0 & 1 \end{bmatrix}$$

Adding row one to row two, we obtain

$$[A \vdots I] \sim \begin{bmatrix} 1 & 1 & 2 & \vdots & 1 & 0 & 0 \\ 0 & 1 & 5 & \vdots & 1 & 1 & 0 \\ 4 & 1 & 2 & \vdots & 0 & 0 & 1 \end{bmatrix}$$

Adding $^-4$ times row one to row three, we obtain

$$[A \vdots I] \sim \begin{bmatrix} 1 & 1 & 2 & \vdots & 1 & 0 & 0 \\ 0 & 1 & 5 & \vdots & 1 & 1 & 0 \\ 0 & {}^-3 & {}^-6 & \vdots & {}^-4 & 0 & 1 \end{bmatrix}$$

Adding $^-1$ times row two to row one, we obtain

$$[A \vdots I] \sim \begin{bmatrix} 1 & 0 & {}^-3 & \vdots & 0 & {}^-1 & 0 \\ 0 & 1 & 5 & \vdots & 1 & 1 & 0 \\ 0 & {}^-3 & {}^-6 & \vdots & {}^-4 & 0 & 1 \end{bmatrix}$$

Adding 3 times row two to row three, we obtain

$$[A \vdots I] \sim \begin{bmatrix} 1 & 0 & {}^-3 & \vdots & 0 & {}^-1 & 0 \\ 0 & 1 & 5 & \vdots & 1 & 1 & 0 \\ 0 & 0 & 9 & \vdots & {}^-1 & 3 & 1 \end{bmatrix}$$

Multiplying row three by 1/9, we obtain

$$[A \vdots I] \sim \begin{bmatrix} 1 & 0 & ^-3 & \vdots & 0 & ^-1 & 0 \\ 0 & 1 & 5 & \vdots & 1 & 1 & 0 \\ 0 & 0 & 1 & \vdots & \dfrac{^-1}{9} & \dfrac{1}{3} & \dfrac{1}{9} \end{bmatrix}$$

Adding $^-5$ times row three to row two, we obtain

$$[A \vdots I] \sim \begin{bmatrix} 1 & 0 & ^-3 & \vdots & 0 & ^-1 & 0 \\ 0 & 1 & 0 & \vdots & \dfrac{14}{9} & \dfrac{^-2}{3} & \dfrac{^-5}{9} \\ 0 & 0 & 1 & \vdots & \dfrac{^-1}{9} & \dfrac{1}{3} & \dfrac{1}{9} \end{bmatrix}$$

Adding 3 times row three to row one, we obtain

$$[A \vdots I] \sim \begin{bmatrix} 1 & 0 & 0 & \vdots & \dfrac{^-1}{3} & 0 & \dfrac{1}{3} \\ 0 & 1 & 0 & \vdots & \dfrac{14}{9} & \dfrac{^-2}{3} & \dfrac{^-5}{9} \\ 0 & 0 & 1 & \vdots & \dfrac{^-1}{9} & \dfrac{1}{3} & \dfrac{1}{9} \end{bmatrix}$$

$$= [I \vdots B]$$

Hence the inverse matrix of A is

$$B = \begin{bmatrix} \dfrac{^-1}{3} & 0 & \dfrac{1}{3} \\ \dfrac{14}{9} & \dfrac{^-2}{3} & \dfrac{^-5}{9} \\ \dfrac{^-1}{9} & \dfrac{1}{3} & \dfrac{1}{9} \end{bmatrix} \quad \blacksquare$$

EXERCISES

Determine the inverse of each of the following matrices if it exists.

1. $\begin{bmatrix} 1 & 2 \\ 3 & 4 \end{bmatrix}.$

2. $\begin{bmatrix} 4 & 2 \\ 3 & 1 \end{bmatrix}.$

3. $\begin{bmatrix} 1 & 0 \\ 0 & 1 \end{bmatrix}.$

4. $\begin{bmatrix} 2 & ^-1 \\ 1 & 0 \end{bmatrix}.$

5. $\begin{bmatrix} 1 & 0 & 0 \\ 0 & 1 & 0 \\ 0 & 0 & 1 \end{bmatrix}$.

6. $\begin{bmatrix} 1 & 1 & 0 \\ 0 & 1 & 1 \\ 3 & 0 & 1 \end{bmatrix}$.

7. $\begin{bmatrix} 3 & 0 & 2 \\ 2 & 0 & ^-1 \\ 1 & 1 & 0 \end{bmatrix}$.

8. $\begin{bmatrix} 2 & ^-1 & 0 \\ 0 & 1 & 3 \\ 1 & ^-1 & 0 \end{bmatrix}$.

9. $\begin{bmatrix} 4 & 3 & 2 \\ 1 & 0 & 1 \\ ^-1 & 0 & 2 \end{bmatrix}$.

10. $\begin{bmatrix} 2 & ^-5 & 1 \\ 1 & ^-1 & 0 \\ 0 & 1 & 1 \end{bmatrix}$.

11.8 A LOOK AT THE SYSTEM OF MATRICES

We have been discussing the system of matrices and have examined the various operations on matrices together with their properties. Some of these properties were similar to the corresponding properties involving operations on the system of real numbers. In this section we will summarize the operations on matrices and their properties.

11.81 DEFINITIONS

1. A *matrix* is a rectangular array of entries that are called *elements*.
2. A *scalar* is a real number, or a function of a real number.
3. The horizontal arrays of a matrix are called its *rows*, and the vertical arrays are called its *columns*.
4. A matrix with m rows and n columns is said to be a matrix of *order* $m \times n$.
5. A matrix all of whose elements are zero is called the *zero* or *null* matrix.
6. A square matrix all of whose elements along the main diagonal are 1's and all other elements are 0's is called an *identity matrix*.

11.82 EQUALITY AND INEQUALITY OF MATRICES

1. Two matrices A and B are said to be *equal*, denoted by $A = B$, if and only if they are of the same order and their corresponding elements are equal.
2. A matrix A with real number entries is said to be *greater than* a matrix B with real number entries, denoted by $A > B$, if and only if the matrices are of the same order and each of the entries of A is greater than the corresponding entries of B.
3. A matrix A with real number entries is said to be *less than* a matrix B with real number entries, denoted by $A < B$, if and only if B is greater than A.

11.83 OPERATIONS ON MATRICES

1. MATRIX ADDITION   If $A = [a_{ij}]_{m \times n}$ and $B = [b_{ij}]_{m \times n}$ are two matrices of order $m \times n$, then the *sum* of A and B, denoted by $A + B$, is the matrix $C = [c_{ij}]_{m \times n}$ of order $m \times n$ such that $c_{ij} = a_{ij} + b_{ij}$ for every i and j. Matrices of the same order are said to be *conformable for matrix addition*.

2. SCALAR MULTIPLE OF A MATRIX   If $A = [a_{ij}]_{m \times n}$ is an arbitrary $m \times n$ matrix and c is an arbitrary scalar, then $cA = [ca_{ij}]_{m \times n}$ is an $m \times n$ matrix called a *scalar multiple* of A.

3. DIFFERENCE OF MATRICES   If A and B are arbitrary matrices of the same order, then the *difference* $A - B = A + (-B)$, where $-B$ means $(^-1)B$.

4. MULTIPLICATION OF MATRICES   Let $A = [a_{ij}]$ be an $m \times p$ matrix, and let $B = [b_{ij}]$ be a $p \times n$ matrix. Then the *product* $C = AB$ is the $m \times n$ matrix such that each entry c_{ij} of the matrix $C = [c_{ij}]$ is obtained by multiplying the entries in the ith row of A by the corresponding entries in the jth column of B and adding the results. The element $c_{ij} = a_{i1}b_{1j} + a_{i2}b_{2j} + a_{i3}b_{3j} + \cdots + a_{ip}b_{pj}$ for every c_{ij} in C.

5. INVERSE OF A MATRIX   If A is a square matrix of order n, then the *inverse matrix* of A, denoted by A^{-1}, if it exists is a square matrix of order n such that $AA^{-1} = A^{-1}A = I_n$, where I_n is the identity matrix of order n.

11.84 EQUIVALENCE OF MATRICES

Two matrices A and B are said to be *equivalent*, denoted by $A \sim B$, if and only if the matrix A can be transformed to the matrix B using only elementary row (column) operations.

An *elementary row operation* on a matrix consists of one of the following operations:

1. Interchanging any two rows.

2. Multiplying any row by a nonzero scalar.

3. Adding to any row a nonzero scalar multiple of another row.

11.85 COMMUTATIVE PROPERTY

1. FOR MATRIX ADDITION   If A and B are any two matrices that are conformable for matrix addition, then $A + B = B + A$.

2. FOR MATRIX MULTIPLICATION   If A and B are matrices that are conformable for matrix multiplication, then in general AB is *not* equal to BA. There are, however, certain matrices such that $AB = BA$.

11.86 ASSOCIATIVE PROPERTY

1. FOR MATRIX ADDITION   If A, B, and C are matrices that are conformable for matrix addition, then $(A + B) + C = A + (B + C)$.
2. FOR MATRIX MULTIPLICATION   If $A = [a_{ij}]$, $B = [b_{ij}]$, and $C = [c_{ij}]$ are matrices that are conformable for the indicated matrix multiplications, then $(AB)C = A(BC)$.

11.87 DISTRIBUTIVE PROPERTIES

If $A = [a_{ij}]$, $B = [b_{ij}]$, and $C = [c_{ij}]$ are matrices that are conformable for the indicated matrix additions and matrix multiplications, then
1. $A(B + C) = AB + AC$.
2. $(A + B)C = AC + BC$.

11.88 IDENTITY PROPERTIES

1. MATRIX ADDITION IDENTITY   There exists a square zero matrix of order n, denoted by $\mathbf{0}_n$, such that $A + \mathbf{0}_n = \mathbf{0}_n + A = A$ for *all* square matrices A of order n.
2. MATRIX MULTIPLICATION IDENTITY   There exists a square identity matrix of order n, denoted by I_n, such that $AI = IA = A$ for *all* square matrices A of order n.

11.89 INVERSE PROPERTIES

1. FOR MATRIX ADDITION   If A is a matrix of order $m \times n$, then there exists a matrix B of order $m \times n$ such that $A + B = B + A = \mathbf{0}_{m \times n}$. $B = -A$ and is called the inverse of A for matrix addition. *Every* matrix has such an inverse.
2. FOR MATRIX MULTIPLICATION   If A is a square matrix of order n and if there exists a square matrix B of order n such that $AB = BA = I_n$, then B is called the inverse of A for matrix multiplication and $B = A^{-1}$. Such an inverse *does not* always exist.

11.810 PROPERTIES FOR SCALAR MULTIPLES OF MATRICES

1. If A is an arbitrary matrix and c is an arbitrary scalar, then $cA = Ac$, where $Ac = [a_{ij}c]$.
2. If A and B are matrices conformable for matrix addition and c and d are arbitrary scalars, then
   a. $c(A + B) = cA + cB$,
   b. $(c + d)A = cA + dA$.

3. If A and B are matrices conformable for matrix multiplication and c and d are arbitrary scalars, then

a. $c(dA) = (cd)A$,
b. $c(AB) = (cA)B$,
c. $(Ac)B = A(cB)$.

11.811 DEFICIENCIES ON THE SYSTEM OF MATRICES

1. COMMUTATIVE PROPERTY FOR MATRIX MULTIPLICATION   If A and B are matrices conformable for matrix multiplication, AB is not always equal to BA.

2. CANCELLATION PROPERTY   If A, B, and C are matrices conformable for matrix multiplication and $AB = AC$, then B is not necessarily equal to C even if $A \neq 0$.

3. MULTIPLICATION PROPERTY OF ZERO   If A and B are matrices conformable for matrix multiplication and $AB = 0$, then it does not follow that at least one of the matrices A, B is a zero matrix.

SUMMARY

In this chapter we introduced a new mathematical system known as the system of matrices. We defined terms such as a matrix, row and column matrices, equality and equivalence of matrices, the zero or null matrix, and the identity matrix. We introduced and discussed operations on matrices. These included matrix addition, scalar multiples of a matrix, matrix multiplication, and finding an inverse of a matrix if it exists. Various properties associated with these operations also were discussed. For matrices whose elements are real numbers, we discussed the relation of order. Similarities and lack of similarities were noted between the system of matrices and other systems discussed earlier in the text. As an application of the use of matrices we considered a brief discussion of the solution of systems of linear equations.

During our discussions the following symbols were introduced.

SYMBOL	INTERPRETATION
$A = [a_{ij}]_{m \times n}$	A matrix A of order $m \times n$
$A_{m \times n}$	Matrix A of order $m \times n$
A_n	Square matrix of order n
$0_{m \times n}$	Zero matrix of order $m \times n$
0_n	Square zero matrix of order n
$A = B$	Equality of two matrices of the same order
$A > B$	Matrix A is greater than the matrix B
$A < B$	Matrix A is less than the matrix B

$A \sim B$ Matrix A is equivalent to matrix B, if A and B are of the same order

$A + B$ The sum of two matrices A and B

cA A scalar multiple of the matrix A

AB The product of the matrices A and B

$A - B$ The difference of the matrices A and B

I_n Square matrix of order n called the identity matrix

$AB \neq BA$ The product of the matrices A and B is not equal to the product of the matrices B and A

A^{-1} The inverse of the matrix A if it exists

A^* The augmented matrix of A

REVIEW EXERCISES FOR CHAPTER 11

1. Let $A = \begin{bmatrix} 2 & -3 \\ 1 & 0 \end{bmatrix}$, $B = \begin{bmatrix} 1 & 2 \\ 3 & 4 \end{bmatrix}$, $C = \begin{bmatrix} 2 & -1 & 0 \\ 1 & 3 & 2 \end{bmatrix}$, $D = \begin{bmatrix} 1 & -2 \\ 3 & -1 \\ 0 & 2 \end{bmatrix}$,

$E = \begin{bmatrix} 4 & -1 \\ 5 & 0 \\ 3 & 4 \end{bmatrix}$, $F = \begin{bmatrix} 2 & -1 & -3 \\ 1 & -1 & 0 \\ -1 & 2 & -2 \end{bmatrix}$, $G = \begin{bmatrix} 3 & 0 & -1 \\ 1 & 2 & 1 \\ 4 & 2 & 0 \end{bmatrix}$, and

$H = \begin{bmatrix} 2 & 1 & -1 & 3 \\ 4 & 0 & 2 & 1 \\ 0 & 1 & -2 & 1 \end{bmatrix}$.

Perform each of the following operations or indicate clearly why the operation(s) cannot be performed.

a. $A + B$.
     b. BC.
c. $2D - 3E$.
     d. $(2E)F$.
e. $A(D + F)$.
     f. $(2A - 3B)C$.
g. $G(2H)$.
     h. $(AC)H$.
i. $E(CF)$.
     j. $(GD)H$.

2. Consider the matrices given in Exercise 1.
   a. Is $A > B$? Why or why not?
   b. Is $A < B$? Why or why not?
   c. Is $A = B$? Why or why not?
   d. Is $D < E$? Why or why not?
   e. Is $D > E$? Why or why not?
   f. Is $G > F$? Why or why not?
   g. Is $G \geq F$? Why or why not?
   h. Is $H \leq H$? Why or why not?

3. Is the set of all matrices of order $m \times n$ closed under the operation of matrix addition? Explain.

4. Is the set of all square matrices of order p closed under the operation of matrix multiplication? Explain.

5. What if any is the additive identity for the set of all matrices of order 4×3?

6. What if any is the multiplicative identity for the set of all square matrices of order 5?

7. Does every square matrix have a multiplicative inverse? Explain.

8. If A is a square matrix of order p and B is the multiplicative inverse of A, what is AB?

9. Determine a value for each of the unknowns so that the following statements are true.

a. $\begin{bmatrix} 2 & x \\ y & 4 \end{bmatrix} = 3 \begin{bmatrix} u & 2 \\ 1 & v \end{bmatrix}$.

b. $\begin{bmatrix} x & 1 \\ 2 & y \end{bmatrix} - 2 \begin{bmatrix} 1 & u \\ v & 2 \end{bmatrix} = \begin{bmatrix} 1 & 0 \\ 0 & 1 \end{bmatrix}$.

c. $2 \begin{bmatrix} x & 1 \\ 2 & y \end{bmatrix} \geq 3 \begin{bmatrix} 1 & 0 \\ -1 & 2 \end{bmatrix}$.

d. $\begin{bmatrix} 2x & 3 \\ 0 & y \end{bmatrix} - 3 \begin{bmatrix} 1 & u \\ v & -1 \end{bmatrix} \leq \begin{bmatrix} 2 & 1 \\ -1 & 0 \end{bmatrix}$.

10. Rewrite the following system of equations in matrix form.

$$\begin{array}{rcl} 2x - 3y + z & = & 1 \\ x + y - z + 4 & = & 0 \\ 3x - 2y + 3z & = & 2 \end{array}$$

11. Solve the system of equations given in Exercise 10.

12. Determine the inverse of the following matrix if it exists.

$$A = \begin{bmatrix} 1 & 2 & -3 \\ 0 & 1 & 0 \\ -1 & 0 & 2 \end{bmatrix}$$

CHAPTER 12
OTHER MATHEMATICAL SYSTEMS

In this text we have considered various mathematical systems. For most of these systems the nonempty set involved was a subset of the real numbers. In the previous chapter we considered the system of matrices where the elements involved were matrices. In this chapter we will consider mathematical systems that may seem to be of a more abstract nature.

12.1 MATHEMATICAL SYSTEMS

DEFINITION 12.1 A *mathematical system* consists of a nonempty set, at least one operation defined on the elements of the set, at least one relation defined among these elements, and properties concerning these elements, operation(s) and relation(s).

EXAMPLE The system of propositions introduced in Chapter 2 is an example of a mathematical system where the nonempty set consisted of elements called propositions and the operations were the connectives of negation, conjunction, disjunction, conditional, and the biconditional. The relations defined among these elements were implication and equivalence. The properties included the commutative and associative properties of disjunction and conjunction, the distributive properties of disjunction (conjunction) over conjunction (disjunction), the identity properties, and the idempotent properties, among others.

EXAMPLE The system of sets introduced in Chapter 3 is an example of a mathematical system whose nonempty set consisted of elements called sets, and the operations were those of union, intersection, complementation, difference, and cartesian product. The relations defined among these elements were "is a subset of" and "is equal to." The properties included the commutative and associative properties of union and intersection, the distributive property of union (intersection) over intersection (union), the identity properties, and the complement properties, among others.

EXAMPLE The system of integers introduced in Chapter 7 is an example of a mathematical system whose nonempty set I consisted of elements called integers, and the operations were addition, multiplication, subtraction, and division. The relations defined among these elements were the order relation and equality. The properties included the commutative and associative properties of addition and multiplication, the distributive property of multiplication over addition, the identity properties, and the additive inverse property, among others.

EXAMPLE The system of matrices introduced in Chapter 11 is an example of a mathematical system where the nonempty set consists of elements called matrices (plural of matrix), and the operations were matrix addition, the scalar multiple of a matrix, and matrix multiplication. The relations defined among these elements were equality and equivalence. The properties included the commutative and associative properties for matrix addition, the associative property for matrix multiplication, the identity properties, and the inverse properties, among others.

From the four examples given above we see that the operations for one mathematical system may be quite different from those of another mathematical system. For instance, ordinary multiplication of integers is quite different from matrix multiplication.

DEFINITION 12.2 An *operation* $*$ over a nonempty set S is a rule or procedure by which two elements of S may be combined to produce a unique third element which may or may not belong to S. If $a, b \in S$ and $a * b \in S$ for all $a, b \in S$, then we say that the set S is *closed* under the operation $*$.

EXAMPLE
a. The set of all integers I is closed under the operation of ordinary addition; that is, if $a, b \in I$, then $a + b \in I$ for all $a, b \in I$.
b. The set of all whole numbers W is *not* closed under the operation of subtraction; if $a, b \in W$, then $a - b \in W$ if and only if $a \geq b$.

The operations used in mathematical systems are subject to certain properties. For instance, the operation of ordinary multiplication of real numbers is a commutative property whereas the multiplication of matrices is, in general, not commutative. We will now define properties used in previous chapters but for arbitrary operations.

DEFINITION 12.3 Commutative Property If S is an arbitrary nonempty set and $*$ is an arbitrary operation, then $*$ is *commutative* on S if and only if $a * b = b * a$ for all $a, b \in S$.

DEFINITION 12.4 Associative Property If S is an arbitrary nonempty set and $*$ is an arbitrary operation, then $*$ is *associative* on S if and only if $(a * b) * c = a * (b * c)$ for all $a, b, c \in S$.

If two operations are defined on a set, then we may discuss the distributive property of one of these operations over the other.

DEFINITION 12.5 Distributive Property If S is an arbitrary nonempty set and $*$ and $\#$ are arbitrary operations, then
1. If $a * (b \# c) = (a * b) \# (a * c)$ for all $a, b, c \in S$, we say that the operation $*$ can be distributed over the operation $\#$ on the set S.
2. If $a \# (b * c) = (a \# b) * (a \# c)$ for all $a, b, c \in S$, we say that the operation $\#$ can be distributed over the operation $*$ on the set S.

EXAMPLE
a. The operation of ordinary addition $+$ is both commutative and associative on the set of integers.
b. The operation of matrix addition is both commutative and associative on the set of matrices that are conformable for the matrix addition.
c. The operation of matrix multiplication is not commutative on the set of matrices, but the operation is associative when the matrices are conformable for matrix multiplication.
d. The operation of ordinary multiplication can be distributed over the operation of ordinary addition on the set of rational numbers.
e. The operation of disjunction can be distributed over the operation of conjunction on the set of all propositions.

EXERCISES

1. Consider the set Re of all real numbers. Is the set Re all by itself a mathematical system? Explain.

2. Consider the set P of all propositions together with the operation of disjunction. Is the set P together only with the operation of disjunction a mathematical system? Explain.

3. Which of the following sets are closed relative to the indicated operations?
   a. The set of all integers under the operation of addition.
   b. The set of all whole numbers under the operation of subtraction.
   c. The set of all rational numbers under the operation of division.
   d. The set of all irrational numbers under the operation of multiplication.
   e. The set of all 2×2 matrices under the operation of matrix addition.
   f. The set of all 3×4 matrices under the operation of matrix multiplication.
   g. The set $S = \{1\}$ under the operation of ordinary addition.
   h. The set $T = \{0\}$ under the operation of ordinary multiplication.
   i. The set of all odd integers under the operation of ordinary addition.
   j. The set of all odd integers under the operation of ordinary multiplication.

4. Which of the following statements are true and which are false?
   a. The operation of addition is a commutative operation on the set of all integers.
   b. The operation of subtraction is an associative operation on the set of all whole numbers.
   c. The operation of matrix multiplication is a commutative operation on the set of all 3×3 matrices.
   d. The operation of matrix addition is an associative operation on the set of all square matrices of order n.
   e. The operation of division is an associative operation on the set of all nonzero rational numbers.
   f. The operation of multiplication can be distributed over the operation of subtraction on the set of all rational numbers.
   g. The operation of addition can be distributed over the operation of multiplication on the set of all integers.
   h. The operation of matrix multiplication can be distributed over the operation of matrix addition on the set of all matrices conformable for the indicated operations.

5. Let $S = Re$. Define the operation $*$ as follows: If $a, b \in S$, then $a * b = a + b + 1$.
   a. Is the set S closed under the operation of $*$? Why or why not?
   b. If the operation of $*$ a commutative operation on the set S? Why or why not?
   c. Is the operation of $*$ an associative operation on the set S? Why or why not?
   d. Compute (1) $2 * 3$, (2) $3 * 4$, (3) $\frac{1}{2} * \frac{1}{4}$, (4) $(2 * 1) * 2$, and (5) $3 * (2 * 3)$.

6. Let T be the set of all integers. Define the operation $\#$ as follows: If $a, b \in T$, then $a \# b = a$.

a. Is the set T closed under the operation of $\#$? Why or why not?
b. Is the operation of $\#$ a commutative operation on the set T? Why or why not?
c. Is the operation of $\#$ an associative operation on the set T? Why or why not?
d. Compute (1) $2 \# 3$, (2) $3 \# 4$, (3) $(2 \# 1) \# 2$, (4) $3 \# (2 \# 3)$, and (5) $(^-2 \# 3) \# (^-3 \# 4)$.

12.2 GROUP—A MATHEMATICAL SYSTEM

A simple mathematical system that involves only one operation is known as a group.

DEFINITION 12.6 A *group* is a mathematical system consisting of a non-empty set, S, together with an operation $*$ and satisfying the following properties:

1. The set S is closed under the operation $*$; that is, if $a, b \in S$, then $a * b \in S$ for all $a, b \in S$.
2. The operation $*$ is associative on S; that is, if $a, b, c \in S$, then $(a * b) * c = a * (b * c)$ for all $a, b, c \in S$.
3. The set S contains an identity e for the operation $*$; that is, there exists $e \in S$ such that $a * e = e * a = a$ for *all* $a \in S$.
4. Each element of the set S has an inverse element in S relative to the operation $*$; that is, if $a \in S$, then there exists $b \in S$ such that $a * b = b * a = e$.

Notice that the commutative property for the operation $*$ on S does not have to be satisfied for a group.

DEFINITION 12.7 A group that also satisfies the property that the operation $*$ is commutative on S is called a *commutative group* or an *abelian group*.

A commutative group is also called an abelian group in honor of the famous Norwegian mathematician Niels Henrik Abel (1802–1829), whose career was very productive but short; he died at the age of 26.

EXAMPLE Example of groups are
a. The set Re together with the operation $+$ is a commutative group.
b. The set of all $m \times n$ matrices whose elements are scalars, together with the operation of matrix addition, is a commutative group.
c. The set $\{0, 1, 2, 3, 4\}$ together with the operation of addition (mod 5) is a commutative group.

EXERCISES

1. Determine which of the following form a group. For those which do not, indicate all of the properties for a group which fail to hold.

   a. The set of all integers together with the operation of ordinary addition.

   b. The set of all even integers together with the operation of subtraction.

   c. The set of all odd integers together with the operation of ordinary multiplication.

   d. The set of all 2×2 matrices with real elements under the operation of matrix addition.

   e. The set of all 2×2 matrices of the form $\begin{bmatrix} a & 1 \\ 0 & b \end{bmatrix}$, where $a, b \in Re$, under the operation of matrix addition.

   f. The set $S = \{a, b, c\}$ under the operation $*$ where the operation $*$ is defined according to the following table:

$*$	a	b	c
a	a	b	c
b	b	c	a
c	c	a	b

   g. The set $T = \{^-1, 0, 1\}$ under the operation of ordinary addition.

   h. The set $T = \{^-1, 0, 1\}$ under the operation of ordinary multiplication.

2. Determine which of the groups given in Exercise 1 are commutative groups.

3. If the set of elements in a group is finite, then the group is said to be a *finite group*. Determine which of the groups given in Exercise 1 are finite groups.

12.3 FIELD—A MATHEMATICAL SYSTEM

Another mathematical system which involves two operations is that called a field.

DEFINITION 12.8 A *field* is a mathematical system consisting of a nonempty set S together with two operations $\oplus$ and $\odot$ called an "addition" and "multiplication" and satisfying the following properties:

1. The set S is closed under the operation of $\oplus$.
2. The operation $\oplus$ is commutative on the set S.
3. The operation $\oplus$ is associative on the set S.
4. There exists $e_1 \in S$ which is the identity element relative to the operation $\oplus$.
5. Each element in S has an inverse element in S relative to the operation $\oplus$.

6. The set S is closed under the operation $\odot$.

7. The operation $\odot$ is commutative on S.

8. The operation $\odot$ is associative on S.

9. There exists $e_2 \in S$ such that $e_1 \neq e_2$, which is the identity element relative to the operation $\odot$.

10. Every element in S, except e_1, has an inverse element in S relative to the operation $\odot$.

11. The operation $\odot$ can be distributed over the operation $\oplus$ on the set S.

It should be somewhat obvious that a field is a more sophisticated mathematical system compared to a group since substantially more properties must be satisfied for a field. Consequently, it should be easier to find examples of groups than it is to find examples of fields. A careful examination of Definition 12.8, however, should reveal that a field may be defined in terms of groups. Notice that the first five properties listed for a field are those for a commutative group consisting of S and the operation $\oplus$. The next five properties are those for a commutative group consisting of $S - \{e_1\}$ and the operation $\odot$. Finally, the last property is simply the distributive property of $\odot$ over $\oplus$ on the set S.

DEFINITION 12.9 (Alternate) A *field* is a mathematical system consisting of a nonempty set S together with the two operations $\oplus$ and $\odot$ and satisfying the following properties:

1. S together with the operation $\oplus$ is a commutative group. The identity for $\oplus$ is $e_1 \in S$.

2. $S - \{e_1\}$ together with the operation $\odot$ is a commutative group. The identity for $\odot$ is $e_2 \in S$ such that $e_1 \neq e_2$.

3. The operation $\odot$ can be distributed over the operation $\oplus$ on S.

EXAMPLE

a. The set Re together with the operations of ordinary addition and ordinary multiplication is an example of a field.

b. The set of complex numbers together with the operations of ordinary addition and ordinary multiplication is an example of a field.

c. The set of integers together with the operations of ordinary addition and ordinary multiplication is *not* an example of a field. (Why not?)

If we delete property 10 in Definition 12.8 and add the cancellation property for the operation $\odot$, we have a new mathematical system called an *integral domain*.

EXAMPLE The set of all integers together with the operations of ordinary addition and ordinary multiplication is an example of an integral domain.

EXERCISES

1. Determine which of the following form a field. For those which do not, indicate all of the properties for a field that fail to hold.
   a. The set of all integers together with the operations $\oplus$ and $\odot$ being ordinary addition and ordinary multiplication.
   b. The set of all rational numbers together with the operations $\oplus$ and $\odot$ being ordinary addition and ordinary multiplication.
   c. The set $T = \{^-1, 0, 1\}$ together with the operations $\oplus$ and $\odot$ being ordinary addition and ordinary multiplication.
   d. The set of all even integers together with the operations $\oplus$ and $\odot$ being ordinary addition and ordinary multiplication.
   e. The set of all 2×2 matrices with real elements together with the operations $\oplus$ and $\odot$ being matrix addition and matrix multiplication.
   f. The set of all 2×2 matrices of the form $\begin{bmatrix} a & 0 \\ 0 & b \end{bmatrix}$, where $a, b \in Re$, together with the operations $\oplus$ and $\odot$ being matrix addition and matrix multiplication.
   g. The set $S = \{a, b, c\}$ together with the operations $\oplus$ and $\odot$ defined according to the following tables:

$\oplus$	a	b	c
a	a	b	c
b	b	c	a
c	c	a	b

$\odot$	a	b	c
a	a	a	a
b	a	b	c
c	a	c	b

2. Determine which of the sets given in Exercise 1 together with the indicated operations are integral domains. For those which are not, indicate all the properties for an integral domain that fail to hold.

12.4 RING—A MATHEMATICAL SYSTEM

Another interesting mathematical system is called a ring.

DEFINITION 12.10 A *ring* is a mathematical system consisting of a non-empty set, S, together with two operations $\oplus$ and $\odot$, called an "addition" and a "multiplication" and satisfying the following properties:
   1. The set S is closed under the operation $\oplus$.
   2. The operation $\oplus$ is a commutative operation on S.

3. The operation ⊕ is an associative operation on *S*.
4. There exists an element $e_1 \in S$ which is the identity element relative to the operation ⊕.
5. Every element in *S* has an inverse element in *S* relative to the operation ⊕.
6. The set *S* is closed under the operation ⊙.
7. The operation ⊙ is associative on *S*.
8. The operation ⊙ can be distributed over the operation ⊕ on *S*.

DEFINITION 12.11 (Alternate) A *ring* is mathematical system consisting of a nonempty set, *S*, together with two operations ⊕ and ⊙ and satisfying the following properties:
1. The set *S* together with the operation ⊕ forms a commutative group.
2. The set *S* is closed under the operation ⊙.
3. The operation ⊙ is associative on *S*.
4. The operation ⊙ can be distributed over the operation ⊕ on *S*.

EXAMPLE
a. The set of all even integers together with the operation of ordinary addition and ordinary multiplication is a ring.
b. The set of all $m \times n$ matrices with scalar entries together with the operations of matrix addition (for ⊕) and matrix multiplication (for ⊙) is a ring.

DEFINITION 12.12 A ring for which the operation ⊙ is commutative is called a *commutative ring*.

EXERCISES

1. Consider Exercise 1 of Section 12.3. Determine which of the sets listed together with the indicated operations are a ring. For those that are not, indicate all of the properties of a ring that fail to hold.
2. Which of the rings determined in Exercise 1 are commutative rings?

12.5 CLOCK ARITHMETIC

In this section we will consider an example of another mathematical system. However, this system will contain only a finite number of elements. Consider an ordinary 12-hour clock such as the one depicted in Figure 12.1. If the clock were to "run" forever, the hour hand of the clock would never be over a numeral greater than 12.

Figure 12.1

Let $S = \{1, 2, 3, 4, 5, 6, 7, 8, 9, 10, 11, 12\}$ be the set consisting of the numbers on the face of the clock, and let $\oplus$ represent the operation of clock addition with the clock addition being considered in the normal direction as the hour hand of the clock moves. For instance, if the hour hand is on 4, then 6 hours later it will be on 10. If the hour hand is on 9, then 6 hours later it will be on 3. We can now form the "addition" table for this operation as indicated in Table 12.1.

Table 12.1 12-hour clock addition

$\oplus$	1	2	3	4	5	6	7	8	9	10	11	12
1	2	3	4	5	6	7	8	9	10	11	12	1
2	3	4	5	6	7	8	9	10	11	12	1	2
3	4	5	6	7	8	9	10	11	12	1	2	3
4	5	6	7	8	9	10	11	12	1	2	3	4
5	6	7	8	9	10	11	12	1	2	3	4	5
6	7	8	9	10	11	12	1	2	3	4	5	6
7	8	9	10	11	12	1	2	3	4	5	6	7
8	9	10	11	12	1	2	3	4	5	6	7	8
9	10	11	12	1	2	3	4	5	6	7	8	9
10	11	12	1	2	3	4	5	6	7	8	9	10
11	12	1	2	3	4	5	6	7	8	9	10	11
12	1	2	3	4	5	6	7	8	9	10	11	12

We will now consider the system consisting of the nonempty set S and the operation $\oplus$. Clearly, from Table 12.1 we note that S is closed relative to the operation $\oplus$ since all of the entries in the table are elements of S. We also note from Table 12.1 that the operation of $\oplus$ is commutative since if the rows of the table were interchanged with their respective columns, an identical table would result. Hence, although the set S is finite, we conclude that if $a, b \in S$, then $a \oplus b \in S$ and $a \oplus b = b \oplus a$.

EXAMPLE
  a. $3 \oplus 7 = 7 \oplus 3 = 10$.
  b. $4 \oplus 9 = 9 \oplus 4 = 1$.
  c. $6 \oplus 8 = 8 \oplus 6 = 2$.
  d. $12 \oplus 9 = 9 \oplus 12 = 9$.

We also note from Table 12.1 that $12 \in S$ is the identity element for this clock addition. That is, $12 \oplus a = a \oplus 12 = a$ for *every* $a \in S$. With the identity element established, we can now look for inverse elements. Again, from Table 12.1 we note that the inverse element of 3 under the clock addition is 9 since $3 \oplus 9 = 9 \oplus 3 = 12$. Similarly, the inverse element of 8 is 4 and the inverse element of 11 is 1.

EXAMPLE   Consider the operation of 12-hour clock addition. The element $12 \in S$ is the identity element for this operation.
  a. The inverse element of 7 is 5 since $7 \oplus 5 = 5 \oplus 7 = 12$.
  b. The inverse element of 2 is 10 since $2 \oplus 10 = 10 \oplus 2 = 12$.
  c. The inverse element of 12 is 12 since $12 \oplus 12 = 12$.

The operation being discussed is also an associative operation on the set S, but this is not easy to prove. The proof would involve showing that for all a, b, $c \in S$, $(a \oplus b) \oplus c = a \oplus (b \oplus c)$, which would be very time consuming since there would be 12^3 or 1728 cases to consider. (Do you agree?) We will, then, assume this property without proof.

EXAMPLE   Verify the associative property for 12-hour clock addition for
  a. $a = 2, b = 3, c = 8$.
  b. $a = 7, b = 11, c = 2$.
  c. $a = 8, b = 12. c = 9$.
  d. $a = 9, b = 10, c = 4$.
  Solution
  a. $(2 \oplus 3) \oplus 8 = 5 \oplus 8 = 1$ and $2 \oplus (3 \oplus 8) = 2 \oplus 11 = 1$.
  b. $(7 \oplus 11) \oplus 2 = 6 \oplus 2 = 8$ and $7 \oplus (11 \oplus 2) = 7 \oplus 1 = 8$.
  c. $(8 \oplus 12) \oplus 9 = 8 \oplus 9 = 5$ and $8 \oplus (12 \oplus 9) = 8 \oplus 9 = 5$.
  d. $(9 \oplus 10) \oplus 4 = 7 \oplus 4 = 11$ and $9 \oplus (10 \oplus 4) = 9 \oplus 2 = 11$.   ∎

We now have a mathematical system consisting of the nonempty set S and the operation $\oplus$ such that S is closed under the operation $\oplus$, $12 \in S$ is the identity for the operation $\oplus$, and each element of S has an inverse element in S for the operation $\oplus$. Further, the operation $\oplus$ on S is both commutative and associative. Clearly, then, this mathematical system is an *example of a commutative group*.

We can also define 12-hour clock multiplication as repeated 12-hour clock addition. The operation will be denoted by $\otimes$, and the "multiplication" table is given in Table 12.2 which can easily be verified.

Again, an examination of Table 12.2 will reveal that S is closed under the operation $\otimes$ since all entries in the table are elements of S. The element $1 \in S$ is the identity element for this operation since $a \otimes 1 = 1 \otimes a$ for *all* $a \in S$. The operation $\otimes$ is a commutative operation on S since interchanging the rows of the table with their respective columns produces an identical table.

Table 12.2 12-hour clock multiplication

⊗	1	2	3	4	5	6	7	8	9	10	11	12
1	1	2	3	4	5	6	7	8	9	10	11	12
2	2	4	6	8	10	12	2	4	6	8	10	12
3	3	6	9	12	3	6	9	12	3	6	9	12
4	4	8	12	4	8	12	4	8	12	4	8	12
5	5	10	3	8	1	6	11	4	9	2	7	12
6	6	12	6	12	6	12	6	12	6	12	6	12
7	7	2	9	4	11	6	1	8	3	10	5	12
8	8	4	12	8	4	12	8	4	12	8	4	12
9	9	6	3	12	9	6	3	12	9	6	3	12
10	10	8	6	4	2	12	10	8	6	4	2	12
11	11	10	9	8	7	6	5	4	3	2	1	12
12	12	12	12	12	12	12	12	12	12	12	12	12

Although $1 \in S$ is the identity element for the operation $\otimes$, not all elements of S have an inverse element in S for the operation $\otimes$. For instance, there is no $b \in S$ such that $2 \otimes b = b \otimes 2 = 1$. The only elements of S that have inverse elements relative to the operation $\otimes$ are 1, 5, 7, and 11, which are their respective inverses. We will assume the associative property of the operation $\otimes$ on S. The distributive property of $\otimes$ over $\oplus$ on the set S also can be verified.

EXAMPLE
a. $7 \otimes 2 = 2 \otimes 7 = 2.$
b. $8 \otimes 6 = 6 \otimes 8 = 12.$
c. $9 \otimes 11 = 11 \otimes 9 = 3.$
d. $10 \otimes 7 = 7 \otimes 10 = 10.$

EXAMPLE Verify that $(a \otimes b) \otimes c = a \otimes (b \otimes c)$ for $a, b, c \in S$, if
a. $a = 2, b = 8,$ and $c = 10.$
b. $a = 5, b = 7,$ and $c = 9.$
Solution
a. $(2 \otimes 8) \otimes 10 = 4 \otimes 10 = 4$ and $2 \otimes (8 \otimes 10) = 2 \otimes 8 = 4.$
b. $(5 \otimes 7) \otimes 9 = 11 \otimes 9 = 3$ and $5 \otimes (7 \otimes 9) = 5 \otimes 3 = 3.$ ∎

EXAMPLE Verify that $a \otimes (b \oplus c) = (a \otimes b) \oplus (a \otimes c)$ for $a, b, c \in S$, if
a. $a = 4, b = 7,$ and $c = 9.$
b. $a = 2, b = 11,$ and $c = 8.$
Solution
a. $4 \otimes (7 \oplus 9) = 4 \otimes 4 = 4$ and $(4 \otimes 7) \oplus (4 \otimes 9) = 4 \oplus 12 = 4.$
b. $2 \otimes (11 \oplus 8) = 2 \otimes 7 = 2$ and $(2 \otimes 11) \oplus (2 \otimes 8) = 10 \oplus 4 = 2.$ ∎

Observe that the nonempty set S together with the operation $\otimes$ is not a group. However, S together with the operations $\oplus$ and $\otimes$ form a ring that is a commutative ring.

We will conclude this section by commenting that not all clock systems are 12-hour clock systems. For instance, the military uses the 24-hour clock system where 0900 hours means 9:00 A.M. and 1740 hours means 5:40 P.M. Athletic officials use stopwatches or clocks such as a 5-minute clock. Timers on various home appliances may be 3-hour clocks or 10-minute clocks or similar clock systems. We will look at some of these via the exercises that follow.

EXERCISES

1. Using 12-hour clock addition, solve the following equations for x.
   a. $x + 2 = 7$. b. $3 + x = 1$.
   c. $x - 3 = 8$. d. $5 = x - 7$.
   e. $6 = x - 9$. f. $7 + x = 2$.
   g. $12 + x = 2$. h. $x + 1 = 1$.
   i. $x + 9 = 8$. j. $7 + x = 5$.
2. Using 12-hour clock multiplication, solve the following equations for all permissible values of u. If no solution exists, indicate this.
   a. $2 \times 10 = u$. b. $6 \times u = 12$.
   c. $3 \times u = 4$. d. $5 \times u = 1$.
   e. $9 \times u = 3$. f. $8 \times u = 7$.
   g. $11 \times u = 4$. h. $10 \times u = 3$.
   i. $(3 \times 4) \times 5 = u$. j. $11 \times (10 \times 3) = u$.
3. Construct the addition and multiplication tables for the 5-minute clock, and solve the following equations for all permissible values of y. If no solutions exist, indicate this.
   a. $2 + y = 4$. b. $3 + 4 = y$.
   c. $4 - y = 1$. d. $2 - y = 4$.
   e. $3 \times y = 1$. f. $4 \times y = 2$.
   g. $y \times 5 = 2$. h. $y - 4 = 2$.
   i. $(2 + 4) \times 3 = y$. j. $3 \times (4 + 3) = y$.
4. Consider the 5-minute clock of Exercise 3.
   a. What if any is the additive identity?
   b. What if any is the multiplicative identity?
   c. What if any is the additive inverse of 4?
   d. What if any is the multiplicative inverse of 4?
   e. What if any is the additive inverse of 2?
   f. What if any is the multiplicative inverse of 2?
5. Consider the 24-hour clock system as used by the military. Write the equivalent A.M. or P.M. times for the following.
   a. 0930 hours. b. 1040 hours.
   c. 1400 hours. d. 1610 hours.
   e. 0001 hours. f. 1345 hours.
   g. 2215 hours. h. 2400 hours.
   i. 0800 hours. j. 0350 hours.

6. a. A military convoy departs at 0730 hours and requires 10 hours and 15 minutes to arrive at its destination. What is the estimated time of arrival?

   b. A military convoy is to arrive at its destination by 0430 hours. If it requires 8 hours and 40 minutes traveling time, when should the convoy depart?

7. Consider the 12-hour clock system introduced in this section. Construct a 12-month clock system by replacing 1 with January, 2 with February, and so on. For this new system solve the given equations for all appropriate values of u and answer the questions listed.

   a. What if any is the additive identity?

   b. What if any is the additive inverse for March?

   c. What if any is the additive inverse for August?

   d. What if any is the multiplicative identity?

   e. What if any is the multiplicative inverse for February?

   f. What if any is the multiplicative inverse for May?

   g. February + June = u.

   h. August + u = March.

   i. April $\times$ July = u.

   j. July $\times$ u = January.

8. Using the 12-hour clock system, verify each of the following statements.

   a. $3 + (5 + 9) = (3 + 5) + 9$.

   b. $4 + (11 + 2) = (4 + 11) + 2$.

   c. $10 + (9 + 8) = (10 + 9) + 8$.

   d. $2 \times (5 \times 7) = (2 \times 5) \times 7$.

   e. $11 \times (2 \times 9) = (11 \times 2) \times 9$.

   f. $7 \times (12 \times 6) = (7 \times 12) \times 6$.

12.6 MODULAR ARITHMETIC

In the previous section we examined a mathematical system involving clock arithmetic. We used primarily the 12-hour clock system but indicated that any hour or minute clock system would work. In Section 8.6 we examined congruences, and the reader may have noted similarities between clock arithmetic and congruences. The notable difference involves the fact that with congruences, modulo m, the entries in the so-called addition tables were 0, 1, 2, ..., $m - 1$ whereas with the m-hour clock system, the entries in the addition table were 1, 2, 3, ..., m. It was noted that the relation "is congruent to, modulo m" is an equivalence relation on the set of integers. In this section we will discuss congruences in more detail.

When forming the addition table, modulo 5, we observed that the addition would be the same as ordinary addition except when the sum is equal to or greater than the modulus 5, in which case we will take the remainder when the sum is divided by 5 instead of the actual sum. The only entries, then, in the addition table, modulo 5, are 0, 1, 2, 3, and 4. Now every integer when divided by 5 will have a remainder of 0, 1, 2, 3, or 4. Clearly, there are infinitely many integers which, when divided by 5, will have a remainder of 3; hence each of these integers is congruent to 3, modulo 5. The set of all those integers congruent to 3, modulo 5, is called a *residue class, modulo 5*. Similarly, there are infinitely many integers congruent to 4, mod 5, forming another residue class, mod 5. In fact, there are exactly 5 such residue classes, mod 5, and the set of all integers can be divided into five disjoint equivalence classes that are the five residue classes.

DEFINITION 12.13 A *complete residue system, modulo m*, with $m > 1$, is a set of integers such that

1. No two elements of the set are congruent.

2. If $\alpha \in I$, then α is congruent to exactly one element of the set.

EXAMPLE

a. The set $\{0, 1, 2, 3\}$ is a complete residue system, mod 4.

b. The set $\{1, 2, 3, 4, 5, 6\}$ is a complete residue system, mod 6.

c. The set $\{3, 4, 5\}$ is a complete residue system, mod 3.

d. The set $\{0, 1, 2, \ldots, m - 1\}$ is a complete residue system, mod m.

e. The set $\{1, 2, 3, \ldots, m\}$ is a complete residue system, mod m.

f. The set $\{m + 1, m + 2, \ldots, 2m\}$ is a complete residue system, mod m.

We will now look at some interesting theorems pertaining to congruences.

THEOREM 12.1   If $a \equiv b \pmod{m}$ and $c \equiv d \pmod{m}$, then

a. $a + c \equiv b + d \pmod{m}$.

b. $a - c \equiv b - d \pmod{m}$,

c. $ac \equiv bd \pmod{m}$,

d. $ka \equiv kb \pmod{m}$, where $k \in W$.

We will prove part (a) and leave the remainder of the proof as an exercise.

Proof of part (a)   Since $a \equiv b \pmod{m}$ and $c \equiv d \pmod{m}$, then $m|(a - b)$ and $m|(c - d)$ by Definition 8.13. Since $m|(a - b)$ and $m|(c - d)$, $m|[(a + c) - (b + d)]$, by Theorem 8.3. Hence, by Definition 8.13, $a + c \equiv b + d \pmod{m}$.

EXAMPLE   Given $3 \equiv 7 \pmod{4}$ and $5 \equiv 1 \pmod{4}$. Hence

a. $3 + 5 \equiv 7 + 1 \pmod{4}$.

b. $3 - 5 \equiv 7 - 1 \pmod{4}$.

c. $(3)(5) \equiv (7)(1) \pmod{4}$.

d. $(2)(3) \equiv (2)(7) \pmod{4}$.

EXAMPLE   Given $4 \equiv {}^-5$ (mod 9) and $6 \equiv 6$ (mod 9). Then
a. $4 + 6 \equiv {}^-5 + 6$ (mod 9).
b. $4 - 6 \equiv {}^-5 - 6$ (mod 9).
c. $(4)(6) \equiv ({}^-5)(6)$ (mod 9).
d. $({}^-7)(4) \equiv ({}^-7)({}^-5)$ (mod 9).

In part (d) of Theorem 12.1 we note that if $a \equiv b$ (mod m) and k is an arbitrary number, then $ka \equiv kb$ (mod m). Suppose, however, that $ka \equiv kb$ (mod m), where $k \in W$. Can we conclude that $a \equiv b$ (mod m)? The answer to this question is given in the next theorem.

THEOREM 12.2   If $ka \equiv kb$ (mod m) and if g.c.d.$(k, m) = 1$, then $a \equiv b$ (mod m).
Proof   The proof is left as an exercise.

EXAMPLE
a. $(2)(3) \equiv (2)({}^-2)$(mod 5) and g.c.d.$(2, 5) = 1$; hence $3 \equiv {}^-2$ (mod 5).
b. $(3)(4) \equiv (3)(11)$ (mod 7) and g.c.d.$(3, 7) = 1$; hence $4 \equiv 11$ (mod 7).

A more general case of Theorem 12.2 is given in the following theorem, which will be stated without proof.

THEOREM 12.3   If $ap \equiv aq$ (mod m) and if $d = $ g.c.d.(a, m), then $a \equiv b$ (mod m/d).

EXAMPLE
a. $(3)(4) \equiv (3)({}^-2)$ (mod 6) and g.c.d.$(3, 6) = 3$; hence, $4 \equiv {}^-2$ (mod 6/3) or $4 \equiv {}^-2$ (mod 2).
b. $(8)(3) \equiv (8)(7)$ (mod 32) and g.c.d.$(8, 32) = 8$; hence $3 \equiv 7$ (mod 32/8) or $3 \equiv 7$ (mod 4).

An application of congruences can be found in what is referred to as "casting out 9's." This is a check which may be used in computations involving the basic arithmetic operations. For instance, if 4 numbers were being added, each number would be replaced by one that is congruent to it, mod 9. The four resulting congruent numbers would be added, and their sum would be replaced by a number congruent to it, mod 9. The sum of the original four numbers must then be congruent to this result obtained. We will illustrate this in the following examples.

EXAMPLE   Check the following addition by casting out 9's.

$$
\begin{array}{r}
4672 \\
3546 \\
1928 \\
\hline
10146
\end{array}
$$

Solution $4672 \equiv 4 + 6 + 7 + 2 \pmod{9}$ by the use of the test for divisibility by 9. Hence $4672 \equiv 19 \pmod{9}$. But $19 \equiv 1 \pmod{9}$ and, hence, $4672 \equiv 1 \pmod{9}$. In similar manner we obtain

$$4672 \equiv 4 + 6 + 7 + 2 \equiv 19 \equiv 1 \pmod{9}$$
$$3546 \equiv 3 + 5 + 4 + 6 \equiv 18 \equiv 0 \pmod{9}$$
$$1928 \equiv 1 + 9 + 2 + 8 \equiv 20 \equiv 2 \pmod{9}$$

Adding $\qquad\qquad\qquad\qquad\qquad\qquad 3 \pmod{9}$

$$10146 \equiv 1 + 0 + 1 + 4 + 6 \equiv 12 \pmod{9}$$

Since $12 \equiv 3 \pmod{9}$, the answer appears to check. However, note that if we wrote the answer as 11046 with a transposition of 2 digits it would still appear to check since $1 + 1 + 0 + 4 + 6 \equiv 3 \pmod{9}$. Hence our answer is only probably correct by checking in this manner.   ∎

EXAMPLE   Check the following multiplication by casting out 9's.

$$\begin{array}{r} 2364 \\ \times\ 578 \\ \hline 1376492 \end{array}$$

Solution

$$2364 \equiv 2 + 3 + 6 + 4 \equiv 15 \equiv 6 \pmod{9}$$
$$578 \equiv 5 + 7 + 8 \qquad \equiv 20 \equiv 2 \pmod{9}$$

Multiplying $\qquad\qquad\qquad\qquad\qquad 12 \equiv 3 \pmod{9}$

$$1376492 \equiv 1 + 3 + 7 + 6 + 4 + 9 + 2 \equiv 32 \equiv 5 \pmod{9}$$

Since $5 \not\equiv 3 \pmod{9}$, the answer is definitely not correct. Actually, the correct answer is 1366392.   ∎

EXAMPLE   Check the following subtraction by casting out 9's.

$$\begin{array}{r} 71{,}692 \\ -\ 27{,}984 \\ \hline 43{,}708 \end{array}$$

Solution

$$71{,}692 \equiv 7 + 1 + 6 + 9 + 2 \equiv 25 \equiv 7 \pmod{9}$$
$$27{,}984 \equiv 2 + 7 + 9 + 8 + 4 \equiv 30 \equiv 3 \pmod{9}$$

Subtracting $\qquad\qquad\qquad\qquad\qquad\quad 4 \pmod{9}$

$$43{,}708 \equiv 4 + 3 + 7 + 0 + 8 \equiv 22 \pmod{9}$$

Since $22 \equiv 4 \pmod{9}$, the answer is probably correct.   ∎

In summary, the method of casting out 9's can be used as a check in arithmetic operations with the following guidelines:

1. If the results obtained as in the above examples are not congruent, the answer is definitely not correct.

2. If the results obtained are congruent, the answers are only probably correct. Any errors in computations that involve transpositions of digits, inclusion, or exclusion of zeros or multiples of 9 will not show up in the check.

In a similar manner, using the test for divisibility by 11, we could check computational results by casting out 11's.

EXAMPLE Check the following addition by casting out 11's.

$$
\begin{array}{r}
4672 \\
5987 \\
6321 \\
\hline
15880
\end{array}
$$

Solution

$$4672 \equiv {}^{-}4 + 6 - 7 + 2 \equiv {}^{-}3 \ (\text{mod } 11)$$
$$5987 \equiv {}^{-}5 + 9 - 8 + 7 \equiv \ \ 3 \ (\text{mod } 11)$$
$$6321 \equiv {}^{-}6 + 3 - 2 + 1 \equiv {}^{-}4 \ (\text{mod } 11)$$

Adding $\qquad\qquad\qquad\qquad\qquad {}^{-}4 \ (\text{mod } 11)$

$$15880 \equiv 1 - 5 + 8 - 8 + 0 \equiv {}^{-}4 \ (\text{mod } 11)$$

Since ${}^{-}4 \equiv {}^{-}4 \ (\text{mod } 11)$, the result seems to be correct. However, the correct answer is 16980. ∎

EXAMPLE Check the following multiplication by casting out 11's.

$$
\begin{array}{r}
798 \\
\times \ 354 \\
\hline
283492
\end{array}
$$

Solution

$$798 \equiv 7 - 9 + 8 \equiv \ 6 \ (\text{mod } 11)$$
$$354 \equiv 3 - 5 + 4 \equiv \ 2 \ (\text{mod } 11)$$

Multiplying $\qquad\qquad\qquad\qquad\qquad 12 \ (\text{mod } 11)$

$$283492 \equiv {}^{-}2 + 8 - 3 + 4 - 9 + 2 \equiv 0 \ (\text{mod } 11)$$

Since $12 \neq 0 \ (\text{mod } 11)$, the answer is not correct. What should it be? ∎

EXERCISES

1. Determine which of the following statements are true and which are false.

a. The set $\{0, 1, 2, 3, 4\}$ is a complete residue system, mod 5.

b. The set $\{11, 12, 13, 14, 15\}$ is a complete residue system, mod 5.

c. The set $\{0, 1, 2, 3, \ldots, p\}$ is a complete residue system, mod p.

d. The set $\{2m + 1, 2m + 2, 2m + 3, \ldots, 3m\}$ is a complete residue system, mod m.

e. The set $\{1, 2, 3, 4, 5, 6, 7\}$ is a complete residue system, mod 6.

f. If u is an integer greater than 1, then a complete residue system, mod u, is a set which could contain $u + 2$ elements.

g. Let v be an integer greater than 1 and S be a set of integers which forms a complete residue system, mod v. Then, if $w \in I$, w will be congruent to exactly one element of S.

2. Given $4 \equiv 8 \pmod 4$, $6 \equiv 2 \pmod 4$, $5 \equiv 2 \pmod 3$, and $7 \equiv 10 \pmod 3$, determine which of the following statements are true and which are false.

a. $4 + 6 \equiv 8 + 2 \pmod 4$.

b. $5 - 7 \equiv 2 - 10 \pmod 3$.

c. $4 - 7 \equiv 8 - 10 \pmod 4$.

d. $6 \times 5 \equiv 2 \times 2 \pmod{4 \times 3}$.

e. $^-3 \times 7 \equiv {}^-3 \times 10 \pmod 3$.

f. $2 \times 5 \equiv 2 \times 2 \pmod{2 \times 3}$.

3. How many different integers are congruent to a particular element of a set that is a complete residue system?

4. If A is a finite subset of the set of integers and A has cardinality 9, will each element of A be congruent to distinct elements of a set that is a complete residue system, mod 8? Explain.

5. Verify that if $ka \equiv kb \pmod m$ and g.c.d.$(k, m) = 1$, then $a \equiv b \pmod m$ for each of the following.

a. $3 \times 2 \equiv 3 \times 7 \pmod 5 \Rightarrow 2 \equiv 7 \pmod 5$.

b. $4 \times 3 \equiv 4 \times 10 \pmod 7 \Rightarrow 3 \equiv 10 \pmod 7$.

c. $7 \times 2 \equiv 7 \times {}^-4 \pmod 6 \Rightarrow 2 \equiv {}^-4 \pmod 6$.

d. $8 \times 5 \equiv 8 \times 2 \pmod 3 \Rightarrow 5 \equiv 2 \pmod 3$.

6. Verify that if $ap \equiv aq \pmod m$ and g.c.d.$(a, m) = d$, then $a \equiv b \pmod{m/d}$ for each of the following.

a. $2 \times 3 \equiv 2 \times {}^-1 \pmod 4 \Rightarrow 3 \equiv {}^-1 \pmod 2$.

b. $3 \times {}^-2 \equiv 3 \times 1 \pmod 9 \Rightarrow {}^-2 \equiv 1 \pmod 3$.

c. $2 \times {}^-3 \equiv 2 \times 2 \pmod{10} \Rightarrow {}^-3 \equiv 2 \pmod 5$.

d. $2 \times {}^-2 \equiv 2 \times 6 \pmod{16} \Rightarrow {}^-2 \equiv 6 \pmod 8$.

7. Check each of the following computations by casting out 9's, and indicate whether the result is probably correct or definitely incorrect.

a. $234 + 467 = 701$.

b. $9,621 + 2,176 = 11,779$.

c. $247 \times 159 = 28,284$.

d. $862 \times 203 = 147,086$.

e. $4,628 \times 2,169 = 9,146,192$.

f. $6,329 - 2,747 = 3,582$.

g. $8,121 - 3,989 = 1,423$.

h. $63,129 + 26,123 - 47,259 = 41,993$.
i. $92,234 - 13,689 - 30,167 = 43,889$.
j. $6,239 \times 2,765 = 17,250,835$.

8. Repeat Exercise 7 by casting out 11's.
9. Perform each of the operations indicated in Exercise 7, and determine the correct result.
10. Prove Part (b) of Theorem 12.1.
11. Prove Part (c) of Theorem 12.1.
12. Prove Part (d) of Theorem 12.1.
13. Prove Theorem 12.2.

12.7 MOVEMENTS OF A SQUARE

In this section we will consider a mathematical system wherein the nonempty set does not have numbers for its elements but rather has elements that are movements of a square. Consider a square piece of paper or cardboard and label its four corners 1, 2, 3, and 4 as indicated in Figure 12.2. Also label the back of the paper or cardboard in the same manner

Figure 12.2

as the front was labeled. Further, label one side front and the other side back. We will now define the elements of a set as movements of the square using the symbols I, V, H, R_1, R_2, R_3, D_1, and D_2 as in Figure 12.3.

We now have the set $S = \{I, V, H, R_1, R_2, R_3, D_1, D_2\}$, and we will introduce the operation $\circ$ to be successive movements of the square. For instance, $V \circ D_1$ would be illustrated as in Figure 12.4.

We can now construct the so-called "multiplication" table for the operation $\circ$.

Table 12.3

$\circ$	I	V	H	R_1	R_2	R_3	D_1	D_2
I	I	V	H	R_1	R_2	R_3	D_1	D_2
V	V	I	R_2	D_2	H	D_1	R_3	R_1
H	H	R_2	I	D_1	V	D_2	R_1	R_3
R_1	R_1	D_1	D_2	R_2	R_3	I	H	V
R_2	R_2	H	V	R_3	I	R_1	D_2	D_1
R_3	R_3	D_2	D_1	I	R_1	R_2	V	H
D_1	D_1	R_1	R_3	V	D_2	H	I	R_2
D_2	D_2	R_3	R_1	H	D_1	V	R_2	I

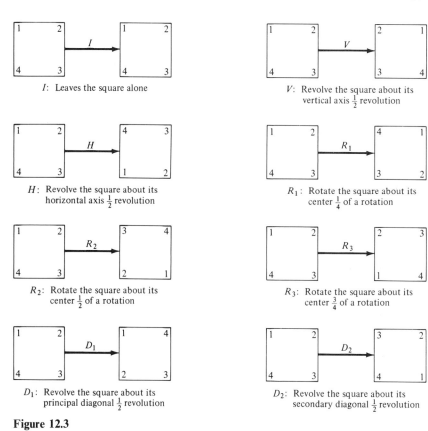

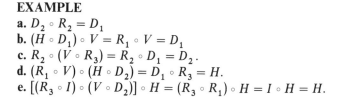

Figure 12.3

EXAMPLE

a. $D_2 \circ R_2 = D_1$

b. $(H \circ D_1) \circ V = R_1 \circ V = D_1$

c. $R_2 \circ (V \circ R_3) = R_2 \circ D_1 = D_2.$

d. $(R_1 \circ V) \circ (H \circ D_2) = D_1 \circ R_3 = H.$

e. $[(R_3 \circ I) \circ (V \circ D_2)] \circ H = (R_3 \circ R_1) \circ H = I \circ H = H.$

From an examination of Table 12.3 we readily determine that I is the identity in S for the operation $\circ$, the set S is closed under the operation $\circ$,

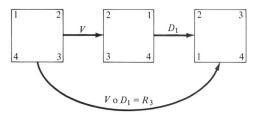

Figure 12.4

the operation ∘ is a commutative operation on S, and each element of S has an inverse element in S relative to the operation ∘. We will assume the associativity property for ∘ on the set S. (The author together with several of his classes have verified all 512 cases to substantiate this property. The reader may wish to do likewise.)

Hence the set S together with the operation ∘ forms a commutative group.

EXERCISES

Using Table 12.3 determine each of the following.

1. $D_1 \circ R_3$.
2. $R_2 \circ H$.
3. $V \circ H$.
4. $D_2 \circ R_1$.
5. $R_1 \circ V$.
6. $D_2 \circ D_1$.
7. $(V \circ H) \circ R_2$.
8. $(D_2 \circ R_3) \circ H$.
9. $(R_1 \circ I) \circ D_1$.
10. $(R_3 \circ D_1) \circ V$.
11. $D_1 \circ (V \circ H)$.
12. $R_2 \circ (D_2 \circ V)$.
13. $H \circ (I \circ R_3)$.
14. $D_2 \circ (D_1 \circ R_1)$.
15. $(H \circ D_1) \circ (V \circ D_2)$.
16. $(R_3 \circ H) \circ (R_2 \circ D_1)$.
17. $(V \circ D_1) \circ (R_2 \circ H)$.
18. $(D_1 \circ D_2) \circ (H \circ V)$
19. $[V \circ (H \circ D_1)] \circ D_2$.
20. $D_1 \circ [H \circ (V \circ R_3)]$.

SUMMARY

In this chapter we defined what is meant by a mathematical system and discussed several such systems. Among these were group, field, ring, clock arithmetic, complete residue systems, and movements of a square. There were no new symbols introduced in this chapter other than those associated with the movements of the square.

REVIEW EXERCISES FOR CHAPTER 12

1. Identify each of the following statements as being true or false.
   a. The operation associated with a group must be commutative.
   b. Every ring is an integral domain.
   c. A group may be finite.
   d. An integral domain is also a field.
   e. If a binary operation is defined on a set S, then S is closed relative to the operation.
   f. If $a \in S$ and a has an inverse in S, then the inverse of the inverse of a is a.
   g. A complete residue system, mod m, contains exactly m elements.

h. There is exactly one integer that is congruent to a particular element of a complete residue system.

i. Casting out 9's always can be used as an absolute check on computations involving arithmetic operations.

j. The set $T = \{^-1, 0, 1\}$ is closed under the operation of ordinary addition.

2. Construct addition and multiplication tables for a complete residue system, mod 9.

3. Consider the complete residue system $\{0, 1, 2, 3\}$ (mod 4). Does this system together with the operation of addition form a group? Why or why not?

4. Consider the complete residue system $\{0, 1, 2, 3\}$ (mod 4). Does this system together with the operations of addition and multiplication form a ring?

5. Repeat Exercise 4 for an integral domain.

6. Repeat Exercise 4 for a field.

CHAPTER 13
AN INTRODUCTION TO
INFORMAL GEOMETRY

Geometry is the branch of mathematics concerned with the study of shapes and sizes together with their relationships. The word *geometry* comes from a combination of two Greek words meaning "earth measurement." Geometry is sometimes referred to as the more practical branch of mathematics insofar as, over the last 2000 or more years, it has helped people find out much about the planet on which they live.

The reader is probably most familiar with geometry from studying Euclidean geometry at the high school level. About 300 B.C. the Greek mathematican Euclid wrote the famous *Elements* in which he collected and organized into an orderly system many geometrical discoveries made by the early Greeks. Contrary to what many high school students believe, Euclid did not "invent" all of the geometry that appears in the *Elements*. It is conjectured, however, that in addition to collecting and organizing the material, he did perfect some of the proofs and also added some original materials.

In this text we have neither the space nor the intent to develop Euclidean geometry in the logical and deductive manner as is done at the high school level. Rather we will use an informal and intuitive approach to describe certain geometric concepts including relationships existing between figures and their sizes.

The language and concepts of sets will be used throughout the chapter in an attempt to provide a unifying approach with topics considered in previous chapters and to provide the foundations for topics to be considered in subsequent chapters.

13.1 UNDEFINED TERMS AND DEFINITIONS

To make more meaningful what is meant by a particular geometric shape, an exact definition must be given using only terms that also have meaning. For instance, if we wish to define the geometric figure we know as a square, we may say that "a square is a figure with four equal sides." However, this definition would not be precise enough because a rhombus also is a figure having four equal sides. To remove this ambiguity, we may wish to say that "a square is a figure with four equal sides and four right angles," assuming, of course, that meaning was already given to the terms "equal sides" and "right angles." Now this second definition for a square is overly precise since it involves facts that can be proven from a more basic definition for a square. The reader may recall a definition that says just enough about a square but not too much. Such a definition is "a square is a rectangle with two adjacent sides equal."

Of course the last definition given for a square depends upon the meaning of a rectangle which is a particular type of parallelogram. But a parallelogram is defined as a special type of quadrilateral. In turn, a quadrilateral is defined as a particular type of polygon, and so on. Continuing in this fashion, we would eventually get to simple geometric figures which cannot be defined in terms of simpler things since none exist. These simple figures are called undefined terms and include such things as point, line, and plane.

Almost everyone has an intuitive idea of what a point is, and to define the word would be unnecessary. For instance, to say that "a point is that thing which has no part" does not make the meaning of the word any clearer. In this text we will leave undefined the words *point*, *line*, *plane*, *space*, and *between*.

We generally represent a point by a pencil dot on a piece of paper or a chalk dot on the board. Obviously, the smaller the dot is, the better. In this respect the dot represents the point just like a numeral represents a number. Both points and numbers are abstractions. They can be represented but not seen! We will name points using capital letters such as A, B, C, P, Q, or R.

By the word *space* we shall consider the totality of all points or the set of all points being discussed. In space we shall consider various geometric figures since each geometric figure will be a set of points. One such figure is a *line*. For the geometry we will be concerned with in this text, all lines are considered to be *straight* lines. For instance, consider two distinct points, P and Q, in space. Then the line determined by these two points can be considered as the set of all the points in space that make up a straight row of dots including dots for P and Q and such that as more and more dots are added, we get a solid figure we may think of as being a line. A line has no thickness but does extend indefinitely in two directions.

The line determined by the points P and Q will be denoted by the symbol $\overleftrightarrow{PQ}$, where the double-headed arrow above P and Q indicates that the line passes through the points P and Q and extends indefinitely in two directions. Lines may also be denoted by lowercase letters such as a, b, c, l, etc.

Since many paths in space can pass through the two distinct points P and Q but not all such paths are lines, we give the following axiom. (Recall that an axiom is a statement that is accepted as being true and requires no proof. An axiom does not have to be self-evident.)

AXIOM 13.1 If P and Q are any two distinct points in space, then there is exactly one line determined by the two points and containing them.

The above axiom guarantees that two distinct points in space determine a unique line and, also, that for any two distinct points in space there is exactly one line containing them.

In Figure 13.1 we have the line $\overleftrightarrow{AB}$ which also contains the point C. This line can also be symbolized by $\overleftrightarrow{AC}$ or by $\overleftrightarrow{BC}$.

If a set of points are all on the same line, then we say that the points are collinear.

Figure 13.1

DEFINITION 13.1 *Collinear points* are points all of which are on or are contained in the same line.

If three distinct points are collinear, then one of them is *between* the other two. In Figure 13.1 point B is between points A and C.

We are considering lines as being sets of points and, hence, we may think of intersecting lines in terms of the intersection of sets. Suppose, then, that we have two lines considered as sets with a nonempty intersection. What is this intersection?

THEOREM 13.1 If two distinct lines have a nonempty intersection, then the intersection is exactly one point.

Proof The proof will be by contradiction. Let the two distinct lines be denoted by a and b and assume that $a \cap b$ contains at least two distinct points A and B. But by Axiom 13.1 only one line can contain both A and B. Hence a and b must be the same line; that contradicts the assumption that they were distinct lines. Thus if two distinct lines have a nonempty intersection, the intersection contains exactly one point.

Consider the line l illustrated in Figure 13.2 which contains the point P.

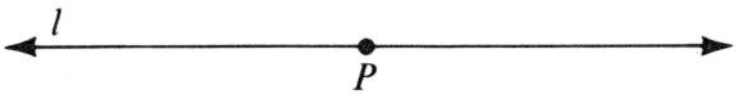

Figure 13.2

The point P divides or separates the line l into three disjoint subsets of the line: the set containing only the point P, and two disjoint subsets called *half-lines*. The point P, which divides the line, is called the *endpoint* for each of the half-lines. If a half-line contains its endpoint, it is called a *closed half-line*; if it does not, it is called an *open half-line*.

DEFINITION 13.2 A *ray* is a closed half-line.

In Figure 13.3 we have a line containing several points. The point A separates the line, and we may denote the ray from A in the direction of B as $\overrightarrow{AB}$. The arrow in the symbol $\overrightarrow{AB}$ starts over the endpoint and extends in the direction of the half-line being considered. Similarly, the symbol $\overrightarrow{AC}$

Figure 13.3

denotes the ray from A in the direction of C. The symbols $\overrightarrow{AC}$ and $\overrightarrow{AD}$ name the same ray. Hence we write $\overrightarrow{AC} = \overrightarrow{AD}$.

Relative to Figure 13.3 we note that $\overrightarrow{AB} \cup \overrightarrow{AC} = \overleftrightarrow{AB} = \overleftrightarrow{BC} = \overleftrightarrow{BD}$ and $\overrightarrow{AB} \cap \overrightarrow{AC} = A$. Also, we note that $\overrightarrow{AC} \cap \overrightarrow{CA}$ is a proper subset of $\overleftrightarrow{AC}$, which we will call a line segment.

DEFINITION 13.3 A *line segment* is a proper subset of a line consisting of two distinct points and the set of all the points between them.

If P and Q are two distinct points on a line, then the line segment consisting of P and Q and the set of all points between them is denoted by $\overline{PQ}$ or by $\overline{QP}$. Hence $\overline{PQ} = \overline{QP}$.

EXAMPLE   Consider the line containing the points indicated below:

Then
a. $\overrightarrow{AB} = \overrightarrow{AC} = \overrightarrow{AD} = \overrightarrow{AE}$.
b. $\overrightarrow{ED} = \overrightarrow{EC} = \overrightarrow{EB} = \overrightarrow{EA}$.
c. $\overrightarrow{BA} \cap \overrightarrow{BE} = B$.
d. $\overrightarrow{BC} \cap \overrightarrow{CB} = \overline{BC}$.
e. $\overrightarrow{DA} \cup \overrightarrow{BC} = \overleftrightarrow{AE}$.
f. $\overrightarrow{CB} \cap \overrightarrow{DE} = \phi$.
g. $\overrightarrow{BC} \cup \overline{DE} = \overrightarrow{BC}$.

h. $\overline{BC} \cup \overrightarrow{CD} = \overrightarrow{BD}.$
i. $\overrightarrow{BA} \cap \overline{BC} = B.$
j. $\overrightarrow{AB} \cap \overrightarrow{ED} = \overline{AE}.$

EXERCISES

1. Identify each of the following figures as representing a line, a ray, or a line segment, and write two different names for each.

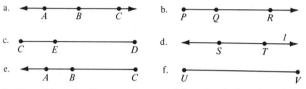

a. A B C b. P Q R

c. C E D d. S T l

e. A B C f. U V

2. Consider the line represented by the following figure:

A B C D

Determine each of the following.
a. $\overrightarrow{AB} \cup \overrightarrow{BC}.$ b. $\overrightarrow{AC} \cup \overline{CD}.$
c. $\overrightarrow{AB} \cap \overrightarrow{CB}.$ d. $\overrightarrow{BC} \cap \overline{CD}.$
e. $\overrightarrow{BC} \cup \overline{BD}.$ f. $\overleftrightarrow{AC} \cap \overline{BD}.$
g. $\overrightarrow{BA} \cup \overrightarrow{AC}.$ h. $\overrightarrow{BA} \cap \overline{BD}.$
i. $(\overrightarrow{AB} \cap \overrightarrow{CA}) \cup \overline{CD}.$ j. $(\overrightarrow{AB} \cup \overrightarrow{CD}) \cap \overrightarrow{CA}.$

3. If P and Q are two distinct points on a line, how many points exist that are collinear with P and Q?
4. If two lines are distinct, must their intersection set be nonempty? Explain.
5. If two distinct lines in the same plane have a nonempty intersection set, how many points must be contained in the intersection set?
6. If a, b, and c represent three distinct lines in the same plane, what are the possibilities for $a \cap b \cap c$? Explain.
7. Draw diagrams with the four points A, B, C, and D on the line l satisfying each of the following.
a. $\overrightarrow{BC} \subset \overrightarrow{AD}.$ b. $\overrightarrow{AD} \cap \overrightarrow{DB} = D.$
c. $\overrightarrow{AC} \cup \overrightarrow{BD} = \overrightarrow{AB}.$ d. $\overrightarrow{AC} \cap \overrightarrow{BD} = \phi.$
e. $\overrightarrow{DB} \cup \overrightarrow{CA} = \overrightarrow{BA}.$ f. $\overrightarrow{AB} \cup \overrightarrow{CD} = \overrightarrow{AD}.$
g. $(\overrightarrow{BD} \cap \overrightarrow{AC}) \subset \overrightarrow{CD}.$ h. $\overrightarrow{BD} \neq \overrightarrow{BC}.$
i. $(\overrightarrow{CB} \cap \overrightarrow{CD}) = (\overrightarrow{CE} \cap \overrightarrow{BC}).$ j. $(\overrightarrow{CD} \cap \overrightarrow{CB}) \cup \overrightarrow{BD} = \overleftrightarrow{AC}.$
8. Consider the diagram below:

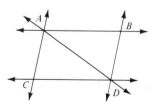

Determine each of the following.
a. $\overleftrightarrow{AD} \cap \overleftrightarrow{BD}$.

b. $(\overleftrightarrow{AC} \cap \overleftrightarrow{AD}) \cup \overline{AB}$.

c. $(\overline{AC} \cup \overline{CD}) \cap \overline{AD}$.

d. $(\overleftrightarrow{AD} \cap \overline{AB}) \cap (\overline{AD} \cup \overleftrightarrow{AC})$.

e. $(\overline{CD} \cup \overline{DB}) \cap \overleftrightarrow{AD}$.

f. $\overrightarrow{AC} \cap (\overrightarrow{DA} \cup \overrightarrow{BA})$.

g. $(\overline{AD} \cup \overline{AC}) \cap (\overrightarrow{AD} \cup \overrightarrow{DB})$.

h. $(\overline{AB} \cup \overline{CD}) \cap (\overline{AC} \cup \overline{BD})$.

9. Identify each of the following statements as being true or false. Give a reason for your answer.

a. A line segment contains a finite number of points.

b. A ray is a closed half-line containing both of its endpoints.

c. A ray can be a proper subset of a line segment.

d. A line segment is a proper subset of a line.

e. A line segment can be a proper subset of an open half-line.

f. A point on a line separates a line into three distinct subsets.

g. The union of the two closed half-lines determined by a point that separates a line is the line.

h. If $\overline{BC}$ is a proper subset of $\overrightarrow{CE}$, then B cannot be the endpoint of $\overrightarrow{CE}$.

i. A point can separate a line segment into two line segments.

j. A point can separate a ray into two rays.

13.2 LINES AND PLANES

As we learned in the previous section, a line is a set of points but the word *line* is left undefined. Likewise, a plane is a set of points but the word *plane* will be left undefined. However, if we try to imagine what a plane is like, we generally think of a set of points all lying on or in a flat surface such as the plane of the chalkboard or the plane of the table top and so forth. Although a chalkboard and a table top have thickness, a plane does not.

Recall that a line has no thickness or width but can be extended indefinitely in two opposite directions. Similarly, a plane has no thickness but can be extended indefinitely in all other directions. Hence the extension of the chalkboard or the table top could be our conception of a plane. We will denote planes by the use of Greek symbols such as α, β, and γ.

In the previous section we learned that any two distinct points will determine a unique line. How many distinct points are necessary to determine a unique plane? We answer this question with the following axiom.

AXIOM 13.2 There is one and only one plane containing three distinct and noncollinear points.

The above axiom guarantees that any three distinct and noncollinear points in space determine a unique plane and, also, that for any three distinct and noncollinear points in space there is exactly one plane containing them. We may represent a plane as illustrated in Figure 13.4 with the understanding that the plane may be extended indefinitely as indicated by the arrows.

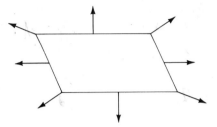

Figure 13.4

In Theorem 13.1 we proved that two distinct lines having a nonempty intersection have exactly one point in common. An analogous statement for the nonempty intersection of two distinct planes is given by the following axiom.

AXIOM 13.3 If two distinct planes have a nonempty intersection, then the intersection is a line.

AXIOM 13.4 If a line contains two distinct points of a plane, then the line lies in the plane (that is, the line is contained in the plane).

Another way of stating Axiom 13.4 is to say that if a line contains two distinct points of a plane, then the line is a subset of the plane.

Just as a point separates a line into three distinct subsets of the line, a line separates a plane into three disjoint subsets of the plane. These are the line itself and two disjoint subsets of points called *half-planes*. The union of a half-plane with the line that separates the plane is called a *closed half-plane*. The half-plane without the line of separation is called an *open half-plane*.

DEFINITION 13.4 If $\overleftrightarrow{AB}$ and $\overleftrightarrow{CD}$ are two distinct lines in the same plane, then $\overleftrightarrow{AB}$ and $\overleftrightarrow{CD}$ are said to be *parallel lines* if and only if $\overleftrightarrow{AB} \cap \overleftrightarrow{CD} = \phi$.

Recall that a ray is a proper subset of a line as is a line segment. In addition to referring to lines as being parallel, we may refer to rays or line segments as being parallel. When we say that two rays (or line segments) are parallel, we shall mean that the lines containing the rays (or line segments) are parallel.

It is possible that two lines in space may be disjoint and yet not be parallel as we will note in the following definition.

DEFINITION 13.5 If $\overleftrightarrow{AB}$ and $\overleftrightarrow{CD}$ are two distinct lines in space such that $\overleftrightarrow{AB}$ and $\overleftrightarrow{CD}$ are not parallel and such that $\overleftrightarrow{AB} \cap \overleftrightarrow{CD} = \phi$, then the two lines are said to be *skew lines*.

Clearly, from the preceding definition we note that skew lines cannot lie in the same plane.

DEFINITION 13.6 If α and β are two distinct planes in space such that $\alpha \cap \beta = \phi$, then α and β are said to be parallel *planes*.

We will onclude this section with some useful and interesting theorems stated without proof.

THEOREM 13.2   If l is a line and A is a point not on the line, then there is one and only one plane which contains l and A.

THEOREM 13.3   If two distinct lines have a nonempty intersection, then the two lines determine one and only one plane.

THEOREM 13.4   If a line intersects a plane not containing it, then the intersection set is a single point.

EXERCISES

Identify each of the following statements as being true or false and give a reason for your answer.
1. Two distinct lines in space either intersect or are parallel.
2. A ray separates a plane into two half-planes.
3. A closed half-plane is a half-plane together with the line that separates the plane.
4. Skew lines in space cannot be coplanar.
5. Three distinct points in space determine a unique plane.
6. If A, B, C, and D are four distinct and noncollinear points in space, then any three of the points must determine a unique plane.
7. If two distinct planes in space have a nonempty intersection set, the intersection set is a line.
8. If three distinct planes in space have a nonempty intersection set, the intersection set could be a point.
9. Any plane that contains two distinct points of a line must contain at least three distinct points of the line.
10. Any line and a point not on the line determine a unique plane in space.

13.3 ANGLES

In this section we will define and discuss the concept of an angle. As we will note, an angle will be a set of points. However, the reader may be tempted to think of an angle in terms of its measure such as $45°$ or $90°$ etc. We will not be concerned with the measure of an angle in this section.

DEFINITION 13.7 An *angle* is the union of two rays that have a common endpoint. The endpoint is called the *vertex* of the angle, and the rays are called its *sides*.

We note that an angle is a set of points and, as such, an angle is an abstraction. We will illustrate the concept of an angle, however, as in Figure 13.5, which represents the angle with vertex A and sides which are the rays $\overrightarrow{AB}$ and $\overrightarrow{AC}$. This angle may be denoted by the symbols $\angle BAC$ or $\angle CAB$. Using these symbols, note that the vertex symbol is written between the other two symbols, which represents points on the rays or sides of the angle.

If no ambiguity exists we also may denote an angle by using the symbol together with its vertex. Hence the angle illustrated in Figure 13.5 may be symbolized by $\angle A$.

A ray does not separate a plane but an angle does. An angle separates a plane into three disjoint subsets of the plane. These are the angle itself, a set of points in the plane called the *interior* of the angle, and a set of points in the plane called the *exterior* of the angle.

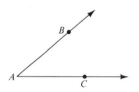

Figure 13.5

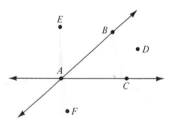

Figure 13.6

Now consider Figure 13.6. The rays $\overrightarrow{AB}$ and $\overrightarrow{AC}$ are the sides of $\angle BAC$. However, the ray $\overrightarrow{AB}$ is contained in the line $\overleftrightarrow{AB}$ which divides the plane into two half-planes. There is the half-plane determined by $\overleftrightarrow{AB}$ and in the direction of point C. Also, there is the half-plane determined by $\overleftrightarrow{AB}$ and in the direction of point E. Similarly, the ray $\overrightarrow{AC}$ is contained in the line $\overleftrightarrow{AC}$. The line $\overleftrightarrow{AC}$ separates the plane into two half-planes, the half-plane in the direction of E and the half-plane in the direction of F.

DEFINITION 13.8 The *interior* of $\angle BAC$, denoted by int($\angle BAC$), is the intersection of the half-plane determined by $\overleftrightarrow{AB}$ and in the direction of C together with the half-plane determined by $\overleftrightarrow{AC}$ and in the direction of B.

The interior of $\angle BAC$ is the shaded portion of the plane in Figure 13.7.

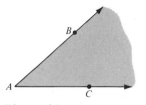

Figure 13.7

DEFINITION 13.9 The *exterior* of $\angle BAC$, denoted by ext($\angle BAC$), is the set of all the points in the plane containing $\angle BAC$ that are neither points of $\angle BAC$ nor the interior of $\angle BAC$.

The exterior of $\angle BAC$ is the shaded portion of the plane in Figure 13.8.

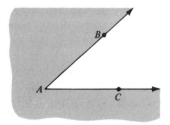

Figure 13.8

Next we will consider two distinct lines $\overleftrightarrow{AB}$ and $\overleftrightarrow{CD}$ which intersect at the point O as illustrated in Figure 13.9. Basically, there are four angles

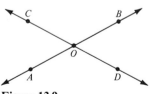

Figure 13.9

determined by these two intersecting lines: $\angle AOD$, $\angle BOD$, $\angle BOC$, and $\angle AOC$. All four angles have a common vertex, O. Also, $\angle AOC$ and $\angle AOD$ have a common ray $\overrightarrow{OA}$. Similarly, $\angle AOD$ and $\angle BOD$ have a common ray, $\overrightarrow{OD}$. However, $\angle AOC$ and $\angle BOD$ do not have a common ray.

DEFINITION 13.10 Two angles in the same plane which have a common vertex and a common ray are said to be *adjacent angles* if the intersection of their interiors is the empty set.

DEFINITION 13.11 Two angles in the same plane which are formed by intersecting lines and which are not adjacent angles are said to be *vertical angles*.

In Figure 13.9 $\angle AOD$ and $\angle AOC$ are adjacent angles as are $\angle BOD$ and $\angle AOD$.

Just as two lines intersect and form certain types of angles, two distinct planes may also intersect and form angles.

DEFINITION 13.12 A *dihedral angle* is the union of two distinct half-planes together with a common line.

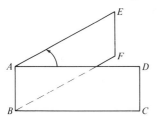

Figure 13.10

A dihedral angle is illustrated in Figure 13.10. It is to be noted that $\overleftrightarrow{AB}$ belongs to neither half-plane but is a part of the angle. The common line of a dihedral angle is also referred to as the *edge* of the angle.

We will conclude this section by stating an axiom and two theorems.

AXIOM 13.5 Let P and l be a point and a line in a plane such that the line l does not contain the point P. Then there is one and only one line in the plane which contains P and is parallel to l.

THEOREM 13.5 Let u and v be two distinct and parallel lines in a plane. Let w be another line contained in the same plane as u and v. Then if w intersects u, w intersects v.

Proof The proof follows immediately from Axiom 13.5 and is left as an exercise.

THEOREM 13.6 If a, b, and c are three distinct lines in a plane and if a is parallel to b and b is parallel to c, then a and c are parallel.

Proof Assume that a is not parallel to c. Then c intersects a. By Theorem 13.5, then, c intersects b. But this contradicts the hypothesis that b and c are parallel. Hence the assumption is false and a is parallel to c.

EXERCISES

1. Write two different names for each of the following angles indicated.

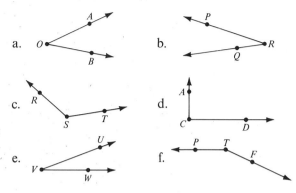

a. b.

c. d.

e. f.

2. For each of the following identify all vertical angles.

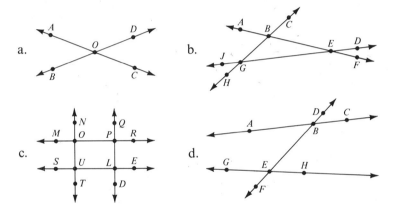

a. b.

c. d.

3. For each of the diagrams in Exercise 2 identify all adjacent angles.
4. Sketch two angles such that their intersection set is
   a. A line segment.
   b. A point.
   c. A ray.
   d. Exactly two points.
   e. Exactly three points.
   f. Exactly four points.
   g. A line segment and a point not on the line segment.
   h. A ray and a point not on the ray.
5. Sketch two angles such that
   a. The interiors of the angles are overlapping sets.
   b. The interior of one is a proper subset of the interior of the other.
   c. The interior of one is a proper subset of the exterior of the other.
   d. The interiors of the angles are disjoint sets.

e. The union of the two angles is a pair of intersecting lines.

f. The union of the two angles is a line and a ray.

g. The vertex of each angle lies in the exterior of the other.

h. The vertex of each angle lies in the interior of the other.

i. The angles have the same vertex and the interior of one intersection with the exterior of the other is the null set.

j. The angles have a ray and vertex in common but such that their interiors are disjoint.

6. Determine each of the following using the diagram below.

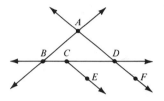

a. $\angle ABD \cap \overrightarrow{CE}$.

b. $\angle ABD \cap \angle ADB$.

c. $\overleftrightarrow{AB} \cap \angle CBA$.

d. $\angle BCE \cap \angle DCE$.

e. $(\overrightarrow{CB} \cap \overrightarrow{CD}) \cup \overrightarrow{CE}$.

f. $\angle ECD \cap \angle ADB$.

g. $\angle BCE \cap \angle FDC$.

h. $\angle BCE \cup \angle FDC$.

i. $(\overleftrightarrow{AF} - \overleftrightarrow{DF}) \cup \overrightarrow{DC}$.

j. $(\angle BCE - \overrightarrow{CE}) \cup \overrightarrow{CD}$.

7. Prove Theorem 13.5.

13.4 CURVES

In this section we will introduce and discuss the concept of a curve. Very precise definitions can be given for a curve, but we will give a very intuitive definition of a curve in terms of a set of points.

DEFINITION 13.13 A *curve* is defined to be a set of all points that can be represented by a pencil (or chalk) drawing without lifting the pencil (or chalk).

Clearly, then, a line segment is a curve. Two line segments with a common endpoint also would be a curve. The graphs of certain types of relations and functions, as were considered in Chapter 4, also are curves.

As a set of points a curve may be represented as a pencil drawing that passes through no point more than once, or it may pass through a single point several times. Also, some curves may start and end at the same point.

DEFINITION 13.14 A *simple curve* is a curve that passes through no point more than once.

DEFINITION 13.15 A *closed curve* is a curve represented by a drawing that begins and ends at the same point.

EXAMPLE The following drawings are examples of simple curves:

EXAMPLE The following drawings are examples of closed curves:

EXAMPLE The following drawings are examples of simple, closed curves:

An important property associated with a simple closed curve which lies in a plane is that it separates a plane into three disjoint subsets of the plane. These subsets are the curve itself and two disjoint subsets called the *interior* of the curve and the *exterior* of the curve. The *interior of a closed curve* is that set of points in the plane which is bounded by the curve and is also called a *bounded region* of the plane. The *exterior* of a closed curve is that set of points of the plane which are not contained on the curve itself or in the interior of the curve. In Figure 13.11 we illustrate a simple, closed curve in a plane and indicate its interior and exterior. Point A is a point in the interior of the curve, and point B is a point in the exterior of the curve. Observe that any curve in the same plane from point A to point B must intersect the closed curve.

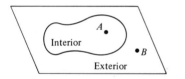

Figure 13.11

The above discussion involves a curve in a plane. Our definition for a curve does not, however, restrict the curve to a plane. We can have a curve in space such that not all of its points are in the same plane.

DEFINITION 13.16 A closed curve in a plane is said to be *convex* if and only if for any two points A and B in its interior, the line segment $\overline{AB}$ is contained completely in the interior of the curve (that is, $\overline{AB}$ is a subset of the interior of the closed curve).

EXAMPLE The following drawings are examples of convex closed curves, since for any two points A and B, $\overline{AB}$ is a subset of the interior of the curve.

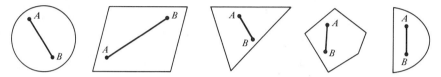

EXAMPLE The following drawings are not examples of convex closed curves since there exist points A and B in the interior of the curve such that $\overline{AB}$ is not a subset of the interior of the curve.

EXERCISES

1. Identify each of the following as being a curve or not being a curve.

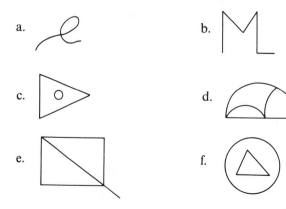

g.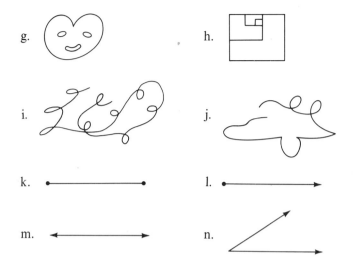

h.

i.

j.

k.

l.

m.

n.

2. Identify each of the following curves as being simple or not being simple.

a.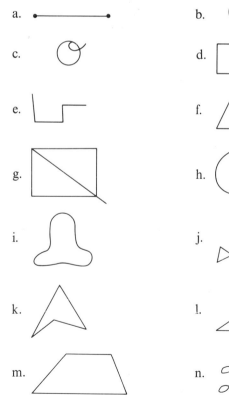

b.

c.

d.

e.

f.

g.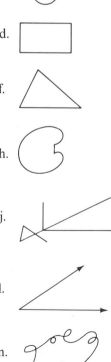

h.

i.

j.

k.

l.

m.

n.

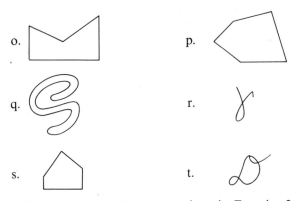

o.   p.

q.   r.

s.   t.

3. Identify each of the curves given in Exercise 2 as being closed or not being closed.

4. Identify each of the closed curves determined in Exercise 3 as being convex or not being convex.

5. Identify each of the following statements as being true or false.

a. Every closed curve is simple.

b. The union of two line segments is always an example of a simple curve.

c. The interior of a circular region is a bounded region.

d. The interior of an angle is a bounded region.

e. The union of three line segments can be a simple closed curve.

f. If A and B are two distinct points in the interior of a simple closed curve that is not convex, then $\overline{AB}$ cannot be a subset of the interior of the curve.

g. If C and D are two distinct points in the interior of a simple closed curve such that $\overline{CD}$ does not lie completely in the interior of the curve, then infinitely many points of $\overline{CD}$ must lie in the exterior of the curve.

h. Every capital letter of the English alphabet (written in printed form) is an example of a simple curve.

13.5 POLYGONS

In the examination of the curves illustrated in the examples of the previous section, we observe that some curves contain line segments and others do not. Further, some of those curves are made up entirely of line segments while others consist of line segments only as part of the curve. In this section we will be concerned with those curves which are made up entirely of line segments; such curves are called broken lines.

DEFINITION 13.17 A curve is said to be a *broken line* if and only if it is the union of line segments, no more than two of which have a common endpoint and such that not all of the line segments are subsets of the same line.

EXAMPLE The following drawings are examples of curves that are broken lines:

EXAMPLE The following drawings are examples of curves none of which are broken lines:

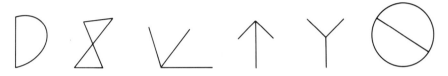

A broken line by definition is basically a curve. However, the curve may be simple or closed, neither or both. We will be especially interested in special types of simple, closed curves that are broken lines.

DEFINITION 13.18 A *polygon* is defined to be a simple, closed broken line. Each line segment of the broken line is called a *side* of the polygon. The common endpoints of the sides of the polygon are called its *vertices*.

Since a polygon is defined in terms of a broken line and since a broken line is the union of line segments, then a polygon can have three or more sides. Also, as a special type of a simple closed curve, a polygon may or may not be convex. Polygons, in general, are named by their vertices (plural of vertex). For instance, in Figure 13.12 the first polygon may be called polygon *ABCD* and the second polygon may be called polygon *EFGHIJ*.

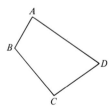 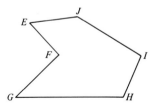

Figure 13.12

DEFINITION 13.19 A polygon having exactly three sides is called a *triangle*. If *A*, *B*, and *C* are its vertices, the triangle is denoted by △ *ABC* (read "triangle *A*, *B*, *C*").

Consider the various triangles in Figure 13.13 with vertices A, B, and C. The sides of each triangle are $\overline{AB}$, $\overline{BC}$, and $\overline{AC}$. Also, $\triangle ABC = \overline{AB} \cup \overline{BC} \cup \overline{AC}$. In addition, to the sides of a triangle, we also refer to

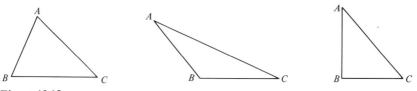

Figure 13.13

the angles of a triangle. For instance, $\angle ABC$, $\angle BAC$, $\angle ACB$ are called the angles of each of the triangles illustrated in Figure 13.13. Each side of a triangle is a subset of the triangle, but no angle of a triangle is a subset of the triangle. An angle is the union of two rays, but there are no rays in a triangle.

DEFINITION 13.20 A polygon having exactly four sides is called a *quadrilateral*. If A, B, C, and D are its vertices, then the quadrilateral is denoted by quad $ABCD$ (read "quadrilateral A, B, C, D").

There are many types of quadrilaterals such as parallelograms, rectangles, and squares which will be discussed later in this chapter. To distinguish one of these from the others would require a knowledge of the concept of measure which we will discuss in the next section.

In addition to triangles and quadrilaterals, we have other special cases of polygons. For instance, a polygon with exactly five sides is called a *pentagon*; a polygon with exactly six sides is called an *hexagon*; a polygon with exactly seven sides is called an *heptagon*; a polygon with exactly eight sides is called an *octagon*; a polygon with exactly nine sides is called a *nonagon*; a polygon with exactly ten sides is called a *decagon*; a polygon with exactly twelve sides is called a *dodecagon*; and names can be given to other polygons.

In this chapter we will be interested primarily in polygons that are triangles or quadrilaterals. Also, we will be interested in the interiors of these polygons.

DEFINITION 13.21 The interior of a polygon is a bounded region called a *polygonal region*. The polygon itself is called the *boundary* of the polygonal region. The union of the polygonal region together with its boundary is called a *closed bounded polygonal region*.

EXERCISES

1. Identify each of the following diagrams as being a broken line or not being a broken line.

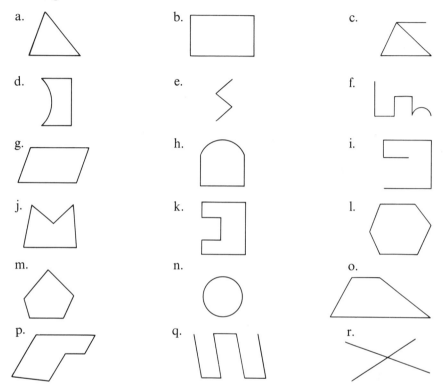

a.   b.   c.

d.   e.   f.

g.   h.   i.

j.   k.   l.

m.   n.   o.

p.   q.   r.

2. Which of the diagrams in Exercise 1 represent polygons?

3. Which of the polygons identified in Exercise 2 are convex?

4. What is the least number of sides that a polygon may have?

5. What is the least number of sides that a convex polygon may have?

6. Can the intersection set of two triangles be exactly one point? If so, sketch an appropriate diagram.

7. Can the intersection set of two quadrilaterals be exactly two points? If so, sketch an appropriate diagram.

8. Can the intersection set of two polygons be a line segment? If so, sketch an appropriate diagram.

9. Is it possible to sketch two triangles such that the interior of one is a proper subset of the interior of the other? If so, sketch two such triangles.

10. Does a triangle separate a plane? If so, into how many parts? What are they?

11. Does every polygon separate a plane? If so, into how many parts? What are they?

12. Is it possible for a ray to be a subset of the interior of a polygon? Explain.
13. If the interior of one polygon is a proper subset of the interior of another polygon, is the intersection set of the two interiors bounded by a polygon? Explain.
14. Sketch if possible a triangle that is not convex.
15. Sketch if possible a pentagon that is not convex.

13.6 CONGRUENCE AND MEASURE OF LINE SEGMENTS AND ANGLES

Suppose that we consider two points A and B on the real number line corresponding to the real numbers 2 and 5, respectively. We may conclude that the line segment from A to B, then, is 3 units long. If the points C and D also are placed on the real number line corresponding to the real numbers 6 and 11, respectively, we may conclude that the line segment from C to D, then, is 5 units long. Further, we may conclude that $\overline{CD}$ is longer than $\overline{AB}$. The real numbers 3 and 5 associated with the length of the line segments $\overline{AB}$ and $\overline{CD}$ are called the *measures* of these line segments.

DEFINITION 13.22 If A and B are any two distinct points on the real number line corresponding to the real numbers a and b, respectively, with $a < b$, then the *measure* of $\overline{AB}$, denoted by $m(\overline{AB})$, is the real number $b - a$. Since $\overline{BA} = \overline{AB}$, then the measure of $\overline{BA}$, denoted by $m(\overline{BA})$, is also the real number $b - a$.

EXAMPLE Consider points located on the real line as indicated below:

$$
\begin{array}{ccccccccccccc}
A & B & C & D & E & F & G & H & I & J & K & L \\
\bullet & \bullet & \bullet & \bullet & \bullet & \bullet & \bullet & \bullet & \bullet & \bullet & \bullet & \bullet \\
^-5 & ^-4 & ^-3 & ^-2 & ^-1 & 0 & 1 & 2 & 3 & 4 & 5 & 6
\end{array}
$$

a. $m(\overline{DH}) = 2 - (^-2) = 4$.
b. $m(\overline{BJ}) = 4 - (^-4) = 8$.
c. $m(\overline{HK}) = 5 - 2 = 3$.
d. $m(\overline{AD}) = {}^-2 - (^-5) = 3$.
e. $m(\overline{GD}) = 1 - (^-2) = 3$.
f. $m(\overline{FL}) = 6 - 0 = 6$.
g. $m(\overline{CF}) = 0 - (^-3) = 3$.

In the above example we note that $\overline{AD}$ and $\overline{CF}$ have the same measure. Clearly, $\overline{AD} \neq \overline{CF}$ since a line segment is a set of points and in order for the two line segments to be equal they must be the same set of points. However, two line segments that have equal measures are said to be *congruent*.

DEFINITION 13.23 Two line segments $\overline{AB}$ and $\overline{CD}$ are said to be *congruent*, denoted by $\overline{AB} \simeq \overline{CD}$, if and only if $m(\overline{AB}) = m(\overline{CD})$.

Congruence of line segments is a special case of congruence of geometric figures, in general. Intuitively, two geometric figures are congruent if one figure could be superimposed upon the other with exact fit. That is, the two figures would have the identical shape and size.

THEOREM 13.7 The relation "is congruent to" on the set of line segments is an equivalence relation.

Proof The proof of this theorem depends upon Definition 13.23 and the fact that the relation "is equal to" is an equivalence relation and is left as an exercise.

We defined an angle to be the union of two rays with a common endpoint. Just as line segments can be congruent, we can also consider the congruence of angles. Basically, two angles are congruent if one of them can be super-imposed upon the other with exact fit. That is, if the two angles have the identical shape and size. To illustrate our intuitive meaning of congruence of angles, consider the angles $\angle BAC$ and $\angle EDF$ given in Figure 13.14.

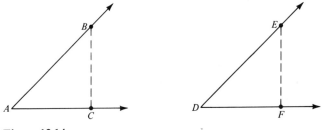

Figure 13.14

Suppose that $\overline{AB} \simeq \overline{DE}$ and $\overline{AC} \simeq \overline{DF}$. Then in order for the two angles to have the same size or measure, it would be necessary that $\overline{BC} \simeq \overline{EF}$. If $\angle BAC$ is congruent to $\angle EDF$, we symbolize this relation as $\angle BAC \simeq \angle EDF$. Then the measures of these two angles would also be equal, and we symbolize this relation as $m(\angle BAC) = m(\angle EDF)$.

The unit of measure most generally given to the measure of an angle is the *degree* that is usually denoted by a small circular superscript. For instance, if A is an angle whose measure is 42 degrees, we would write $m(\angle A) = 42°$. How large is an angle whose measure is $1°$? To answer this question consider an angle whose rays lie on the same line such as $\angle AOB$ illustrated in Figure 13.15. The measure of $\angle AOB$ is $180°$ and $\angle AOB$ is called a *straight angle*. Now if we could construct 180 congruent adjacent angles all with a vertex at the point O and such that the outside rays on the first and last of these 180 angles are the rays $\overrightarrow{OA}$ and $\overrightarrow{OB}$, then each one of those 180 angles would have a measure of $1°$.

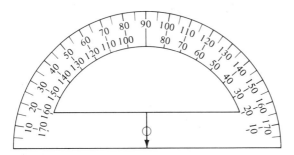

Figure 13.15

The measure of an angle also can be given in what is called *radian* measure. However, in this text we will use the degree measure exclusively. As an aid in measuring angles in degrees, we use an instrument called a *protractor* as illustrated in Figure 13.16.

Figure 13.16

Now consider the angle ∠PRQ given in Figure 13.17. If the point of the arrow on the base of the protractor is placed at the vertex R of the angle and the edge of the protractor is placed along the ray $\overrightarrow{RQ}$, then the measurement of the angle can be read by noting where the protractor intersects the other ray of the angle. We now read the measure of ∠PRQ to be 55°.

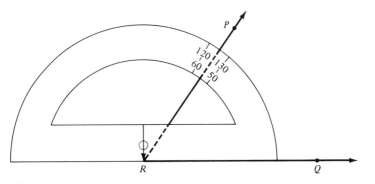

Figure 13.17

Although the measure of a line segment is always positive, the measure of an angle may be positive, zero, or negative. Further, there is no upper bound or lower bound on the possible measurements of an angle. However, in this chapter we shall consider only angles such that for any angle ∠ABC, $0° \le m(\angle ABC) \le 180°$.

We will now classify angles according to their measures.

DEFINITION 13.24 An angle whose measure is 180° is called a *straight angle*. An angle whose measure is 90° is called a *right angle*. An angle whose measure in degrees is greater than or equal to zero but less than 90 is called an *acute angle*. An angle whose measure in degrees is greater than 90 and less than 180 is called an *obtuse angle*.

EXAMPLE Classify each of the following angles in accordance with Definition 13.24:
  a. $\angle A$, if $m(\angle A) = 42°$.
  b. $\angle B$, if $m(\angle B) = 90°$.
  c. $\angle C$, if $m(\angle C) = 102°$.
  d. $\angle D$, if $m(\angle D) = 179°$.
  e. $\angle E$, if $m(\angle E) = 180°$.
  f. $\angle F$, if $m(\angle F) = 89°$.

Solution
  a. $\angle A$ is an acute angle since $0 < 42 < 90$.
  b. $\angle B$ is a right angle.
  c. $\angle C$ is an obtuse angle since $90 < 102 < 180$.
  d. $\angle D$ is an obtuse angle since $90 < 179 < 180$.
  e. $\angle E$ is a straight angle.
  f. $\angle F$ is an acute angle since $0 < 89 < 90$. ■

The measure of an angle may be given in degrees or in parts of a degree. If we divide the unit of 1° into 60 equal parts, then each part is called a *minute* and is denoted by ′. Likewise, if we divide a minute into 60 equal parts, each part is called a *second* and is denoted by ″. For instance, 43°19′ is read "43 degrees, 19 minutes." Similarly, 86°17′56″ is read "86 degrees, 17 minutes, 56 seconds."

EXAMPLE Express 23°12′ in terms of degrees.
Solution Since $1° = 60′$, then $1′ = (\frac{1}{60})°$ and $12′ = 12(\frac{1}{60})° = 0.2°$. Hence $23°12′ = 23.2°$. ■

EXAMPLE Express 123.8° in terms of degrees and minutes.
Solution Since $0.8° = 0.8(60)′ = 48′$, then $123.8° = 123°48′$. ■

In addition to classifying certain angles according to their measures, we may also classify pairs of angles.

DEFINITION 13.25 If two angles have measures whose sum is 90°, then the angles are said to be *complementary*. If two angles have measures whose sum is 180°, then the angles are said to be *supplementary*.

DEFINITION 13.26 Two lines in the same plane are said to be *perpendicular lines* if and only if they intersect and form adjacent angles that are right angles.

Now that we have introduced the measure of a line segment and the congruence of line segments and angles, we can return to the discussion of quadrilaterals and consider their classifications.

DEFINITION 13.27 Certain quadrilaterals are defined as follows:

1. A quadrilateral whose opposite sides lie on parallel lines is called a *parallelogram*. That is, a parallelogram is a quadrilateral both of whose pairs of opposite sides are parallel.
2. A quadrilateral having exactly one pair of opposite sides parallel is called a *trapezoid*.
3. A parallelogram having one right angle is called a *rectangle*.
4. A rectangle whose adjacent sides are congruent is called a *square*.
5. A quadrilateral all of whose sides are congruent is called a *rhombus*.

Relative to a parallelogram, we state the following properties:

1. A parallelogram has two pairs of parallel sides.
2. The opposite sides of a parallelogram are congruent.
3. The opposite angles of a parallelogram are congruent.
4. The adjacent angles of a parallelogram are supplementary.

We defined a rectangle as being a parallelogram with one right angle. From the properties of a parallelogram given above it is easy to prove the following theorem.

THEOREM 13.8   A rectangle has exactly four right angles.
Proof   The proof is left as an exercise.

We will conclude this section by noting that the sum of the measures of the sides of a polygon is called its *perimeter*.

EXERCISES

1. Consider points located on the real number line as indicated below:

Determine each of the following:

a. $m(\overline{CK})$.

b. $m(\overline{KP})$.

c. $m(\overline{HN})$.

d. $m(\overline{EL})$.

e. $m(\overline{GB})$.

f. $m(\overline{NJ})$.

g. $m(\overline{AI})$.

h. $m(\overline{MD})$.

i. $m(\overline{FO})$.

j. $m(\overline{BE} \cup \overline{MP})$.

k. $m(\overline{HB} \cup \overline{KN})$.

l. $m(\overline{CP} \cap \overline{AK})$.

m. $m(\overline{LB} \cap \overline{FN})$.

n. $m(\overline{DJ}) + m(\overline{FM})$.

o. $m(\overline{CL}) - m(\overline{JO})$.

2. Using the diagram in Exercise 1, determine two distinct values for X such that each of the following will be true.

a. $m(\overline{GX}) = 3$.

b. $m(\overline{JX}) = 4$.

c. $m(\overline{FX}) = 5$.

d. $m(\overline{XI}) = 2$.

e. $m(\overline{EX}) + m(\overline{JX}) = 5$.

f. $m(\overline{XD}) + m(\overline{XK}) = 1$.

3. Using the diagram in Exercise 1, determine which of the following statements are true and which are false. Give an appropriate reason for your answer.

a. $\overline{FJ} \simeq \overline{AE}$.

b. $\overline{DH} \simeq \overline{KH}$.

c. $\overline{EK} = \overline{HN}$.

d. $\overline{BG} \simeq \overline{IF}$.

e. $\overline{CK} > \overline{LP}$.

f. $m(\overline{AG}) > m(\overline{OP})$.

g. $m(\overline{JD}) - m(\overline{EH}) = 3$.

h. $\overline{DF} \cup \overline{FL} = \overline{LD}$.

i. $m(\overline{BH} \cup \overline{FK}) = m(\overline{BH}) + m(\overline{FK})$.

j. $m(\overline{CH} - \overline{FJ}) = m(\overline{CH}) - m(\overline{FJ})$.

4. Classify each of the following angles as being a straight angle, a right angle, an acute angle, or an obtuse angle.

a. $\angle A$, if $m(\angle A) = 36°$.

b. $\angle B$, if $m(\angle B) = 92°$.

c. $\angle C$, if $m(\angle C) = 163°$.

d. $\angle D$, if $m(\angle D) = 180°$.

e. $\angle E$, if $m(\angle E) = 90°$.

f. $\angle F$, if $m(\angle F) = 18°$.

g. $\angle G$, if $m(\angle G) = 16°23'$.

h. $\angle H$, if $m(\angle H) = 141°19'$.

i. $\angle I$, if $m(\angle I) = 0°1'1''$.

j. $\angle J$, if $m(\angle J) = 179°59'23''$.

5. Rewrite each of the following in degrees only.

a. $22°12'$.

b. $39°6'$.

c. $56°24'$.

d. $102°30'$.

e. $68°42'$.

f. $100°54'$.

g. $86°36'$.

h. $57°18'$.

i. $179°60'$.

6. Express each of the following in terms of degrees and minutes.

a. $32.6°$.

b. $47.3°$.

c. $117.25°$.

d. $79.35°$.

e. $18.9°$.

f. $17.15°$.

7. Identify each of the following as being true or false and give an appropriate reason for your answers.
   a. Two line segments that have the same measure are equal.
   b. Equal line segments are congruent.
   c. Congruent angles have the same measures.
   d. The opposite sides of a parallelogram are both parallel and equal.
   e. The opposite sides of a rectangle are both parallel and congruent.
   f. A square is a parallelogram with all four sides being congruent.
   g. If two angles are complementary, then the sum of their measures is 90°.
   h. If two angles are supplementary, then their union is a straight line.
   i. If two lines in the same plane are perpendicular, they intersect to form right angles.
   j. A rhombus is a special case of a square.
8. Prove that the relation "is congruent to" on the set of line segments is an equivalence relation.
9. Prove that a rectangle has exactly four right angles.
10. Consider a polygon $ABCDE$ with $m(\overline{AB}) = 2$, $m(\overline{BC}) = 4$, $m(\overline{CD}) = 3$, $m(\overline{DE}) = 4$, and $m(\overline{EA}) = 5$. Determine the perimeter of the polygon.

13.7 CONGRUENCE OF TRIANGLES

If two triangles are congruent, then if one triangle is superimposed upon the other there should be a perfect fit. That is, the two triangles would be identical in shape and size. Intuitively it follows that if the corresponding vertices of the two triangles were paired, the corresponding sides should be congruent and the corresponding angles also should be congruent.

DEFINITION 13.28 Two triangles $\triangle ABC$ and $\triangle DEF$ are said to be *congruent*, denoted by $\triangle ABC \simeq \triangle DEF$, if and only if the vertices A, B, and C of $\triangle ABC$ are paired, respectively, with the vertices D, E, and F of $\triangle DEF$ such that $\overline{AB} \simeq \overline{DE}$, $\overline{AC} \simeq \overline{DF}$, $\overline{BC} \simeq \overline{EF}$, $\angle ABC \simeq \angle DEF$, $\angle BCA \simeq \angle EFD$, and $\angle CAB \simeq \angle FDE$.

From Definition 13.28 we conclude that two triangles are congruent if and only if the three sides of one triangle are congruent, respectively, to the three sides of the other and the three angles of one triangle are congruent, respectively, to the three angles of the other. Hence it appears that there are six congruence relations that may be satisfied. However, it is not necessary to check out all six of these congruence relations, as we will note in the following basic axiom for congruence of triangles.

AXIOM 13.6 If two sides and the included angle of one triangle are congruent, respectively, to two sides and the included angle of another triangle, then the two triangles are congruent.

We also can prove that two triangles are congruent in accordance with the following two theorems which will be stated without proof.

THEOREM 13.9   If two angles and the included side of one triangle are congruent, respectively, to two angles and the included side of another triangle, then the two triangles are congruent.

THEOREM 13.10   If three sides of one triangle are congruent, respectively, to three sides of another triangle, then the triangles are congruent.

Just as the relation "is congruent to" on the set of line segments is an equivalence relation, the relation "is congruent to" on the set of triangles also is an equivalence relation.

THEOREM 13.11   The relation "is congruent to" on the set of triangles is an equivalence relation.

Proof   The proof of this theorem follows immediately from the definition of congruence of triangles and the fact that the relation "is equal to" is an equivalence relation and is left as an exercise.

Triangles can be classified as to the number of sides that are congruent.

DEFINITION 13.29 If $\triangle ABC$ is an arbitrary triangle, then $\triangle ABC$ is
1. An *isosceles triangle*, if exactly two of its sides are congruent.
2. An *equilateral triangle*, if all three of its sides are congruent.
3. A *scalene triangle*, if no sides are congruent.

Triangles also can be classified as to their angles.

DEFINITION 13.30 If $\triangle ABC$ is an arbitrary triangle, then $\triangle ABC$ is
1. A *right triangle*, if one of its angles is a right angle.
2. An *acute triangle*, if all three of its angles are acute angles.
3. An *obtuse triangle*, if one of its angles is an obtuse angle.
4. An *equiangular triangle*, if all three of its angles are congruent.

Relative to Definition 13.30 we make the following observations:
1. There can be *at most one* right angle in a triangle.
2. There can be *at most one* obtuse angle in a triangle.
3. If a triangle has a right angle, the other two angles are acute angles.
4. If a triangle has an obtuse angle, the other two angles are acute angles.
5. If a triangle is equiangular, the measure of each of its angles is 60°.

DEFINITION 13.31 In a right triangle the side opposite the right angle is called the *hypothenuse* and the other two sides are called its *legs*.

Since right triangles are special kinds of triangles, it should not be surprising that we also give special attention to the congruence of right triangles. The various cases of such congruence are given in the following theorem which will be stated without proof.

THEOREM 13.12 Two right triangles are congruent if any one of the following conditions is satisfied:
a. Two legs of one are congruent to the corresponding two legs of the other.
b. The hypothenuse and leg of one are congruent to the hypothenuse and corresponding leg of the other.
c. The hypothenuse and an acute angle of one are congruent to the hypothenuse and corresponding acute angle of the other.
d. A leg and an acute angle of one are congruent to the corresponding leg and acute angle of the other.

EXERCISES

1. Using the given information, classify each of the triangles $\triangle ABC$ and $\triangle DEF$ as being congruent or not being congruent. Give an appropriate reason for your answer. For those which are not congruent, sketch an appropriate diagram to substantiate your answer.
   a. $\overline{AB} \simeq \overline{DE}$, $\overline{BC} \simeq \overline{EF}$, and $\angle ABC \simeq \angle DEF$.
   b. $\overline{AB} \simeq \overline{EF}$, $\overline{BC} \simeq \overline{FE}$, and $\overline{AC} \simeq \overline{DF}$.
   c. $\angle ABC \simeq \angle DFE$, $\overline{AB} \simeq \overline{DF}$, and $\overline{BC} \simeq \overline{EF}$.
   d. $m(\angle ABC) = m(\angle DEF) = 90°$, $\overline{AC} \simeq \overline{DF}$.
   e. $m(\overline{AB}) = m(\overline{DE})$, $m(\angle ABC) = m(\angle DEF)$, $\angle ACB \simeq \angle DFE$.
   f. B and E are vertices of right angles, $m(\overline{AB}) = m(\overline{ED})$.
   g. $\angle ABC \simeq \angle DEF$, $\angle BCA \simeq \angle EFD$, $\angle CAB \simeq \angle FDE$.
   h. $AB \simeq \overline{DE}$, $m(\overline{AC}) = m(\overline{BC}) + 2$, $m(\overline{DF}) = m(\overline{EF}) + 2$.
2. Using the given information, classify each of the triangles $\triangle ABC$ indicated below as being isosceles, equilateral, or scalene.
   a. $\overline{AB} \simeq \overline{AC}$, $m(\overline{BC}) < m(\overline{AB})$.

b. $m(\overline{AB}) = m(\overline{BC})$, $m(\angle ABC) = 90°$.
c. $m(\overline{AC}) > m(\overline{BC})$, $m(\angle ACB) = 90°$.
d. $m(\overline{AB}) < m(\overline{BC}) < m(\overline{AC})$.
e. $\overline{AB} \simeq \overline{AC}$, $m(\overline{BC}) = m(\overline{AB})$.
f. $\angle ABC \simeq \angle BCA$, $\angle BCA \simeq \angle BAC$.

3. Using the given information, classify each of the triangles $\triangle ABC$ indicated below as being right, acute, obtuse, or equiangular.
   a. $m(\angle ABC) = m(\angle BCA)$, $m(\angle BAC) = 70°$.
   b. $m(\angle BAC) = 108°$.
   c. $m(\angle ABC) = m(\angle CAB) = 60°$.
   d. $m(\angle BAC) = 72°$, $\overline{AB} \simeq \overline{AC}$.
   e. $\angle ABC \simeq \angle ACB$, $m(\angle ABC) = 44°$.
   f. $\triangle ABC$ is an equilateral triangle.
   g. $\triangle ABC \simeq \triangle DEF$, $\angle DEF$ is a right angle.
   h. $\triangle ABC \simeq \triangle PQR$, $\triangle PQR$ is an isosceles triangle with $\overline{PQ} \simeq \overline{QR}$, $m(\angle QPR) = 16°$.

4. Identify each of the following statements as being true or false. Give an appropriate reason for your answers.
   a. If the two acute angles of one right triangle are congruent, respectively, to the two acute angles of another triangle, then the two triangles are congruent.
   b. If an angle of a triangle has a measure of $2°$, then the other two angles can both be obtuse angles.
   c. If two angles of a triangle are both acute angles, then the third angle also must be an acute angle.
   d. It is not possible for a triangle to have more than one obtuse angle.
   e. An isosceles triangle also can be an equiangular triangle.
   f. A scalene triangle can have an obtuse angle.
   g. A scalene triangle can have a right angle.
   h. A right triangle cannot be an isosceles triangle.
   i. The relation "is congruent to" on the set of triangles is a reflexive relation.
   j. The relation "is congruent to" on the set of triangles is a symmetric relation.

5. Prove that the relation "is congruent to" on the set of triangles is an equivalence relation.

13.8 AREA OF POLYGONAL REGIONS

We have defined triangles and quadrilaterals as special types of polygons and also have defined the interiors of these polygons to be polygonal regions. Both the polygon and its interior are sets of points, but these are disjoint sets of points and should be treated as such.

We frequently hear reference being made to the "area of a rectangle" or the "area of a triangle" when, in fact, what is probably meant is the "area of the rectangular region" or the "area of the triangular region." A rectangle is the union of four line segments, and each line segment has length. Hence the rectangle has length. But the line segments forming the rectangle have no width and, hence, the rectangle has no width. We conclude, then, that the area of the rectangle must be zero. Likewise, the area of a triangle must be zero and, in general, the area of any polygon must be zero.

DEFINITION 13.32 By the word *area* we shall mean the measure of the region in a plane that is the interior of a closed curve.

Now consider a square such that each side has a measure of 1 unit, as illustrated in Figure 13.18. We shall define this square to be a *unit square*

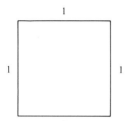

Figure 13.18

such that its interior has a measure that is 1 *square unit.* If the unit of measurement is in feet, then the unit square would have 4 sides of measure 1 foot each and the measure of its interior would be 1 square foot. If the unit of measurement is in centimeters, then the unit square would have 4 sides of measure 1 centimeter each and the measure of its interior would be 1 square centimeter.

We will now attempt to determine the area or the measure of the interior of a rectangle such that two adjacent sides have measures of 4 and 6 units, as illustrated in Figure 13.19. To determine the area of this rectangular

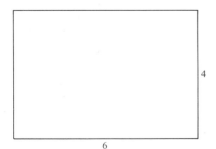

Figure 13.19

region, we would have to determine the number of unit squares contained in the region or the number of unit squares that would be necessary to completely cover the region. To determine the number, we can divide the side of the rectangle whose length is 4 units into 4 congruent line segments, divide the side of the rectangle whose length is 6 units into 6 congruent line segments, and then divide the interior of the rectangle as illustrated in Figure 13.20. Each of the subdivisions of the region is a unit square.

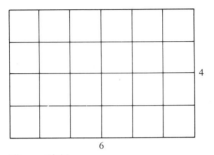

4

6

Figure 13.20

Counting, we determine that there are 24 such unit squares, and since each unit square has an area of 1 square unit, the rectangular region in question will have an area of 24 square units. Observe that $24 = 6 \times 4$ and the area of the rectangular region could have been determined by considering the product of the measures of the two adjacent sides of the rectangle. This procedure would be valid even if the measures of the adjacent sides are not whole numbers.

DEFINITION 13.33 Consider the rectangle $ABCD$ such that $m(\overline{AB}) = b$ and $m(\overline{BC}) = h$ (see Figure 13.21). Then the area A of the rectangular region is given by the formula $A = bh$, provided that b and h are given in the same units of measurement. The area A is given in square units.

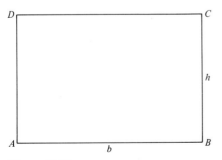

Figure 13.21

DEFINITION 13.34 If the rectangle $ABCD$ is a square such that the measure of each side is given by s, then the area of the square region is given by $A = s^2$ in square units.

EXAMPLE Determine the area of the rectangular region that measures 9 ft on one side and 4 ft on the other side.
Solution $A = (9 \text{ ft})(4 \text{ ft}) = 36$ sq ft. ∎

EXAMPLE Determine the area of the rectangular region that measures $4\frac{1}{2}$ cm on one side and 18 cm on the other side.
Solution $A = (4\frac{1}{2} \text{ cm})(18 \text{ cm}) = 81$ sq cm. ∎

EXAMPLE Determine the area of the rectangular region that measures 9 ft on one side and 4 yd on the other side.
Solution The units of measurement for the two sides are different and, hence, we must get them in terms of the same unit.
a. If we convert 9 ft to 3 yd, then $A = (3 \text{ yd})(4 \text{ yd}) = 12$ sq yd.
b. If we convert 4 yd to 12 ft, then $A = (9 \text{ ft})(12 \text{ ft}) = 108$ sq ft.
Either answer is acceptable. ∎

EXAMPLE Determine the area of the square region whose sides are 3.5 units long.
Solution $A = (3.5 \text{ units})^2 = 12.25$ sq units. ∎

Now consider the region bounded by the parallelogram $ABCD$ in Figure 13.22. If we construct a perpendicular line segment $\overline{DE}$ from point D to line segment $\overline{AB}$, we would form a right triangle $\triangle ADE$. Further, if we

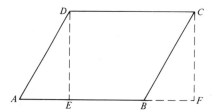

Figure 13.22

remove the region bounded by $\triangle ADE$ from one end of the parallelogram and attach it to the other end, represented by $\triangle BFC$, then the parallelogram $ABCD$ would be transformed into the rectangle $DEFC$. We also would observe that the area of the interior of the rectangle formed is equal to the area of the interior of the given parallelogram. But the area of the rectangular region is $m(\overline{DC}) \cdot m(\overline{DE})$ and, therefore, the area of the interior of the parallelogram $ABCD$ is also $m(\overline{DC}) \cdot m(\overline{DE})$.

The line segment $\overline{DE}$ constructed from the vertex D of the parallelogram $ABCD$ perpendicular to the side $\overline{AB}$ is called an *altitude* of the parallelogram; the side $\overline{AB}$ is called a *base* of the parallelogram. Since $\overline{AB} \simeq \overline{DC}$, $m(\overline{DC}) = m(\overline{AB})$. Hence the area of the interior of parallelogram $ABCD$ is equal to the product $m(\overline{AB}) \cdot m(\overline{DE})$.

DEFINITION 13.35 Consider the parallelogram $ABCD$ with altitude $\overline{DE}$ (and base $\overline{AB}$ such that $m(\overline{AB}) = b$) and $m(\overline{DE}) = h$ (see Figure 13.23). The area, A, of the interior of the parallelogram $ABCD$ is given by the formula $A = bh$, provided b and h are given in the same units of measurement. The area, A, is given in square units.

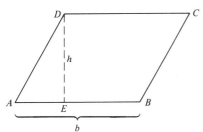

Figure 13.23

EXAMPLE Determine the area of the interior of the parallelogram that has a base whose measure is 9 in. and an altitude whose measure is 6 in.

Solution $A = (9 \text{ in.})(6 \text{ in.}) = 54 \text{ sq in,}$ ∎

EXAMPLE Determine the area of the interior of the parallelogram that has a base whose measure is 2 cm and an altitude whose measure is 3 mm.

Solution   Since the measurements of the base and altitude are given in different units, we must first convert them to the same unit of measurement.

a. If we convert 2 cm to 20 mm, then $A = (20 \text{ mm})(3 \text{ mm}) = 60 \text{ sq mm}$.

b. If we convert 3 mm to 0.3 cm, then $A = (2 \text{ cm})(0.3 \text{ cm}) = 0.6 \text{ sq cm}$.

Either answer is acceptable.   ∎

Now let us consider the area of a triangular region. We can start with the parallelogram $ABCD$ with altitude $\overline{AE}$, as in Figure 13.24, and consider the line segment $\overline{AC}$, called a *diagonal* of the parallelogram. Since $ABCD$ is a parallelogram, we have $\overline{AD} \simeq \overline{BC}$, $\overline{AB} \simeq \overline{CD}$ and $\angle ADC \simeq \angle ABC$. Hence $\triangle ABC \simeq \triangle ADC$ and, therefore, the measure of the region bounded by $\triangle ABC$ must be equal to the measure of the region bounded by $\triangle ADC$. Hence the measure of the region bounded by $\triangle ABC$ is equal to one-half the measure of the region bounded by the parallelogram $ABCD$. Again, if we let $m(\overline{BC}) = b$ and $m(\overline{AE}) = h$, the area of the interior of the parallelogram $ABCD$ is equal to bh and the area of the interior of $\triangle ABC = \frac{1}{2}bh$.

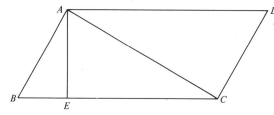

Figure 13.24

DEFINITION 13.36 Consider the triangle ABC with altitude AD such that $m(\overline{BC}) = b$ and $m(AD) = h$ (see Figure 13.25). The *area* A of the interior of triangle ABC is given by the formula $A = \frac{1}{2}bh$, provided b and h are given in the same units of measurement. The area, A, is given in square units.

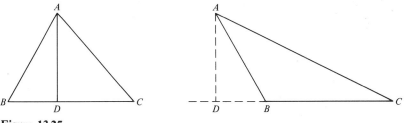

Figure 13.25

EXAMPLE   Determine the area of the triangular region that has a base with a measure of 8 yd and an altitude with a measure of 4 yd.
   Solution $A = \frac{1}{2}(8 \text{ yd})(4 \text{ yd}) = 16$ sq yd.   ■

The formula for finding the area of a triangular region is very useful since every polygonal region can be divided into disjoint triangular regions and the area of the polygonal region can be formed by taking the sum of the triangular regions so formed. For instance, to find the area of the trapezoidal region given in Figure 13.26, consider the region as the two disjoint triangular regions bounded by $\triangle ABC$ and $\triangle ADC$. Then the area of the trapezoidal region would be the sum of the areas of the two triangular regions. The area of the triangular region bounded by $\triangle ABC$

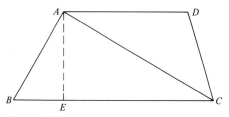

Figure 13.26

is $\frac{1}{2} \cdot m(\overline{BC}) \cdot m(\overline{AE})$, and the area of the triangular region bounded by $\triangle ADC$ is $\frac{1}{2} \cdot m(\overline{AD}) \cdot m(\overline{AE})$. Hence the area of the trapezoidal region is $\frac{1}{2} \cdot m(\overline{BC}) \cdot m(\overline{AE}) + \frac{1}{2} \cdot m(\overline{AD}) \cdot m(\overline{AE}) = \frac{1}{2} \cdot m(\overline{AE})(m(\overline{BC}) + m(\overline{AD}))$.

The sides of a trapezoid that are parallel are called its *bases*, and a line segment from a vertex of the trapezoid perpendicular to the opposite side is called an *altitude* of the trapezoid.

DEFINITION 13.37 Consider the trapezoid $ABCD$ with bases $\overline{AD}$ and $\overline{BC}$ and altitude $\overline{AE}$ and such that $m(\overline{AD}) = a$, $m(\overline{BC}) = b$, and $m(\overline{AE}) = h$ (see Figure 13.27). The area, A, of the interior of the trapezoid $ABCD$ is given by the formula $A = \frac{1}{2}h(a + b)$, provided that a, b, and h are all given in the same units of measurement. The area, A, is given in square units.

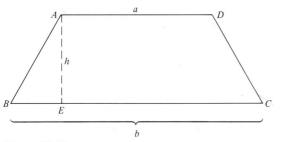

Figure 13.27

EXAMPLE Determine the area of the interior of the trapezoid whose bases have measures of 8 cm and 10 cm, respectively, and whose altitude has a measure of 5 cm.

Solution

$$A = \frac{1}{2}(5 \text{ cm})(8 \text{ cm} + 10 \text{ cm})$$
$$= \frac{1}{2}(5 \text{ cm})(18 \text{ cm})$$
$$= \frac{1}{2}(90 \text{ sq cm})$$
$$= 45 \text{ sq cm} \quad \blacksquare$$

We will conclude this section with a discussion of the Pythagorean theorem.

THEOREM 13.13 The Pythagorean Theorem If a triangle is a right triangle, then the square of the measure of the hypothenuse of the triangle is equal to the sum of the squares of the measures of the two legs.

Given right triangle $\triangle ABC$ with right angle at C, $m(\overline{BC}) = a$, $m(\overline{AC}) = b$, and $m(\overline{AB}) = c$ as illustrated in Figure 13.28, then the Pythagorean theorem becomes associated with the formula $c^2 = a^2 + b^2$.

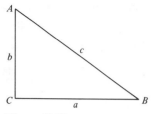

Figure 13.28

Also, if we consider a^2, b^2, and c^2 to be the measures of square regions whose sides have measures a, b, and c, respectively, as illustrated in Figure 13.29, then we could state the Pythagorean theorem as the measure of the square region on the hypothenuse of a right triangle is equal to the sum of the measures of the square regions on its legs.

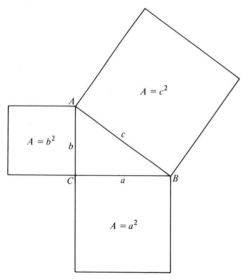

Figure 13.29

EXAMPLE   Given right triangle ABC with right angle at C, $m(\overline{AC}) = b = 4$, and $m(\overline{BC}) = a = 2$, determine $m(\overline{AB}) = c$.

Solution   Using the formula $c^2 = a^2 + b^2$ with $a = 2$ and $b = 4$, we have

$$c^2 = 2^2 + 4^2$$
$$= 4 + 16$$
$$= 20$$

If $c^2 = 20$, then $c = \pm 2\sqrt{5}$. But since c represents the length of a line segment, it cannot be negative. Hence $c = m(\overline{AB}) = 2\sqrt{5}$. ∎

The Pythagorean theorem gives us the relationship existing among the measures of the sides of a right triangle. Equally important is the converse of the theorem.

THEOREM 13.14 Converse of the Pythagorean Theorem If the square of the measure of one side of a triangle is equal to the sum of the squares of the measures of the other two sides, then the triangle is a right triangle.

EXAMPLE A triangle has three sides whose measures are 7, 24, and 25 units. Determine if the triangle is a right triangle.
Solution Since $7^2 = 49$, $24^2 = 576$ and $25^2 = 625$, we readily determine that $625 = 49 + 576$ or that $25^2 = 7^2 + 24^2$. Hence the triangle is a right triangle. ∎

EXAMPLE A triangle has three sides whose measures are 4, 7, and 11. Determine if the triangle is a right triangle.
Solution Since $4^2 = 16$, $7^2 = 49$, and $11^2 = 121$ and since $121 \neq 16 + 49$, then $11^2 \neq 7^2 + 4^2$ and the triangle is not a right triangle. ∎

EXERCISES

1. Determine the area of each of the triangular regions indicated.

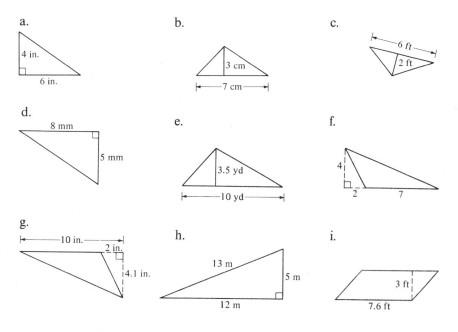

a.
4 in.
6 in.

b.
3 cm
7 cm

c.
6 ft
2 ft

d.
8 mm
5 mm

e.
3.5 yd
10 yd

f.
4
2
7

g.
10 in.
2 in.
4.1 in.

h.
13 m
5 m
12 m

i.
3 ft
7.6 ft

2. Determine the area of each of the polygonal regions indicated.

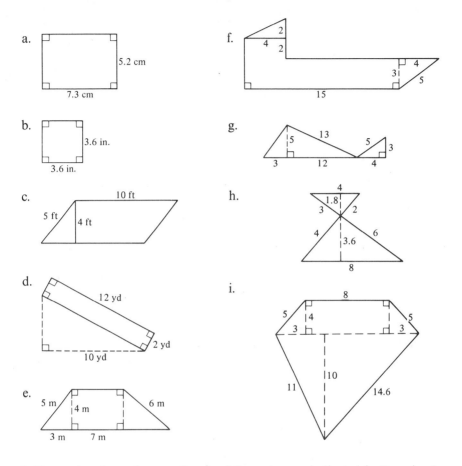

3. Determine the perimeter of each of the polygons indicated in Exercise 2.
4. Each of the right triangles indicated below has its right angle at C. Determine the measure of the indicated side using the given data and the theorem of Pythagoras.
   a. $m(\overline{AC})$, given $m(\overline{BC}) = 4$, $m(\overline{AB}) = 5$.
   b. $m(\overline{AB})$, given $m(\overline{AC}) = m(\overline{BC}) = 7$.
   c. $m(\overline{BC})$, given $m(\overline{AB}) = 13$, $m(\overline{AC}) = 8$.
   d. $m(\overline{BC})$, given $m(\overline{AB}) = 7$, $m(\overline{AC}) = 2 + m(\overline{AB})$.
   e. $m(\overline{AB})$, given $m(\overline{AC}) = 6$, $\overline{AC} \simeq \overline{BC}$.
5. Determine which of the following sets of measurements correspond to the measures of sides of right triangles.
   a. 6, 8, 10.   b. 7, 9, 16.
   c. 5, 12, 13.   d. 8, 15, 17.
   e. 9, 12, 15.   f. 30, 40, 50.
   g. 12, 19, 24.   h. 2, 4, $2\sqrt{3}$.

6. Identify each of the following as being true or false, and give an appropriate reason for your answer.
   a. If two polygons are congruent, then their polygonal regions have the same area.
   b. If two polygonal regions have the same area, then the polygons are congruent.
   c. A diagonal of a rectangle separates the rectangular region into two triangular regions.
   d. An altitude of a rectangle separates the rectangular region into two square regions.
   e. An altitude of a rectangle separates the rectangular regions into two rectangular regions.
   f. The measure of a polygonal region is always given in square units.
   g. The area of a polygon is zero.
   h. The area of a polygonal region always can be determined by summing the areas of the triangular regions comprising the polygonal region.
   i. All altitudes of the same triangle have the same measure.
   j. All altitudes of the same rectangle have the same measure.
   k. It is not possible to construct a square region with measure of $\sqrt{21}$.
   l. The perimeter of a trapezoid is given by one half the sum of the measures of its bases times the measure of its altitude.
   m. A diagonal of a parallelogram separates the parallelogram into two equal triangles.
   n. The interior of a polygon is a set of points all of which are coplanar.
   o. The measure of the union of a polygon with its interior is equal to the measure of the polygonal region.

13.9 SIMILAR TRIANGLES

So far in our discussions we have observed that if two geometric figures are congruent, then they have the same shape and size. Now consider the two triangular regions given in Figure 13.30. They appear to have the same shape but, quite clearly, their measures are not equal.

DEFINITION 13.38 Two geometric figures are said to be *similar* if and only if they have the same shape.

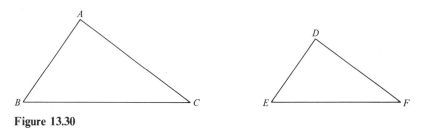

Figure 13.30

In this section we will be concerned with the similarity of triangles.

DEFINITION 13.39 Two triangles $\triangle ABC$ and $\triangle DEF$ are said to be *similar*, denoted by $\triangle ABC \sim \triangle DEF$ if and only if their corresponding angles are congruent; that is, $\angle ABC \simeq \angle DEF$, $\angle BCA \simeq \angle EFD$, and $\angle CAB \simeq \angle FDE$.

If two triangles are similar, then the measures of their corresponding sides are in proportion. For instance, if the triangles represented in Figure 13.30 are similar, then

$$\frac{m(\overline{AB})}{m(\overline{DE})} = \frac{m(\overline{BC})}{m(\overline{EF})}, \frac{m(\overline{AB})}{m(\overline{DE})} = \frac{m(\overline{AC})}{m(\overline{DF})} \quad \text{and} \quad \frac{m(\overline{BC})}{m(\overline{EF})} = \frac{m(\overline{AC})}{m(\overline{DF})}$$

EXAMPLE   Given $\triangle ABC$ and $\triangle DEF$ with sides of measures as indicated in Figure 13.31. Determine if the triangles are similar.

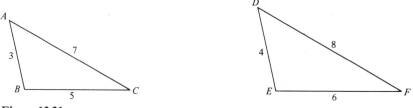

Figure 13.31

Solution   If the two triangles are similar, then the shortest side of $\triangle ABC$ will correspond to the shortest side of $\triangle DEF$ and the longest side of $\triangle ABC$ will correspond to the longest side of $\triangle DEF$. Now

$$\frac{m(\overline{AB})}{m(\overline{DE})} = \frac{3}{4} \quad \text{and} \quad \frac{m(\overline{AC})}{m(\overline{DF})} = \frac{7}{8}$$

but

$$\frac{3}{4} \neq \frac{7}{8} \quad \text{since } (3)(8) = 24 \neq 28 = (4)(7)$$

Hence, the triangles are not similar.   ∎

EXAMPLE   If $\triangle PQR \sim \triangle STW$ with $m(\angle P) = 50°$ and $m(\angle W) = 80°$, determine $m(\angle Q)$ and $m(\angle T)$.

Solution   Since $\triangle PQR \sim \triangle STW$, $\angle P \simeq \angle S$, $\angle Q \simeq \angle T$, and $\angle R \simeq \angle W$.

$$m(\angle Q) = 180° - [m(\angle P) + m(\angle R)]$$
$$= 180° - [(50° + m(\angle R))]$$
$$= 130° - m(\angle R)$$

But $\angle R \simeq \angle W$ and, therefore, $m(\angle R) = m(\angle W) = 80°$. Hence $m(\angle Q) = 130° - m(\angle R) = 130° - 80° = 50°$.

Similarly,

$$m(\angle T) = 180° - [m(\angle S) + m(\angle W)]$$
$$= 180° - [m(\angle S) + 80°]$$
$$= 100° - m(\angle S)$$

But $\angle S \simeq \angle P$ and, therefore, $m(\angle S) = m(\angle P) = 50°$. Hence $m(\angle T) = 100° - m(\angle S) = 100° - 50° = 50°$. ∎

The concepts of similarity of triangles can be extended to the similarity of other polygons.

EXERCISES

1. Determine which of the triangles $\triangle ABC$ and $\triangle DEF$ are similar and which are not similar. Give an appropriate reason for your answer.
   a. $\angle A \simeq \angle D$, $\angle B \simeq \angle E$, $m(\overline{AB}) = m(\overline{DE})$.
   b. $m(\overline{AB}) = \frac{1}{2}m(\overline{DE})$, $m(\overline{AC}) = \frac{1}{2}m(\overline{DF})$, $m(\overline{EF}) = 2m(\overline{BC})$.
   c. $\triangle ABC \simeq \triangle DEF$.
   d. $m(\overline{AB}) = 5$, $m(\overline{AC}) = 6$, $m(\overline{DE}) = 10$, $m(\overline{DF}) = 11$.
   e. $\angle B \simeq \angle E$, $\angle C \simeq \angle F$, $m(\overline{BC}) = 3$, $m(\overline{EF}) = 4$.
   f. $m(\overline{DE}) = m(\overline{AB})$, $m(\overline{EF}) = 6$, $m(\overline{BC}) = 5$.

2. For each of the triangles indicated below determine the measure of the unknown sides if the triangles are assumed to be similar.

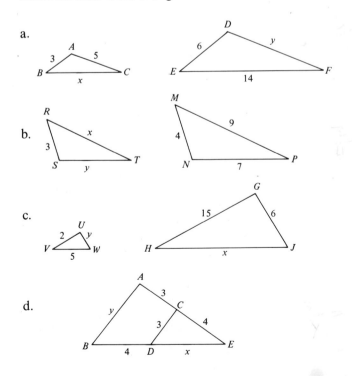

3. If $\triangle RST \sim \triangle UVW$ with $m(\angle R) = 42°$ and $m(\angle W) = 63°$, determine $m(\angle S)$.

4. If $\triangle MNP \sim \triangle GHJ$ with $\angle M \simeq \angle P$ and $m(\angle H) = 72°$, determine $m(\angle G)$.

5. Consider $\triangle RST$ with U being the midpoint of $\overline{RS}$ and V being the midpoint of $\overline{RT}$. What is the relationship between $m(\overline{UV})$ and $m(\overline{ST})$? Explain.

6. Determine the perimeter of each of the triangles given in Exercise 2.

7. Determine the height of the flagpole indicated below, using the given information:

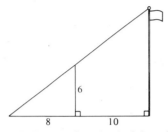

8. Determine the height of the building indicated below, using the given information:

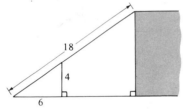

9. Similarity of figures is used in map construction. For instance, for a particular map the given scale may be 1 : 25,000. That is, each unit of length on the map will correspond to 25,000 units actually on the ground. If the distance between two points on a map is 3.4 in., what is the actual distance between the two corresponding locations on the ground? (Ignore differences in elevation between the two locations.)

10. If two triangles are isosceles, are they also similar? Explain.

11. If two triangles are equilateral, are they also similar? Explain.

12. Prove or disprove that the relation "is similar to" on the set of triangles is an equivalence relation.

13.10 SOME SIMPLE CLOSED SURFACES

A surface is a set of points in space analogous to a curve in a plane. Just as some curves are simples curves, some surfaces are simple surfaces, and just as some curves are closed curves, some surfaces are closed surfaces.

Finally, just as some curves in a plane are simple and closed, some surfaces in space also are simple and closed.

A line in a plane is an example of a simple curve that is not closed. The analogous extension would be a plane in space. A plane in space is an example of a simple surface that is not closed.

A simple closed curve separates a plane into three disjoint subsets of the plane: the curve itself and the two disjoint subsets of the plane called the interior and the exterior of the curve. Similarly, a simple closed surface separates space into three disjoint subsets of points: the surface itself and the two disjoint subsets of points called the interior of the surface and the exterior of the surface.

A circle is an example of a simple closed curve that separates the plane containing it into three disjoint sets: the circle, the interior of the circle, and the exterior of the circle. The analogous extension would be the sphere that separates the space containing it into three disjoint subsets of points: the sphere itself, the interior of the sphere, and the exterior of the sphere.

In this section we will be primarily concerned with simple closed surfaces formed by polygonal regions.

DEFINITION 13.40 A simple closed surface that is formed by polygons and polygonal regions is called a *polyhedron*. The polygonal regions are called the *sides* or *faces* of the polyhedron. A *regular polyhedron* is a polyhedron formed by congruent polygons.

Polyhedrons may be classified according to the number of sides:
   A polyhedron with 4 sides is called a *tetrahedron*.
   A polyhedron with 6 sides is called an *hexahedron*.
   A polyhedron with 8 sides is called an *octahedron*.
   A polyhedron with 12 sides is called a *dodecahedron*.
   A polyhedron with 20 sides is called an *icoscahedron*.
   A regular hexahedron is called a *cube*.
We will now consider a special type of simple polyhedron called a *prism*.

DEFINITION 13.41 A simple closed surface formed by two congruent polygonal regions in parallel planes, together with three or more quadrilateral regions connecting the two parallel polygonal regions, is called a *prism*. The parallel polygonal regions of a prism are called its *bases*, and the quadrilateral polygonal regions are called its *lateral sides* or, simply, its *sides*.

For some prisms the planes containing the sides are perpendicular to the planes containing the bases. Such a prism is called a *right prism*. Illustrations of right prisms are given in Figure 13.32.

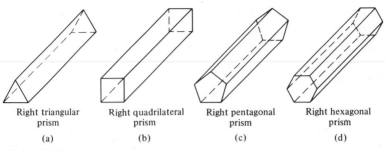

Right triangular prism

(a)

Right quadrilateral prism

(b)

Right pentagonal prism

(c)

Right hexagonal prism

(d)

Figure 13.32

For a right prism the sides of the quadrilateral that form the faces of the prism are called the *edges* of the prism and the vertices of the quadrilaterals are called the *vertices* of the prism. There is an interesting relationship existing between the number of vertices, V, the number of edges, E, and the number of sides, S, of a polyhedron. This relationship is given by the formula

$$V + S - E = 2$$

which was discovered by the famous Swiss mathematician Leonard Euler (1707–1783). This formula can easily be verified for the different polyhedrons. For instance, consider the pentagonal prism in Figure 13.33. There are 10 vertices, 15 edges, and 7 sides (the bases are also sides). Hence $V + S - E = 10 + 7 - 15 = 2$. Euler's formula may be verified for other polyhedrons in a similar manner.

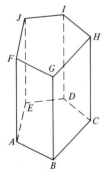

Figure 13.33

Another interesting polyhedron is the pyramid.

DEFINITION 13.42 A simple closed surface formed by a simple closed polygonal region, a point not in the plane containing the region, and the triangular regions joining the point and the sides of the polygonal region is called a *pyramid*. The simple closed polygonal region of the pyramid is called its *base*.

Pyramids are classified according to their base. If the base is a triangular region, the pyramid is called a triangular pyramid; if the base is a quadrilateral region, the pyramid is called a quadrilateral pyramid; and so forth. Figure 13.34 illustrates certain pyramids.

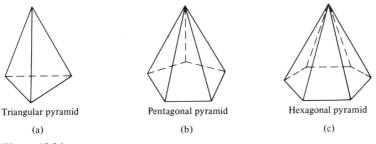

Triangular pyramid   Pentagonal pyramid   Hexagonal pyramid

(a)   (b)   (c)

Figure 13.34

It is easy to verify that Euler's formula is valid for pyramids also.

EXERCISES

1. Identify each of the following statements as being true or false and give an appropriate reason for your answer.
   a. A circle separates space into three disjoint subsets of points.
   b. A sphere separates space into three disjoint subsets of points.
   c. The interior of a simple closed surface and the exterior of a simple closed surface have the simple closed surface as their intersection set.
   d. The least number of faces for a polyhedron is three.
   e. Euler's formula is valid for pyramids as well as for prisms.
   f. A pentagonal pyramid is a polyhedron with exactly six sides.
   g. All prisms are right prisms.
   h. Prisms and pyramids are classified according to their bases.
2. Sketch and verify Euler's formula for each of the following:
   a. A triangular prism.   b. A triangular pyramid.
   c. A quadrilateral prism.   d. A quadrilateral pyramid.
   e. A hexagonal prism.   f. A hexagonal pyramid.
   g. A tetrahedron.   h. A dodecahedron.
3. Is it possible to construct a pyramid with eight sides if its base is an hexagon? Explain.
4. Sketch a pentagonal pyramid in the interior of a sphere.

13.11 SURFACE AREA AND VOLUME OF POLYHEDRONS

In the previous section we considered polyhedrons as examples of simple closed surfaces. A polyhedron is a set of points that has a measure called *surface area,* and the interior of the polyhedron is a space region that also has a measure called *volume*. To determine the surface area of a polyhedron we determine the area of each of its polygonal regions and sum these areas. For instance, consider the rectangular polyhedron given in Figure 13.35 having edges that measure 4 in., 6 in. and 10 in.

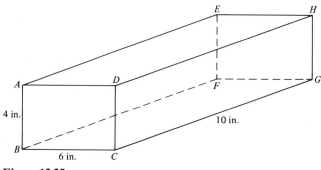

Figure 13.35

The measure of $\overline{AB}$ is called the *height* of the polyhedron, the measure of $\overline{BC}$ is called its *width*, and the measure of $\overline{CG}$ is called its *length*. This rectangular polyhedron has two sides, the rectangular polygonal regions *ABCD* and *EFGH* that are congruent and such that each of these regions has an area of (4 in.)(6 in.) = 24 sq in. The polyhedron also has two sides that are the congruent polygonal regions *DCGH* and *ABFE* each of which has an area of (4 in.)(10 in.) = 40 sq in. The polyhedron also has two sides that are the congruent polygonal regions *ADHE* and *BCGF* each of which has an area of (6 in.)(10 in.) = 60 sq in. Hence the *surface area* of this rectangular polyhedron is equal to 2(24 sq in.) + 2(40 sq in.) + 2(60 sq in.) = 248 sq in.

DEFINITION 13.43 Consider the rectangular polyhedron illustrated below such that $m(\overline{AB}) = h$ (called *height*), $m(\overline{BC}) = w$ (called *width*), and $m(\overline{CG}) = l$ (called *length*). Then the *surface area*, *S*, of this rectangular polyhedron is given by the formula $S = 2(hw + hl + lw)$, provided *l*, *w*, and *h* are given in the same units of measurement. The surface area, *S*, is given in square units (see Figure 13.36).

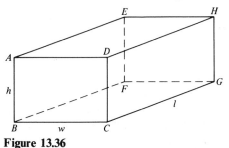

Figure 13.36

EXAMPLE   Determine the surface area of the rectangular polyhedron whose height is 6 ft, width is 3 yd, and length is 11 ft.

Solution   Converting $w = 3$ yd to feet, we have $w = 9$ ft. Hence

$$S = 2(lw + hw + hl)$$
$$= 2[(11 \text{ ft})(9 \text{ ft}) + (6 \text{ ft})(9 \text{ ft}) + (6 \text{ ft})(11 \text{ ft})]$$
$$= 2[(99 \text{ sq ft}) + (54 \text{ sq ft}) + (66 \text{ sq ft})]$$
$$= 2(219 \text{ sq ft})$$
$$= 438 \text{ sq ft} \quad ■$$

To determine the volume of the interior of a rectangular polyhedron we use a cube (that is, a regular hexahedron) such that each edge has a measure of 1 unit as illustrated in Figure 13.37. We shall define this cube to be a *unit cube* such that its interior has a measure that is 1 *cubic unit*. If the unit of measure is in feet, then the unit cube would have 12 edges of measure 1 ft each and the measure of its interior would be 1 cubic foot.

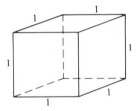

Figure 13.37

We will now attempt to determine the volume or the measure of the interior of a rectangular polyhedron with $w = 4$ units, $l = 8$ units, and $h = 3$ units, as illustrated in Figure 13.38. To determine the volume of this

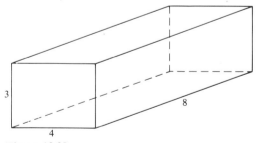

Figure 13.38

space region, we would have to determine the number of unit cubes contained in the interior or the number of unit cubes that would be necessary to completely cover the interior. To do so we could place 12-unit cubes on one face of the polyhedron as illustrated in Figure 13.39.

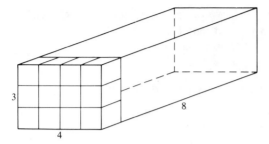

Figure 13.39

Also observe that we can place 8 such "layers" of 12-unit cubes to completely cover the interior as illustrated in Figure 13.40. Hence there would be $8 \times 12 = 96$ unit cubes, and these would completely cover the

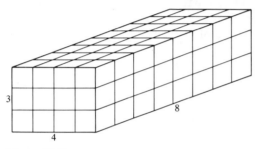

Figure 13.40

interior of the polyhedron. Since each unit cube has a volume of 1 cubic unit, the volume of this interior would be 96 cubic units. Since $96 = 3 \times 4 \times 8$, it is easy to formulate the following definition.

DEFINITION 13.44 Consider the rectangular polyhedron with height $= h$ units, width $= w$ units, and length $= l$ units. Then the volume, V, of the interior of the rectangular polyhedron is given by the formula $V = lwh$, provided l, w, and h are given in the same units of measurement. The volume, V, is given in cubic units.

EXAMPLE A rectangular polyhedron is such that $l = 9$ cm, $w = 7$ cm, and $h = 2\frac{1}{2}$ cm. Determine the volume of its interior.
Solution

$$V = lwh$$
$$= (9 \text{ cm})(7 \text{ cm})(2\frac{1}{2} \text{ cm})$$
$$= 157\frac{1}{2} \text{ cu cm} \quad \blacksquare$$

A rectangular polyhedron is a special case of a right prism. In general, to find the surface area of a prism, we determine the area of each of its bases and the area of each of its sides and then form the sum of these areas. To determine the volume of the interior of a prism, determine the number of square units in the area of one of its bases and multiply this number by the height of the prism, considering the height to be a perpendicular distance between the bases of the prism.

EXAMPLE Determine (a) the surface area and (b) the volume associated with the right triangular prism in Figure 13.41, where all measurements are given in the same units.

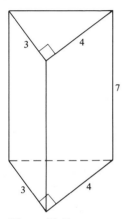

Figure 13.41

Solution The bases of this triangular prism are right triangles as indicated, and the legs of each triangular base have measures of 3 and 4 units, respectively. Hence, by the Pythagorean theorem, we determine that the measure of the hypothenuse of each triangular base is 5 units. Therefore

a. $S = [\frac{1}{2}(3 \times 4) + \frac{1}{2}(3 \times 4) + (3 \times 7) + (4 \times 7) + (5 \times 7)]$ sq units

$= (6 + 6 + 21 + 28 + 35)$ sq units

$= 96$ sq units

b. $V = [\frac{1}{2}(3 \times 4)] \times 7$ cu units

$= (6 \times 7)$ cu units

$= 42$ cu units ∎

EXAMPLE Determine the volume of the interior of the trapezoidal prism in the accompanying figure where each base has parallel edges with measures of 7 and 9 units, respectively, the height of the trapezoid is 4 units, and the height of the prism is 8 units (see Figure 13.42).

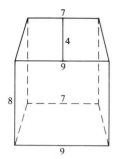

Figure 13.42

Solution

$V = Ah$ where $A = $ the area of a base and h is the height of the prism

$= [\frac{1}{2}(7 + 9) \times 4] \times 8$ cu units

$= (32 \times 8)$ cu units

$= 256$ cu units ■

In general we have the following formulas associated with prisms:

1. The *volume* of the interior of a prism is given by the formula $V = Bh$, where B is the area of a base of the prism (in square units) and h is the height of the prism (in units). V is given in cubic units.

2. The *lateral surface* area, L, of a prism is the sum of all the areas of the polygonal regions (excluding the bases) that comprise the prism. L is given in square units.

3. The *total surface area*, S, of a prism is given by the formula $S = 2B + L$, where B represents the area (in square units) of a base of the prism and L (in square units) is the lateral surface area of the prism. S is given in square units.

4. Unless otherwise indicated, by surface area we shall mean total surface area.

It is not immediately obvious, but it can be proved, that the volume of the region bounded by a pyramid is equal to one-third of the interior of a prism having the same base and height as the prism. Hence the volume, V, of a pyramid is given by the formula $V = \frac{1}{3}Bh$, where B is the area of the base of the pyramid and h is its height. V is given in cubic units.

EXERCISES

1. Consider the rectangular polyhedron indicated below.

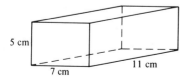

a. Determine the total surface area of the polyhedron.
b. Determine the lateral surface area of the polyhedron.
c. Determine the volume of the interior of the polyhedron.
d. Determine the total length of the edges of the polyhedron.
2. Consider the triangular prism indicated below.

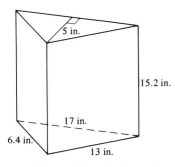

a. Determine the lateral surface area of the prism.
b. Determine the total surface area of the prism.
c. Determine the volume of the interior of the prism.
d. Determine the total length of the edges of the prism.
3. Consider the trapezoidal prism indicated below.

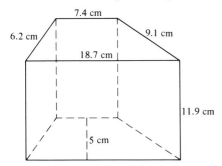

a. Determine the lateral surface area of the prism.
b. Determine the total surface area of the prism.
c. Determine the volume of the interior of the prism.
d. Determine the total length of the edges of the prism.
4. A pentagonal prism has a height of 12 in. and bases as indicated below.

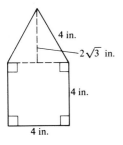

a. Determine the lateral surface area of the prism.
b. Determine the total surface area of the prism.
c. Determine the volume of the interior of the prism.
d. Determine the total length of the edges of the prism.

13.12 BASIC CONSTRUCTIONS

In this section we will consider some simple constructions requiring the use of a compass, a straightedge, and a pencil only.

We considered two line segments to be congruent if and only if they have the same measure. We will now consider "constructing" a line segment congruent to a given line segment.

CONSTRUCTION 13.1 Construct a line segment congruent to a given line segment.

Given   Line segment $\overline{AB}$ and line l (see Figure 13.43).

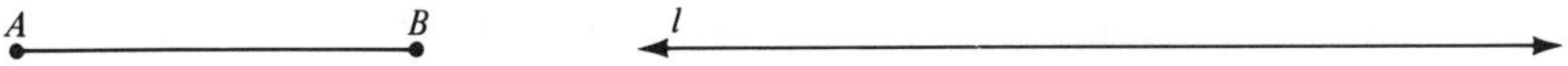

A ●————————● B l

Figure 13.43

To construct   Line segment $\overline{CD}$ on line l such that $\overline{AB} \simeq \overline{CD}$.
Construction   (Figure 13.44):
a. Set your compass on line segment $\overline{AB}$ so that one point of the compass is at A and the other is at B.
b. Select a point C on line l and without changing the setting of your compass, place one point of the compass at C and describe an arc intersecting line l at the point D.
c. The line segment $\overline{CD}$ so constructed is the required line segment.

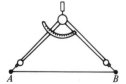
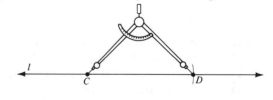

Figure 13.44

CONSTRUCTION 13.2 Divide a given line segment into two congruent parts. (This is known as *bisecting* the line segment.)

Given   Line segment $\overline{AB}$ (Figure 13.45).

A ●————————● B

Figure 13.45

To construct   Line segments $\overline{AC}$ and $\overline{CB}$ such that C is a point on $\overline{AB}$, $\overline{AC} \simeq \overline{CB}$, and $\overline{AC} \cup \overline{CB} = \overline{AB}$.

Construction (Figure 13.46):

a. Set one point of your compass at point A, and open the compass so that the distance between its legs is greater than one-half $m(\overline{AB})$.

b. With this setting and one point of the compass at point A describe arcs above and below the line segment $\overline{AB}$.

c. Without changing this setting place one point of the compass at point B and describe arcs above and below the line segment $\overline{AB}$ intersecting the previous arcs described at points P and Q.

d. With a straightedge, connect the points P and Q with the line segment $\overline{PQ}$.

e. The point C of intersection of the line segments $\overline{AB}$ and $\overline{PQ}$ is the required point such that $\overline{AC} \simeq \overline{CB}$.

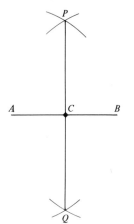

Figure 13.46 Line segment $\overline{PQ}$ is called a perpendicular bisector of $\overline{AB}$.

CONSTRUCTION 13.3 Construct a line segment perpendicular to a given line from a point not contained on the given line.

Given   Line l and point C not on l. (Figure 13.47).

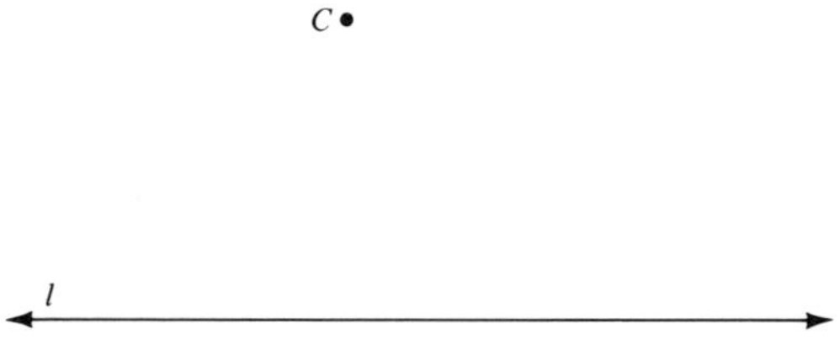

$C \bullet$

l

Figure 13.47

To construct   Line segment $\overline{CD}$ such that $\overline{CD}$ is perpendicular to l.

Construction (Figure 13.48):

a. With one point of your compass placed at point C and a setting greater than the shortest distance from C to l, describe an arc that will intersect the line l at two distinct points A and B.

b. With one point of your compass now placed at point A and a setting greater than one-half $m(\overline{AB})$ describe an arc in the half-plane determined by l that does not contain C.

c. Without changing the setting but with one point of the compass now at point B describe another arc in the same half-plane intersecting the previous arc described at the point E.

d. With a straightedge connect the points C and E.

e. The line segment $\overline{CE}$ intersects the line segment $\overline{AB}$ at the point D.

f. The line segment $\overline{CD}$ is the required line segment.

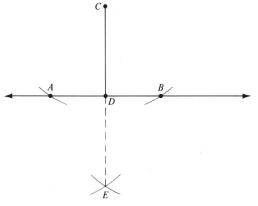

Figure 13.48

CONSTRUCTION 13.4 Construct an angle congruent to a given angle.

Given Angle $\angle ABC$ (Figure 13.49).

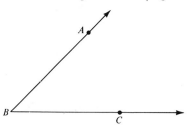

Figure 13.49

To construct Angle $\angle DEF$ such that $\angle ABC \simeq \angle DEF$.

Construction (Figure 13.50):

a. Start with ray $\overrightarrow{EP}$.

b. Place one point of your compass at vertex B of $\angle ABC$, and describe an arc through the interior of the angle intersecting ray $\overrightarrow{BA}$ at point M and ray $\overrightarrow{BC}$ at point N.

c. Without changing the setting of the compass but with one point now at E on ray $\overrightarrow{EP}$, describe an arc intersecting ray $\overrightarrow{EP}$ at point F.

d. Now place one point of the compass at point M and the other point at point N.

e. Without changing this setting, place one point of the compass at point F on ray $\overrightarrow{EP}$ and describe an arc intersecting the arc described in (c) at point D.

f. With a straightedge connect points D and E forming ray $\overrightarrow{ED}$.

g. $\angle DEF$ is the required angle.

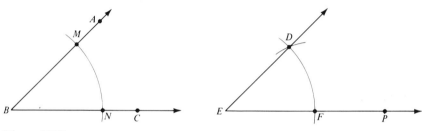

Figure 13.50

CONSTRUCTION 13.5 To bisect an angle.

Given   Angle $\angle ABC$ (Figure 13.51).

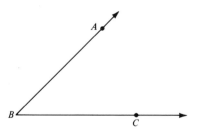

Figure 13.51

To construct   A ray $\overrightarrow{BD}$ in the interior of $\angle ABC$ such that $\angle ABD \simeq \angle DBC$.

Construction (Figure 13.52):

a. With one point of your compass at the vertex B of $\angle ABC$ describe an arc through the interior of the angle and intersecting ray $\overrightarrow{BA}$ at P and ray $\overrightarrow{BC}$ at Q.

b. With one point of the compass now at point P and a setting greater than one-half $m(\overline{PQ})$, describe an arc in the interior of $\angle ABC$.

c. Without changing the setting of the compass but with one point now at point Q, describe an arc in the interior of $\angle ABC$, intersecting the arc described in (b) at the point D.

d. With a straightedge connect the points B and D to form ray $\overrightarrow{BD}$.

e. $\angle ABD \simeq \angle DBC$; $\angle ABC$ is said to have been bisected.

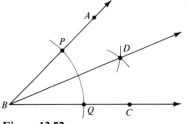

Figure 13.52

CONSTRUCTION 13.6 To construct a triangle congruent to a given triangle.

Given Triangle $\triangle ABC$ (Figure 13.53).

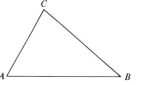

Figure 13.53

To construct $\triangle DEF \simeq \triangle ABC$.

Construction (Figure 13.54):

a. Start with ray $\overrightarrow{DQ}$.

b. Place one point of your compass at point D and with a setting equal to the measure $\overline{AB}$, describe an arc intersecting ray $\overrightarrow{DQ}$ at point E.

c. Now place one point of your compass at point E and with a setting equal to the measure of $\overline{BC}$, describe an arc in the half-plane determined by $\overleftrightarrow{DE}$ and in the direction of M.

d. Now place one point of your compass at point D and with a setting equal to the measure of $\overline{AC}$, describe an arc in the same half-plane and intersecting the arc described in (c) at point F.

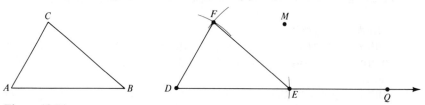

Figure 13.54

e. With a straightedge connect the points E and F to form line segment $\overline{EF}$.

f. With a straightedge connect the points D and F to form line segment $\overline{DF}$.

g. The triangle $\triangle DEF$ is the required triangle.

SUMMARY

In this chapter we considered some geometric concepts from an intuitive point of view. Basically, we examined the geometry of shape, size, and measure. We started with the undefined terms of *point, line, plane, space,* and *between*; definitions were then given in terms of these undefined terms. The definitions introduced included those for collinear points, half-line, open half-line, ray, line segment, half-plane, open half-plane, closed half-plane, intersecting lines, skew lines, parallel lines, parallel planes, intersecting planes, angle, interior and exterior of an angle, curve, broken line, polygon, polyhedron, area, and volume.

Angles were classified as being adjacent, vertical, complementary, supplementary, right, straight, acute, and obtuse. Dihedral angles also were defined. Polygons were classified according to the number of sides, and polyhedrons were classified according to the number of sides. Congruence of figures was discussed as was similarity of geometric figures.

The concept of measure was introduced and discussed and included length, area, surface area, lateral surface area, volume, and perimeter.

We concluded the chapter with six basic constructions as applications of congruence of line segments, angles, and triangles.

During our discussion the following symbols were introduced.

SYMBOL	INTERPRETATION
$A, B, C, \ldots$	Symbols representing points
$\overleftrightarrow{AB}$	Symbol for line determined by the distinct points A and B
$\overrightarrow{AB}$	The ray from point A and in the direction of point B
$\overline{AB}$	The line segment from A to B
$\angle ABC$	Angle A, B, C
$\text{int}(\angle ABC)$	Interior of $\angle ABC$
$\text{ext}(\angle ABC)$	Exterior of $\angle ABC$
$\triangle ABC$	Triangle A, B, C
$m(\overline{AB})$	Measure (or length) of line segment $\overline{AB}$
$\overline{AB} \simeq \overline{CD}$	Congruence of line segments $\overline{AB}$ and $\overline{CD}$
$m(\angle ABC)$	Measure of angle $\angle ABC$
$\angle ABC \simeq \angle DEF$	Congruence of angles $\angle ABC$ and $\angle DEF$
$\triangle ABC \simeq \triangle DEF$	Congruence of triangles $\triangle ABC$ and $\triangle DEF$
$\triangle ABC \sim \triangle DEF$	Similarity of triangles $\triangle ABC$ and $\triangle DEF$

REVIEW EXERCISES FOR CHAPTER 13

1. Identify each of the following as being true or false and given an appropriate reason for your answer.
   a. A ray separates a plane.
   b. An angle separates a plane.
   c. If $\overrightarrow{AB} = \overrightarrow{AC}$, then the points A, B, and C are collinear.
   d. If four distinct points in space are noncollinear, then any three of them must determine a unique plane.
   e. A ray can be a proper subset of a line segment.
   f. The area of a triangle is zero.
   g. The area of a trapezoidal region is given by the formula $A = \frac{1}{2}h(a + b)$, where h refers to the height and a and b are the measures of the bases.
   h. A sphere is an example of a simple closed surface.
   i. A pyramid is a special case of a prism.
   j. All prisms are right prisms.
   k. Euler's formula is valid for both prisms and pyramids.
   l. Similar triangles always are congruent.
   m. All equilateral triangles are similar.
   n. The relation "is congruent to" on the set of all triangles is an equivalence relation.
   o. The relation "is similar to" on the set of all triangles is an equivalence relation.
   p. Volume is a term used to refer to the measure of a space region.
   q. If three angles of one triangle are congruent, respectively, to three angles of a second triangle, then the triangles are congruent.
   r. The lateral surface area of a prism is equal to its total surface area.
   s. The volume of the interior of a prism is given by the formula $V = Bh$, where B refers to the measure of its base and h refers to its height.
   t. If $m(\overline{AB}) = m(\overline{CD})$, then $\overline{AB} = \overline{CD}$.
   u. If $m(\overline{AB}) = m(\overline{BC})$, then $B = C$.
   v. It is not possible for two angles to intersect in exactly two points.
   w. If the intersection set of two angles is a line segment, then the angles must have a common ray.
   x. If $m(\angle A) = 62°$ and $m(\angle B) = 28°$, then angles A and B are complementary.
   y. Two intersecting lines form both vertical and adjacent angles.
   z. All closed curves are simple.
2. Give an example for each of the following. If this is not possible, indicate why not.
   a. Two angles whose intersection set is a ray.
   b. Two angles whose intersection set is a line segment.
   c. Two rays whose union is a line.
   d. Two triangles that are similar but not congruent.

e. Four collinear points A, B, C, and D such that $\overline{AC} \cap \overline{BD} = \varnothing$.

f. A scalene triangle that is obtuse.

g. A right triangle that is isosceles.

h. A right triangle that is equilateral.

i. Two angles such that their interiors are overlapping sets.

j. A pyramid with exactly three polygonal regions.

k. Two angles whose intersection set is a ray but such that they have no ray in common.

l. A triangle with two obtuse angles.

m. The measure of a side of a square such that the area of the square region is 17.

n. A real value between 5 and 7 satisfying the theorem of Pythagoras.

o. A rational value between 10 and 13 satisfying the theorem of Pythagoras.

CHAPTER 14

AN INTRODUCTION TO COORDINATE GEOMETRY

We have considered various number systems in this text that involve subsets of the real numbers. We also have considered some concepts of informal geometry in the previous chapter. There are relationships between geometry and the arithmetic of numbers that are studied at the elementary level under the heading of *coordinate geometry*.

14.1 SPECIAL LINES AND PLANES

In Chapter 5 we referred to the so-called number line to represent whole numbers, order relations between whole numbers, and operations with whole numbers. In Chapter 13 we referred to a line as a set of points that can be extended indefinitely in either direction. Hence the whole numbers really cannot be represented by a number line if by a number line we refer to our geometric conception of a line, since the whole numbers begin with 0 and increase indefinitely. Hence the so-called whole number line would begin with the 0-point and extend indefinitely in only one direction.

To further complicate matters, recall that a geometric line is dense; that is, between any two distinct points of the line there always exists another point of the line. However, the set of whole numbers is not dense. Hence on the whole number line there would be spaces between the points representing the whole numbers. For all practical purposes, then, the whole number line should be represented by a set of collinear points equally spaced as in Figure 14.1.

Figure 14.1

In Chapter 7 we extended the system of whole numbers and introduced the system of integers. The whole numbers were now considered as integers and, further, for each whole number, n, there was now an additive inverse, ^-n. Further, on the so-called number line the integer ^-a was located the same number of units to the left of the integer 0 as the integer a was to the right of 0. The set of integers is not dense and, hence, the integer number line should be represented by a set of collinear points equally spaced as in Figure 14.2 such that the points extend indefinitely in either direction from 0.

Figure 14.2

In Chapter 9 we introduced the system of rational numbers and learned that the set of rational numbers was dense. However, not every point on the number line represented a rational number. The rational number line would have many "holes" in it, and it would be difficult to represent it as a set of collinear points.

In Chapter 10 we introduced the system of real numbers. The set of real numbers is dense and also is complete. Hence there are not "holes" in the so-called real number line and this real number line does conform to our concept of a geometric line. In Figure 14.3, to represent the real numbers on the real number line, we locate the real number 0 on the line at point O,

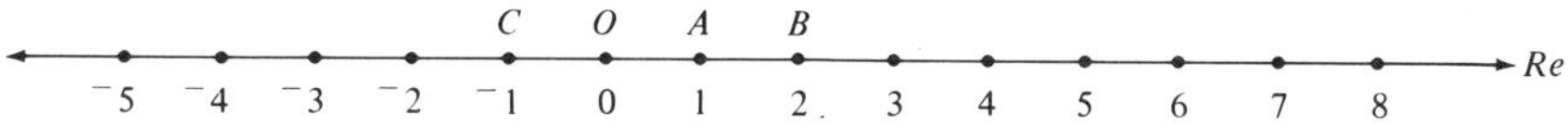

Figure 14.3

called the *origin*, and the real number 1 on the line and to the right of O at the point A. A *scale* is now determined which can be used to locate any real number on this line. The real number 2 would be located at the point B such that $m(\overline{OA}) = m(\overline{AB})$. Similarly, the real number $^-1$ would be located at the point C such that $m(\overline{OC}) = m(\overline{OA})$.

Finally, in Chapter 4 we introduced the concept of a relation as a cartesian product, and graphed relations in cartesian planes. If the ordered pairs in the relation had real components, the cartesian plane was the graph of $Re \times Re$ illustrated in Figure 14.4 with two such real number lines being perpendicular to each other and intersecting at the point with coordinates $(0, 0)$. The horizontal number line is called the *horizontal axis*, the vertical number line is called the *vertical axis*, and the point of intersection of the two

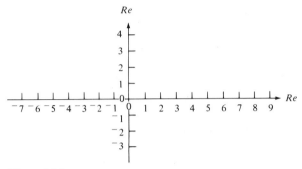

Figure 14.4

axes (plural of axis) is called the *origin*. The plane containing these axes is called the *real plane* or the *euclidean plane*.

The cartesian product $Re \times Re$ contains infinitely many elements that are ordered pairs of real numbers. Each such element can be graphed in the euclidean plane as a single point. The totality of all such points would be the graph of $Re \times Re$. It is impossible to label each point in this plane and, consequently, we generally label only certain integer values on the horizontal and vertical axes. All points of the plane are then given in terms of these coordinates. In Figure 14.5 we graph the ordered pairs of real numbers $(^-3, 2)$, $(0, 2)$, $(2, 5)$, $(4, \ ^-2)$, and $(^-5, \ ^-3)$ as the points A, B, C, D, and E, respectively.

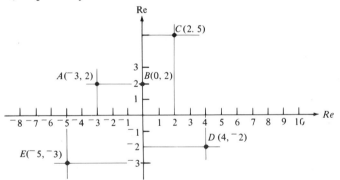

Figure 14.5

EXERCISES

1. Locate each of the following indicated real numbers on the real number line.

    a. 3.   b. $\frac{1}{2}$.   c. 0.

    d. $\sqrt{2}$.   e. $^-3$.   f. $\sqrt[-]{3}$.

    g. $^-2/3$.   h. $^-1\frac{1}{4}$.   i. $2 + \sqrt{3}$.

j. $3 - \sqrt{2}$.   k. 0.66$\bar{6}$.   l. $^-$0.3.

m. 2.33$\bar{3}$.   n. $^-$7.4.   o. 6.9.

2. Locate each of the following indicated ordered pairs of real numbers in the real plane.

a. (2, 1).   b. $(^-3, 2)$.   c. (0, 2).

d. $(^-3, 0)$.   e. $(^-4, ^-1)$.   f. $(5, ^-3)$.

g. $(0, ^-3)$.   h. (0.5, 0).   i. (0, 0).

j. $(\frac{1}{2}, \frac{1}{3})$.   k. $(^-\frac{1}{4}, \frac{1}{2})$.   l. $(^-\frac{1}{3}, ^-\frac{1}{3})$.

m. $(0, \frac{1}{2})$.   n. $(^-\frac{1}{3}, 0)$.   o. $(\sqrt{2}, \sqrt{3})$.

14.2 DISTANCE

If we label the horizontal axis of a cartesian plane as the x axis and choose a definite coordinate system or scale on it, we observe that the distance from the origin of the plane to the point A with x coordinate of 2 is 2, which corresponds to the measure of the line segment $\overline{OA}$. (See Figure 14.6.) Also, the distance from the origin to the point B on the x axis with x coordinate $^-2$ is 2, which corresponds to the measure of the line segment $\overline{OB}$. If we label the vertical axis of the cartesian plane as the y axis and choose the scale on this axis as the same as was used for the x axis, we observe that the distance from the origin to the point C with y coordinate 4 is 4, which corresponds to the measure of the line segment $\overline{OC}$. Also, the distance from the origin to the point D on the y axis with y coordinate of $^-4$ is 4, which corresponds to the measure of line segment $\overline{OD}$.

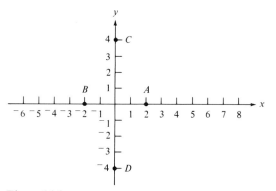

Figure 14.6

DEFINITION 14.1 On a cartesian plane the *distance from the origin O to a point A on the horizontal axis* is equal to $m(\overline{OA})$ and the distance from the origin to a point B on the vertical axis is equal to $m(\overline{OB})$.

EXAMPLE

a. The point A on the horizontal x axis of the xy plane has an x-coordinate of 3. (See Figure 14.7.) Then the distance from 0 to A is $m(\overline{OA}) = 3 - 0 = 3$.

b. The point B on the horizontal x axis of the xy plane has an x coordinate $^-5$. (See Figure 14.7.) Then the distance from O to B is $m(\overline{OB}) = 0 - (^-5) = 5$.

c. The point C on the vertical y axis of the xy plane has a y coordinate of 4. (See Figure 14.7.) Then the distance from O to C is $m(\overline{OC}) = 4 - 0 = 4$.

d. The point D on the vertical y axis of the xy plane has a y coordinate of $^-3\frac{1}{2}$. (See Figure 14.7.) Then the distance from O to D is $m(\overline{OD}) = 0 - (^-3\frac{1}{2}) = 3\frac{1}{2}$.

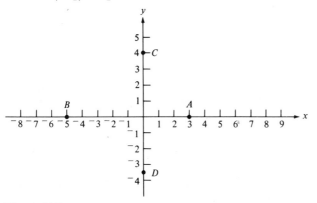

Figure 14.7

Let us now consider determining the distance between any two distinct points on the horizontal or on the vertical axis. For instance, to determine the distance between the points $A(1, 0)$ and $B(^-4, 0)$, respectively, we observe (Figure 14.8) that the distance from O to A is $m(\overline{OA}) = 1$, and the

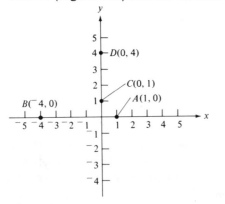

Figure 14.8

distance from O to B is $m(\overline{OB}) = 4$. Since $\overline{BA} = \overline{BO} \cup \overline{OA}$, $m(\overline{BA}) = m(\overline{BO}) + m(\overline{OA})$. Hence $m(\overline{AB}) = 4 + 1 = 5$, which represents the distance between the points A and B.

To determine the distance between the two points $C(0, 1)$ and $D(0, 4)$ on the y axis, we observe that the distance from O to D is $m(\overline{OD}) = 4$ and the distance from O to C is $m(\overline{OC}) = 1$. Since $\overline{CD} = \overline{OD} - \overline{OC}$, $m(\overline{CD}) = m(\overline{OD}) - m(\overline{OC})$. Hence $m(\overline{CD}) = 4 - 1 = 3$, which represents the distance between the two points $C(0, 1)$ and $D(0, 4)$.

DEFINITION 14.2 On the xy plane the *distance d between two distinct points* on the x axis with coordinates $A(x_1, 0)$ and $B(x_2, 0)$ such that $x_1 < x_2$ is equal to $m(\overline{AB}) = x_2 - x_1$. The distance d between two distinct points on the y axis with co-ordinates $C(0, y_1)$ and $D(0, y_2)$ such that $y_1 < y_2$ is equal to $m(\overline{CD}) = y_2 - y_1$.

EXAMPLE Find the distance d between the points $P(^-4, 0)$ and $Q(3, 0)$.
Solution The points P and Q both lie on the horizontal axis. Hence by Definition 14.2 we have $d = m(\overline{PQ}) = 3 - (^-4) = 7$. ∎

EXAMPLE Find the distance d between the points $S(0, ^-2)$ and $T(0, ^-7)$.
Solution The points S and T both lie on the vertical axis. Hence $d = m(\overline{ST}) = ^-2 - (^-7) = ^-2 + 7 = 5$. ∎

To determine the distance between two distinct points in a cartesian plane, we proceed as follows. Let P_1 and P_2 be any two distinct points in the xy plane with coordinates (x_1, y_1) and (x_2, y_2), respectively. There are three possible cases we will now discuss separately.

CASE 1   Suppose that $x_1 = x_2$ and $y_1 \neq y_2$. Let $y_1 < y_2$. Then P_1 and P_2 are on the same vertical line (see Figure 14.9). Construct perpendiculars from P_1 and P_2 to the y axis as indicated. Clearly, then, the distance between P_1 and P_2 is equal to the distance between the points of inter-section of the perpendiculars with the y axis. Hence $d = m(\overline{P_1P_2}) = m(\overline{AB}) = y_2 - y_1$.

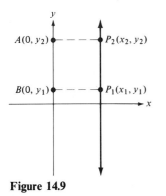

Figure 14.9

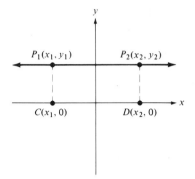

Figure 14.10

CASE 2   Suppose that $x_1 \neq x_2$ and $y_1 = y_2$. Let $x_1 < x_2$. Then P_1 and P_2 are on the same horizontal line (see Figure 14.10). Construct perpendiculars from P_1 and P_2 to the x axis as indicated. Clearly, then, the distance between P_1 and P_2 is equal to the distance between the points of intersection of the perpendicular with the x axis. Hence $d = m(\overline{P_1 P_2}) = m(\overline{CD}) = x_2 - x_1$.

CASE 3   Suppose $x_1 \neq x_2$ and $y_1 \neq y_2$. Let $x_1 < x_2$ and $y_1 < y_2$. Then P_1 and P_2 lie on an oblique line (that is, a line that is neither horizontal nor vertical) as shown in Figure 14.11. Construct a line through P_1 perpendicular to the y axis and construct a line through P_2 perpendicular to the x axis. Label the point of intersection of these lines P_3, as indicated in Figure 14.11. Since P_1 and P_3 lie on the same horizontal line, the two points must have the same y coordinate. Similarly, since P_2 and P_3 lie on the same vertical line, the two points must have the same x coordinate. Hence the coordinates of P_3 are (x_2, y_1). Now the distance between the points P_1 and P_3 is $d_1 = m(\overline{P_1 P_3}) = x_2 - x_1$, by case 2, and the distance between the points P_2 and P_3 is $d_2 = m(\overline{P_2 P_3}) = y_2 - y_1$, by case 1. But

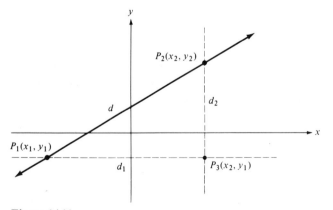

Figure 14.11

P_1, P_2, and P_3 are vertices of a right triangle whose hypothenuse is the line segment $\overline{P_1P_2}$. Therefore by the theorem of Pythagoras we have $d = m(\overline{P_1P_2})$ and

$$d^2 = d_1{}^2 + d_2{}^2$$

where $d_1{}^2 = (x_2 - x_1)^2$ and $d_2{}^2 = (y_2 - y_1)^2$. Substituting these values in the Pythagorean equation we obtain

$$d^2 = (x_2 - x_1)^2 + (y_2 - y_1)^2$$

or,

$$d = \sqrt{(x_2 - x_1)^2 + (y_2 - y_1)^2}$$

This last equation is referred to as the *undirected distance formula*. It may be used to determine the distance between any two points in the same plane. The formula involves only the coordinates of the two given points. Further, it should be noted that $(x_2 - x_1)^2 = (x_1 - x_2)^2$ and that $(y_2 - y_1)^2 = (y_1 - y_2)^2$. We can now combine all three cases into the following theorem.

THEOREM 14.1   Let $P_1(x_1, y_1)$ and $P_2(x_2, y_2)$ be any two points in the xy plane. Then the undirected distance d between P_1 and P_2 is given by

$$d = \sqrt{(x_1 - x_2)^2 + (y_1 - y_2)^2}$$

EXAMPLE   Determine the distance between the points $A(2, {}^-3)$ and $B({}^-1, 4)$, with both points lying in the same plane.
   Solution   Using Theorem 14.1, we have

$$d = \sqrt{(2 - ({}^-1))^2 + ({}^-3 - 4)^2}$$
$$= \sqrt{(3)^2 + ({}^-7)^2}$$
$$= \sqrt{9 + 49}$$
$$= \sqrt{58} \quad \blacksquare$$

Let x_1 and x_2 be the coordinates of two distinct points on a horizontal line. Let $\bar{x}$ be the coordinate of the point midway between them or the *midpoint* of the line segment joining the two distinct points. If $x_1 < x_2$, then $x_1 < \bar{x} < x_2$; otherwise $x_1 > \bar{x} > x_2$. Suppose that $x_1 < \bar{x} < x_2$. Then since $\bar{x}$ is the midpoint, the distance from x_1 to $\bar{x}$ is equal to the distance from $\bar{x}$ to x_2 or

$$\bar{x} - x_1 = x_2 - \bar{x}$$

Hence

$$\bar{x} + \bar{x} = x_2 + x_1 \quad \text{or} \quad 2\bar{x} = x_1 + x_2$$

It follows that

$$\bar{x} = \tfrac{1}{2}(x_1 + x_2)$$

If $x_1 > \bar{x} > x_2$, we obtain the same results.

For two points on a vertical line, the y coordinate of the midpoint of the line segment joining the points can be found in a similar manner. If y_1 and y_2 are the coordinates of the points and $\bar{y}$ is the coordinate of the point midway between them, then the following formula may be used to find $\bar{y}$:

$$\bar{y} = \tfrac{1}{2}(y_1 + y_2)$$

EXAMPLE   Determine the coordinates of the midpoint of the line segment in the xy plane whose endpoints are $A(5, {}^-1)$ and $B(5, 4)$.

Solution   The line segment $\overline{AB}$ is vertical, so we use the formula

$$\begin{aligned}
\bar{y} &= \tfrac{1}{2}(y_1 + y_2) \\
&= \tfrac{1}{2}({}^-1 + 4) \\
&= \tfrac{3}{2}
\end{aligned}$$

The coordinates of the midpoint of the segment $\overline{AB}$, then, are $(5, \tfrac{3}{2})$.   ■

Now let us consider the coordinates of the midpoint of any line segment in the xy plane. Let $P_1(x_1, y_1)$ and $P_2(x_2, y_2)$ be two distinct points in the plane as in Figure 14.12.

Let $P(\bar{x}, \bar{y})$ be the midpoint of the line segment joining P_1 and P_2. Through P_1, P, and P_2 construct lines perpendicular to and meeting the x axis at A, B, and C, respectively, with coordinates $A(x_1, 0)$, $B(\bar{x}, 0)$, and

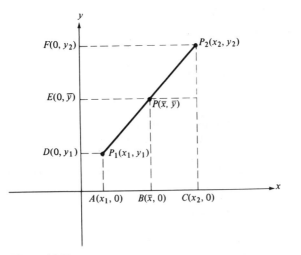

Figure 14.12

$C(x_2, 0)$. Clearly, $B(\bar{x}, 0)$ is the midpoint of the line segment joining A to C. Hence $\bar{x} = \frac{1}{2}(x_1 + x_2)$. Similarly, construct lines through P_1, P, and P_2 perpendicular to and meeting the y axis at D, E, and F, respectively, with coordinates $D(0, y_1)$, $E(0, \bar{y})$, and $F(0, y_2)$. Again, $E(0, \bar{y})$ is the midpoint of the line segment joining the points D and F and $\bar{y} = \frac{1}{2}(y_1 + y_2)$. The abscissa (or first coordinate) of the midpoint of a line segment, then, is the average of the abscissas of the endpoints; the ordinate (or second coordinate) of the midpoint is the average of the ordinate of the endpoints.

THEOREM 14.2   Let $P_1(x_1, y_1)$ and $P_2(x_2, y_2)$ be two distinct points in the xy plane. Then the coordinates of the midpoint $P(\bar{x}, \bar{y})$ of the line segment joining P_1 and P_2 are $\bar{x} = \frac{1}{2}(x_1 + x_2)$ and $\bar{y} = \frac{1}{2}(y_1 + y_2)$.

EXAMPLE   Determine the coordinates of the midpoint of the line segment whose endpoints are $A(^-2, ^-3)$ and $B(1, ^-5)$.
   Solution   Using Theorem 14.2, we have $\bar{x} = \frac{1}{2}(^-2 + 1) = \frac{1}{2}(^-1) = ^-1/2$ and $\bar{y} = \frac{1}{2}(^-3 - 5) = \frac{1}{2}(^-8) = ^-4$. Hence $P(\bar{x}, \bar{y}) = P(^-1/2, ^-4)$.   ∎

EXERCISES

1. For each of the following pairs of points in the same plane, determine whether the points lie on the same horizontal line. For those that do, find the distance between them.
   a. $(2, ^-3)$, $(^-3, ^-3)$.
   b. $(-\sqrt{2}, 1)$, $(1, \sqrt{2})$.
   c. $(3, ^-4)$, $(6, ^-4)$.
   d. $(^-5, 6)$, $(^-5, ^-7)$.
   e. $(\sqrt{2}, ^-3)$, $(2, ^-3)$.
   f. $(\frac{1}{2}, ^-3)$, $(\frac{1}{3}, ^-3)$.
   g. $(\frac{2}{3}, 4)$, $(\frac{2}{3}, 0)$.
   h. $(0, 5)$, $(^-5, 5)$.
   i. $(^-4, 0)$, $(4, 0)$.
   j. $(-\frac{1}{2}, ^-16)$, $(\frac{2}{7}, ^-16)$.
2. For each of the following pairs of points in the same plane determine whether the points lie on the same vertical line. For those that do, find the distance between them.
   a. $(2, ^-3)$, $(^-3, ^-3)$.
   b. $(\sqrt{3}, ^-1)$, $(^-2, ^-1)$.
   c. $(0, 3)$, $(0, ^-3)$.
   d. $(^-1/2, 1/3)$, $(2/5, 1/3)$.
   e. $(1/3, ^-1/2)$, $(1/3, 2/5)$.
   f. $(5, 0)$, $(^-5, 0)$.
   g. $(0, 5)$, $(0, ^-5)$.
   h. $(\sqrt{3}, \sqrt{2})$, $(\sqrt{3}, ^-\sqrt{2})$.
   i. $(2, 1 - \sqrt{3})$, $(2, 5)$.
   j. $(^-7, ^-3)$, $(7, ^-9)$.
3. For each pair of points in the same plane determine if they lie on the same horizontal line and find the distance between those that do. Also, determine which pairs lie on the same vertical line, and find the distance between them.
   a. $(2, ^-3)$, $(4, ^-3)$.
   b. $(0, 5)$, $(0, ^-6)$.
   c. $(\sqrt{3}, ^-2)$, $(^-\sqrt{3}, ^-2)$.
   d. $(4, 7)$, $(6, 5)$.
   e. $(0, 0)$, $(2, 0)$.
   f. $(^-2, ^-1)$, $(3, ^-2)$.
   g. $(0, 0)$, $(0, 2)$.
   h. $(^-3, 2)$, $(^-3, 4)$.
   i. $(^-6, 0)$, $(5, 0)$.
   j. $(1, 2 + \sqrt{5})$, $(1, 3 - \sqrt{2})$.

4. Using the formula for the undirected distance between two points in the same plane, find the distance between the following pairs of points.
   a. $(^-3, 4)$, $(0, 2)$.
   b. $(2, 0)$, $(4, ^-3)$.
   c. $(3, 7)$, $(3, ^-2)$.
   d. $(^-4, 5)$, $(7, 6)$.
   e. $(0, ^-7)$, $(0, 8)$.
   f. $(\sqrt{2}, \sqrt{3})$, $(^-2, 0)$.
   g. $(0, 0)$, $(8, ^-3)$.
   h. $(1, 1)$, $(4, 4)$.
   i. $(^-5, \sqrt{5})$, $(2, ^-1)$.
   j. $(^-3, ^-3)$, $(^-2, ^-1)$.

5. Determine the coordinates of the midpoint of each of the line segments in the same plane whose endpoints are given by the following pairs of points.
   a. $(^-3, 4)$, $(0, 2)$.
   b. $(2, 0)$, $(4, ^-3)$.
   c. $(3, 7)$, $(3, ^-2)$.
   d. $(^-4, 5)$, $(7, 6)$.
   e. $(0, ^-7)$, $(0, 8)$.
   f. $(\sqrt{2}, \sqrt{3})$, $(^-2, 0)$.
   g. $(0, 0)$, $(8, ^-3)$.
   h. $(1, 1)$, $(4, 4)$.
   i. $(^-5, \sqrt{5})$, $(2, ^-1)$.
   j. $(^-3, ^-3)$, $(^-2, ^-1)$.
   k. $(4, ^-6)$, $(9, ^-8)$.
   l. $(\frac{1}{2}, \frac{1}{2})$, $(\frac{1}{4}, \frac{1}{4})$.
   m. $(1/3, 2/7)$, $(2/5, 1/9)$.
   n. $(0, 1/2)$, $(^-1/2, 0)$.
   o. $(2 + \sqrt{3}, ^-4)$, $(^-3 - \sqrt{3}, 7)$.

14.3 LINES

In the previous chapter we discussed the concept of a line as a set of points. We also learned that a line is contained completely in a plane. In this chapter we introduced the concept of the real number line and the real plane. There are times when we would like to assign names to various lines contained in a plane; those names are given as equations. Hence we refer to the equation of a line in a plane. In this section we shall discuss the naming of lines.

DEFINITION 14.3 If P_1 and P_2 are two distinct points in the xy plane with coordinates (x_1, y_1) and (x_2, y_2), respectively, and such that $x_1 < x_2$, then P_1 and P_2 determine a unique line. The difference $x_2 - x_1$ is called the *run* associated with the line and the difference $y_2 - y_1$ is called the *rise*.

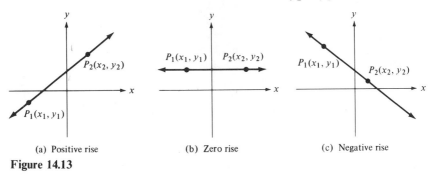

(a) Positive rise   (b) Zero rise   (c) Negative rise

Figure 14.13

In the above definition we qualified that $x_1 < x_2$. Hence the run given by $x_2 - x_1$ is always positive. However, we did not qualify y_1 and y_2. Therefore the rise given by $y_2 - y_1$ may be positive (if $y_1 < y_2$), zero (if $y_1 = y_2$), or negative (if $y_1 > y_2$). In Figure 14.13 we illustrate lines with positive rise, zero rise, and negative rise.

DEFINITION 14.4 Let P_1 and P_2 be two distinct points in the xy plane with coordinates (x_1, y_1) and (x_2, y_2), respectively, and such that $x_1 < x_2$. Then the *slope*, denoted by m, of the line determined by P_1 and P_2 is defined to be the rise divided by the run; that is,

$$m = \frac{y_2 - y_1}{x_2 - x_1}$$

EXAMPLE   Determine the slope of the line that passes through the points $A(2, {}^-3)$ and $B(5, 2)$ in the xy plane.
Solution   Letting $x_1 = 2 < 5 = x_2$, the run is $x_2 - x_1 = 5 - 2 = 3$; the rise is $y_2 - y_1 = 2 - ({}^-3) = 5$. Hence the slope m is equal to $\frac{3}{5}$. ■

EXAMPLE   Determine the slope of the line that passes through the points $P_1({}^-3, 5)$ and $P_2(4, 5)$ in the xy plane.
Solution

$$m = \frac{y_2 - y_1}{x_2 - x_1} = \frac{5 - 5}{4 - ({}^-3)} = \frac{0}{7} = 0 \quad ■$$

DEFINITION 14.5 For the point $P_1(x_1, y_1)$ in the xy plane the x_1 value is called the *abscissa* of the point and the y_1 value is called the *ordinate* of the point.

Combining Definitions 14.4 and 14.5, we may now conclude that the slope of a line in a given plane is simply the quotient of the difference in the ordinates of the two distinct points on the line divided by the *corresponding* difference in the abscissas. Hence if $P_1(x_1, y_1)$ and $P_2(x_2, y_2)$ are two distinct points in the xy plane, we have

$$m = \frac{y_2 - y_1}{x_2 - x_1} \quad \text{or} \quad m = \frac{y_1 - y_2}{x_1 - x_2}$$

provided that $x_1 \neq x_2$. If $x_1 = x_2$, the two points P_1 and P_2 would lie on the same vertical line. (Do you agree?) In this case we say that the slope of the line is not defined.

Using the concept of the slope of a line, we may now consider the various forms of the equation of a line.

14.31 POINT-SLOPE FORM

Let L be the line in the xy plane determined by the two distinct points $P_1(x_1, y_1)$ and $P_2(x_2, y_2)$ such that $x_1 \neq x_2$, and let $P(x, y)$ be an arbitrary point on L. Since $P_1(x_1, y_1)$ and $P_2(x_2, y_2)$ are two distinct points on L, we can determine the slope of L as follows:

$$m = \frac{y_2 - y_1}{x_2 - x_1}$$

Further, if the point $P(x, y)$ lies on L, the slope of the line through the points $P(x, y)$ and $P_1(x_1, y_1)$ must also be equal to m since all line segments of the same line have the same slope. Therefore we have

$$m = \frac{y - y_1}{x - x_1}$$

Since $x - x_1 \neq 0$, we can multiply both sides of this equation by $x - x_1$ to obtain

$$y - y_1 = m(x - x_1) \tag{14.1}$$

Equation (14.1) is called the *point-slope form* of the equation of the line L since it is given in terms of the coordinates of the point $P_1(x_1, y_1)$, which lies on the line L and the slope m of the line.

EXAMPLE Determine the equation of the line in the xy plane that passes through the point $A(2, {}^-3)$ and has a slope $m = {}^-2$.
  Solution Using Equation (14.1) with $x_1 = 2$, $y_1 = {}^-3$, and $m = {}^-2$, we have

$$y - y_1 = m(x - x_1)$$

or

$$y - ({}^-3) = {}^-2(x - 2)$$
$$y + 3 = {}^-2x + 4$$
$$y = {}^-2x + 1$$

as the required equation. ▪

It should be somewhat obvious that Equation (14.1) can also be used if we do not know the slope of the line but do know the coordinates of two distinct points through which it passes. To illustrate this fact consider the following example.

EXAMPLE Determine the equation of the line in the xy plane that passes through the two points $A(2, {}^-3)$ and $B({}^-4, 5)$.

Solution   Let $P_1(x_1, y_1) = A(2, {}^-3)$ and $P_2(x_2, y_2) = B({}^-4, 5)$. Since we defined

$$m = \frac{y_2 - y_1}{x_2 - x_1}$$

we have

$$m = \frac{5 - ({}^-3)}{{}^-4 - 2}$$

$$= \frac{5 + 3}{{}^-6}$$

$$= \frac{8}{{}^-6}$$

$$= \frac{{}^-4}{3}$$

We may now use the coordinates of either point A or point B together with the computed value of m. Using the coordinates of $A(2, {}^-3)$ and Equation (14.1), we now have

$$y - y_1 = m(x - x_1)$$

or

$$y - ({}^-3) = \frac{{}^-4}{3}(x - 2)$$

$$y + 3 = \frac{{}^-4}{3}x + \frac{8}{3}$$

$$y = \frac{{}^-4}{3}x - \frac{1}{3}$$

as the required equation. ∎

14.32 SLOPE-INTERCEPT FORM

Equation (14.1) may be written as

$$y = m(x - x_1) + y_1$$

$$= mx - mx_1 + y_1$$

Substituting b for $y_1 - mx_1$, we get

$$y = mx + b \tag{14.2}$$

In the form of Equation (14.2) the b is the y intercept or the ordinate of the point where the line crosses the y axis. Equation (14.2) is called the *slope-intercept form* of the equation of the line L, since the equation is given in terms of the slope, m, of the line and its y intercept, b.

EXAMPLE   Determine the equation of the line in the xy plane that has a slope of $^-1/3$ and a y intercept of 2.

Solution   Using Equation (14.2) with $m = {}^-1/3$ and $b = 2$, we obtain

$$y = mx + b$$

or

$$y = \frac{^-1}{3} x + 2$$

as the required equation.   ∎

EXAMPLE   Determine the slope of the line with equation $2x - 7y = 6$.

Solution   Solving the above equation for y, we have

$$y = \frac{2}{7} x - \frac{6}{7}$$

Comparing this equation with Equation (14.2), we have that $m = 2/7$ and $b = {}^-6/7$. Hence the slope is $m = 2/7$.   ∎

14.33 THE TWO-INTERCEPTS FORM

Another form of the equation of a line that is useful in mathematics is the *two-intercepts form*. Rewriting Equation (14.2), we have

$$y - mx = b$$

If $b \neq 0$, then we may divide both sides of this equation by b, obtaining

$$\frac{y - mx}{b} = 1$$

or

$$\frac{y}{b} - \frac{mx}{b} = 1$$

or

$$\frac{y}{b} + \frac{x}{(^-b/m)} = 1$$

If we substitute a for $^-b/m$, we get

$$\frac{x}{a} + \frac{y}{b} = 1 \tag{14.3}$$

Observe that if $y = 0$, then $x = a$, which is called the x intercept or the abscissa of the point where the line crosses the x axis. If $x = 0$, then $y = b$, the y intercept. Equation (14.3) is called the *two-intercept form* of the equation of the line L, since it is given in terms of the x intercept and the y intercept.

14.34 THE GENERAL FORM

Equations (14.1), (14.2), and (14.3) may each be written in the form

$$Ax + By + C = 0 \tag{14.4}$$

where A, B, and C are constants such that at least one of the values of A and B is not equal to zero. Equation (14.4) is called the *general form* of the equation of a line.

EXAMPLE Determine the general form of the equation of the line in the xy plane that passes through the points $A(0, 2)$ and $B(^-3, 0)$.

Solution Since the line passes through the point $A(0, 2)$, the y intercept is $b = 2$. Since the line passes through the point $B(^-3, 0)$, the x intercept is $a = {}^-3$. Hence using the two intercepts form of the equation of a line, in the xy plane we have

$$\frac{x}{a} + \frac{y}{b} = 1$$

or

$$\frac{x}{^-3} + \frac{y}{2} = 1$$

Multiplying both sides of the last equation by $^-6$, we obtain

$$2x - 3y = {}^-6$$

or

$$2x - 3y + 6 = 0$$

This last equation is now in the general form and is the required solution. ∎

We will conclude this section by noting that lines can be classified according to their slopes.

DEFINITION 14.6 Two lines, L_1 and L_2, in the same plane, with slopes m_1 and m_2, respectively, are said to be *parallel* if and only if $m_1 = m_2$. Vertical lines have no slope, but all vertical lines in the same plane are parallel.

EXAMPLE Determine the equation of the line in the xy plane which passes through the point $A(^-1, 3)$ and is parallel to the line L with the equation $2y - 3x = 7$.

Solution If the line in question is parallel to L, then its slope must be equal to the slope of L. We rewrite the equation for the line L, obtaining $y = \frac{3}{2}x + \frac{7}{2}$. The slope of L, then, is $\frac{3}{2}$, and this is also the slope of our required line. We know that the required line passes through the point $A(^-1, 3)$, so we use the point-slope form for the equation of the line. Hence

$$y - y_1 = m(x - x_1)$$

$$y - 3 = \frac{3}{2}(x + 1)$$

$$2(y - 3) = 3(x + 1)$$

$$2y - 6 = 3x + 3$$

$$2y = 3x + 9$$

which is the required equation. ■

DEFINITION 14.7 Two lines, L_1 and L_2, in the same plane, with slopes m_1 and m_2, respectively, and such that $m_1 \neq 0$ are said to be *perpendicular* if and only if $m_2 = {}^-1/m_1$.

EXERCISES

1. Determine the slope of the line that passes through each of the following pairs of points in the same plane.
   a. $(^-3, 4)$, $(0, 2)$.   b. $(2, 0)$, $(4, ^-3)$.   c. $(3, 7)$, $(3, ^-2)$.
   d. $(^-4, 5)$, $(7, 6)$.   e. $(0, ^-7)$, $(0, 8)$.   f. $(\sqrt{2}, \sqrt{3})$, $(^-2, 0)$.
   g. $(0, 0)$, $(8, ^-3)$.   h. $(1, 1)$, $(4, 4)$.   i. $(^-5, \sqrt{5})$, $(2, ^-1)$.
   j. $(^-3, ^-3)$, $(^-2, ^-1)$.
2. For each of the equations given below determine the slope and the y intercept of the lines represented. Graph the lines.
   a. $y = 3x - 4$.   b. $y = 1 - 5x$.   c. $2y = 3x + 2$.
   d. $3x = y + 7$.   e. $x - 2y = 3$.   f. $y - 4x = ^-2$.
   g. $y = 2(x - 3)$.   h. $2x - 3y = 1$.   i. $4x + 3y - 2 = 0$.
   j. $^-5x = 2y - 3$.

3. For each of the following write an appropriate equation with integral coefficients for the line in the xy plane with slope and y intercept as given. Draw the graph of each line.

a. $m = {}^-3, b = 2$.   b. $m = \frac{1}{2}, b = 0$.   c. $m = 2, b = {}^-1/3$.
d. $m = 0, b = {}^-4$.   e. $m = {}^-2/3, b = 1$ f. $m = 2c, b = 3p$.
g. $m = 0, b = 0$.   h. $m = {}^-1/7, b = 2/9$.

4. a. Determine the slope of the line in the xy plane with equation $y = 4$.
   b. Determine the slope of the line in the xy plane with equation $x = {}^-2$.

5. Determine the equation of the line in the xy plane satisfying the following conditions.

a. Slope $= \frac{1}{2}$, y intercept $= {}^-3$.
b. Slope $= {}^-\frac{1}{2}$, x intercept $= 3$.
c. Passes through the points $(3, 0)$ and $(3, {}^-2)$.
d. Slope $= 2$ and passes through the point $(2, {}^-3)$.
e. Passes through the points $(3, {}^-2)$ and $(0, 4)$.
f. Passes through the points $({}^-2, 4)$ and $(0, 4)$.
g. Slope $= 0$ and passes through the point $(2, {}^-3)$.
h. Slope not defined and passes through the point $(2, {}^-3)$.
i. x intercept $= {}^-3$, y intercept $= 2$.
j. x intercept $= 2$, and passes through the point $(2, {}^-1)$.

6. If two lines in the same plane are parallel, then their slopes are equal. For each of the following determine which pairs of equations represent parallel lines. (The independent variable precedes the dependent variable in alphabetical order.)

a. $y = 3x + 2; 2y = 3x + 4$.   b. $2r = s - 3; r = 2s + 7$.
c. $2u - 3v + 4 = 0; 3v = 3 + 2u$.   d. $2t - 3s = 4; 3t = 3 + 2s$.
e. $z = 3; z = \sqrt{2}$.   f. $2x = 3y + 4; y = \frac{2}{3}(5 + x)$.

7. Determine the equation of the line in the xy plane satisfying the following conditions.

a. Parallel to the line with the equation $2x - 3y + 4 = 0$ and passes through the point $(2, {}^-5)$.
b. Parallel to the line with the equation $3x - 2y = 4$ and has a y intercept of 3.
c. Has a slope of $\frac{2}{3}$ and passes through the point of intersection of the two lines with equations $y = {}^-3$ and $x = {}^-2$.

8. If three or more points in the same plane lie on the same line, we say that the points are collinear. Show that the following sets of points are collinear. [Hint: If $P_1(x_1, y_1)$, $P_2(x_2, y_2)$, and $P_3(x_3, y_3)$ are collinear, then the slope of the line between P_1 and P_2 equals the slope of the line between P_1 and P_3.]

a. $({}^-1, {}^-1), (0, 0), (2, 2)$.   b. $(1, 2), (2, 4), (5, 10)$.
c. $({}^-2, {}^-4), (0, 2), (1, 5)$.   d. $(1, {}^-3), ({}^-3, {}^-11), (4, 3)$.
e. $(\frac{1}{2}, 2), ({}^-1, {}^-1), (3, 7)$.   f. $(3, 11), (1, 7), ({}^-3, {}^-1)$.
g. $({}^-2, 5), ({}^-1/3, 0), (4, {}^-13)$.   h. $(1, {}^-1), (3, 5), (2, 2)$.

9. Rewrite each of the equations in Exercise 2 in the general form of the equation of a line.

10. Rewrite each of the equations in Exercise 2 in the two-intercepts form of the equation of a line.

11. Determine which of the following pairs of equations represent perpendicular lines.

   a. $y = 2x + 3$; $x = 2y + 4$.   b. $2x - 3y + 4 = 0$; $2y = 1 + 3x$.
   c. $2r + 5s + 1 = 0$; $2s = 5r - 2$.   d. $u - v = 7$; $v + u = 9$.
   e. $y = 4$; $x = {}^-3y + 4$.   f. $3a = 2b$; $2a + 3b = 1$.

12. Determine the equation of the line in the xy plane satisfying the following conditions.

   a. Perpendicular to the line with the equation $4x = 5y + 1$ and passes through the point $({}^-2, 3)$.
   b. Perpendicular to the line with the equation $x - 6y + 2 = 0$ and has an x intercept of $^-2$.

14.4 CIRCLES

In the previous chapter we indicated that a circle is an example of a simple closed curve. In this section we will define a circle as a particular set of points and derive equations for circles.

DEFINITION 14.8 A *circle* is a set of points in a plane that are a given distance from a fixed point in the plane. The fixed point is called the *center* of the circle, and the given distance is called its *radius*.

In Figure 14.14 we illustrate a circle in the xy plane with its center at the point $C(h, k)$ and a radius, r.

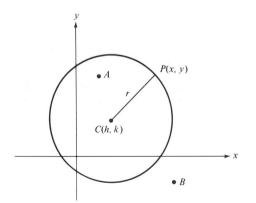

Figure 14.14

As a simple closed curve the circle separates the plane containing it into three disjoint subsets of the plane. They are the circle itself, the interior of the circle, and the exterior of the circle. It should be noted that the center of the circle is a point in the interior of the circle and, hence, does not belong to the circle.

The interior of the circle can be thought of as the set of all points in the plane containing the circle whose distances from the center of the circle are less than the radius of the circle. Likewise, if a point lies in the exterior of the circle, then the distance from the point to the center of the circle is greater than the radius. The point A in Figure 14.14 lies in the interior of the circle, and the point B lies in its exterior. Any path from point A to point B that lies in the plane must necessarily cross the circle.

To determine an algebraic specification or equation for a circle, let $C(h, k)$ denote the given point or center of the circle in the xy plane and let r be the given distance or the radius of the circle. (See Figure 14.14.) For an arbitrary point $P(x, y)$ on the circle we require that $m(\overline{CP}) = r$. Since the undirected distance between the points C and P is given by

$$\sqrt{(x - h)^2 + (y - k)^2}$$

we now have

$$\sqrt{(x - h)^2 + (y - k)^2} = r$$

Squaring both sides of this equation, we get

$$(x - h)^2 + (y - k)^2 = r^2 \tag{14.5}$$

Equation (14.5) is called the *standard form* of the equation of a circle in the xy plane with its center at $C(h, k)$ and its radius equal to r.

In the special case where $h = k = 0$, Equation (14.5) becomes

$$x^2 + y^2 = r^2 \tag{14.6}$$

Equation (14.6) is called the standard form of the equation of the circle in the xy plane with its center at the origin $C(0, 0)$ and its radius equal to r.

EXAMPLE   Determine the equation of a circle in the xy plane whose center is at the point with coordinates $C(^-3, 2)$ and whose radius is 3.

Solution   Using Equation (14.5) with $h = -3$, $k = 2$, and $r = 3$, we have

$$(x - h)^2 + (y - k)^2 = r^2$$

or

$$(x - (^-3))^2 + (y - 2)^2 = (3)^2$$

or

$$(x + 3)^2 + (y - 2)^2 = 9$$

as the required equation.   ∎

EXAMPLE Determine the equation of a circle in the st plane whose center is at the point $C(0, {}^-4)$ and whose radius is 4.

Solution We use Equation (14.5) with $h = 0$, $k = {}^-4$, and $r = 4$ and replace x by s and y by t. Hence we have

$$(s - h)^2 + (t - k)^2 = r^2$$

or

$$(s - 0)^2 + (t - ({}^-4))^2 = (4)^2$$
$$s^2 + (t + 4)^2 = 16$$

as the required equation. ∎

If we expand the terms in Equation (14.5) and combine similar terms, we get

$$(x - h)^2 + (y - k)^2 = r^2$$
$$x^2 - 2hx + h^2 + y^2 - 2ky + k^2 = r^2$$
$$x^2 + y^2 - 2hx - 2ky + (h^2 + k^2 - r^2) = 0$$

This last equation can be written in the more general form

$$x^2 + y^2 + Dx + Ey + F = 0 \tag{14.7}$$

which is called the *general form of the equation of a circle* in the xy plane.

Even though the equation of each circle in the xy plane can be written in the form of Equation (14.7), we must be cautious since not all equations in this form necessarily describe a circle. It can be shown that the equation

$$x^2 + y^2 + Dx + Ey + F = 0$$

can be rewritten in the equivalent form

$$\left(x + \frac{D}{2}\right)^2 + \left(y + \frac{E}{2}\right)^2 = \frac{D^2 + E^2 - 4F}{4} \tag{14.8}$$

Now since D, E, and F are arbitrary real values, $\frac{1}{4}(D^2 + E^2 - 4F)$ may be positive, negative, or zero. Hence we shall consider the three cases separately.

CASE 1   If $\frac{1}{4}(D^2 + E^2 - 4F) > 0$, then Equation (14.8) is in the standard form of the equation of a circle with $r^2 > 0$, and we have a *circle* with $C(h, k) = C({}^-D/2, {}^-E/2)$ and $r = \frac{1}{2}\sqrt{D^2 + E^2 - 4F}$.

CASE 2   If $\frac{1}{4}(D^2 + E^2 - 4F) = 0$, then Equation (14.8) is in the standard form of a circle with $r^2 = 0$, and we have a *point* at $C({}^-D/2, {}^-E/2)$.

CASE 3   If $\frac{1}{4}D^2 + E^2 - 4F) < 0$, then Equation (14.8) would correspond to the standard form of the equation of a circle with $r^2 < 0$, which is not possible. Hence we would have *no graph* for this equation.

EXAMPLE If the equation $x^2 + 2y^2 - 3x + 4y + 5 = 0$ describes a circle, determine its center and its radius.

Solution Since the coefficients of x^2 and y^2 are not equal, the equation does not describe a circle. ∎

EXAMPLE If the equation $2x^2 + 2y^2 - 4x - 6y = -6$ describes a circle, determine its center and its radius.

Solution Dividing both sides of the equation by 2 and transposing the constant term, we obtain

$$x^2 + y^2 - 2x - 3y + 3 = 0$$

Hence $D = {}^-2$, $E = {}^-3$, and $F = 3$ and

$$r^2 = \tfrac{1}{4}(D^2 + E^2 - 4F) = \tfrac{1}{4}[({}^-2)^2 + ({}^-3)^2 - 4(3)]$$
$$= \tfrac{1}{4}(4 + 9 - 12)$$
$$= \tfrac{1}{4}(1)$$
$$= \tfrac{1}{4}$$

Since $r^2 = \tfrac{1}{4} > 0$, using Equation (14.8) we determine that the given equation describes a circle with center at $C({}^-D/2, \ {}^-E/2) = C(1, \ 3/2)$ and radius $r = \tfrac{1}{2}\sqrt{D^2 + E^2 - 4F} = \tfrac{1}{2}$. ∎

We will conclude this section with a brief discussion of measure associated with a circle. Since a circle is a simple closed curve, we can consider the length of this circle called its *circumference*.

DEFINITION 14.9 The perimeter or *circumference*, C, of a circle is given by the formula $C = 2\pi r$, where r is the radius of a circle and π (read "pi") is an irrational number approximately equal to 3.14. (Both r and C are given in units.)

EXAMPLE Determine the circumference of the circle whose equation is $(x + 1)^2 + (y - 2)^2 = 4$.

Solution From the given equation we determine that $r^2 = 4$ and, hence, the radius, r, of the circle is equal to 2. Using Definition 14.9, we now have

$$C = 2\pi r$$

or

$$C = 2\pi \ (2)$$
$$= 4\pi \ \text{units}$$

The *exact* value of C is 4π units. However, using 3.14 for the value of π, we can get the *approximate* value of C to be $4(3.14) = 12.56$ units. ∎

DEFINITION 14.10 The *area*, A, of the interior of the circle whose radius is r is given by the formula $A = \pi r^2$, where r is given in units and A is given in square units.

    EXAMPLE The area of the region bounded by the circle whose equation is $x^2 + y^2 = 25$ is given by

$$A = \pi r^2$$

or

$$A = \pi(25)$$

or

$$A = 25\pi \text{ sq units}$$

since $r^2 = 25$.

EXERCISES

1. Write the standard form of the equation of the circle in the coordinate plane indicated with the given centers and radii (plural of radius).
   a. $C(2, {}^-3)$, $r = 3$, xy plane.   b. $C({}^-4, 1)$, $r = 2$, st plane.
   c. $C(0, {}^-3)$, $r = \sqrt{3}$, uv plane.   d. $C(\sqrt{2}, 1)$, $r = \frac{2}{3}$, yz plane.
   e. $C(2, 0)$, $r = \frac{3}{7}$, pq plane.   f. $C({}^-3, \sqrt{5})$, $r = 4$, mn plane.
2. Write the general form of the equation for each of the circles described in Exercise 1.
3. Using Equation (14.8), determine whether each of the following equations describes a circle, a point, or no graph in the xy plane.
   a. $(x - 2)^2 + (y + 3)^2 = 16$.   b. $x^2 + y^2 - 6x = 3 + 4y$.
   c. $2x^2 + 2y^2 - 4x - 8y + 10 = 0$.   d. $3x^2 + 3y^2 + 4x + 2y + 6 = 0$.
   e. $4(x + 2)^2 + 4(y - 5)^2 = 12$.   f. $(2x - 3)^2 + (2y + 5)^2 = 4$.
4. Find the circumference for each of the circles described in Exercise 1.
5. Find the circumference for each of the circles described in Exercise 3.
6. Find the area for each of the circular regions bounded by the circles described in Exercise 1.

14.5 SYSTEMS OF EQUATIONS

In Section 14.3 we considered finding the equation of a line. A line is a set of points in a plane, and each point is given by an ordered pair of coordinates. Actually, a line is the graph of a relation as was introduced in Chapter 4 and the equation of the line is the specification for the relation.

    For instance, if the line L_1 in the xy plane has the equation $y = 2x + 3$, we may write

$$L_1 = \{(x, y) | y = 2x + 3\}$$

to indicate that the line L_1 is the set of all points in the xy plane whose coordinates are (x, y) and such that $y = 2x + 3$.

Now we may wish to determine if the point A in the xy plane with coordinates $(4, 7)$ lies on the line. If it does, then the coordinates $x = 4$ and $y = 7$ must satisfy the equation $y = 2x + 3$. By this we mean that substitution of 4 for x and 7 for y must make $y = 2x + 3$ a true statement. Substituting, we determine that $7 \neq 2(4) + 3$ and, hence, the point A does not belong to the line L_1. However, the point B with coordinates $(2, 7)$ does belong to the line since $7 = 2(2) + 3$.

Obviously, there are infinitely many points belonging to the line L_1. Knowing the abscissa, x, of the point, we can determine the ordinate, y, by substituting the x value in the equation $y = 2x + 3$ and determining the y value.

To graph the line in the xy plane, we really need to know the coordinates of only two distinct points of the line since any two distinct points determine a unique line in a plane. However, we should use at least three points for a check of our calculations. To graph the line L_1 with the equation $y = 2x + 3$ we could let $x = {}^-1$, 0, and 3 and then determine the corresponding values for y:

$$y = 2x + 3$$

If $x = {}^-1$ $y = 2({}^-1) + 3 = 1$
If $x = 0$ $y = 2(0) + 3 = 3$
If $x = 3$ $y = 2(3) + 3 = 9$

We now have the points $A({}^-1, 1)$, $B(0, 3)$ and $C(3, 9)$ that lie on the line

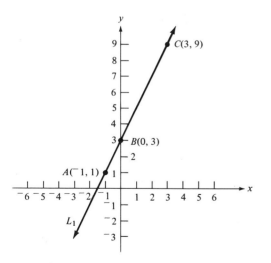

Figure 14.15

L_1. Plotting these points in the xy plane and passing a line through them, we obtain the graph as illustrated in Figure 14.15.

Now consider the line L_2 in the xy plane with equation $y = {}^-3x + 8$. We may write

$$L_2 = \{(x, y) \mid y = {}^-3x + 8\}$$

to indicate that the line L_2 is the set of all points in the xy plane whose coordinates are (x, y) and such that $y = {}^-3x + 8$. To graph L_2 we may select three values for x and compute the corresponding values for y:

$$y = {}^-3x + 8$$
$$\text{If } x = {}^-2 \qquad y = {}^-3({}^-2) + 8 = 14$$
$$\text{If } x = 0 \qquad y = {}^-3(0) + 8 = 8$$
$$\text{If } x = 1 \qquad y = {}^-3(1) + 8 = 5$$

We now have the points $D({}^-2, 14)$, $E(0, 8)$ and $F(1, 5)$ that lie on L_2. Plotting these points in the xy plane and passing a line through them, we obtain the graph as illustrated in Figure 14.16.

Next consider the graphs of L_1 and L_2 together in the xy plane as in Figure 14.17. An examination of Figure 14.17 reveals that the two lines

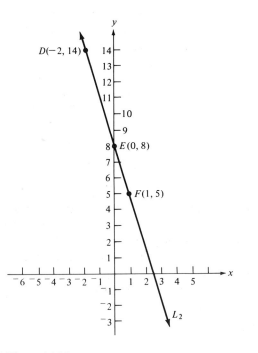

Figure 14.16

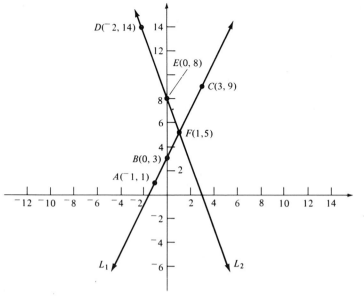

Figure 14.17

probably intersect at the point $F(1, 5)$. This is readily checked by substituting $x = 1$ and $y = 5$ in *each* of the equations for L_1 and L_2 and, indeed, F is the point of intersection of the two lines. Hence we have

$$L_1 = \{(x, y)\,|\,y = 2x + 3\}$$
$$L_2 = \{(x, y)\,|\,y = {}^-3x + 8\}$$

and

$$L_1 \cap L_2 = \{(x, y)\,|\,y = 2x + 3\} \cap \{(x, y)\,|\,y = {}^-3x + 8\}$$
$$= \{(x, y)\,|\,y = 2x + 3 \quad and \quad y = {}^-3x + 8\}$$
$$= \{(1, 5)\}$$

Because there are infinitely many points on L_1 there are infinitely many solutions for the equation $y = 2x + 3$ since the coordinates of each of the points on L_1 must satisfy this equation. Similarly, the equation $y = {}^-3x + 8$ also has infinitely many solutions. However, the two equations have only one solution in common which we found using a graphical method.

We also can find the solution of the system of equations

$$\begin{cases} y = 2x + 3 \\ y = {}^-3x + 8 \end{cases}$$

using algebraic methods. To solve the system means to find solutions common to the two equations. Now if we substitute $2x + 3$ for y in the equation $y = {}^-3x + 8$, we obtain

$$y = {}^-3x + 8$$

or

$$2x + 3 = {}^-3x + 8$$

or

$$2x + 3x = 8 - 3$$

or

$$5x = 5$$

or

$$x = 1$$

But when $x = 1$, $y = 2x + 3$ becomes $y = 2(1) + 3 = 5$ and $y = -3x + 8$ becomes $y = {}^-3(1) + 8 = 5$. Clearly, then, $(1, 5)$ is a solution for the system of these two equations. The method used to find the solution is known as the *method of substitution*.

Another method is the *method of elimination*, which we will now illustrate. Starting with

$$\begin{cases} y = 2x + 3 \\ y = {}^-3x + 8 \end{cases}$$

we may subtract the second equation from the first, obtaining

$$0 = 5x - 5$$

Solving this last equation for x, we obtain

$$5x = 5$$

or

$$x = 1$$

Substituting 1 for x in *either* of the original equations, say the first, we obtain

$$y = 2x + 3$$

or

$$y = 2(1) + 3$$

or

$$y = 5$$

Hence $(1, 5)$ is our required solution.

Equations of the form $Ax + By + C = 0$ such that not both A and B are zero are called linear equations since the graphs associated with them are lines. There are three possibilities:

1. If $A \neq 0$ and $B = 0$, the line is *vertical*.
2. If $A = 0$ and $B \neq 0$, the line is *horizontal*.
3. If $A \neq 0$ and $B \neq 0$, the line is *oblique*; that is, the line is neither vertical nor horizontal.

Now if we attempt to solve a system of linear equations consisting of two equations, remembering that each equation is associated with a line, it should be clear that

1. We may have a *unique solution*, if the lines are intersecting lines.

2. We may have *no solution*, if the lines are parallel lines.

3. We may have *infinitely many solutions*, if the lines are the same.

EXAMPLE   Solve the following system of equations:

$$\begin{cases} 2x - 3y = 7 \\ \quad x + y = 2 \end{cases}$$

Solution   Solving for y in the second equation, we obtain $y = 2 - x$. Substituting $2 - x$ for y in the first equation, we get

$$2x - 3y = 7$$

or

$$2x - 3(2 - x) = 7$$

$$2x - 6 + 3x = 7$$

$$5x = 7 + 6$$

$$5x = 13$$

$$x = \frac{13}{5}$$

Substituting $x = 13/5$ in the equation $y = 2 - x$, we obtain

$$y = 2 - x$$

or

$$y = 2 - \frac{13}{5}$$

$$y = \frac{10 - 13}{5}$$

$$y = \frac{^-3}{5}$$

Checking, we determine

$$2x - 3y = 2\left(\frac{13}{5}\right) - 3\left(\frac{^-3}{5}\right)$$

$$= \frac{26}{5} + \frac{9}{5}$$

$$= \frac{35}{5}$$

$$= 7$$

and

$$x + y = \frac{13}{5} + \frac{^-3}{5}$$

$$= \frac{13 - 3}{5}$$

$$= \frac{10}{5}$$

$$= 2$$

Therefore $x = 13/5$, $y = {}^-3/5$ is the unique solution. ■

EXAMPLE Solve the following system of equations:

$$\begin{cases} x - 2y = 6 \\ 2x + 3y = 4 \end{cases}$$

Solution Multiplying the first equation by 2, we obtain the equivalent system

$$\begin{cases} 2x - 4y = 12 \\ 2x + 3y = 4 \end{cases}$$

Subtracting the second equation from the first, we eliminate the x term and obtain

$$^-7y = 8$$

or

$$y = \frac{^-8}{7}$$

Substituting $^-8/7$ for y in either of the original equations, say the first, we obtain

$$x - 2y = 6$$

or

$$x - 2\left(\frac{^-8}{7}\right) = 6$$

$$x + \frac{16}{7} = 6$$

$$x = 6 - \frac{16}{7}$$

$$x = \frac{42 - 16}{7}$$

$$x = \frac{26}{7}$$

Checking, we determine that

$$x - 2y = \frac{26}{7} - 2\left(\frac{^-8}{7}\right)$$

$$= \frac{26}{7} + \frac{16}{7}$$

$$= \frac{42}{7}$$

$$= 6$$

and

$$2x + 3y = 2\left(\frac{26}{7}\right) + 3\left(\frac{^-8}{7}\right)$$

$$= \frac{52}{7} - \frac{24}{7}$$

$$= \frac{28}{7}$$

$$= 4$$

Therefore $x = 26/7$, $y = {}^-8/7$ is the unique solution. ∎

Similarly, since a circle is a set of points also, systems of equations of the form

$$\begin{cases} x^2 + y^2 = 16 \\ x + y = 4 \end{cases}$$

also can be solved using a graphical method or the algebraic method of substitution.

Since the algebraic method requires a knowledge of the solution of a quadratic equation of the form $ax^2 + bx + c = 0$ $(a \neq 0)$, we will use only the graphical method in this chapter.

The first equation describes a circle with its center at $C(0, 0)$ and radius $r = 4$. The second equation describes a line. Graphing both on the same set of coordinate axes, we obtain the graph as illustrated in Figure 14.18. The points of intersection are A and B, which appear to have the coordinates $A(0, 4)$ and $B(4, 0)$. Checking $A(0, 4)$, we determine that

$$x^2 + y^2 = (0)^2 + (4)^2 = 0 + 16 = 16$$

and

$$x + y = 0 + 4 = 4$$

which checks.

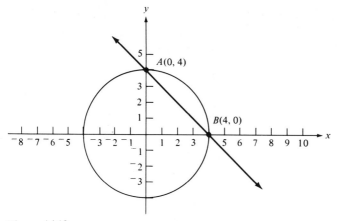

Figure 14.18

Checking $B(4, 0)$, we determine that

$$x^2 + y^2 = (4)^2 + (0)^2 = 16 + 0 = 16$$

and

$$x + y = 4 + 0 = 4$$

which also checks.

Hence the given system has two solutions $(x = 0, \ y = 4)$ and $(x = 4, \ y = 0)$.

Clearly, a line and a circle in the same plane may have two, one, or no points of intersection, and the corresponding systems of equations, then, may have two, one, or no solutions.

EXERCISES

1. Using the method of substitution, solve each of the following systems of linear equations.

   a. $2x - 3y = 1$ b. $x + 2y = 3$
      $y = 2x + 1$ $2x - \ y = 0$
   c. $3x - 2y = 2$ d. $4x - \ y = 5$
      $x + 2y = 1$ $2x + 3y = 1$
   e. $5x + 2y = {}^-2$ f. $3x - 4y = 0$
      $x - \ y = 1$ $2x + 3y = 5$

2. Using the method of elimination, solve each of the following systems of linear equations.

   a. $2x - y = 4$ b. $3x + 2y = 5$
      $x + y = 1$ $x - 3y = 1$
   c. $3x + 4y = 1$ d. $5x - 4y = 2$
      $2x - 3y = 4$ $2x + 3y = {}^-1$

3. Using graphical methods, determine the number of solutions for each of the following systems of equations.

a. $x^2 + y^2 = 16$
   $x - y = 2$

b. $x^2 + y^2 = 25$
   $2x - 3y = 4$

c. $x^2 + y^2 = 9$
   $x + y = {}^-7$

d. $x^2 + y^2 = 16$
   $y = 3$

e. $x^2 + y^2 = 49$
   $x - 7 = 0$

f. $x^2 + y^2 = 36$
   $y = x$

SUMMARY

In this chapter we introduced the concept of coordinate geometry as a unifying relationship between geometry and the arithmetic of numbers. We discussed lines and circles together with their equations and considered certain systems of equations and their solutions.

Standard forms of the equations for lines and circles were derived, and the general forms of the equation for lines and circles also were introduced.

Distance between two points in a plane was introduced as the measure of the line segment determined by the two points. Various formulas for computing distances were derived culminating with the general undirected distance formula.

During our discussions the following symbols were introduced.

SYMBOL	INTERPRETATION
$(\bar{x}, \bar{y})$	The coordinates of the midpoint of a line segment
m	The slope of a line
b	The y intercept of a line in the xy plane
a	The x intercept of a line in the xy plane

REVIEW EXERCISES FOR CHAPTER 14

1. Determine the distance between the two points in the same plane with the indicated coordinates.

a. $({}^-2, 3), (1, {}^-2)$.

b. $({}^-3, 0), ({}^-3, 4)$.

c. $(1, 0), (2, {}^-3)$.

d. $({}^-4, {}^-1), ({}^-5, {}^-2)$.

e. $(0, 2), ({}^-3, 2)$.

f. $({}^-2, {}^-2), ({}^-4, {}^-4)$.

g. $(2, {}^-3), ({}^-2, 3)$.

h. $(4, {}^-5), (0, 1)$.

i. $(1, {}^-7), ({}^-2, 6)$.

j. $({}^-5, 0), (2, {}^-3)$.

2. Determine the slope of the line that passes through each pair of indicated points given in Exercise 1. If the slope is undefined, indicate this.

3. Determine the equation of the line in the xy plane satisfying the following conditions:

   a. $m = 2$, $b = 3$.

   b. Passes through the points with coordinates $(2, {}^-3)$ and $({}^-1, 2)$.

   c. Has an x intercept of 3 and a y intercept of ${}^-4$.

   d. Passes through the point with coordinates $(1, 2)$ and is parallel to the line with equation $y = 2x + 3$.

   e. Passes through the point with coordinates $({}^-2, 3)$ and is perpendicular to the line with equation $y = 4$.

4. Determine the coordinates of the midpoint of the line segment whose endpoints have the indicated coordinates.

   a. $({}^-2, 3)$, $(1, 0)$.   b. $(2, {}^-3)$, $({}^-4, {}^-3)$.

   c. $(1, 1)$, $(3, {}^-3)$.   d. $(3, {}^-1)$, $(3, 5)$.

   e. $({}^-4, 7)$, $(2, {}^-3)$.   f. $(1, {}^-3)$, $(2, 2)$.

5. Determine the equation of the circle in the xy plane with center at $({}^-2, 3)$ and radius equal to 3.

6. If the equation $2x^2 + y^2 - 3x + 4y = 7$ describes a circle, determine its center and radius. If it does not, indicate why not.

7. If the equation $x^2 + y^2 - 4x + 6y + 9 = 0$ describes a circle, determine its center and radius. If it does not, indicate why not.

8. Solve each of the following systems of equations.

   a. $x + y = 2$ b. $3x + 2y = {}^-5$ c. $x^2 + y^2 = 16$

       $2x - 3y = 4$ $2x - 3y = 4$ $y = x + 4$

CHAPTER 15
EQUATIONS, INEQUALITIES, AND LINEAR PROGRAMMING

In this chapter we will present a brief discussion of the solution of equations associated with various word problems and of the solution of various inequalities. An an application of inequalities we will also discuss an interesting concept used commonly in the areas of the social sciences and business. This concept is called linear programming.

15.1 WORD PROBLEMS AND EQUATIONS

In the previous chapter we discussed the solution of equations of the form $ax + by + c = 0$ such that not both a and b are zero. We noted that such equations are called *linear* equations because their graphs are lines.

EXAMPLE Solve the equation $2x + 7 = 12$.

Solution Starting with $2x + 7 = 12$, we could add $(^-7)$ to both sides of the equation, obtaining

$$(2x + 7) + (^-7) = 12 + (^-7)$$

or

$$2x = 5$$

Multiplying both sides of the last equation by $\frac{1}{2}$, we obtain

$$\tfrac{1}{2}(2x) = \tfrac{1}{2}(5)$$

493

or

$$x = \tfrac{5}{2}$$

Hence $\tfrac{5}{2}$ is a solution or a *root* of the equation. We can check this result by substituting $\tfrac{5}{2}$ for x in the original equation:

$$2x + 7 = 2(\tfrac{5}{2}) + 7$$
$$= 5 + 7$$
$$= 12$$

It does check.   ∎

An equation such as the one in the example above is fairly easy to solve. However, we often encounter mathematical problems in everyday experiences that are stated in verbal form rather than in the form of an equation. To solve such problems, it becomes necessary first to translate the verbal statement into a mathematical statement involving an equation. The next step is to solve the mathematical statement. Finally, we examine the result to determine if it represents a meaningful solution.

EXAMPLE   If 4 is added to a number, the sum is 9. What is the number?

Solution   This obvious example is used to illustrate the procedure employed in solving a word problem. Since the unknown in this problem is the number, we could let $y =$ the number. Reading the statement carefully, we have the sum of 4 and the number is 9. The verbal statement, then, is

4 plus the number equals 9

Substituting $+$ for plus, y for the number, and $=$ for equals, we can translate the verbal statement into the mathematical statement

$$4 + y = 9$$

Solving the equation, we get $y = 5$. Finally, we substitute the answer in the equation to get $4 + 5 = 9$, which is a true statement. Hence our answer is the solution to the problem.   ∎

EXAMPLE   John's weight is double that of Mary's. Three times Mary's weight less one-fourth of John's weight is equal to 100 pounds. How much does each of them weigh?

Solution   John's weight is double that of Mary's. Since we do not know the weight of either, we could let $u =$ Mary's weight (in pounds). Then John's weight becomes $2u$, satisfying the verbal statement. Writing down the next verbal statement, we have

Three times Mary's weight less one-fourth of
John's weight is equal to 100 pounds

Substituting symbols for the verbal statement, we have the following mathematical statement:

$$3(u) - \tfrac{1}{4}(2u) = 100$$

This is equivalent to

$$3u - \tfrac{1}{2}u = 100$$

Solving for u, we have

$$(\tfrac{5}{2})u = 100$$

$$5u = 200$$

$$u = 40$$

Hence Mary weighs 40 lb and John weighs 80 lb. These are reasonable results. ■

EXAMPLE A butcher has two grades of ground meat, one that sells for 79¢ a pound and another that sells for 99¢ a pound. How much of each should the butcher mix together in order to have a 100-lb mixture he can sell during a weekend special at 87¢ a pound?

Solution We do not know how many pounds of each grade are to be combined to form the 100-lb mixture. However, we can let y represent the number of pounds of ground meat selling for 79¢ a pound and the balance, or $100 - y$, represent the number of pounds of ground meat selling for 99¢ a pound. The verbal statement associated with this problem is

> The value of the 79¢ a pound ground meat plus the value of the 99¢ a pound ground meat is equal to the value of the 87¢ a pound ground meat

The value of the 79¢ a pound ground meat is equal to the cost per pound ($.79) times the number of pounds y. Similarly, the value of the 99¢ a pound ground meat is equal to the cost per pound ($.99) times the number of pounds $100 - y$. The value of the 87¢ a pound ground meat is equal to the cost per pound ($.87) times the number of pounds 100. The mathematical statement associated with the verbal statement becomes

$$(\$.79)(y) + (\$.99)(100 - y) = (\$.87)(100)$$

which is equivalent to the equation

$$79y + 99(100 - y) = 87(100)$$

Solving for y, we have

$$79y + 9900 - 99y = 8700$$

$$-20y = {}^{-}1200$$

$$y = 60$$

Therefore the butcher should mix 60 lb of 79¢ a pound ground meat and 100 − 60, or 40, pounds of 99¢ a pound ground meat to form the 87¢ a pound mixture.

Checking these results, we find that the value of the 79¢ a pound ground meat is ($.79)(60) or $47.40; the value of the 99¢ a pound ground meat is ($.99)(40) or $39.60; and $47.40 + $39.60 = $87.00 = ($.87)(100), the value of the total mixture. ∎

The above example represents what is known as a mixture problem. Instead of ground meat we may have different types of candy, grades of gasolines, kinds of grass seed, and so forth. In problems of this type the following "box" format may be useful.

	Unit price	Number of units	Value
Unit 1			
Unit 2			
Mixture			

The mathematical statement associated with the problem then would be

Value of item 1 + value of item 2 = value of total mixture

For the last example we would have the following.

	Price per pound	Number of pounds	Value
Ground meat 1	$.79	y	$.79y$
Ground meat 2	$.99	$100 - y$	$.99(100 - y)$
Mixture	$.87	100	$.87(100)$

$$\$.79y + \$.99(100 - y) = \$.87(100)$$

EXAMPLE Michael has a collection of coins consisting of pennies, nickels, dimes, and quarters. He has $\frac{1}{2}$ as many quarters as dimes and 2 more nickels than dimes. The number of pennies in his collection is 3 less than twice the number of dimes. If the total collection of coins is worth $3.02, how many of each kind of coin does Michael have?

Solution The number of pennies, nickels, and quarters is given in terms of the number of dimes, which we could represent by b. Since Michael has $\frac{1}{2}$ as many quarters as dimes, he would have $\frac{1}{2}b$ quarters. He has 2 more nickels than dimes, or $b + 2$. Finally, the number of pennies is 3 less than twice the number of dimes, or $2b - 3$. The values of the pennies, nickels, dimes, and quarters is given in the following table:

	Number	Unit worth	Value
Pennies	$2b - 3$	$.01	$.01(2b - 3)$
Nickels	$b + 2$	$.05	$.05(b + 2)$
Dimes	b	$.10	$.10b$
Quarters	$\frac{1}{2}b$	$.25	$.25(\frac{1}{2}b)$

The verbal statement associated with the problem is

The value of the pennies plus the value of the nickels plus the values of the dimes plus the value of the quarters is equal to the total value of the collection.

The corresponding mathematical statement, then, would be

$$\$.01(2b - 3) + \$.05(b + 2) + \$.10(b) + \$.25(\tfrac{1}{2}b) = \$3.02$$

which is equivalent to

$$(2b - 3) + 5(b + 2) + 10(b) + 25(\tfrac{1}{2}b) = 302$$

Solving for b, we have

$$2b - 3 + 5b + 10 + 10b + \left(\frac{25}{2}\right)b = 302$$

$$\left(\frac{59}{2}\right)b = 295$$

$$59b = 590$$

$$b = 10$$

Hence Michael has b or 10 dimes, $2b - 3$ or 17 pennies, $b + 2$ or 12 nickels, and $\frac{1}{2}b$ or 5 quarters in his collection. Check to determine that the collection is worth $3.02. ∎

EXERCISES

1. Solve each of the following linear equations for the indicated variable.
   a. $3x - 2 = 2x + 7$.
   b. $5 - 3y = 7 - 4y$.

   c. $2u - 3 = 4u + 1$.
   d. $3w + 7 = {}^{-}2(w + 1)$.

   e. $\frac{1}{2}p + 3 = 2 - 4p$.
   f. $6t + 1 = (\frac{2}{3})(t + 5)$.

   g. $3.2x + 7 = 3.9 - 5x$.
   h. $6.2 + 7y = 2(4.9 - 1.8y)$.

   i. $(\frac{2}{3})u - \frac{1}{2} = (\frac{3}{4}) - (\frac{1}{5})u$.
   j. $\dfrac{m - 2}{3} = \dfrac{m - 3}{2}$.

k. $\dfrac{2x}{3} - 3 = \dfrac{5x}{4} + 2.$ l. $9.8t - 2.9 = 5.1 - 3.7t.$

m. $6(1.2 - 2.3r) = {}^-3(4.2r + 1.7).$ n. $\dfrac{5.3x}{4} = \dfrac{2.7x}{3} + 1.9.$

o. $\dfrac{2m - 1}{7} = 3 - \dfrac{4 - 3m}{5}.$ p. $\dfrac{w + 5}{6} - \dfrac{w - 6}{9} = 2.$

q. $\dfrac{3x + 5}{4} = \dfrac{2x - 1}{3} + 4.$ r. $\dfrac{3y - 2}{4} + \dfrac{10y - 8}{12} = \dfrac{13}{4} + \dfrac{y - 2}{3}.$

s. $\dfrac{4u}{5} + \dfrac{2 - 3u}{7} = \dfrac{5u + 1}{6} - \dfrac{u}{2}.$ t. $\dfrac{9s + 1}{2} + 4 = 7 - \dfrac{3 - 4s}{5}.$

2. Solve each of the following equations for the indicated variable.

a. $P = 2L + 2W$ for L.   b. $A = bh$ for h.

c. $C = 2\pi r$ for r.   d. $T = p + vr$ for v.

e. $5F = 9C + 160$ for C.   f. $A = P(1 + rt)$ for t.

g. $2a - 3b = 4c$ for b.   h. $F = ma$ for a.

i. $A = \dfrac{B - C}{CD}$ for D.   j. $\dfrac{r - st}{u - s} = k$ for s.

3. Find two consecutive odd natural numbers whose sum is 20.
4. A plane rectangular region is such that its length is 7 ft more than its width. If the perimeter of the region is 182 ft, determine its dimensions.
5. A piece of string 24 in. long is cut into two pieces such that one piece is 4 in. less than 3 times the length of the other piece. Determine the length of each piece of string.
6. Working alone, Joe can paint a house in 8 days. Working alone, Jim can paint the same house in 6 days. How long would it take both men to paint the house working together? (Hint: Consider the portion of the house that each paints in one day.)
7. Liz has a collection of coins consisting of pennies, dimes, and quarters. She has 7 more pennies than dimes, and the number of quarters is 3 less than $\frac{1}{2}$ the number of dimes. If the total collection is worth $2.14, how many of each kind of coin does she have?
8. Diane has a collection of coins consisting of pennies, nickels, dimes, and quarters. She has 1 more dime than 3 times the number of nickels; she has 3 less quarters than twice the number of nickels; and she has 9 more pennies than she has quarters. If the total collection is worth $6.37, how many of each kind of coin does Diane have?
9. Temperature readings may be taken using either a Fahrenheit scale or a centigrade scale thermometer. If F denotes Fahrenheit and C denotes

centigrade, the basic relation between the two can be represented as $5F - 9C = 160$. Solve for C in terms of F and compute the centigrade temperatures corresponding to the following Fahrenheit temperatures: (a) $^-31°$, (b) $^-12°$, (c) $0°$, (d) $32°$, and (e) $90°$.

Lever principle: In physics a *lever* is defined as a rigid rod supported at one point called the *fulcrum*. (A familiar example of a lever is a teeterboard or seesaw.) If a mass of x pounds is placed on a lever at a distance of d units from its fulcrum, then the product of x and d is called the *moment* of x about the fulcrum. The lever principle states that when the lever is in a position of equilibrium, the sum of the moments for all masses on one side of the fulcrum is equal to the sum of the moments for all masses on the other side.

10. If two boys weighing 120 lb and 70 lb, respectively, sit at opposite ends of a teeterboard which is 19 ft long, where should the board be supported to maintain a position of equilibrium?
11. A boy weighing 85 lb sits 6 ft from the fulcrum on a teeterboard. How far from the other side of the fulcrum should a 70-lb boy sit in order to balance him?
12. Two boys weighing 60 lb and 80 lb sit on the same side of a teeterboard 5 ft and 9 ft, respectively, from the fulcrum. A third boy weighing 90 lb sits 7 ft from the fulcrum on the other side. Where should a fourth boy weighing 100 lb sit to balance the board?

Uniform motion and distance: Uniform motion involves an object traveling at a constant rate of speed along a particular path. The speed of the object refers to the distance traveled along the path during a specified period of time such as hour, day, or second. If rate of speed is denoted by r, time by t, and distance by d, the relationship existing among the three variables is given by the equation $d = rt$.

13. Determine the speed of a snowmobile that travels 150 mi in 4 hr.
14. How long will it take a jet to travel 2300 mi if its rate of speed is 700 mph?
15. Two men in automobiles travel toward each other from points 600 mi apart at speeds of 45 mph and 60 mph, respectively. When will they meet (theoretically, that is) if they start at the same instant?
16. Two runners start from the same place and travel in opposite directions at speeds of 3 mph and 4 mph, respectively. How long after they start will they be 50 mi apart?
17. One boy can run the 100 yd in 7 sec, and a second boy can run the same distance in 6.5 sec. In a race between them, how long will it take the faster boy to gain a lead of 2 yd over the other boy?
18. A businessman has $15,000 to invest at simple interest rates of 5 percent and 6.5 percent. What amounts should be invested at each rate to obtain an annual cash dividend of $840?

19. In a particular school district 11 percent, or 2,167, of the students ride a bus to school. How many students are there in the school district?
20. Determine the selling price for a particular item if the cost of the item is $12.20 and the profit is 20 percent of the cost.
21. Determine the selling price for a particular item that cost $10.22 if the profit is 30 percent of the selling price.
22. How many gallons of a solution of antifreeze and water with 45 percent antifreeze should be added to 22 gallons of a solution with 25% antifreeze to obtain a solution with 38 percent antifreeze?
23. A football game was attended by 4730 people, some of whom paid $1.50 each for reserved seats and the rest of whom paid 90¢ general admission. If the total receipts for the game were $5,172, how many tickets of each kind were sold?

15.2 LINEAR INEQUALITIES

In Chapter 13 we learned that a line separates a plane containing it into three disjoint subsets of the plane. They are the line itself and two disjoint subsets called half-planes. We also learned that the set of points in the plane whose graph is the line can be given by an equation of the form $ax + by + c = 0$ such that not both a and b are zero.

Now consider the line L in the xy plane described by the equation $y = x + 3$, illustrated in Figure 15.1. The line separates the xy plane into two half-planes: the half-plane determined by L and in the direction of A and the half-plane determined by L and in the direction of B.

If we consider $x = 1$, then any point on L with an abscissa of 1 must be such that its coordinates satisfy the equation $y = x + 3$. Hence $y = 1 + 3 = 4$ and the point $C(1, 4)$ lies on the line. (See Figure 15.1). Clearly, if we take a y value less than 4 associated with the x value equal to 1, the corresponding

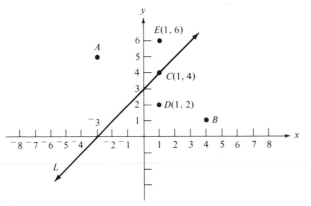

Figure 15.1

point must lie in the half-plane determined by L and in the direction of B, such as the point $D(1, 2)$ in Figure 15.1. In fact all points in this half-plane may be represented by

$$T = \{(x, y) \,|\, y < x + 3\}$$

In a similar manner, if we take a y value greater than 4 associated with the x value equal to 1, the corresponding point must lie in the half-plane determined by L and in the direction of A, such as the point $E(1, 6)$ in Figure 15.1. In fact all points in this half-plane may be represented by

$$V = \{(x, y) \,|\, y > x + 3\}$$

The statements $y < x + 3$ and $y > x + 3$ are called *inequalities* and involve the order relation as was introduced earlier in the text for the various systems of numbers. Throughout this chapter we will be concerned with the set of all real numbers. The sets T and V are relations (as defined in Chapter 4), and their graphs are half-planes.

EXAMPLE   Graph the relation $P = \{(x, y) \,|\, y < 2x + 5\}$.
Solution   First we will graph the relation $L = \{(x, y) \,|\, y = 2x + 5\}$, which will be a line. To determine two points on the line, we could take $x = 0$ and $x = {}^-3$ and find that $y = 5$ and $y = {}^-1$, respectively. Hence the line L passes through the points $(0, 5)$ and $({}^-3, {}^-1)$. We will graph L by means of a dashed line as illustrated in Figure 15.2. This line divides the xy plane into two half-planes, one of which will be the graph of P. We now can take two distinct points in the plane, one from each half-plane, and test their coordinates in the inequality $y < 2x + 5$. For instance, take $A({}^-3, 5)$ and $B(0, 0)$. Testing for $A({}^-3, 5)$, we determine that $5 \not< 2({}^-3) + 5 = {}^-1$. Testing for $B(0, 0)$, we determine that $0 < 2(0) + 5 = 5$. Hence the required graph is the half-plane in the direction of B, indicated by the shaded region.   ■

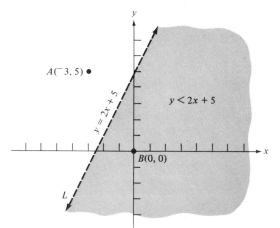

Figure 15.2

Sometimes we are interested in the portion of a plane that is the union of a half-plane together with the line separating the plane or, simply, the closed half-plane.

EXAMPLE Graph the relation $Q = \{(x, y)|2x - 3y \geq 4\}$.

Solution We observe that Q is the union of two sets, $Q_1 = \{(x, y)|2x - 3y = 4\}$ and $Q_2 = \{(x, y)|2x - 3y > 4\}$. The graph of Q_1 is a line, and the graph of Q_2 is a half-plane determined by the graph of Q_1. To graph Q_1 we consider the two points with coordinates $(2, 0)$ and $(^-1, ^-2)$, both lying on the graph of Q_1. We graph this as indicated in Figure 15.3, using a solid line. To obtain the graph of Q_2 we consider the test points $A(0, 0)$ and $B(2, ^-3)$. Testing at $A(0, 0)$, we have $2(0) - 3(0) = 0 \not> 4$. Testing at $B(2, ^-3)$, we have $2(2) - 3(^-3) = 13 > 4$, which does check. Hence the required half-plane for Q_2 is in the direction of B. The required graph of Q, then, is the half-plane together with the line as indicated by the shaded portion of the plane. ■

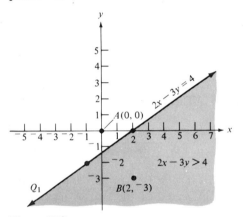

Figure 15.3

To determine whether the graph of a linear inequality is an open half-plane or a closed half-plane we observe the following convention:
1. If the graph is an open half-plane, represent the line that separates the plane by a *dashed* line.
2. If the graph is a closed half-plane, represent the line that separates the plane by a *solid* line.

EXERCISES

Determine the graph of each of the following relations.
1. $\{(x, y)|y > 2x + 1\}$.
2. $\{(x, y)|y < 3x - 2\}$.
3. $\{(x, y)|2x + 3y > 0\}$.

4. $\{(x, y)\,|\,3x - 4y < 1\}$.
5. $\{(x, y)\,|\,3x - 7y > {}^-2\}$.
6. $\{(x, y)\,|\,2x + 3y \geq 1\}$.
7. $\{(x, y)\,|\,3x - 4y \leq {}^-2\}$.
8. $\{(x, y)\,|\,3 + 2x \geq y\}$.
9. $\{(x, y)\,|\,2 - 3y \geq x\}$.
10. $\{(x, y)\,|\,3x + 2 \leq 3 - 2y\}$.

15.3 SYSTEMS OF LINEAR INEQUALITIES

The solution of a linear inequality is a half-plane that is either open or closed. If we consider the solution of two such inequalities in the same plane, then the solution would be the intersection set of their half-planes. We will now consider such solutions by means of the following examples.

EXAMPLE   Determine the graph in the xy plane associated with the following system of inequalities:

$$\begin{cases} y - 2x < 5 \\ 2x - 3y \geq 4 \end{cases}$$

Solution   The graph associated with the inequality $y - 2x < 5$ was determined in the first example in the previous section and the graph of $2x - 3y \geq 4$ was determined in the second example in the same section. If we represent the graph of the first inequality by the shaded half-plane in Figure 15.4(a) and the graph of the second inequality by the shaded half-plane in Figure 15.4(b), then the required graph would be the intersection of the two half-planes indicated by the shaded portion of the plane in Figure 15.4(c) (see page 504).   ■

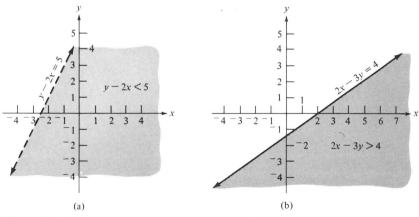

(a)   (b)

Figure 15.4

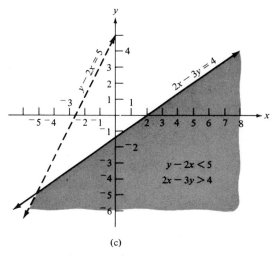

(c)

Figure 15.4 (*Continued*)

EXAMPLE Determine the graphical solution of the following system of linear inequalities

$$\begin{cases} 3x - 2y \le {}^-6 \\ 7x + 4y \ge {}^-28 \end{cases}$$

Solution The graph of $3x - 2y \le {}^-6$ is the closed half-plane given in Figure 15.5(b). The graph of $7x + 4y \ge {}^-28$ is the closed half-plane given in Figure 15.5(a). The required graphical solution of the given system of inequalities is given in Figure 15.5(c). The reader should verify this. ■

If a system of inequalities contains more than two linear inequalities, the graphical solution may be a polygonal region with part or all of its boundary. We will now consider such a system.

EXAMPLE Determine the graphical solution of the following system of linear inequalities

$$\begin{cases} x - y \le 3 \\ 3x + 2y \ge {}^-6 \\ x \ge {}^-2 \\ y \le 5 \end{cases}$$

Solution The graph of $x - y \le 3$ is the closed half-plane given in Figure 15.6(a). The graph of $3x + 2y \ge {}^-6$ is the closed half-plane given in Figure 15.6(b). The graph of $x \ge {}^-2$ is the closed half-plane given in Figure 15.6(c). The graph of $y \le 5$ is the closed half-plane given in Figure 15.6(d). Taking the intersection set of all four of these closed half-planes,

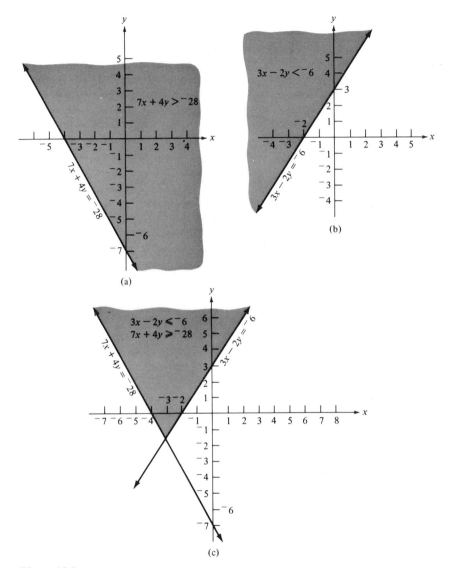

Figure 15.5

we obtain the required graphical solution as given in Figure 15.6(e) as the closed polygonal region bounded by the polygon *ABCD*. ■

If the graphical solution of a system of linear inequalities is a polygonal region, it may be a closed region, or the graphical solution may be a polygonal region with only part of its boundary. Further, if the graphical solution is a closed polygonal region, it may also be a convex region. Recall from Chapter 13 that if a polygonal region is convex, then a line segment

(a)

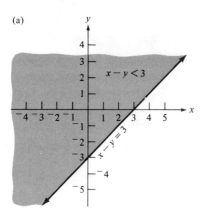

(b)

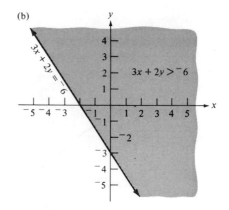

(c)

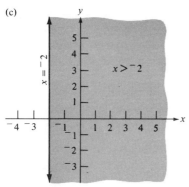

(d)

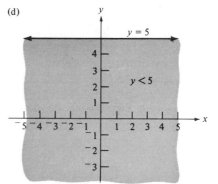

(e)

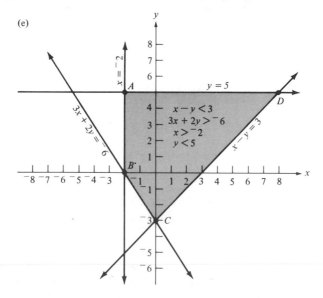

Figure 15.6

connecting any two distinct points in the region must lie completely in the region. A closed convex polygonal region is required for the discussion of linear programming which will be taken up in the next section.

EXERCISES

Determine the graphical solution for each of the following systems of inequalities. If the graphical solution is a polygonal region, determine if it is closed. Also determine if it is convex.

1. $x > 2.$
   $y < 3$

2. $x \geq {}^-3$
   $x < 4$

3. $x + y \leq 2$
   $2x - y > 3$

4. $2x - 3y < {}^-2$
   $3x + 4y \leq 3$

5. $3x + 5y \leq {}^-1$
   $x + 2y \geq 1$

6. $2 - 3x \leq 4y$
   $1 + 2y > 3x$

7. $x \geq {}^-1$
   $x \leq 3$
   $y \geq {}^-2$
   $y \leq 4$

8. $y \geq x$
   $y \leq 4$
   $y \geq {}^-2$
   $y \leq x + 3$

9. $2y \geq x - 4$
   $2y \leq x + 12$
   $y \leq {}^-2x + 12$
   $x \leq 3$

10. $y \geq x - 1$
   $y \leq x + 6$
   $y \leq 6x + 6$
   $3x + 4y \leq 31$

15.4 LINEAR PROGRAMMING

A linear programming problem basically involves maximizing or minimizing a linear function, F, of the form $F(x_1, x_2, x_3, \ldots, x_n) = c_1 x_1 + c_2 x_2 + \cdots + c_n x_n$ subject to certain restrictions given in the form of linear inequalities. In this chapter we will consider linear programming problems involving only two variables.

Let F be a linear function in the two variables x and y of the form $F(x, y) = c_1 x + c_2 y$. Then F is called the *objective function*. Also consider the system of linear inequalities of the form

$$a_1 x + b_1 y \leq d_1$$
$$a_2 x + b_2 y \leq d_2$$
$$\vdots$$
$$a_n x + b_n y \leq d_n$$

Each of the above inequalities is called a *constraint* or a restriction.

If the graphical solution of the above system of linear inequalities is a closed convex polygonal region, then we would be interested in maximizing (or minimizing) the objective function F over his region.

Every point in the closed convex region is called a *feasible solution* of the problem. However, a point in the closed convex polygonal region at which $F(x, y)$ becomes maximum (or minimum) is called an *optimum solution.*

For purposes of our discussion we will accept without proof the fact that every linear function F of the form $F(x, y) = c_1 x + c_2 y$ defined over a closed convex region in a plane takes on its maximum value at a vertex point of the region and also takes on its minimum value at another vertex point of the region.

Hence the following steps may be followed in the solution of a linear programming problem:

1. Determine the closed convex region that is the graphical solution of the system of constraints.
2. Determine the coordinates of the vertices of the boundary of the polygonal region.
3. Evaluate the objective function at each of these vertex point values.
4. The smallest of these values obtained in (3) will be the minimum value of the objective function and, the largest of the values will be the maximum value.

We will illustrate this procedure with the following example.

EXAMPLE Using the procedure outlined above, determine the maximum value of the objective function F defined by $F(x, y) = 2x - 3y$ over the closed convex region of the xy plane determined by the following constraints:

$$\begin{cases} x - y \leq 3 \\ 3x + 2y \geq {}^-6 \\ x \geq {}^-2 \\ y \leq 5 \end{cases}$$

Solution

Step 1 In this step we are to determine the graphical solution of the above system of linear inequalities. However, this was done in the last example in the previous section, and we have the closed convex polygonal region $ABCD$ given in Figure 15.7.

Step 2 Determine the coordinates of the vertices of the boundary of the region.

a. The coordinates of vertex A may be found by solving the system of equations

$$\begin{cases} x = {}^-2 \\ y = 5 \end{cases}$$

from which we determine $A({}^-2, 5)$.

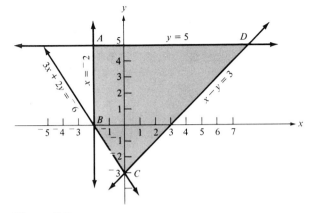

Figure 15.7

    b. The coordinates of vertex B may be found by solving the system of equations

$$\begin{cases} x = {}^-2 \\ 3x + 2y = {}^-6 \end{cases}$$

from which we determine $B({}^-2, 0)$.

    c. The coordinates of vertex C may be found by solving the system of equations

$$\begin{cases} 3x + 2y = {}^-6 \\ x - y = 3 \end{cases}$$

from which we determine $C(0, {}^-3)$.

    d. The coordinates of vertex D may be found by solving the system of equations

$$\begin{cases} x - y = 3 \\ y = 5 \end{cases}$$

from which we determine $D(8, 5)$.

Step 3   Evaluate the objective function $F(x, y) = 2x - 3y$ at each of the vertex points.

$A({}^-2, 5)$ $F(x, y) = F({}^-2, 5) = 2({}^-2) - 3(5) = {}^-4 - 15 = {}^-19$
$B({}^-2, 0)$ $F(x, y) = F({}^-2, 0) = 2({}^-2) - 3(0) = {}^-4 - 0 = {}^-4$
$C(0, {}^-3)$ $F(x, y) = F(0, {}^-3) = 2(0) - 3({}^-3) = 0 + 9 = 9$
$D(8, 5)$ $F(x, y) = F(8, 5) = 2(8) - 3(5) = 16 - 15 = 1$

Step 4   Determine that $F(x, y) = 2x - 3y$ attains its maximum value at vertex $C(0, {}^-3)$ and that this value is 9. ■

EXAMPLE Suppose that a firm is engaged in the manufacture of lats (a fictitious product formed by using the author's initials) and has them stored in warehouses A and B in different cities. Warehouses A and B contain 40 lats and 50 lats, respectively. Further, suppose that a customer in city M orders 30 lats and that a customer in city N orders 40 lats. The cost of shipping each lat from warehouse A to cities M and N is $27 and $36, respectively. The cost of shipping each lat from warehouse B to cities M and N is $33 and $42, respectively. How should the manufacturer ship the lats so as to minimize his shipping costs and, hence, maximize his profits? Obviously, the manufacturer has many possible ways of filling the orders.

Solution Suppose we let x represent the number of lats shipped from warehouse A to city M and y represent the number of lats shipped from warehouse A to city N. Then $(30 - x)$ and $(40 - y)$ represent the number of lats shipped from warehouse B to cities M and N, respectively. The total cost of shipping would then be $P = 27x + 36y + 33(30 - x) + 42(40 - y)$ in dollars. We wish to minimize the value of P subject to the following constraints:

a. $x \geq 0$.
b. $y \geq 0$.
c. $30 - x \geq 0$.
d. $40 - y \geq 0$.
e. $x + y \leq 40$.
f. $(30 - x) + (40 - y) \leq 50$.

The first four constraints simply state that the number of lats shipped from each warehouse A and B cannot be negative. The fifth constraint implies that the combined number of lats shipped from warehouse A to M and N cannot exceed 40, the number stored. Similarly, the last constraint implies that the combined number of lats shipped from warehouse B to cities M and N cannot exceed 50. The graph of the above system of inequalities is the shaded portion of the xy plane given in Figure 15.8.

The graphical solution of the given system of inequalities is a closed convex polygonal region bounded by the polygon $ABCDE$ with the coordinates for its vertices being $A(0, 40)$, $B(0, 20)$, $C(20, 0)$, $D(30, 0)$, and $E(30, 10)$. The values for $P(x, y) = 27x + 36y + 33(30 - x) + 42(40 - y)$ or equivalently $P(x, y) = {}^-6x - 6y + 2670$ at each of the vertex points are given in the table below:

SHIPPING COSTS (IN DOLLARS)

VERTEX	$P(x, y) = 2670 - 6x - 6y$
$A(0, 40)$	2430 (minimum value)
$B(0, 20)$	2550
$C(20, 0)$	2550
$D(30, 0)$	2490
$E(30, 10)$	2430 (minimum value)

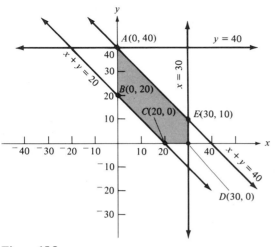

Figure 15.8

From the table we note that the minimum shipping costs occur at vertex *A* with coordinates (0, 40) or at vertex *E* with coordinates (30, 10). Hence the shipping costs would be minimized (and profits maximized) by shipping 40 lats from warehouse A to city N and 30 lats from warehouse B to city M. Or the shipping costs would be minimized by shipping 30 lats from warehouse A to city M, 10 lats from warehouse A to city N, and 30 lats from warehouse B to city N. ∎

EXAMPLE A manufacturer makes two different models of a boat, A and B. During the manufacture of these boats each model must be processed in different areas of the factory C and D. For each unit of model A, area C must work 5 labor-days and area D must work 2 labor-days. For each unit of model B, area C must work 3 labor-days and area D must work 3 labor-days. Work in area C is limited to 180 labor-days per week and in area D, to 135 labor-days per week. The manufacturer can sell all of the boats of each model he can make. The profit on each boat of model A is $300 and is $200 on each boat of model B. How many boats of each model should he produce weekly in order to maximize his profit?

Solution If we let *x* represent the number of boats of model A and *y* the number of boats of model B that he makes, then the manufacturer's profit would be denoted by $P(x, y) = 300x + 200y$, where *P* is in dollars. We wish to determine the values of *x* and *y* for which $P(x, y)$ will be maximized subject to the following constraints:

a. $x \geq 0$.
b. $y \geq 0$.
c. $5x + 2y \leq 180$.
d. $3x + 3y \leq 135$.

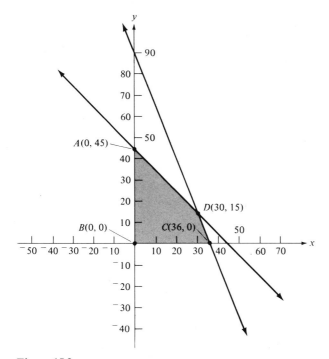

Figure 15.9

The first two constraints simply state that it is impossible to produce a negative number of each model of the boats. The third constraint states that area C cannot be operated for more than 180 labor-days per week and the last constraint states that area D cannot be operated for more than 135 labor-days per week.

We now graph the above system of inequalities and observe that the set of all feasible points lie within the shaded region and on the boundary of the region in Figure 15.9.

Since the optimal value will occur at one of the vertices of the above region, we form the table of values of $P(x, y)$ for each vertex value.

| | Profit (in Dollars) |
Vertex	$P(x, y) = 300x + 200y$
$A(0, 45)$	9,000
$B(0, 0)$	0
$C(36, 0)$	10,800
$D(30, 15)$	12,000 (maximum value)

From the above table we see that the maximum profit, $12,000, occurs at the vertex $D(30, 15)$ and, hence, the manufacturer should produce 30 units of model A and 15 units of model B per week to maximize his profit. ■

EXERCISES

1. Determine the closed convex polygonal region that is the graphical solution for each of the following systems of inequalities.

a. $x \leq 5$
   $x \geq {}^{-}2$
   $y \geq 0$
   $y \leq 4$

b. $y \geq x - 1$
   $y \leq x + 6$
   $y \leq 8$
   $y \geq {}^{-}2$

c. $2y \leq 5x + 10$
   $y \leq 5$
   $2y \geq {}^{-}x - 2$
   $x \leq 6$
   $6y \geq x - 6$

d. $x \geq {}^{-}3$
   $y \geq {}^{-}3$
   $x \leq 5$
   $y \leq x + 5$
   $5y \leq {}^{-}3x + 35$

e. $\quad y \geq {}^{-}2x - 6$
   $\quad y \geq {}^{-}4$
   $2x - y \leq 12$
   $3y \leq 5x + 15$
   $\quad y \leq 5$
   $\quad x \leq 6$

f. $\quad x \geq {}^{-}4$
   $\quad y \geq {}^{-}4$
   $5y \geq 2x - 20$
   $\quad x \leq 5$
   $2y + x \leq 11$
   $\quad y \leq 5$

2. Determine the coordinates of the vertices for each of the closed convex polygonal regions indicated in Exercise 1.

3. Evaluate the objective function $F(x, y) = 3x + y - 2$ at each of the vertex points of the polygonal region determined in Exercises 1 and 2.

4. Referring to Exercise 3, determine the maximum value for each objective function and also determine where this value occurs.

5. Referring to Exercise 3, determine the minimum value for each objective function and also determine where this value occurs.

6. Repeat Exercises 3 to 5 using the objective function $G(x, y) = 2x - 3y$.

7. Repeat Exercises 3 to 5 using the objective function $H(x, y) = x - y + 5$.

8. A manufacturer produces two types of lawn mowers, rotary and reel, requiring skilled, semiskilled, and unskilled labor. Each rotary mower requires 1 hr of skilled labor, 2 hr of semiskilled labor, and 3 hr of unskilled labor. Each reel mower requires 2 hr of skilled labor, 1 hr of semiskilled labor, and 1 hr of unskilled labor. Each day the manufacturer has available, 200, 160, and 270 hours of skilled, semiskilled, and unskilled labor, respectively. If the profit on a rotary mower is $30 and the profit on a reel mower is $25, how many of each type of mower should the manufacturer produce each day to receive a maximum profit? What is the maximum profit?

9. The Roast-O-Fresh Coffee Company sells two types of coffee, regular grind and drip grind. The regular grind coffee consists of 6 parts Brazilian coffee, 2 parts Mexican coffee, and 3 parts Venezuelan coffee. The drip grind coffee consists of 8 parts Brazilian coffee, 6 parts of Mexican coffee and 5 parts of Venezuelan coffee. The coffee company has in inventory 2500 lb of Brazilian coffee, 1500 lb of Mexican coffee, and

1500 lb of Venezuelan coffee. The profit for the regular grind coffee is 35¢ per pound, and for the drip grind the profit is 40¢ per pound. How many pounds of each type of coffee should be company sell in order to maximize its profit?

SUMMARY

In this chapter we discussed the solution of simple linear equations and linear inequalities. Systems of linear inequalities also were discussed. We observed that the graph of a linear inequality is a half-plane and that the graph of a system of linear inequalities is the intersection of all the half-planes corresponding to the system.

As an application of a system of linear inequalities, we considered some basic linear programming problems.

There were no new symbols introduced during our discussions.

REVIEW EXERCISES FOR CHAPTER 15

1. Solve each of the following equations for the indicated variable.

a. $2x - 3 = 3x + 2$.

b. $3y + 4 = 5y - 7$.

c. $\frac{1}{2}x = (\frac{2}{3})x - 4$.

d. $3u - \frac{1}{2} = \frac{1}{2}u + \frac{1}{4}$.

e. $\dfrac{t + 3}{4} = \dfrac{4 - 2t}{5}$

f. $\dfrac{v - 3}{2} - \dfrac{2 - v}{3} = 4$

2. Solve each of the following equations for the indicated variable.

a. $2x + 3y = 5$ for y.

b. $3r - 2 = 3 - 2s$ for s.

c. $A = P(1 + rt)$ for r.

d. $T = p + vr$ for r.

e. $A = \dfrac{B - C}{CD}$ for B.

f. $\dfrac{r - st}{u - s} = k$ for u.

3. Determine the graphical solution for each of the following inequalities.

a. $y \leq 3$.

b. $x \geq {}^-2$.

c. $2x - y \leq 4$.

d. $3x > 2y - 5$.

e. $\frac{1}{2}x + (\frac{2}{3})y \geq 1$.

f. $3 - (\frac{2}{5})x \leq (\frac{1}{3})y$.

4. Determine the graphical solution for each of the following systems of inequalities.

a. $x \geq {}^-1$
$x \leq 4$

b. $y \leq 2$
$y > {}^-3$

c. $y \geq x$
    $y < 5$
    $x < 3$

e. $y \geq x$
    $y < 2x + 1$
    $y < 2 - x$

d. $y \leq 5$
    $y > {}^-2$
    $x + y \geq 1$
    $x + y \leq 6$

f. $x \geq 2$
    $y \geq 1$
    $x - 2y < 2$
    $x + 4y < 14$

CHAPTER 16
INTRODUCTION TO PROBABILITY

The word probability is used frequently in everyday life experiences. For instance, we hear statements such as the following: The probability of precipitation today is 20 percent this afternoon and 40 percent this evening; The probability that the price of a loaf of bread will exceed $1 is very likely; It is equiprobable that our basketball team will capture the league title and that our wrestling team will tie for second place; There is a greater probability that one will pass this course by studying on a regular basis than by cramming before examinations.

The study of probability deals with the study of uncertainties, and the above examples involve a degree of confidence or belief that one has in the occurrence of the particular event. However, probability also may be considered in terms of expectation. For instance, if a coin is tossed repeatedly a large number of times, then we would expect that the coin will land "heads" half of the time. Hence we would say that the probability that a head will show when a coin is tossed is 1/2.

In this chapter we will discuss some of the theory of probability but, first, we will introduce some of the basic terminology necessary to make our discussions more precise.

16.1 PRELIMINARIES

DEFINITION 16.1 We define an *experiment* to be the process by which an observation or measurement is made. The result of the experiment is called its *outcome*.

An experiment that is very easy for a young child to understand involves the use of a spinner such as that illustrated in Figure 16.1. Such spinners are found in many games for young children and consist of a needle and a circular piece of cardboard. The needle is attached to the cardboard in such a way that it is allowed to spin freely. Also, the needle is equally likely to stop at any one numbered place as at any other numbered place. Using the spinner, the experiment would be spinning the needle and the outcome would be the numbered position where it stops.

Figure 16.1

For the spinner illustrated in Figure 16.1 the set of all the possible outcomes is $\{1, 2, 3, 4, 5, 6\}$. If the needle stops on a line separating two numbered regions, we agree to determine the outcome as the numbered region the needle would move to if it were rotated in a clockwise direction.

For the experiment outlined above the only possible outcomes are 1, 2, 3, 4, 5, or 6. The outcome of this experiment is one and only one of these elements and each element is a possible outcome.

DEFINITION 16.2 The set of all possible outcomes of an experiment is called the *sample space* of the experiment if the set is finite and has the following properties:
   1. Every element of the set is a possible outcome of the experiment.
   2. The outcome of every experiment is exactly one of the elements of the set.

DEFINITION 16.3 Every element of a sample space is called a *sample point*.

EXAMPLE Consider the spinner illustrated in Figure 16.1. The *experiment* consists of spinning the needle. The *outcomes* would be the numbered positions where it stops. The *sample space* is the set $\{1, 2, 3, 4, 5, 6\}$, and each element of this set is a *sample point*.

EXAMPLE Consider the *experiment* of tossing a coin and observing the possible *outcomes* which would be a head or a tail. Assume that the coin does not land on its edge or roll away. The *sample space* is {*H*, *T*}, where *H* denotes head and *T* denotes tail and each element of this set is a *sample point*.

EXAMPLE Consider the *experiment* of tossing two coins and observing the possible *outcomes* which are a head on both coins, a tail on both coins, a head on the first coin and a tail on the second coin, and a tail on the first coin and a head on the second coin. The *sample space* is {*HH*, *TT*, *HT*, *TH*}, and each element of this set is a *sample point*.

EXAMPLE Consider the *experiment* of rolling a pair of dice and observing the possible *outcomes*. Each die (singular of dice) will show 1, 2, 3, 4, 5, or 6. Determine the sample space.
  Solution The first die may show 1, and the second die may show 1. We will write this possible outcome as the ordered pair (1, 1). Similarly, the ordered pair (4, 5) will denote that the first die shows 4 and the second die shows 5. Do you agree that the *sample space* is the set {(1, 1), (1, 2), (1, 3), (1, 4), (1, 5), (1, 6), (2, 1), (2, 2), (2, 3), (2, 4), (2, 5), (2, 6), (3, 1), (3, 2), (3, 3), (3, 4), (3, 5), (3, 6), (4, 1), (4, 2), (4, 3), (4, 4), (4, 5), (4, 6), (5, 1), (5, 2), (5, 3), (5, 4), (5, 5), (5, 6), (6, 1), (6, 2), (6, 3), (6, 4), (6, 5), (6, 6)}, which consists of 36 *sample points*? ■

For some experiments we may be interested in only some but not all of the sample points. For instance, in the last example above we may be interested only in the sample points representing the sum of the faces of the two dice being equal to 8. Hence the sample points would be (2, 6), (3, 5), (4, 4), (5, 3), and (6, 2) and the set {(2, 6), (3, 5), (4, 4), (5, 3), (6, 2)} containing these sample points is a proper subset of the sample space for the experiment.

DEFINITION 16.4 An *event* is a subset of a sample space. When an event contains only one sample point, it is called a *simple event*. When an event contains more than one sample point, it is called a *compound event*. When an event contains no sample points, the event is the *null set*.

EXAMPLE Consider the experiment of tossing three coins and observing the possible outcomes. The sample space is the set {*HHH*, *HHT*, *HTH*, *THH*, *HTT*, *THT*, *TTH*, *TTT*} consisting of eight sample points. Some of the events associated with this sample space are
  a. {*HHH*}, all coins heads.
  b. {*HTT*, *THT*, *TTH*}, one coin heads and two coins tails.
  c. {*HHH*, *TTT*}, all coins alike.
  d. ∅, more than three heads.

EXAMPLE   Consider the experiment of rolling a single die and observing the outcomes. The sample space is $S = \{1, 2, 3, 4, 5, 6\}$. Let A be the event that the die shows an even number. Then $A = \{2, 4, 6\}$. Since $A \subseteq S$, the set $\{1, 3, 5\}$, which consists of all those sample points in S that are not in A, is also an event and is called the complement of A and denoted by A'.

DEFINITION 16.5 Let S be the sample space associated with an experiment, and let A be an event such that $A \subseteq S$. Then the set of all sample points in S that are not in A is called the *complement event* of A and is denoted by A'.

In the example following Definition 16.4 we note that events (a) and (b) are disjoint; events (a) and (c) are not disjoint but have the sample point HHH in common.

DEFINITION 16.6 Let S be the sample space associated with an experiment, and let A and B be events associated with S. If $A \cap B = \varnothing$, then the events A and B are said to be *mutually exclusive events*.

Since an event is defined as a subset of a sample space and since more than one event can be associated with some sample spaces, it is meaningful to consider the union and intersection of events.

DEFINITION 16.7 Let A and B be events associated with a sample space S. Then by the symbol $A \cap B$ we shall mean the set of all sample points common to A and also B; that is, A occurs *and* B occurs. Also, by the symbol $A \cup B$ we shall mean the set of all sample points that are found in A or in B or in both; that is, A occurs *or* B occurs *or* both A and B occur.

EXAMPLE   Consider the experiment of drawing a card from a deck of 8 cards numbered 1 through 8 and let

   A be the event that the card drawn is even-numbered; that is, $A = \{2, 4, 6, 8\}$.

   B be the event that the card drawn is odd-numbered; that is, $B = \{1, 3, 5, 7\}$.

   C be the event that the card drawn is at least a 4; that is, $C = \{4, 5, 6, 7, 8\}$.

   D be the event that the card drawn is at most a 2; that is, $D = \{1, 2\}$.

Then

   $A \cup B = \{1, 2, 3, 4, 5, 6, 7, 8\}$, the sample space.

   $A \cap B = \varnothing$, representing the events A and B as being mutually exclusive events.

$A \cap C = \{4, 6, 8\}.$
$B \cap D = \{1\}.$
$C \cup D = \{1, 2, 4, 5, 6, 7, 8\}.$

EXERCISES

1. Consider the spinners illustrated below and determine the sample space for each spinner.

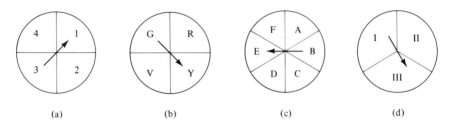

(a)   (b)   (c)   (d)

2. An experiment consists of spinning each of the spinners in Exercise 1 twice. List the sample space for each spinner.
3. An experiment consists of spinning two of the spinners in Exercise 1. List the sample space if the two spinners are
   a. (a) and (b).   b. (a) and (b).
   c. (a) and (d).   d. (b) and (c).
   e. (b) and (d).   f. (c) and (d).
4. An experiment consists of spinning three of the spinners in Exercise 1. List the sample space if the three spinners are
   a. (a), (b), and (c).   b. (a), (b), and (d).
   c. (a), (c), and (d).   d. (b), (c), and (d).
5. A coin is tossed three times.
   a. Describe the sample space using H for heads and T for tails.
   b. If A is the event that the first toss is a head, list A.
   c. If B is the event that there are exactly two heads, list B.
   d. If C is the event that there are at most two heads, list C.
   e. If D is the event that the first and third toss are the same, list D.
   f. If E is the event that there are no tails, list E.
   g. List A'.
   h. List D'.
   i. Are the events A and B mutually exclusive? Explain.
   j. Are the events D and E mutually exclusive? Explain.
6. A coin is tossed four times.
   a. Describe the sample space for the experiment.
   b. If A is the event that the second toss is a head, list A.

    c. If *B* is the event that the number of heads equals the number of tails, list *B*.

    d. If *C* is the event that there are at most three heads, list *C*.

    e. If *D* is the event that there are at least two tails, list *D*.

    f. If *E* is the event that heads and tails alternate, list *E*.

    g. List *B'*.

    h. List *E'*.

    i. Which events if any are mutually exclusive with *C*?

    j. Which events if any are mutually exclusive with *D*?

7. An experiment consists of tossing a coin and drawing a card from a set of cards numbered 1 through 8. Describe the sample space for the experiment.

8. The mathematics faculty at Matrix College consists of members A, B, C, D, and E. Two members of this faculty are to be selected to serve on the College's Promotion Committee. This selection can be accomplished in ten ways.

    a. Describe the sample space for this experiment.

    b. If *F* is the event that member A is selected, list *F*.

    c. If *G* is the event that C or D is selected, list *G*.

    d. If *H* is the event that B and F are selected, list *H*.

    e. If *I* is the event that neither A nor E is selected, list *I*.

    f. If *J* is the event that C is selected but D is not selected, list *J*.

    g. If *K* is the event that B or D but not both are selected, list *K*.

    h. List *H'*.

    i. List *K'*.

    j. Which of the above events if any are mutually exclusive with *J*?

9. In a survey of families having exactly three children, the sex of each child was recorded in the order of their births.

    a. Describe the sample space for this experiment using *G* for girl and *B* for boy.

    b. If *A* is the event that the second child is a girl, list *A*.

    c. If *B* is the event that there are exactly two boys in the family, list *B*.

    d. If *C* is the event that the second child is a girl and that there are exactly two girls in the family, list *C*.

    e. If *D* is the event that the third child is a boy and that there are exactly two girls, list *D*.

    f. If *E* is the event that the children are all of the same sex, list *E*.

    g. If *F* is the event that there are more boys than girls in the family, list *F*.

    h. If *G* is the event that there are at least two girls in the family, list *G*.

    i. If *H* is the event that there are at most two boys in the family, list *H*.

    j. If *I* is the event that the first and third children are of the same sex, list *I*.

10. Consider the events *A* through *I* indicated in Exercise 9. Which of these events if any are mutually exclusive?

16.2 PROBABILITY

When an experiment is performed we observe the outcome associated with the experiment. If the experiment were repeated under the identical conditions, the outcome may or may not be the same as the outcome when the experiment was initially performed. For instance, if a coin is tossed, the outcome may be a head. If the coin is tossed again, the outcome may be a tail. If the coin is tossed repeatedly, say 10 times, the outcomes may be 7 heads and 3 tails. For the next 10 tosses of the coin, the outcomes may be 6 tails and 4 heads. For the first 10 tosses we may conclude that the chance of obtaining a head on the following tosses would be 7/10. The ratio 7/10 is called a *relative frequency* and, in general, the relative frequency may be denoted by the symbol f/N, where N represents the number of times the experiment is performed and f represents the number of favorable outcomes (in this case heads showing on the toss of a coin).

If the same experiment is repeated many times, we may find that the relative frequency changes and, hence, this ratio is not very useful for predicting outcomes. However, as N, the number of times the experiment is performed, becomes very large, the relative frequency tends to approach a specific value and stabilizes around that value. For instance, if a coin were tossed a very large number of times, the relative frequency associated with obtaining a head would be 1/2.

The above discussion suggests that if A is an event, then the relative frequency associated with A as N becomes larger and larger approaches a value p. This value p associated with the event A is called the *probability of the event A* and is denoted by the symbol $p(A)$. This definition of probability is called *empirical probability*. Another definition of probability, known as the *classical* definition of probability, concerns equally likely outcomes.

If each of the outcomes of an experiment is equally likely, then we may define the probability of an event as the ratio of the number of favorable outcomes of the experiment to the total number of possible outcomes.

DEFINITION 16.8 If S is the sample space and A is an event, then the *probability of A*, denoted by $p(A)$, is equal to the number of favorable outcomes of the experiment divided by the total number of outcomes, provided the outcomes are all equally likely. If f represents the favorable outcomes and t represents the total number of outcomes, then $p(A) = f/t$.

From Definition 16.8 it should be clear that the probability of any event must be a numerical value greater than or equal to 0 and less than or equal to 1. The total number of outcomes must be positive, the number of favorable outcomes must be 0 or positive, and f must be less than or equal to t. Hence we have

$$0 \leq p(A) \leq 1$$

for all events A of any sample space S. If $p(A) = 0$, we say that the event A is impossible, and if $p(A) = 1$, we say that the event A is certain to occur. The more positive the value of $p(A)$, the more likely that event A will occur.

EXAMPLE An experiment consists of rolling a single die and observing the value associated with the face of the die that shows. The sample space $S = \{1, 2, 3, 4, 5, 6\}$, and we will assume that all sample points are equally likely. Let A be the event of rolling a 5, B be the event of rolling an even number, and C be the event of rolling at least a 3. Determine (a) $p(A)$, (b) $p(B)$, and (c) $p(C)$.
Solution
a. $p(A) = 1/6$ since there are a total of 6 possible outcomes and the the only favorable outcome is rolling a 5. Hence $f/t = 1/6$.
b. $p(B) = 1/2$ since $t = 6$, and $f = 3$ because the favorable outcomes are rolling a 2, 4, or 6.
c. $p(C) = 2/3$ since $t = 6$, and $f = 4$ because the favorable outcomes are rolling a 3, 4, 5, or 6. ■

EXAMPLE A fair coin is tossed 3 times, and the outcomes are observed. The sample space $S = \{HHH, HHT, HTH, THH, HTT, THT, TTH, TTT\}$. Let A be the event of obtaining exactly 2 heads, B be the event of obtaining at least 2 heads, C be the event of obtaining all like outcomes, and D be the event of obtaining no tails. Determine (a) $p(A)$, (b) $p(B)$, (c) $p(C)$, and (d) $p(D)$.
Solution
a. $p(A) = 3/8$ since $t = 8$ (there are 8 sample points in S), and $f = 3$ because the favorable outcomes are HHT, HTH, and THH.
b. $p(B) = 1/2$ since $t = 8$, and $f = 4$ because the favorable outcomes (at least 2 heads) are HHH, HHT, HTH, and THH.
c. $p(C) = 1/4$ since $t = 8$, and $f = 2$ because the favorable outcomes are HHH and TTT.
d. $p(D) = 1/8$ since $t = 8$, and $f = 1$ because the only favorable outcome is HHH. ■

EXAMPLE An experiment consists of rolling a pair of dice and observing the outcomes. The sample space S consists of the 36 sample points where the first component in each ordered pair denotes the value associated with the first die and the second component denotes the value associated with the second die.

$S = \{(1, 1), (1, 2), (1, 3), (1, 4), (1, 5), (1, 6), (2, 1), (2, 2), (2, 3), (2, 4), (2, 5),$
$\quad (2, 6), (3, 1), (3, 2), (3, 3), (3, 4), (3, 5), (3, 6), (4, 1), (4, 2), (4, 3), (4, 4),$
$\quad (4, 5), (4, 6), (5, 1), (5, 2), (5, 3), (5, 4), (5, 5), (5, 6), (6, 1), (6, 2), (6, 3),$
$\quad (6, 4), (6, 5), (6, 6)\}$

a. The probability that the sum of the values on the two dice will be 9 is 1/9 since $t = 36$, and $f = 4$ because the favorable outcomes are (3, 6), (4, 5), (5, 4), and (6, 3).

b. The probability that the sum of the values on the two dice will be even is 1/2 since $t = 36$, and $f = 18$ because the favorable outcomes are (1, 1), (1, 3), (1, 5), (2, 2), (2, 4), (2, 6), (3, 1), (3, 3), (3, 5), (4, 2), (4, 4), (4, 6), (5, 1), (5, 3), (5, 5), (6, 2), (6, 4), and (6, 6).

EXERCISES

1. A pair of dice is rolled and all sample points are equally probable.
   a. What is the probability of getting a total of 7?
   b. What is the probability of getting a total of 8?
   c. What is the probability of getting a total of at least 6?
   d. What is the probability of getting a total of at most 9?
2. A card is drawn at random from a well-shuffled deck of 52 ordinary playing cards.
   a. What is the probability of drawing a queen of diamonds?
   b. What is the probability of drawing a king, an ace, or a 7?
   c. What is the probability of drawing a black card?
   d. What is the probability of drawing a 3, 5, 7, 8, or 9?
   e. What is the probability of drawing a club or a 5?
3. An urn contains 15 balls of which 8 are white, 4 are red, and 3 are black. A ball is drawn at random from the urn.
   a. What is the probability that the ball is white?
   b. What is the probability that the ball is red or black?
   c. What is the probability that the ball is red or white?
   d. What is the probability that the ball is not red?
4. A coin is tossed 3 times.
   a. What is the probability that the first toss is a head?
   b. What is the probability that there are exactly 2 heads?
   c. What is the probability that there are at most 2 heads?
   d. What is the probability that the first and third toss are the same?
   e. What is the probability that there are no tails?
   f. What is the probability of (a) or (b)?
   g. What is the probability of (c) or (d)?
5. A coin is tossed 4 times.
   a. What is the probability that the second toss is a head?

b. What is the probability that the number of heads equals the number of tails?

c. What is the probability that there are at most 3 heads?

d. What is the probability that there are at least 2 tails?

e. What is the probability that heads and tails alternate?

f. What is the probability of (b) and (c)?

g. What is the probability of (d) or (e)?

6. An experiment consists of tossing a coin and rolling a die.

a. What is the probability of tossing a head and rolling a 3?

b. What is the probability of tossing a tail and rolling at least a 4?

c. What is the probability of tossing a head and rolling at most a 4?

d. What is the probability of tossing a tail and rolling an even number?

e. What is the probability of (b) or (d)?

16.3 INDEPENDENT EVENTS

DEFINITION 16.9 Two events are said to be *independent* if the outcome of one event in no way affects the outcome of the other.

Independent events and mutually exclusive events do not mean the same thing. Recall from Definition 16.6 that two events are mutually exclusive if they are disjoint.

To obtain the probability of event A, denoted by $p(A)$, we have been using the formula $p(A) = f/t$, where f refers to the number of favorable outcomes and t refers to the total number of outcomes in the experiment. Now the total number of outcomes in the experiment is the number of sample points in the sample space S, which we may denote by $n(S)$. Likewise, the number of favorable outcomes would correspond to the number of outcomes in the event A, which we may denote by $n(A)$. Substituting $f = n(A)$ and $t = n(S)$, we now have

$$p(A) = \frac{n(A)}{n(S)}$$

where A is a subset of S and S is the sample space consisting of equally likely sample points.

Also, from Definition 16.7 we recall that the symbol $A \cup B$ means that the event A occurs or the event B occurs or both events occur. Hence if A and B are both events in S, we have

$$p(A \cup B) = \frac{n(A \cup B)}{n(S)}$$

But recall that if A and B are arbitrary sets, then $n(A \cup B) = n(A) + n(B) - n(A \cap B)$. Therefore we have

$$p(A \cup B) = \frac{n(A \cup B)}{n(S)}$$

$$= \frac{n(A) + n(B) - n(A \cap B)}{n(S)}$$

$$= \frac{n(A)}{n(S)} + \frac{n(B)}{n(S)} - \frac{n(A \cap B)}{n(S)}$$

$$= p(A) + p(B) - p(A \cap B)$$

Hence we have the following theorem.

THEOREM 16.1 If A and B are any two events associated with a sample space S, then the probability of A or B, denoted by $p(A \cup B)$, is given by

$$p(A \cup B) = p(A) + p(B) - p(A \cap B)$$

Clearly, if the events A and B are mutually exclusive, then $A \cap B = \varnothing$, and $p(A \cap B) = 0$.

THEOREM 16.2 If A and B are any two mutually exclusive events associated with a sample space S, then $p(A \cup B) = p(A) + p(B)$.

Proof This theorem is a special case of Theorem 16.1, and the proof is left as an exercise.

EXAMPLE An experiment consists of drawing a card from a well-shuffled deck of 6 cards numbered 1 through 6 inclusive. What is the probability of drawing a card with a number greater than or equal to 3 or an odd-numbered card?

Solution Let A be the event of drawing a card with a number greater than or equal to 3, and let B be the event of drawing a card with an odd number. We now determine that $A = \{3, 4, 5, 6\}$, $B = \{1, 3, 5\}$, and $A \cap B = \{3, 5\}$. Since $p(A \cup B) = p(A) + p(B) - p(A \cap B)$, we have $p(A \cup B) = (2/3) + (1/2) - (1/3) = 5/6$. ∎

EXAMPLE An experiment consists of rolling a single die. What is the probability of rolling an odd number or a 6?

Solution Let A be the event of rolling an odd number, and let B be the event of rolling a 6. Then $A = \{1, 3, 5\}$ and $B = \{6\}$. Since $A \cap B = \varnothing$, the events A and B are mutually exclusive and $p(A \cap B) = 0$. Hence $p(A \cup B) = p(A) + p(B)$. Since $p(A) = 1/2$ and $p(B) = 1/6$, we have $p(A \cup B) = (1/2) + (1/6) = 2/3$. ∎

We have defined two events as being independent events if the outcome of one event in no way affects the outcome of the other.

DEFINITION 16.10 (Alternate) Two events A and B are said to be *independent* if and only if $p(A \cap B) = p(A) \cdot p(B)$.

EXAMPLE An experiment consists of tossing three coins. Let A be the event of tossing a head on the first coin and B be the event of tossing a tail on the second coin. Determine (a) $p(A)$, (b) $p(B)$, and (c) $p(A \cap B)$.
   Solution The sample space $S = \{HHH, HHT, HTH, THH, HTT, THT, TTH, TTT\}$, $A = \{HHH, HHT, HTH, HTT\}$, $B = \{HTH, HTT, TTH, TTT\}$, and $A \cap B = \{HTH, HTT\}$.
   a. $p(A) = 1/2$.
   b. $p(B) = 1/2$.
   c. $p(A \cap B) = 1/4$.
Observe that $p(A)p(B) = 1/2 \cdot 1/2 = 1/4 = p(A \cap B)$. Hence the events A and B are independent events. However, since $A \cap B \neq \emptyset$, A and B are not mutually exclusive. ∎

EXAMPLE If A and B are independent events in a sample space such that $p(A) = 3/5$ and $p(B) = 1/5$, find (a) $p(A \cap B)$ and (b) $p(A \cup B)$.
   Solution
   a. Since A and B are independent events, we know that $p(A \cap B) = p(A) \cdot p(B)$. Hence $p(A \cap B) = (3/5)(1/5) = 3/25$.
   b. From Theorem 16.1 we know that for any two events $p(A \cup B) = p(A) + p(B) - p(A \cap B)$. Hence $p(A \cup B) = (3/5) + (1/5) - (3/25) = 17/25$. ∎

In Definition 16.5 we defined the complement event of A, denoted by A', associated with a sample space S to be the set of all sample points in S that are not in A. Clearly, $A \cap A' = \emptyset$ and $A \cup A' = S$.

THEOREM 16.3 For any event A in a sample space S, $p(A') = 1 - p(A)$.
   Proof Since $A \cap A' = \emptyset$, A and A' are mutually exclusive and $p(A \cup A') = p(A) + p(A')$. Since $A \cup A' = S$ and $p(S) = 1$, we have $p(A) + p(A') = 1$. Hence $p(A') = 1 - p(A)$.

EXAMPLE An experiment consists of rolling a single die. Determine the probability of *not* getting a 5.
   Solution Let A be the event of getting a 5. Then $p(A) = 1/6$. Hence A' is the event of not getting a 5, and $p(A') = 1 - p(A) = 1 - (1/6) = 5/6$. ∎

EXERCISES

1. A coin is tossed. Show that the events "tail on the first toss" and "head on the second toss" are independent events.

2. Two coins are tossed. Let A be the event "head on the first toss," B be the event "exactly 1 head," and C be the event "at least 1 tail." Determine which of the following pairs of events are independent.
   a. A and B.
   b. A and C.
   c. B and C.
   d. A' and B.
   e. A' and C.
   f. B' and C.
   g. A' and B'.
   h. A' and C'.
   i. B' and C'.
   j. $(A \cup B)$ and C.

3. Let A and B be independent events defined on the same sample space such that $p(A) = 0.20$ and $p(B) = 0.35$. Determine each of the following probabilities.
   a. $p(A')$.
   b. $p(B')$.
   c. $p(A \cap B)$.
   d. $p(A \cup B)$.
   e. $p(A' \cap B')$.
   f. $p(A' \cup B')$.

4. Let A and B be independent events defined on the same sample space such that $p(A \cap B) = 1/4$ and $p(A \cap B') = 1/2$. Determine each of the following probabilities.
   a. $p(A)$.
   b. $p(A')$.
   c. $p(B)$.
   d. $p(B')$.
   e. $p(A \cup B)$.
   f. $p(A' \cup B')$.
   g. $p(A' \cap B)$.
   h. $p(A' \cap B')$.

5. Two dice, one white and the other red, are rolled. Let A be the event "white die shows an even number," B be the event "red die shows an odd number," and C be the event "the sum of the numbers on the two dice is odd."
   a. Show that A and B are independent events.
   b. Show that A and C are independent events.
   c. Show that B and C are independent events.
   d. Show that $p(A \cap B \cap C) \neq p(A) \cdot p(B) \cdot p(C)$.

6. Two events, A and B, are mutually exclusive with $p(A) = 0.25$ and $p(B) = 0.60$. Determine
   a. $p(A')$.
   b. $p(B')$.
   c. $p(A \cup B)$.
   d. $p(A \cap B)$.
   e. $p(A \cup B)'$.
   f. $p(A \cap B)'$.
   g. $p(A' \cup B')$.
   h. $p(A' \cap B')$.
   i. $p(A' \cap B)$.
   j. $p(A \cup B')$.

7. The probabilities that a student will receive a grade of A, B, C, D, or F in this course are 0.23, 0.47, 0.19, 0.06, and 0.05, respectively. What is the probability that the student will receive
   a. Either an A or a B?
   b. Neither an A nor an F?

    c. At most a C?

    d. At least a C?

8. Professor Smart reported to his dean that the probabilities associated with grades assigned in his course are

    0.20 for A

    0.27 for B

    0.35 for C

    0.19 for D

    0.10 for F

    What's wrong with his report?

9. Grades assigned for this course are A, B, C, D, and F such that all grades other than F are passing grades. If the probability that a particular student will pass this course is 0.82 and the probability that the student will get a grade of C or lower is 0.51, what is the probability that the student will receive a grade of C or D?

10. Based upon previous voting practices of a particular couple, the probability that the husband will vote in an upcoming election is 0.42, the probability that the wife will vote is 0.49, and the probability that both will vote is 0.26. What is the probability that:

    a. The husband will not vote?

    b. At least one of the couple will vote?

    c. Neither will vote?

    d. Exactly one of them will vote?

    e. At most one of them will vote?

11. Prove Theorem 16.2.

16.4 CONDITIONAL PROBABILITY

Now consider an experiment of drawing two cards without replacement from a well-shuffled standard deck of cards. What is the probability that the second card is an ace given that the first card is an ace? If we let A be the event that the second card is an ace and B be the event that the first card is an ace, then $p(A, \text{ given } B)$ can be determined as follows. There are 52 cards in a standard deck of cards, and 4 of these are aces. If the first card drawn is an ace, then there are 3 aces left in the deck, which now has 51 cards, since there is no replacement. Hence the required probability is 3/51 or 1/17.

DEFINITION 16.11 Let A and B be any events in a sample space S. Then the *conditional probability* of event A, given event B, denoted by $p(A \mid B)$, is defined to be

$$p(A \mid B) = \frac{p(A \cap B)}{p(B)} \quad \text{if } p(B) \neq 0$$

EXAMPLE An experiment consists of drawing two cards without replacement from a well-shuffled standard deck of cards. (a) What is the probability that the second card is a diamond given that the first card is a diamond? (b) What is the probability that both of the cards are diamonds?

Solution Let A be the event that the second card is a diamond and B be the event that the first card drawn is a diamond.

    a. Since it is given that the first card drawn is a diamond, there would be 12 diamonds left in the deck, which now contains 51 cards, since there is no replacement. Hence $p(A|B) = 12/51 = 4/17$.

    b. From Definition 16.11 we have $p(A \cap B) = p(B) \cdot p(A|B)$. Since $p(A|B) = 4/17$ from part (a) and since $p(B) = 13/52 = 1/4$, we have $p(A \cap B) = (1/4)(4/17) = 1/17$. ∎

THEOREM 16.4 If A and B are two independent events with nonzero probabilities, then $p(A|B) = p(A)$ and $p(B|A) = p(B)$.

Proof Since A and B are independent events, $p(A \cap B) = p(A) \cdot p(B)$. Hence

$$p(A|B) = \frac{p(A \cap B)}{p(B)} = \frac{p(A) \cdot p(B)}{p(B)} = p(A)$$

since it is given that $p(B) \neq 0$. Similarly,

$$p(B|A) = \frac{p(A \cap B)}{p(A)} = \frac{p(A) \cdot p(B)}{p(A)} = p(B)$$

EXERCISES

1. A man tosses two coins. What is the conditional probability of tossing 2 tails given that he tossed at least 1 tail?

2. A coin is tossed three times. What is the probability of getting 3 tails given that there is at least 1 tail?

3. A family has three children. Under the assumption that each child is as likely to be a boy as it is to be a girl, what is the probability that all three children are girls given at least one child is a girl?

4. Two distinguishable dice are rolled. What is the probability that the sum of the two faces which turn up is greater than 8, given that one turns up a 5?

5. The probability that a customer in a clothing store will buy a sports coat is 0.24, the probability that he will buy a pair of slacks is 0.27, and the probability that he will buy both is 0.15. What is the probability that the customer will buy a pair of slacks given that he already purchased a sports coat?

6. Students in a special college program took only two courses, mathematics and a language. Five percent of the students failed the language course, 4 percent failed mathematics, and 1 percent failed both. Among those who failed mathematics, what percent also failed the language course?

7. In a group of 500 college students, 225 are enrolled in English, 200 are enrolled in statistics, and 120 are not taking either course.
   a. How many students are enrolled in both courses?
   b. What is the probability that a student selected at random is enrolled in exactly one course given that he is enrolled in at least one course?
   c. What is the probability that a student selected at random is enrolled in English given that he is also enrolled in statistics?
   d. What is the probability that a student selected at random is enrolled in neither course?

16.5 EXPECTATION

In this section we will discuss briefly what is commonly called expectation or expected value. In playing various games of chance, a game is said to be a *fair* game if the expected value of winning (or losing) per game is zero. It should be obvious, then, that gambling houses do not operate fair games because to do so would cause them to "go broke." For such games the player has an expected value that is negative. What, then, do we mean by expected value?

DEFINITION 16.12 If an event has several possible outcomes which we will denote by x_i $(i = 1, 2, 3, \ldots, n)$, and these outcomes have corresponding probabilities p_i, then the *expectation* or *expected value* of E is given by

$$E = x_1 p_1 + x_2 p_2 + x_3 p_3 + \cdots + x_n p_n$$

It is to be noted that the expected value is not the value to be expected each time a game is played. Rather it is an average expectation over a long run of repeated game playing.

EXAMPLE A player pays $3 to play a game. For this particular game the probability of winning is 1/5 and the probability of losing is 4/5. If the player wins the game, he receives $12. What is the expected value if he continues to play the game?
Solution Let $x_1 = \$9$, which represents the $12 won minus the $3 to play the game, and let $x_2 = {}^-\$3$, the amount it cost to play the game. We also have that $p_1 = 1/5$ and $p_2 = 4/5$. Hence $E = x_1 p_1 + x_2 p_2 = (\$9)(1/5) + ({}^-\$3)(4/5) = {}^-\$3/5 = {}^-\$.60$. Hence if the player continues to play the game, he would lose 60¢ per game on the average. ∎

EXAMPLE A game consists of drawing a card from a well-shuffled standard deck of cards. The player pays $5 for playing the game and receives $10 if he draws a queen or seven, $7 if he draws a jack, and $6 if he draws a two or king. What is the expected value of this game?

Solution   Let $x_1 = \$5$, which represents the $10 won minus the $5 to play the game, $x_2 = \$2$ ($7 − $5), $x_3 = \$1$ ($6 − $5), and $x_4 = {}^-\$5$, the amount it cost to play the game. We also have $p_1 = 8/52$ since there are 4 queens and 4 sevens in a standard deck of cards and no queen is also a seven, $p_2 = 4/52$ since there are 4 jacks in the deck, $p_3 = 8/52$ since there are 4 kings and 4 twos in the deck and no king is also a two, and $p_4 = 32/52$ since there are 32 cards in the deck which are other than queen, seven, jack, two, or king. Hence $E = x_1 p_1 + x_2 p_2 + x_3 p_3 + x_4 p_4 = (\$5)(8/52) + (\$2)(4/52) + (\$1)(8/52) − (\$5)(32/52) = {}^-\$104/52 = {}^-\$2$. Hence if the player continues to play the game, he will lose $2 per game on the average.   ■

EXERCISES

1. In a particular raffle 1000 tickets are sold at $1 each. The first prize is $500, the second prize is $200, and the three third prizes are $50 each. Is this a fair game? If not, for whom is it favorable?
2. In a game of dice a player wins $5 if he throws a sum of 7 or 11 and loses $1 if he throws any other sum. Is this a fair game? If not, for whom is it favorable?
3. A man has four dimes, three quarters, and two half-dollars in his pocket and plans to purchase an item worth 35¢. The salesclerk offers to sell the item to him in exchange for a coin selected at random from his pocket. Assuming that there are only the nine coins referred to above, is this a fair offer? If not, for whom is it favorable?
4. For a 50-question multiple-choice test, each question has four possible choices, only one of which is correct. Bette took the test without reading the questions and randomly selected the answers.
   a. What is Bette's expected score on the test if each correct answer receives 2 points and there is no penalty for guessing?
   b. What is Bette's expected score on the test if each correct answer receives 2 points and each incorrect answer receives $^-1$ points?
5. Two players A and B play a game of rolling a pair of dice. A collects 11¢ if an even sum appears, and B collects 4¢ if an odd sum appears.
   a. What is A's expected average value per roll of the dice?
   b. What is B's expected average value per roll of the dice?

SUMMARY

In this chapter we introduced some of the basic concepts of probability. We defined and discussed an experiment, an outcome, a sample space, a sample point, an event, mutually exclusive events, relative frequency, probability of an event, independent events, conditional probability, and expectation.

During our discussion the following symbols were introduced.

SYMBOL	INTERPRETATION
S	Sample space
A	An event in S
$p(A)$	Probability of A
A'	Complement event of event A
$p(A \mid B)$	Probability of event A given event B
E	Expectation or expected value

REVIEW EXERCISES FOR CHAPTER 16

1. A coin is tossed. If A is the event of getting a head and B is the event or getting a tail, are A and B mutually exclusive? Explain.
2. Two coins are tossed. If A is the event of getting 2 heads and B is the event of getting at least 1 tail, are A and B mutually exclusive? Explain.
3. A die is rolled. If A is the event that an even number shows and B is the event that a 5 shows, are A and B mutually exclusive? Explain.
4. A die is rolled. If A is the event that an even number shows and B is the event that at least a 5 shows, are A and B mutually exclusive? Explain.
5. The registrar at Big League University indicates that the probabilities associated with students registering for numbers of courses are
   0.07 for 0 courses
   0.28 for exactly 1 course
   0.34 for exactly 2 courses
   0.20 for exactly 3 courses
   0.06 for 4 or more courses
   What's wrong with the Registrar's statement?
6. A student who successfully completed this course was told by a garage mechanic that the probability that her car would pass inspection was 0.40 and that the probability that her car would not pass inspection was 0.50. On the basis of this information, the student decided to have her car inspected elsewhere. Why?
7. A card is drawn at random from a well-shuffled deck of 52 ordinary playing cards. Let A be the event "the card is a nine", B be the event "the card is a diamond", and C be the event "the card is an ace or a diamond." Which of the following pairs of events are independent? Explain.
   a. A and B.   b. A and C.
   c. B and C.   d. A and B'.
   e. A' and C.   f. B' and C.

8. Let A and B be independent events defined on the same sample space such that $p(A) = 0.40$ and $p(A \cap B) = 0.18$. Determine each of the following probabilities.
   a. $p(A')$.
   b. $p(B)$.
   c. $p(B')$.
   d. $p(A \cup B)$.
   e. $p(A' \cap B')$.
   f. $p(A' \cup B')$.
   g. $p(A \cup B')$.
   h. $p(A' \cap B)$.
   i. $p(B \cup A')$.
   j. $p(B' \cap A)$.

9. Mr. Backup has two automobiles. The probability that one will not start up on a cold morning is 0.2, and the probability that the other will not start up is 0.15. On a cold morning, what is the probability that
   a. Both cars will start up?
   b. At least one car will start up?
   c. Neither car will start up?

10. A faculty committee concerned with tenure produced the following results relating to the total faculty at their college:

	WITH DOCTORATE	WITHOUT DOCTORATE
Tenured	42	54
Nontenured	63	41

   One of these professors is chosen at random to be chairman of the committee. If we let A denote the event that the professor has a doctorate and let B denote the event that the professor is tenured, determine each of the following probabilities.
   a. $p(A)$.
   b. $p(B)$.
   c. $p(A')$.
   d. $p(B')$.
   e. $p(A \cap B)$.
   f. $p(A' \cap B')$.
   g. $p(A \cap B')$.
   h. $p(A' \cap B)$.
   i. $p(A \cup B)$.
   j. $p(A' \cup B)$.
   k. $p(A \cup B')$.
   l. $p(A \cup B)'$.
   m. $p(A \cap B)'$.
   n. $p(B|A)$.
   o. $p(A|B)$.

11. On a particular Saturday night the probability that a student will stay in the dorm and study is 0.14, the probability that the student will go out to the movies is 0.29, the probability that the student will stay in and watch television is 0.22, and the probability that the student will go out on a date is 0.35.
   a. What is the probability that the student will not stay in and study?
   b. What is the probability that the student will neither watch television nor have a date?
   c. What is the probability that the student will stay in?

12. A group of 100 people consists of 39 persons who read *New* magazine and 53 people who ride the bus to work. Of these, 20 read *New* magazine and also ride the bus to work. If a person is chosen at random from this group, what is the probability that the person

a. Either reads *New* magazine or rides the bus to work?

b. Neither reads *New* magazine nor rides the bus to work?

c. Does not read *New* magazine but rides the bus to work?

d. Does not ride the bus to work but reads *New* magazine?

13. Of 150 college students, 62 are enrolled in a mathematics course, 93 are enrolled in a social science course, and 48 are enrolled in both. If a student is selected at random from this group, what is the probability that the student is

    a. Enrolled in only one of the courses?

    b. Enrolled in at least one of the courses?

    c. Enrolled in neither of the courses?

14. Two dice are rolled. What is the probability that the sum of the face numbers that turn up is greater than 10 given that one of the dice shows a 6?

15. Consider an urn containing 8 balls of which 6 are black and 2 are white. Two balls are drawn without replacement. What is the conditional probability that the second ball drawn is black, given that the first ball is black?

16. A family has three children. Under the assumption that each child is as likely to be a boy as it is to be a girl, what is the probability that all 3 children are girls given that at least 2 children are girls?

17. The probability that a customer in a clothing store will buy a sports coat is 0.24, the probability that he will buy a pair of slacks is 0.27, and the probability that he will buy both is 0.15. What is the probability that a customer in the clothing store will buy a sports coat, given that he has already purchased a pair of slacks?

18. Students in a special college program took only two courses, mathematics and reading. Five percent of the students failed reading, 4 percent failed mathematics, and 1 percent failed both. Among those who failed reading, what percentage also failed mathematics?

19. A player pays $5 to play a game. For this particular game the probability of winning is 2/7 and the probability of losing is 5/7. If the player wins the game, he recieves $15.

    a. What is the average expected value for this game?

    b. Is this a fair game? Explain.

    c. If the game is not a fair game, who is favored by the game?

20. For a particular raffle 2000 tickets are sold at $3 each. The first prize winner recieves $1000, the second prize winner recieves $500, the third prize winner receives $200, the fourth prize winner receives $100, and the five fifth prize winners receive $20 each. What is the expected value for this raffle?

CHAPTER 17
AN INTRODUCTION TO STATISTICS

This chapter will introduce the reader to the most basic concepts involved in statistics and indicate how statistics can be used. The reader will learn to better understand the use of statistics and statistical terms encountered in various textbooks, newspapers, magazines, and other media. Although the branch of statistics known as inferential statistics is very interesting, this chapter will emphasize the branch known as descriptive statistics.

17.1 DESCRIPTIVE STATISTICS

There are two basic branches of statistics called descriptive statistics and inferential statistics. *Descriptive statistics* is concerned with collecting, tabulating, summarizing, and presenting information known about some particular situation. From this information we may attempt to infer something from this situation to a larger situation about which we do not have complete information. This use of statistics is in the category known as *statistical inference* or *inferential statistics*. Throughout this chapter we will emphasize descriptive statistics.

DEFINITION 17.1 By the word *population* we shall mean the set of all people or objects under consideration for a particular situation. By the word·*sample* we shall mean a subset of the population.

We will be interested in selecting elements from a population, and the words random and randomly are very important. Discussion of these terms could be made very precise and lengthy. However, we will use them from an intuitive point of view. Consider a population in which every element in the population has the same probability or chance of being selected. If the process of selection does not favor any element in the population over any other element in the population in any manner whatsoever, we say we have a random selection. Any sample of elements from this population that has been selected in this manner is called a *random sample.* If we shuffle thoroughly a standard deck of cards before we select cards from the deck we would have an example of a random selection, and the cards selected in this manner would constitute a random sample.

Now consider a set of test grades. Each grade is called a *raw score,* and the collection of all such raw scores is called a *distribution* of scores. If we consider the heights of all students in your class, then each height measurement is a raw score and the set of all height measurements would be a distribution of these scores. Raw scores also are called *measures.*

DEFINITION 17.2 *Parameters* are measures of a population. That is, parameters are measures that are taken of every element of a population. Parameters are generally denoted by the use of Greek symbols such as μ.

DEFINITION 17.3 *Statistics* are measures of a sample from some population. That is, statistics are measures that are taken of only some elements of the population. Statistics are generally denoted by English letters such as m.

EXERCISES

Identify each of the following statements as being true or false, and give an appropriate reason for your answer.
1. Descriptive statistics is concerned with making inferences about populations based upon samples taken from these populations.
2. A population is defined as a complete set of individuals, objects, or measurements having some observable characteristics.
3. Generally, the purpose of calculating statistics from a sample is to understand the characteristics of that sample.
4. The total number of people who voted in the last presidential election is a statistic.
5. A statistic is generally represented by the use of a Greek letter.
6. In a city of 82,000 voters, 500 were carefully selected for study. The 500 represented a sample of the total population of 82,000.
7. A group of 1,000 people selected at random from a total of 1,216,813 were surveyed. Forty-one percent of this group indicated that they

were pleased with the manner in which the president of the United States was performing in office. The figure of 41 percent is a statistic.

8. The total number of pre-school age children in the United States at any given time is a parameter.
9. To estimate the number of defective units in the daily output in a particular factory, 200 items were randomly selected and tested. The 200 items constitute the population.
10. The two branches of statistics are statistical reference and inferential statistics.

17.2 MEASURES OF CENTRAL TENDENCY

Consider a set of test grades for a particular group of 20 students such as those grades listed below.

100	94	88	76
100	91	88	63
98	90	88	58
97	90	88	47
95	89	83	39

Suppose that I. M. Bright received a grade of 90 on the test. How well did he perform? Clearly, 90 was not the highest grade nor was it the lowest grade. There were 8 students who performed at least as well or better than Bright, and 11 students who did not perform as well. From the distribution listed above we also note that 90 was not the most common grade since that distinction belongs to 88. Finally, we note that if all 20 grades were added together and the sum divided by 20, the result would be 83.1, which is called the mean or the arithmetic average. Notice that only 5 of the grades in this distribution fall below the mean.

In the above paragraph we are referring to what are known as measures of central tendency which we shall now discuss.

DEFINITION 17.4 The *mean* or *arithmetic average* of a set of n measurements is the sum of the measurements divided by n, the number of measurements. If the n measurements are denoted by $X_1, X_2, X_3, \ldots, X_n$, then the mean, denoted by $\overline{X}$, is given by

$$\overline{X} = \frac{X_1 + X_2 + X_3 + \cdots + X_n}{n}$$

When working with the mean, it is sometimes useful to introduce the symbol $\sum$, the Greek letter sigma, which is used to denote summation.

For instance, the expression $X_1 + X_2 + X_3 + \cdots + X_n$ may be written in the abbreviated form

$$\sum_{i=1}^{n} X_i$$

The i in the notation is called the *index* and indicates that which is being changed in each successive term of the sum. It is referred to as a dummy index insofar as we could have used j or k or any other symbol not already in the expression. The 1 is called the lower bound and tells us where to start; the n is called the upper bound and tells us where to stop. Hence we may write

$$\overline{X} = \frac{\sum_{i=1}^{n} X_i}{n}$$

EXAMPLE   Determine the value of each of the following:

a. $\displaystyle\sum_{i=1}^{4} i$

b. $\displaystyle\sum_{j=1}^{3} j^2$

c. $\displaystyle\sum_{k=0}^{2} \frac{1}{k+1}$

d. $\displaystyle\sum_{m=4}^{7} \frac{m+3}{m}$

Solution

a. $\displaystyle\sum_{i=1}^{4} i = 1 + 2 + 3 + 4 = 10$

b. $\displaystyle\sum_{j=1}^{3} j^2 = 1^2 + 2^2 + 3^2 = 1 + 4 + 9 = 14$

c. $\displaystyle\sum_{k=0}^{2} \frac{1}{k+1} = \frac{1}{0+1} + \frac{1}{1+1} + \frac{1}{2+1} = 1 + \frac{1}{2} + \frac{1}{3} = \frac{11}{6}$

d. $\displaystyle\sum_{m=4}^{7} \frac{m+3}{m} = \frac{4+3}{4} + \frac{5+3}{5} + \frac{6+3}{6} + \frac{7+3}{7} = \frac{7}{4} + \frac{8}{5} + \frac{9}{6} + \frac{10}{7} = \frac{879}{140}$

∎

EXAMPLE   Determine the mean of 12, 16, 20, 30, and 42.
Solution

$$\overline{X} = \frac{12 + 16 + 20 + 30 + 42}{5} = \frac{120}{5} = 24 \quad \blacksquare$$

EXAMPLE Determine the arithmetic average, $\overline{X}$, of 20, 20, 25, 30, 30, 30, 36, 40, 40, and 40.
Solution

$$\overline{X} = \frac{20 + 20 + 25 + 30 + 30 + 30 + 36 + 40 + 40 + 40}{10} = \frac{311}{10} = 31.1 \quad \blacksquare$$

DEFINITION 17.5 Consider a set of measurements arranged in order of either increasing or decreasing magnitude. If the number of measurements is odd, then the *median* of the set is the middle measurement in the arrangement. If the number of measurements is even, then the *median* of the set is halfway between the two middle measurements.

EXAMPLE Consider the set of seven measurements 8, 10, 4, 1, 9, 7, 11. Determine the median.
Solution Arranging these measurements in increasing order of magnitude, we have 1, 4, 7, 8, 9, 10 and 11, and the median is 8. $\blacksquare$

EXAMPLE Consider the set of measurements 93, 89, 68, 95, 88, 72, 79, 82, 69, and 96. Determine the median.
Solution Arranging these measurements in increasing order of magnitude, we have 68, 69, 72, 79, 82, 88, 89, 93, 95 and 96. There are ten measurements, and the two middle measurements are 82 and 88. Hence the median is 85. $\blacksquare$

DEFINITION 17.6 The *mode* for a set of measurements is the measurement that occurs the greatest number of times in the distribution.

EXAMPLE Determine the mode for the set of measurements 2, 4, 8, 4, 9, 6, 6, 5, 4, 9, and 1.
Solution The mode is 4 since 4 occurs 3 times and all other measurements occur fewer times. $\blacksquare$

Now consider the following salaries of a group of workers: $8,000, $10,000, $14,000, $14,000, $16,000, and $100,000. The mean or arithmetic average of these salaries is $27,000. The median salary is $14,000, and the mode is also $14,000. Which of these measures of central tendency is most meaningful? To claim that the average salary of $27,000 is representative of the data given is misleading due to the extreme value of $100,000 being included in the computation of the mean. Since each measurement is used in the computation of the mean, we readily see that the mean becomes very sensitive to extreme measurements unless, of course, these extreme measurements are balanced on both sides of the mean.

If a measurement differs from the mean, the difference between the measurement and the mean is called a *deviation*. For instance, the mean of the salary data given above is $27,000. The deviation associated with the measurement of $10,000 is $^-$$17,000, and the deviation associated with the measurement of $100,000 is $73,000. An important characteristic of the mean, however, is that the sum of the squares of the deviations from the mean is less than the sum of the squares of the deviations about any other measurement in the distribution.

The median, however, is not sensitive to extreme values. Hence this characteristic makes the median more acceptable for describing central tendency in a distribution where the mean is not favored due to extreme values present.

Of the three measures of central tendency discussed, the mean is generally the preferred statistic because of its value in more advanced statistical analysis. The deviations of scores from the mean provide additional useful information about a distribution. Also, in the language of statistics we say that the mean is more stable or reliable than either the median or mode. It should be noted, however, that in certain distributions which are known as "skewed" distributions, the median is preferred over the mean.

EXERCISES

1. Determine the value of each of the following:

a. $\sum\limits_{j=0}^{4} j.$ b. $\sum\limits_{j=1}^{4} \dfrac{j+1}{j}.$ c. $\sum\limits_{k=2}^{5} \dfrac{k-2}{3k}.$

d. $\sum\limits_{i=3}^{7} \dfrac{1}{i-2}.$ e. $\sum\limits_{t=1}^{3} \dfrac{2t}{t^2}.$ f. $\sum\limits_{k=1}^{4} k(k+1).$

2. For each of the following sets of data determine the mean or arithmetic average.

a. 2, 3, 5, 2, 6, 3, 2, 4, 5, 2, 3.

b. 12, 10, 7, 10, 9, 9, 13, 11, 8, 9, 12, 8.

c. 17, 16, 20, 16, 15, 22, 16, 20, 19.

d. 91, 91, 91, 91.

e. 101, 101, 98, 89, 67, 110, 98, 69.

f. 33, 36, 37, 35, 36, 42, 30.

3. For each set of data given in Exercise 2 determine the median.

4. For each set of data given in Exercise 2 determine the mode.

5. During a labor negotiation session the union representative made the statement, "The median annual wage for my membership is $8300." The management representative, using the same data, made the statement, "The mean annual wage for your membership is $9,100." Assuming that both statements were true, what could account for the "discrepancy"?

17.3 GROUPED DATA AND FREQUENCY DISTRIBUTIONS

In the previous section we considered measures of central tendency for distributions with relatively few measurements and we treated the measurements individually. Now consider Table 17.1 giving a set of 50 scores on a particular examination with a maximum possible score of 200:

Table 17.1

182	173	190	168	142
139	187	178	182	190
173	168	142	173	168
157	193	168	177	145
168	176	174	175	163
190	169	157	169	158
182	158	151	157	182
178	173	145	189	191
128	132	189	163	156
187	190	137	159	160

Examining the scores in the table, we determine that the high score is 193 and the low score is 128. We could now list all the scores from 193 to 128 in decreasing order and record alongside each score the number of times it occurs. The number of times a score appears is called the *frequency* and is denoted by f. When this has been completed, the resulting table so constructed will be what is called a *frequency distribution*. In Table 17.2 we show the frequency distribution for the ungrouped data given in Table 17.1. The symbol X represents individual measurements or scores.

The scores tabulated in Table 17.2 are widely spread and many have 0 frequencies associated with them. An examination of the frequency distribution fails to reveal any clear indication of a measure of central tendency. In situations of this type it is sometimes useful to group the scores into mutually exclusive class intervals such that any particular score will belong to exactly one such class interval. When this is done, the resulting frequency distribution is called a *grouped frequency distribution.*

How many class intervals should there be for a grouped frequency distribution? Although no particular number is universally prescribed, statisticians do agree that most data of the type we are concerned with in this chapter can be accomodated by from 10 to 20 class intervals, aiming for approximately 15 class intervals.

Now that we know about how many class intervals we should have for a grouped frequency distribution, how do we go about determining what they are? The first step would be to determine the difference between the highest and lowest scores which, for the data given in Table 17.2, would be $193 - 128 = 65$. We then add 1 to obtain the total number of scores or

Table 17.2

X	f	X	f	X	f	X	f	X	f	X	f
193	1	182	4	171	0	160	1	149	0	138	0
192	0	181	0	170	0	159	1	148	0	137	1
191	1	180	0	169	2	158	2	147	0	136	0
190	4	179	0	168	5	157	3	146	0	135	0
189	2	178	2	167	0	156	1	145	2	134	0
188	0	177	1	166	0	155	0	144	0	133	0
187	2	176	1	165	0	154	0	143	0	132	1
186	0	175	1	164	0	153	0	142	2	131	0
185	0	174	1	163	2	152	0	141	0	130	0
184	0	173	4	162	0	151	1	140	0	129	0
183	0	172	0	161	0	150	0	139	1	128	1

potential scores. Hence we have $65 + 1 = 66$. The next step would be to divide 66, in this case, by 15 to obtain the number of scores or potential scores in each class interval and round the result to the nearest whole number. Hence we have $66 \div 15 = 4\frac{2}{5}$ or 4, rounded to the nearest whole number. With 4 scores to an interval and a total of 66 scores or potential scores, we would need 17 class intervals for this distribution.

We next consider the lowest score in the distribution and let that be the least value in the lowest class interval. Hence 128 is the minimum value in the lowest class interval in this case. Add 3 to this to obtain the highest value in this class interval. Hence the lowest class interval would be 128–131 which does contain 4 scores or potential scores. The next higher class interval would have 132 for its least value and $132 + 3 = 135$ for its highest value. We would continue this procedure to obtain each higher class interval until all the scores in the distribution have been included.

The resulting *grouped frequency distribution* for the data given in Table 17.1 is shown in Table 17.3.

It should be noted that in an ungrouped frequency distribution the individual scores or measurements are preserved. However, in a grouped

Table 17.3

CLASS INTERVAL	f	CLASS INTERVAL	f
192–195	1	156–159	7
188–191	7	152–155	0
184–187	2	148–151	1
180–183	4	144–147	2
176–179	4	140–143	2
172–175	6	136–139	2
168–171	7	132–135	1
164–167	0	128–131	1
160–163	3		$N = 50$

frequency distribution this is not so. For instance, in the class interval 180–183 there are 4 scores but the identity of the individual scores is not known unless we refer back to the original data.

Relative to the grouped frequency distribution given in Table 17.3, the class intervals such as 164–167 have lower and upper values referred to as *apparent limits* of the class interval. Hence the apparent limits of the lowest class interval are 128 and 131. The measurements recorded in this interval are given in whole numbers, and the interval would actually contain all measurements from 127.5 to 131.5. The *real limits*, then, of the class interval 128–131 are 127.5 and 131.5.

Frequency distributions also can be represented graphically. The horizontal axis of the graph would reflect the measurements, and the vertical scale would reflect the frequencies associated with these measurements. One such graph is the *histogram*; a histogram for the data given in Table 17.3 is illustrated in Figure 17.1. As noted, rectangles are constructed with the widths given in terms of the real limits of the class intervals and the heights given according to the frequencies associated with the class intervals.

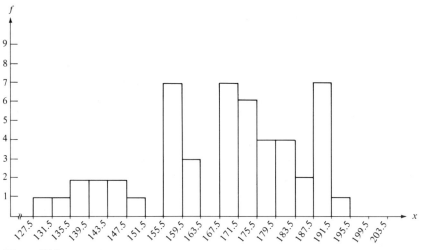

Figure 17.1

Another type of graph that could be used to illustrate the frequency distribution is a *frequency polygon*. To construct a frequency polygon, the midpoints of the "tops" of the rectangles in the corresponding histogram are connected by line segments. Since a frequency polygon always starts and terminates on the horizontal axis, it is necessary to add an extra class interval at each end of the grouped frequency distribution. Of course, each of these added class intervals will have a 0 frequency. A frequency polygon for the data given in Table 17.3 is illustrated in Figure 17.2.

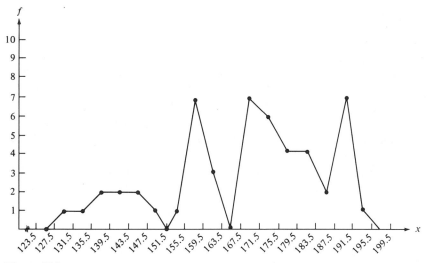

Figure 17.2

EXERCISES

1. The following set of scores were recorded for 60 students for a particular English examination:

82	79	63	42	57	86	98	79	90	75
96	87	72	71	68	84	90	89	84	91
61	83	55	74	77	89	51	71	73	84
97	90	92	80	81	63	59	68	74	77
71	77	88	90	96	85	84	65	69	85
84	79	77	75	60	58	49	88	83	92

  a. Construct a frequency distribution for the above data.
  b. Construct a grouped frequency distribution for the above data taking the lowest class interval as 42–45.
  c. What are the real limits for the lowest class interval in the above-grouped distribution?
  d. What are the real limits for the highest class interval in the above-grouped distribution?
  e. Construct a histogram for the above-grouped frequency distribution.
  f. Construct a frequency polygon for the above-grouped frequency distribution.

2. Scores obtained by 30 high school seniors on the verbal portion of the CEEB examinations are tabulated below:

425	556	551	487	610	621
479	600	586	573	499	530
586	490	386	562	543	490
510	721	597	630	482	609
690	613	510	489	602	463

a. How many scores or potential scores are represented in the above data?

b. Construct a grouped frequency distribution for the above data taking 22 scores or potential scores to a class interval and such that the lowest class interval has a lower apparent limit of 385.

c. What are the real limits for the highest class interval in the above-grouped frequency distribution?

d. Construct a histogram for the above-grouped frequency distribution.

e. Construct a frequency polygon for the above-grouped frequency distribution.

3. Purchases rounded off to the nearest dollar for the first 40 customers in a food store on a particular day are recorded below:

19	2	6	17	13	23	36	1
16	42	30	25	19	7	11	9
31	57	21	6	11	8	17	21
20	30	13	26	22	10	2	10
22	29	17	14	32	63	15	5

a. Construct a frequency distribution for the above data.

b. Construct a grouped frequency distribution for the above data taking the highest class interval to be 62–65.

c. What are the real limits for the lowest class interval of the above-grouped frequency distribution?

d. Construct a histogram for the above-grouped distribution.

e. Construct a frequency polygon for the above-grouped frequency distribution.

4. Identify each of the following statements as being true or false, and give an appropriate reason for your answer.

a. A grouped frequency distribution is obtained when we group scores into class intervals and calculate the frequency associated with each class interval.

b. When grouping data into class intervals, a general rule to follow is to use no more than 10 class intervals.

c. The individual scores are not identified in a grouped frequency distribution.

d. If a class interval for a grouped frequency distribution is 63–66, then there are 3 scores or potential scores represented in that interval.

e. The real limits for the interval 76–80 are 75 and 81.

f. If the real limits for a class interval are 19.5 and 22.5, then the apparent limits are 19 and 23.

g. In a grouped frequency distribution there cannot be class intervals with 0 frequency.

h. In a set of data, if the highest score is 93 and the lowest score is 17, then there are 77 scores or potential scores represented among the data.

i. In an ungrouped frequency distribution the individual scores or measurements are preserved.

j. A histogram is a graphical representation for a grouped frequency distribution.

17.4 GROUPED DATA AND MEASURES OF CENTRAL TENDENCY

Consider the data given in Table 17.4 which represents the amount of money rounded off to the nearest dollar that a group of 20 students had with them on a particular day.

Table 17.4

18	36	8	5	12
42	5	18	12	10
16	29	16	10	5
30	5	29	20	12

To determine the mean associated with these data, we could use the formula

$$\overline{X} = \frac{\sum\limits_{i=1}^{20} X_i}{20}$$

where each X_i corresponds to a particular measurement from Table 17.4 and such that X_i's are not necessarily distinct. For instance, there would be 4 X's with a value of 5 but only 1 X with a value of 42. We determine that $\overline{X} = 16.9$.

To determine the median associated with these data we could rearrange the measurements from highest to lowest values and, since $N = 20$, take the average of the two middle measurements. We determine that the median is 14.

We determine the mode by inspection to be 5.

Now the same data given in Table 17.4 can be displayed as an ungrouped frequency distribution as illustrated in Table 17.5. In this frequency distribution we also include a column headed fX representing the value of the measurement, X, multiplied by its corresponding frequency.

To obtain the mean of this ungrouped frequency distribution, we simply multiply each X value by its corresponding f value, sum all the fX values, and divide by N. Hence

$$\overline{X} = \frac{\sum fX}{N} = \frac{338}{20} = 16.9$$

which is the same value obtained for $\overline{X}$ earlier.

Table 17.5

X	f	fX
42	1	42
36	1	36
30	1	30
29	2	58
20	1	20
18	2	36
16	2	32
12	3	36
10	2	20
8	1	8
5	4	20
	$N = \sum f = 20$	$\sum fX = 338$

From Table 17.5 we also may determine the median to be 14, and the mode is now readily determined to be 5.

EXAMPLE Suppose that on a trip Ms. Traveler buys 10 gal of gasoline at 54¢ per gallon at one station, 12 gal of gasoline at 56¢ per gallon at a second station, 16 gal of gasoline at 53¢ per gallon at a third station, and 14 gallons of gasoline at 59¢ per gallon at a fourth station. What was the average cost per gallon of gasoline purchased at the four gas stations?

Solution The price per gallon of gasoline was different at each of the four stations. To take the average of 54, 56, 53, and 59 (in cents) would be wrong since the amount of gas purchased at each station also varied. We must consider what is called a *weighted mean* as in Table 17.6.

Table 17.6

	GALLONS PURCHASED	PRICE PER GALLON	AMOUNT OF PURCHASE
Station 1	10	$.54	$5.40
Station 2	12	.56	6.72
Station 3	16	.53	8.48
Station 4	14	.59	8.26
	$\sum = 52$		$\sum = \$28.86$

To determine the average price per gallon of gasoline we would divide the total amount spent for gasoline, $28.86, by the total number of gallons of gasoline purchased, 52, and obtain $55\frac{1}{2}$ cents per gallon. ∎

Returning now to the data given in Table 17.4, we could form a grouped frequency distribution as in Table 17.7.

To obtain the mean of the data now given in Table 17.7, we use the same procedure as for determining the mean of the data given in Table 17.5 except that the midpoint of the interval is used in a grouped frequency distribution to represent all of the measurements within that interval. Then

Table 17.7

CLASS INTERVAL	f	MIDPOINT OF INTERVAL	$f \cdot \left(\begin{array}{c} \text{MIDPOINT OF} \\ \text{INTERVAL} \end{array} \right)$
41–43	1	42	42
38–40	0	39	0
35–37	1	36	36
32–34	0	33	0
29–31	3	30	90
26–28	0	27	0
23–25	0	24	0
20–22	1	21	21
17–19	2	18	36
14–16	2	15	30
11–13	3	12	36
8–10	3	9	27
5–7	4	6	24
	$\sum = 20$		$\sum = 342$

the midpoint value of each interval is multiplied by the corresponding frequency of the interval; these values are then summed, and the sum is divided by N. Hence we have that $\overline{X} = 342/20 = 17.1$, which differs slightly from $\overline{X} = 16.9$ as determined for the ungrouped data.

The median for the data given in a grouped frequency distribution is determined in a manner similar to the ungrouped frequency distribution situation with the exception that the values in the class interval containing the median are assumed to be distributed uniformly within the interval. For the particular data in Table 17.7 the median occurs between the two class intervals, 11–13 and 14–16. In this case we take the average of the upper limit of the lower class interval and the lower limit of the upper class interval. Hence the median is $(13 + 14)/2 = 13.5$.

The mode seems to occur in the class interval 5–7.

EXAMPLE Determine the median for the data given in the following grouped frequency distribution.

CLASS INTERVAL	f
27–29	2
24–26	1
21–23	4
18–20	3
15–17	2
12–14	1
9–11	3

Solution Since $N = 16$ we determine that the class interval 18–20 contains the median since the lowest three class intervals contain 6 measurements, the lowest four class intervals contain 9 measurements, and $8 = \frac{1}{2}N$ is between 6 and 9. With the assumption that the 3 measurements in the interval 18–20 are uniformly distributed and since we want only 2 of these measurements to add to 6 for the median value, we consider two-thirds of the "width" of the interval in question. The interval 18–20 has true limits 17.5–20.5, and its width is 20.5–17.5 = 3. Hence $\frac{2}{3} \times 3 = 2$, and the median is 17.5 + 2 = 19.5. ■

EXERCISES

1. Consider the following ungrouped frequency distribution:

X	f
6	4
5	2
4	6
3	9
2	5
1	1

   a. What is the value of N in the above distribution?
   b. What is the mode of the above data?
   c. Compute the mean score of the above data.
   d. Determine the median score of the above data.

2. An investor purchased the following shares of common stock at the indicated prices:
     200 shares of Stock A at $4.50 per share
     150 shares of Stock B at $4.00 per share
     100 shares of Stock C at $3.75 per share
     100 shares of Stock D at $3.50 per share
     50 shares of Stock E at $2.00 per share
   a. What is the weighted mean price per share of stocks purchased?

3. The salaries for the 150 faculty of Community University are tabulated below:

DOLLARS	INSTRUCTOR	ASSISTANT PROFESSOR	ASSOCIATE PROFESSOR	PROFESSOR
28,000–29,999				4
26,000–27,999				3
24,000–25,999				3
22,000–23,999			3	4
20,000–21,999			5	5
18,000–19,999		5	11	4
16,000–17,999	2	8	13	3
14,000–15,999	7	16	9	
12,000–13,999	10	12	1	
10,000–11,999	11	3		
8,000–9,999	8			

a. What is the weighted mean salary for an instructor?
b. What is the weighted mean salary for an assistant professor?
c. What is the weighted mean salary for an associate professor?
d. What is the weighted mean salary for a professor?
e. What is the weighted mean salary for the total faculty?

4. Answer the following questions by referring to Exercise 3.
   a. What class interval contains the mode for instructor's salary?
   b. What class interval contains the mode for assistant professor's salary?
   c. What class interval contains the mode for associate professor's salary?
   d. What class interval contains the mode for professor's salary?
   e. What class interval contains the mode for the total faculty's salary?

5. Answer the following questions by referring to Exercise 3.
   a. What is the median salary for instructor?
   b. What is the median salary for assistant professor?
   c. What is the median salary for associate professor?
   d. What is the median salary for professor?
   e. What is the median salary for the total faculty?

6. Construct a grouped frequency distribution for the data given in Exercise 3 using the indicated class intervals but disregarding academic rank.

7. During a special sale a merchant sold varying numbers of items for $3, $4, and $5.
   a. Was the average value per sale $4? Explain.
   b. If the merchant sold 160 items and the total value of the sales was $608, what was the average value per sale?

17.5 MEASURES OF VARIABILITY

When working with measures of central tendency, we determine a single value from among the scale of values in a distribution and identify it as the mean, the median, or the mode. However, it is more meaningful to be able to determine if the measurements in a distribution are fairly close together or far apart. Such an index of spread is called a *measure of variability* or a *measure of dispersion*. One of the simplest measures of variability is the range.

DEFINITION 17.7 The *range* is a measure of variability defined to be the difference between the largest measurement in a distribution and the smallest measurement in the distribution.

EXAMPLE If for a particular test the highest score was 96 and the lowest score was 57, then the range is $96 - 57 = 39$.

EXAMPLE Temperature readings for a particular 24-hour period varied from a low of 27.3°F to a high of 57.1°F. Then the range in temperature readings in degrees Fahrenheit is $57.1 - 27.3 = 29.8$.

Another measure of variability that involves the mean is the mean deviation.

DEFINITION 17.8 The *mean deviation*, denoted by *MD*, is a measure of variability obtained by subtracting the mean from each measurement, adding the deviations without regard to algebraic sign, and then dividing the sum by the number of measurements.

If X represents a measurement in a distribution and $\overline{X}$ is the mean of the distribution, then

$$MD = \frac{\sum\limits_{i=1}^{N} |X_i - \overline{X}|}{N}$$

where $|X_i - \overline{X}|$ represents the absolute value of the difference between a particular score and the mean; that is, the difference without regard to algebraic sign.

EXAMPLE   Determine the mean deviation for the distribution

$$77, 80, 72, 81, 90, 63, 48$$

Solution

$$\overline{X} = \frac{77 + 80 + 72 + 81 + 90 + 63 + 48}{7} = \frac{511}{7} = 73$$

and

X	$\overline{X}$	$|X - \overline{X}|$
77	73	4
80	73	7
72	73	1
81	73	8
90	73	17
63	73	10
48	73	25
		$\sum = 72$

Hence $MD = \dfrac{72}{7} = 10\,\dfrac{2}{7}$. ■

The mean deviation may be used to compare the dispersion or variability of two or more distributions. The smaller the value of the mean deviation, the less the dispersion in the distribution. However, the mean deviation is not useful in statistics as a measure of variability in the interpretation of measurements within a distribution.

A more useful measure of variability, closely related to the mean deviation, is known as variance.

DEFINITION 17.9 The *variance* of a distribution, denoted by s^2, is a measure of dispersion defined as the sum of the squares of the deviations from the mean, divided by the number of measurements.

Instead of treating each deviation from the mean without regard to sign, as is done with the mean deviation, we square each deviation, sum these squares, and divide by N, the number of measurements. Symbolically, the variance s^2 is given as

$$s^2 = \frac{\sum_{i=1}^{N} (X_i - \overline{X})^2}{N}$$

Alternately, if we let $x = X - \overline{X}$, the variance may be given as

$$s^2 = \frac{\sum_{i=1}^{N} x_i^2}{N}$$

Closely related to the variance is the standard deviation.

DEFINITION 17.10 The *standard deviation* of a distribution, denoted by s, is a measure of dispersion defined as the square root of the variance of the distribution.

Symbolically, then, we have

$$s = \sqrt{\frac{\sum_{i=1}^{N} (X_i - \overline{X})^2}{N}} \quad \text{or} \quad s = \sqrt{\frac{\sum_{i=1}^{N} x_i^2}{N}}$$

The formulas given above for the variance and the standard deviation of a distribution involve the deviations from the mean. Sometimes more useful formulas that involve the raw scores and not their deviations from the mean are

$$s^2 = \frac{\sum_{i=1}^{N} X^2}{N} - \overline{X}^2$$

and

$$s = \sqrt{\frac{\sum_{i=1}^{N} X^2}{N} - \overline{X}^2}$$

These formulas also may be used to determine the variance and the standard deviation of either an ungrouped or a grouped frequency distribution provided we modify the formulas to consider the corresponding frequency of each score or class interval. Remember, for a grouped frequency distribution the scores in any particular class interval are represented by the midpoint value of the interval. In modified form, then, the variance and the standard deviation can be computed using the formulas

$$s^2 = \frac{\sum\limits_{i=1}^{N} f X^2}{N} - X^2$$

and

$$s = \sqrt{\frac{\sum\limits_{i=1}^{N} f X^2}{N} - \overline{X}^2}$$

Earlier we introduced the symbol fX which represented the product of a measurement, X, with its corresponding frequency, f. The symbol fX^2 used in the preceding formulas represents the product of the *square* of the measurements, X^2, and its corresponding frequency, f. The symbols fX^2 and $(fX)^2$ are not to be interchanged.

EXAMPLE Determine the variance, s^2, and the standard deviation, s, for the following distribution, using the mean deviation approach.

11, 23, 17, 26, 19, 20, 22, 28, 32

Solution

$$\overline{X} = \frac{11 + 23 + 17 + 26 + 19 + 20 + 22 + 28 + 32}{9} = 22$$

X	$X - \overline{X}$	$(X - \overline{X})^2$
11	⁻11	121
23	1	1
17	⁻5	25
26	4	16
19	⁻3	9
20	⁻2	4
22	0	0
28	6	36
32	10	100

$$\sum_{i=1}^{9} X_i = 198$$

$$N = 9$$

$$\overline{X} = \frac{198}{9} = 22$$

$$\sum_{i=1}^{9} (X_i - \overline{X})^2 = 312$$

Hence

$$s^2 = \frac{\sum\limits_{i=1}^{9} (X_i - \overline{X})^2}{N}$$

$$= \frac{312}{9} = 34.67$$

and

$$s = \sqrt{34.67} = 5.9 \quad \blacksquare$$

EXAMPLE Determine the variance and the standard deviation for the distribution given in the example above using the raw score approach.

Solution

X	X^2
11	121
23	529
17	289
26	676
19	361
20	400
22	484
28	784
32	1024

$$\sum_{i=1}^{9} X^2 = 4668$$

$\overline{X} = 22$ (from previous example)

Hence

$$s^2 = \frac{\sum\limits_{i=1}^{9} X^2}{N} - \overline{X}^2$$

$$= \frac{4668}{9} - (22)^2 = 518.67 - 484$$

$$= 34.67$$

and

$$s = \sqrt{34.67} = 5.9 \quad \blacksquare$$

EXAMPLE Determine the variance and the standard deviation for the following grouped frequency distribution.

CLASS INTERVAL	f
24–26	2
21–23	3
18–20	5
15–17	1
12–14	4
9–11	6

Solution We will complete the table as follows where the X values represent the midpoint of each class interval.

CLASS INTERVAL	f	X	fX	fX^2
24–26	2	25	50	1250
21–23	3	22	66	1452
18–20	5	19	95	1805
15–17	1	16	16	256
12–14	4	13	52	676
9–11	6	10	60	600

From the above table we determine that

$$N = \sum f = 21$$
$$\sum fX = 339$$
$$\sum fX^2 = 6039$$

Also

$$\overline{X} = \frac{\sum fX}{N} = \frac{339}{21} = 16.14$$

Hence

$$s^2 = \frac{\sum fX^2}{N} - \overline{X}^2$$

$$= \frac{6039}{21} - (16.14)^2 = 287.57 - 260.50$$

$$= 27.07$$

and

$$s = \sqrt{27.07} = 5.2 \quad \blacksquare$$

EXERCISES

1. Consider the following distribution of scores: 2, 2, 2, 4, 6, 8, 10, 12, 12.
   a. What is the mean of the distribution?
   b. What is the median of the distribution?
   c. What is the mode of the distribution?
   d. What is the range of the distribution?
   e. Determine the mean deviation for the distribution?
   f. Determine the variance and the standard deviation for the distribution using the mean deviation approach.
   g. Determine the variance and the standard deviation for the distribution using the raw score method.
2. Consider the following distribution of test scores: 86, 95, 83, 96, 72, 69, 87, 90, 72, 75, 87, 91, 67, 87, 85, 77, 66, 93, 82, 79.
   a. What is the mean of the distribution?

b. What is the median of the distribution?

c. What is the mode of the distribution?

d. What is the range of the distribution?

e. Determine the mean deviation for the distribution.

f. Determine the variance and the standard deviation for the distribution using the mean deviation approach.

g. Determine the variance and the standard deviation for the distribution using the raw score method.

3. Consider the following grouped frequency distribution:

CLASS INTERVAL	f
72–74	2
69–71	3
66–68	2
63–65	4
60–62	3
57–59	3
54–56	2
51–53	1

a. Determine the variance of the above distribution.

b. Determine the standard deviation of the above distribution.

4. Consider the following grouped frequency distribution:

CLASS INTERVAL	f
82–85	2
78–81	6
74–77	4
70–73	3
66–69	1
62–65	5
58–61	2
54–57	3
50–53	5
46–49	4

a. Determine the variance of the above distribution.

b. Determine the standard deviation of the above distribution.

5. Consider the following groups of scores:

(1) 4, 6, 8, 10, 12, 14.

(2) 4, 8, 9, 9, 10, 14.

(3) 4, 5, 6, 12, 13, 14.

(4) 4, 4, 5, 13, 14, 14.

(5) 4, 8, 8, 9, 12, 14.

a. Which of the above groups of scores exhibit the *least* variance?

b. Which of the above groups exhibit the *most* variance?

17.6 MEASURES OF AN INDIVIDUAL SCORE

In the previous sections of this chapter we have discussed measures of central tendency and measures of dispersion or variability. These measures relate to a total distribution, and any score from the distribution taken by itself is meaningless. An individual score from a distribution takes on meaning only when it can be compared relative to other scores in the distribution. In this section we will discuss two measures of an individual score—the percentile rank of the score and a z-score associated with the individual score.

DEFINITION 17.11 The *percentile rank* of a score in a distribution or reference group indicates the percentage of scores in the distribution that are below the given score.

For instance, if a student scored 45 on a particular examination, did he necessarily perform poorly? Suppose that the mean of the scores in the distribution was 40. Then in reference to the mean his score of 45 was above average. Suppose, however, that the median was 53. Then in reference to the median score his score of 45 would put his performance with the lower half of the reference group. If the score of 45 has a percentile rank of 65, then we would know that 65 percent of the reference group scored lower than 45. If the percentile rank of the score of 45 is 65, we would denote this by the symbol $P_{65} = 45$. As a general rule percentile ranks are rounded off to the nearest whole number.

EXAMPLE In a distribution of 90 scores, 63 scores are below the score of 73. Determine the percentile rank of the score of 73.
Solution We determine that 63 scores in a distribution of 90 scores represents 70 percent of the scores in the distribution. Hence 70 percent of scores are below the score of 73; $P_{70} = 73$ for this distribution. ∎

EXAMPLE In a distribution of 85 scores, 58 scores are below the score of 70. Determine the percentile rank of the score of 70.
Solution We determine that 58 scores in a distribution of 85 scores represents 68.2 percent of the scores in the distribution. Hence 68.2 percent of the scores are below the score of 70. But 68.2, rounded off to the nearest whole number, is 68. Hence $P_{68} = 70$ for this distribution. ∎

If a score of 55 in a particular distribution has a percentile rank of 70, we also say that the score of 55 is at the 70th *percentile*. That is, the score of 55 is the score in the distribution below which 70 percent of the scores lie.
Special cases of percentiles are
1. The *median*, denoted by M, or the 50th percentile.

2. The *first quartile*, denoted by Q_1, or the 25th percentile.

3. The *third quartile*, denoted by Q_3, or the 75th percentile.

It should be noted that Q_1, M, and Q_3 occur at P_{25}, P_{50}, and P_{75}, respectively. The 10th percentile and the 90th percentile occur at P_{10} and P_{90}, respectively.

Because of the sensitivity to extreme values in the distribution, we noted that the range is not a very good measure of variability or dispersion. However, better measures that are somewhat related to the range are the decile range and the interquartile range.

DEFINITION 17.12 The *decile range* is defined to be the difference between the 90th percentile and the 10th percentile; that is, decile range $= P_{90} - P_{10}$.

DEFINITION 17.13 The *interquartile range* is defined to be the difference between the 3rd quartile and the 1st quartile; that is, interquartile range $= P_{75} - P_{25}$.

EXAMPLE   Determine the decile range for the following distribution:

6, 19, 27, 32, 8, 11, 43, 52, 47, 83, 72, 19, 25, 29, 7, 17, 56, 63, 22, 37.

Solution   To determine the decile range we need P_{10} and P_{90}. There are 20 scores in the distribution and $(\frac{1}{10})(20) = 2$. Arranging the scores in increasing order, we note that 7 is the second score and 8 is the third score. The value of P_{10} is a score between 7 and 8. We will agree to select a value halfway between 7 and 8 and determine that $P_{10} = 7.5$. In a similar manner, to determine P_{90} we consider $(\frac{9}{10})(20) = 18$. Arranged in increasing order, the 18th score is 63 and the 19th score is 72. Hence $P_{90} = 63 + \frac{1}{2}(72 - 63) = 63 + 4.5 = 67.5$. Thus the decile range is $P_{90} - P_{10} = 67.5 - 7.5 = 60.$ ■

EXAMPLE   For the same distribution given in the example above determine the interquartile range.

Solution   To determine the interquartile range we need P_{25} and P_{75}. Since $\frac{1}{4}(20) = 5$, we note that when the scores are arranged in increasing order the 5th score is 17 and the 6th score is 19. Hence $P_{25} = 17 + \frac{1}{2}(19 - 17) = 17 + 1 = 18$. Similarly, $(\frac{3}{4})(20) = 15$ and the 15th score is 47 and the 16th score is 52. Hence $P_{75} = 47 + \frac{1}{2}(52 - 47) = 47 + 2.5 = 49.5$. Hence the interquartile range is $P_{75} - P_{25} = 49.5 - 18 = 31.5$ ■

Another important measure of an individual score relative to a distribution is called the *z-score*.

DEFINITION 17.14 A *z-score* is a measurement of how many standard deviations a particular score in a distribution is from the mean of the distribution.

Symbolically, we have

$$z = \frac{X - \overline{X}}{s}$$

where X denotes the particular score and s denotes the standard deviation associated with the distribution.

EXAMPLE If a distribution has a mean of 80 and a standard deviation of 5, determine the z-score for a score of 70 in the distribution.
Solution $X = 70$, $\overline{X} = 80$, and $s = 5$. Hence

$$z = \frac{X - \overline{X}}{s} = \frac{70 - 80}{5} = \frac{^-10}{5} = {^-2}.$$

Thus a score of 70 in the given distribution would be 2 standard deviations below the mean. ■

EXAMPLE If a distribution has a mean of 85 and a standard deviation of 6, what score in the distribution will be 1.5 standard deviations above the mean?
Solution $\overline{X} = 85$, $s = 6$, $z = 1.5$. Hence

$$z = \frac{X - \overline{X}}{s}$$

becomes

$$1.5 = \frac{X - 85}{6} \quad \text{or} \quad 9 = X - 85$$

Hence $X = 94$ represents the score in the given distribution which is 1.5 standard deviations above the mean. ■

EXERCISES

1. Determine the percentile rank of each of the following scores in the indicated distributions.
   a. A score of 69, if 81 scores in a distribution of 90 scores are below the score of 69.
   b. A score of 69, if 56 scores in a distribution of 80 scores are below the score of 69.

   c. A score of 81, if 84 scores in a distribution of 112 scores are below the score of 81.

   d. A score of 42, if 50 scores in a distribution of 100 scores are below the score of 42.

   e. A score of 58, if 63 scores in a distribution of 82 scores are below the score of 58.

   f. A score of 61, if 81 scores in a distribution of 113 scores are below the score of 61.

   g. A score of 75, if 78 scores in a distribution of 90 scores are below the score of 75.

   h. A score of 120, if 58 scores in a distribution of 69 scores are below the score of 120.

   i. A score of 80, if 23 scores in a distribution of 90 scores are *above* the score of 80.

   j. A score of 42, if 17 scores in a distribution of 25 scores are *above* the score of 42.

2. Consider the following distribution: 8, 11, 13, 17, 23, 31, 32, 9, 7, 5, 24, 37, 42, 11.

   a. Determine P_{10} for the distribution.

   b. Determine P_{25} for the distribution.

   c. Determine P_{50} for the distribution.

   d. Determine P_{75} for the distribution.

   e. Determine P_{90} for the distribution.

   f. Determine the decile range for the distribution.

   g. Determine the interquartile range for the distribution.

3. Consider the following distribution: 19, 26, 37, 11, 8, 10, 43, 57, 36, 23, 19, 27, 19, 30, 29, 38, 40, 5, 12, 26, 35, 41, 5, 17, 46, 40, 10, 20, 30, 25, 31, 22, 21, 18, 2, 16, 39.

   a. Determine P_7 for the distribution.

   b. Determine P_{10} for the distribution.

   c. Determine P_{18} for the distribution.

   d. Determine P_{25} for the distribution.

   e. Determine P_{32} for the distribution.

   f. Determine P_{50} for the distribution.

   g. Determine P_{62} for the distribution.

   h. Determine P_{71} for the distribution.

   i. Determine P_{75} for the distribution.

   j. Determine P_{82} for the distribution.

   k. Determine P_{90} for the distribution.

   l. Determine P_{95} for the distribution.

  m. Determine P_{99} for the distribution.

   n. Determine the range of the distribution.

   o. Determine the decile range of the distribution.

   p. Determine the interquartile range of the distribution.

4. Determine the z-score for the following:
   a. A score of 62, if $\overline{X} = 70$ and $s = 4$.
   b. A score of 59, if $\overline{X} = 55$ and $s = 3$.
   c. A score of 82, if $\overline{X} = 72$ and $s = 5$.
   d. A score of 96, if $\overline{X} = 77$ and $s = 6$.
   e. A score of 19, if $\overline{X} = 60$ and $s = 17$.
   f. A score of 23, if $\overline{X} = 47$ and $s = 5$.
   g. A score of 60, if $\overline{X} = 60$ and $s = 2$.
   h. A score of 47, if $\overline{X} = 39$ and $s = 4$.
   i. A score of 92, if $\overline{X} = 70$ and $s = 11$.
   j. A score of 80, if $\overline{X} = 75$ and $s = 8$.
5. a. What score in a distribution will be 1.4 standard deviations above the mean if $\overline{X} = 80$ and $s = 6$?
   b. What score in a distribution will be 0.6 standard deviations above the mean if $\overline{X} = 62$ and $s = 5$?
   c. What score in a distribution will be 1.7 standard deviations below the mean if $\overline{X} = 82$ and $s = 4$?
   d. What score in a distribution will be 0.9 standard deviations below the mean if $\overline{X} = 83$ and $s = 4$?
   e. What score in a distribution will be 1.1 standard deviations above the mean if $\overline{X} = 76$ and $s = 9$?
   f. What score in a distribution will be 2.6 standard deviations below the mean if $\overline{X} = 66$ and $s = 7$?
   g. What score in a distribution will be 3.1 standard deviations above the mean if $\overline{X} = 70$ and $s = 4$?
   h. What score in a distribution will be 2.7 standard deviations below the mean if $\overline{X} = 56$ and $s = 5$?
   i. What score in a distribution will be 2.2 standard deviations above the mean if $\overline{X} = 91$ and $s = 2$?
   j. What score in a distribution will be 0.7 standard deviations below the mean if $\overline{X} = 86$ and $s = 3$?

SUMMARY

In this chapter we discussed some very basic concepts of descriptive statistics. Terminology of descriptive statistics we introduced included population, sample, random sample, distribution, parameter, and statistics. Measures of central tendency were introduced and discussed. These included the mean or arithmetic average, the median, and the mode. Ungrouped and grouped frequency distributions were introduced, and measures of central tendency relating to them were discussed. Measures of variability or dispersion also were discussed. These included range, mean deviation, variance, standard deviation, decile range, and interquartile range. Percentiles, percentile rank, and z-scores were introduced as measures of individual scores. Histograms and frequency polygons were introduced as graphical representations of data.

During our discussions the following symbols were introduced.

Symbol	Interpretation
X	A score or measurement
$\overline{X}$	The mean or arithmetic average
$\sum$	Greek letter sigma for summation
f	Frequency
MD	Mean deviation
s^2	Variance
s	Standard deviation
P_{10}	10th percentile
P_{25}	25th percentile
P_{75}	75th percentile
P_{90}	90th percentile
Q_1	1st quartile $(= P_{25})$
M	Median $(= P_{50})$
Q_3	3rd quartile $(= P_{75})$
z	A z-score

REVIEW EXERCISES FOR CHAPTER 17

Identify each of the following statements as being true or false, and give an appropriate reason for your answer.

1. A general rule to follow when grouping data into class intervals is to strive for no more than 6 class intervals.
2. When calculating the median for a grouped frequency distribution, we use the real limits of the class intervals.
3. When we arrange a set of scores in increasing order of magnitude and indicate the frequency associated with each score, we have a frequency distribution.
4. The real limits for the interval 72–76 are 71.5 and 75.5.
5. The midpoint of the interval 27–30 is 28.5.
6. If there are 7 scores in a class interval that has a midpoint of 12, then the apparent upper limit is 15.
7. The frequency of 4 in the class interval 26–29 means that the 4 scores are uniformly spread out throughout the interval.
8. In a grouped frequency distribution all scores in a class interval are represented by the upper real limit of the class interval.
9. A score taken by itself is meaningless.

10. The mean, mode, median, and interquartile range are all measures of central tendency.
11. The standard deviation of a distribution is always equal to the mean deviation of the distribution.
12. A percentile rank taken by itself is meaningless.
13. If a percentile rank of a score of 30 is 55, that means that 30 percent of the scores in the distribution were below the score of 55.
14. If a student obtained a score of 96 on a school examination, you should assume that the student answered 96 questions correctly from a group of 100 questions.
15. The median is a score or potential score that balances all the scores on either side of it.
16. The median of a distribution is also the 50th percentile rank.
17. Consider a distribution of scores with a mean of 50. If 10 points are added to each score in the distribution, then the new mean is 60.
18. If the median of a particular distribution is 43 and 5 points are subtracted from each score in the distribution, then the new median is 43.
19. The standard deviation of the distribution 76, 77, 78 is greater than the standard deviation for the distribution 11, 12, 13.
20. If two distributions have the same mean and range, we may conclude that the two distributions are identical.
21. If each score in a distribution is doubled, the new standard deviation will be equal to twice the old standard deviation.
22. If the mean score of 35 students on a particular examination is 55 and $s = 0$, then all 35 students received a score of 55.
23. Of all the measures of dispersion only the range reflects only the two most extreme scores in a distribution.
24. If $\overline{X} = 20$, $X = 12$, and $s = 6$, then $z = 2.00$.
25. The mean of a set of z-scores is 0.

ANSWERS TO SELECTED ODD-NUMBERED EXERCISES

CHAPTER 1

SECTION 1.2

1. 45   3. 3   5. $.10   7. No   9. 7, 9, 11, indicating the odd
natural numbers being continued; or, 8, 13, 21, obtained by adding two
consecutive entries to obtain the next; or, 7, 11, 13, indicating the prime
numbers being continued, etc.

CHAPTER 2

SECTION 2.1

1. a. Neither   b. True proposition   c. Neither   d. False proposition
   e. True proposition   f. Propositional form   g. True proposition
   h. Neither   i. False proposition   j. True proposition

SECTION 2.2

1. a. A dog does not have four legs.   b. Vitamins are metals.
   c. Integers are not real numbers.   d. A house is not a building.
   e. $3 + 4 = 9$.

3. a. Red is a color or February is a month of the year.
   b. Five is not a rational number or $p \to q$.   c. $\sim p$ or $2 + 2 = 5$.
   d. $p \vee \sim q$ or $\sim(p \to q)$.
5. a. Converse: If you like music, then you like to play the piano.
      Inverse: If you do not like to play the piano, then you do not like
            music.
      Contrapositive: If you do not like music, then you do not like to
                play the piano.
   b. Converse: $u \to w$
      Inverse: $\sim w \to \sim u$
      Contrapositive: $\sim u \to \sim w$
   c. Converse: If a proposition is false, then the negation of the
            proposition is true.
      Inverse: If the negation of a proposition is false, then the
            proposition is true.
      Contrapositive: If a proposition is true, then the negation of the
                proposition is false.
   d. Converse: $q \to \sim p$
      Inverse: $p \to \sim q$
      Contrapositive: $\sim q \to p$
   e. Converse: $\sim s \to r$
      Inverse: $\sim r \to s$
      Contrapositive: $s \to \sim r$
7. Laura's team lost both games.   9. None
11. a. Brian does not have brown eyes and brown hair.   b. Brian does
    not have brown eyes or brown hair.   c. Brian has brown eyes and
    brown hair.   d. Brian has brown eyes or brown hair.   e. Brian does
    not have brown eyes or has brown hair.   f. Brian does not have
    brown eyes and has brown hair.
13. a. $r \wedge \sim s$ b. $r \vee \sim s$ c. $\sim r \wedge s$ d. $\sim r \vee s$ e. $r \vee s$ f. $r \wedge s$

SECTION 2.3

1. Open; closed   3. Negation   5. Parallel

SECTION 2.4

1. Tautology   3. Tautology   5. Neither   7. Neither
9. Neither   11. Neither   13. Tautology   15. Neither
17. Contradiction   19. Tautology

SECTION 2.5

1.

p	q	$p \vee q$	$\sim p \rightarrow q$
T	T	T	T
T	F	T	T
F	T	T	T
F	F	F	F

3. a. Today is Tuesday or it will rain.   b. $b \neq 6$ or $b \neq 4$
   c. $a = 5$ or $a = 7$ d. $\sim q$ or p e. p or q f. q or $\sim p$
5. a. Yes   b. Yes   c. No   d. Yes   e. No   f. Yes   g. No   h. Yes   i. Yes
   j. Yes

SECTION 2.6

1. a. If a grade of B is received, then this course is passed.
   b. If $2 + 2 = 5$, then $3 \times 2 = 4$.   c. If an angle is a right angle, then its
   measure is 90 degrees.   d. If a quadrilateral is a square, then its
   opposite sides are congruent.   e. If $\sim q$, then w.   f. If $u \rightarrow w$, then
   $\sim u \vee w$.   g. If today is Wednesday, then yesterday was Tuesday.
   h. If the judge is impartial, then he is fair.

SECTION 2.7

1. a. p b. t c. p d. c e. p f. t g. c h. p i. $p \vee (q \wedge r)$
   j. $(p \vee r) \wedge q$ k. $\sim q \wedge (\sim q \vee r)$ l. $p \vee \sim q$ m. $\sim (p \wedge q)$ n. $p \wedge (q \wedge r)$
   o. $(p \wedge q) \vee p$

SECTION 2.8

1. Fallacy   3. Fallacy   5. Fallacy   7. Valid   9. Valid
11. Fallacy   13. Valid   15. Valid   17. Valid   19. Valid

SECTION 2.9

1. Tautology   3. Tautology
5. a. q, by the law of detachment   b. $\sim q$, by the law of contrapositive
   inference   c. p (or q), by the law of conjunctive simplification
   d. $q \vee p$, by the commutative property for disjunction of statements
   e. $\sim p$, by the law of detachment   f. $(p \rightarrow q) \wedge (q \rightarrow r)$, by the law of
   conjunctive inference; or, $p \rightarrow r$, by truth tables   g. p, by the law of
   detachment   h. $\sim p$, by the law of contrapositive inference
   i. $(p \rightarrow q) \wedge \sim p$, by the law of conjunctive inference   j. r, by the law of
   detachment

SECTION 2.10

1. *b*   3. a. False   b. True   c. True   d. True   e. False

SECTION 2.11

1. a. $(p \wedge q) \vee (p \wedge r)$ b. $[(q \wedge r) \vee \sim q] \wedge r$ c. $(p \vee q) \vee \sim q \vee (\sim p \wedge q)$
   d. $[(p \wedge \sim q) \vee q \vee (\sim p \wedge q)] \wedge \sim q$ 3. a. ⊣ p ⊢   b. ⊣ p ⊢

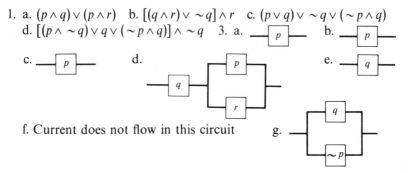

c. ⊣ p ⊢   d.   e. ⊣ q ⊢

f. Current does not flow in this circuit   g.

REVIEW EXERCISES FOR CHAPTER 2

1. Refer to the following definitions: a. 2.1   b. 2.2   c. 2.3   d. 2.12   e. 2.4
   f. 2.5   g. 2.6   h. 2.7   i. 2.8   j. 2.9   k. 2.10   l. 2.11   m. 2.15   n. 2.16
   o. 2.13   p. 2.14   q. 2.17   r. 2.18
3. None
5. a. Some women are not blondes.   b. There exists a quadrilateral
   which is not a rectangle.   c. All quadrilaterals are not rectangles.
   d. For all *y* in the set *X*, $y \neq 13$.   e. Some professors are intelligent.
   f. Apples are not candy or roses are not yellow.   g. Men are not
   mortal and dogs do not have three legs.   h. Today is Monday and
   tomorrow is not Tuesday.
·7. c, d, l, m, o, p, q

CHAPTER 3

SECTION 3.1

1. a. The set of all the letters in the English alphabet. The set of all U.S.
   Senators from 1900 to the present.   b. The set of all real numbers
   which are greater than 10. The set of all real numbers between 0 and 1.
   c. The set of all real numbers that are greater than 0 and also less than 0.
   The set of all matriculated college freshmen who are at least 9 ft. 10 in.
   tall.
3. a, c, d   5. a. See Definition 3.5.   b. See Definition 3.2.
7. No, {*A*} is the set that contains *A*.
9. a. $\{y \mid y$ is a natural number that is less than 7$\}$.
   b. $\{u \mid u$ is the name of the day of the week$\}$.

c. $\{z \mid z$ is one of the first five letters of the English alphabet$\}$.
d. $\{t \mid t$ is the name of the month of the year having exactly 30 days$\}$.
e. $\{s \mid s$ is a positive multiple of 3$\}$.

SECTION 3.2

1. a. $\{a, b, c, d\}, \{a, b, c\}, \{a, b, d\}, \{a, c, d\}, \{b, c, d\}, \{a, b\}, \{a, c\}, \{a, d\},$
   $\{b, c\}, \{b, d\}, \{c, d\}, \{a\}, \{b\}, \{c\}, \{d\}, \varnothing$
   b. $\{1, 2, 3\}, \{1, 2\}, \{1, 3\}, \{2, 3\}, \{1\}, \{2\}, \{3\}, \varnothing$
   c. $\{^-2, 0, 3\}, \{^-2, 0\}, \{^-2, 3\}, \{0, 3\}, \{^-2\}, \{0\}, \{3\}, \varnothing$
   d. $\{a, e, i, o, u\}, \{a, e, i, o\}, \{a, e, i, u\}, \{a, e, o, u\}, \{a, i, o, u\},$
   $\{e, i, o, u\}, \{a, e, i\}, \{a, e, o\}, \{a, e, u\}, \{a, i, o\}, \{a, i, u\}, \{a, o, u\},$
   $\{e, i, o\}, \{e, i, u\}, \{e, o, u\}, \{i, o, u\}, \{a, e\}, \{a, i\}, \{a, o\}, \{a, u\}, \{e, i\},$
   $\{e, o\}, \{e, u\}, \{i, o\}, \{i, u\}, \{o, u\}, \{a\}, \{e\}, \{i\}, \{o\}, \{u\}, \varnothing$
   e. $\{\varnothing\}, \varnothing$
   f. $\{5, 7, ?, \#\}, \{5, 7, ?\}, \{5, 7, \#\}, \{5, ?, \#\}, \{7, ?, \#\}, \{5, 7\}, \{5, ?\},$
   $\{5, \#\}, \{7, ?\}, \{7, \#\}, \{?, \#\}, \{5\}, \{7\}, \{?\}, \{\#\}, \varnothing$
3. a. $\{a, b, c\}, \{a, b, d\}, \{a, c, d\}, \{b, c, d\}, \{a, b\}, \{a, c\}, \{a, d\}, \{b, c\}, \{b, d\},$
   $\{c, d\}, \{a\}, \{b\}, \{c\}, \{d\}, \varnothing$
   b. $\{1, 2\}, \{1, 3\}, \{2, 3\}, \{1\}, \{2\}, \{3\}, \varnothing$
   c. $\{^-2, 0\}, \{^-2, 3\}, \{0, 3\}, \{^-2\}, \{0\}, \{3\}, \varnothing$
   d. $\{a, e, i, o\}, \{a, e, i, u\}, \{a, e, o, u\}, \{a, i, o, u\}, \{e, i, o, u\}, \{a, e, i\},$
   $\{a, e, o\}, \{a, e, u\}, \{a, i, o\}, \{a, i, u\}, \{a, o, u\}, \{e, i, o\}, \{e, i, u\}, \{e, o, u\},$
   $\{i, o, u\}, \{a, e\}, \{a, i\}, \{a, o\}, \{a, u\}, \{e, i\}, \{e, o\}, \{e, u\}, \{i, o\}, \{i, u\},$
   $\{o, u\}, \{a\}, \{e\}, \{i\}, \{o\}, \{u\}, \varnothing$
   e. $\varnothing$
   f. $\{5, 7, ?\}, \{5, 7, \#\}, \{5, ?, \#\}, \{7, ?, \#\}, \{5, 7\}, \{5, ?\}, \{5, \#\}, \{7, ?\},$
   $\{7, \#\}, \{?, \#\}, \{5\}, \{7\}, \{?\}, \{\#\}, \varnothing$
5. a, b, f, g, h
7. a. See Definition 3.6.   b. See Definition 3.7.   c. See Definition 3.8.
   d. See Definition 3.9.

SECTION 3.3

1. a. $\{1, 3, 6, 7, 8, 9\}$ b. $\{8, 9\}$ c. $\{1, 2, 3, 6, 7, 8, 9\}$ d. $\{8, 9\}$
   e. $\{1, 2, 3, 6, 7, 8, 9\}$ f. $\{6, 8, 9\}$ g. $\{8, 9\}$ h. $\{6, 8, 9\}$ i. $\{6, 8, 9\}$
   j. $\{2, 6, 8, 9\}$
3. e, f

SECTION 3.4

1. a. $\{1, 2, 3, 4, 5, 6\}$ b. $\{4, 5, 6\}$ c. $\varnothing$ d. $\{1, 2\}$
   e. $\{(4, 1), (4, 2), (4, 3), (4, 4), (5, 1), (5, 2), (5, 3), (5, 4), (6, 1), (6, 2),$
   $(6, 3), (6, 4), (7, 1), (7, 2), (7, 3), (7, 4)\}$

f. $A \cup (B \cap C) = \{1, 2, 3, 4\} \cup [\{3, 4, 5, 6\} \cap \{4, 5, 6, 7\}] =$
$\{1, 2, 3, 4\} \cup \{4, 5, 6\} = \{1, 2, 3, 4, 5, 6\}.$
$(A \cup B) \cap (A \cup C) = [\{1, 2, 3, 4\} \cup \{3, 4, 5, 6\}] \cap$
$[\{1, 2, 3, 4\} \cup \{4, 5, 6, 7\}] = \{1, 2, 3, 4, 5, 6\} \cap \{1, 2, 3, 4, 5, 6, 7\} =$
$\{1, 2, 3, 4, 5, 6\}$

5. a. Yes, the null set is a subset of *every* set.
  b. No, since S could be a nonempty set.
  c. $S \cup \varnothing = S$ and $S \cup \varnothing = \varnothing$ only if $S = \varnothing$.
  d. Yes, by the definition of the intersection of two sets.
  e. Yes, if $S = \varnothing$
  f. Only if $S \neq \varnothing$, since $\varnothing$ is a proper subset of every nonempty set.

7. $\varnothing$ 9. They could be, but either could also be a subset of the other.

11. a. $\{1, 2, 3, 4, 5, 6\}$ b. $\{1, 2, 7, 8\}$ c. $\{1, 2\}$ d. $\{7, 8\}$
   e. $\{5, 6, 7, 8\}$ f. $\{3, 4\}$ g. $\{7, 8\}$ h. $\{1, 2, 5, 6, 7, 8\}$ i. $\{1, 2, 5, 6, 7, 8\}$
   j. $\{3, 4, 5, 6, 7, 8\}$

13. a. $\{2, 3\}$ b. $\{1, 2, 3, 4, 5\}$ c. $\{2, 3\}$ d. $\{1, 2, 3, 4, 5\}$

SECTION 3.5

1. 70   3. a. 39   b. 30   c. 35   d. 195   e. 165

REVIEW EXERCISES FOR CHAPTER 3

1. Refer to the following definitions: a. 3.1   b. 3.4   c. 3.4   d. 3.5   e. 3.6
   f. 3.7   g. 3.8   h. 3.12   i. 3.15   j. 3.9   k. 3.10   l. 3.11   m. 3.17   n. 3.14
   o. 3.13   p. 3.2   q. 3.2   r. 3.3   s. 3.16

11. a. $\{21, 22, 23, 24, 25, 26, 27, 28\}$ b. $\{^-4, \ ^-3, \ ^-2, \ ^-1, 0, 1, 2\}$
   c. $\{a, e, i, o, u\}$ d. $\{\alpha, \beta, \gamma, \delta, \varepsilon, \zeta, \eta, \theta, \iota, \kappa, \lambda, \mu, \nu, \xi, o, \pi, \rho, \sigma, \tau,$
   $\upsilon, \phi, \chi, \psi, \omega\}$ e. $\{\triangledown, \triangleleft\}$ f. $\{I, V, X, L, C, D, M\}$

13. a. True, $x \in E \Rightarrow x \in B$ b. True, a $1:1$ correspondence between their
   elements can be established.   c. False, $1 \in A$ but $1 \notin C$.
   d. False, $F \nsubseteq C$.   e. True, $n(D) = n(E) = 8$.   f. False, $n(E) = 8$,
   $n(F) = 7$.   g. True, there are no elements common to both B and C.
   h. True, $B \cap D \neq \varnothing$, $B \nsubseteq D$, and $D \nsubseteq B$.   i. False, $B \cup C = A - \{1\}$.
   j. True, $E \cap F \neq \varnothing$, $E \nsubseteq F$, $F \nsubseteq E$.

15. a. $\{1, 2\}$ b. $\{6, 7\}$ c. $\{(2, 1), (2, 5), (2, 7), (3, 1), (3, 5), (3, 7)\}$
   d. $\{(1, 3), (1, 4), (1, 5), (1, 6), (1, 7), (5, 3), (5, 4), (5, 5), (5, 6), (5, 7),$
   $(7, 3), (7, 4), (7, 5), (7, 6), (7, 7)\}$ e. $\{1, 4, 5\}$
   f. Not possible, $D \nsubseteq B$ g. $\{2, 3\}$ h. $\{3, 4, 6\}$
   i. $\{(1, 2), (1, 3), (5, 2), (5, 3), (7, 2), (7, 3)\}$
   j. $\{(1, 3), (1, 4), (1, 5), (1, 6), (1, 7), (2, 3), (2, 4), (2, 5), (2, 6), (2, 7),$
   $(3, 3), (3, 4), (3, 5), (3, 6), (3, 7), (4, 3), (4, 4), (4, 5), (4, 6), (4, 7),$
   $(5, 3), (5, 4), (5, 5), (5, 6), (5, 7)\}$

CHAPTER 4

SECTION 4.1

5. a. $D_1 = \{^-3, 2, 1, ^-5, 4\}$, $R_1 = \{2, ^-3, 0, 7\}$, $D_2 = \{^-2, 0, 4, \sqrt{3}, ^-1\}$,
$R_2 = \{3, 0, ^-7, 2, ^-1\}$, $D_3 = \{^-5, ^-3, 0, 1, \sqrt{2}, 5, 9\}$, $R_3 = \{2\}$,
$D_4 = \{^-7, ^-5, ^-3, ^-1, 1, 3, 5, 7, 9\}$, $R_4 = \{^-5, ^-4, ^-3, ^-2, ^-1, 0, 1, 2, 3\}$,
$D_5 = \{^-3, 1, 0, \sqrt{2}, 5\sqrt{3}\}$, $R_5 = \{^-3, 1, 0, \sqrt{2}, 5\sqrt{3}\}$
b.

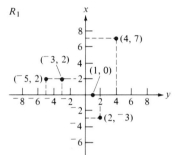

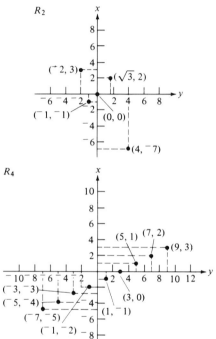

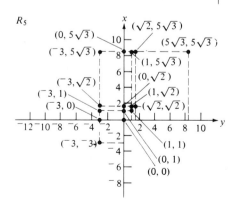

c. $\{(2, ^-3), (^-3, 2), (0, 1), (2, ^-5), (7, 4)\}$,
$\{(3, ^-2), (0, 0), (^-7, 4), (2, \sqrt{3}), (^-1, ^-1)\}$,
$\{(2, ^-5), (2, ^-3), (2, 0), (2, 1), (2, \sqrt{2}), (2, 5), (2, 9)\}$,
$\{(a, b)\,|\,b = 2a + 3, a \in M\}$ such that $M = \{^-5, ^-4, ^-3, ^-2, ^-1, 0, 1, 2, 3\}$,
$\{(x, y)\,|\,x, y \in A, y \le x\}$ such that $A = \{^-3, 1, 0, 2, 5\sqrt{3}\}$.

d.

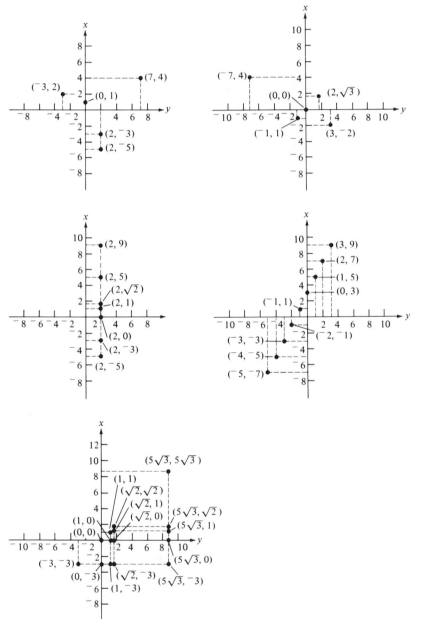

SECTION 4.2

1. a. Yes, $R \neq \varnothing$, $R \subseteq S \times S$ b. No, $(5, 2) \in R$ but $5 \notin S$
   c. Yes, $R \neq \varnothing$, $R \subseteq S \times S$ d. No, $(2, {}^-2) \in R$ but ${}^-2 \notin S$
3. b   7. a, c   b. d

SECTION 4.3

1. a. (1), (2), (4), (6), (8), (9), (12)   b. (1), (4), (8), (12)
   c.

(1)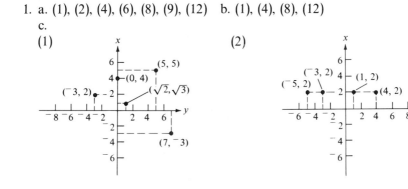

(2)

(3)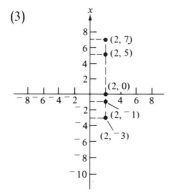

(4)

(5)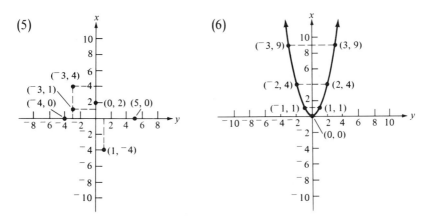

(6)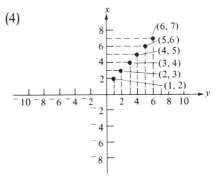

(7) 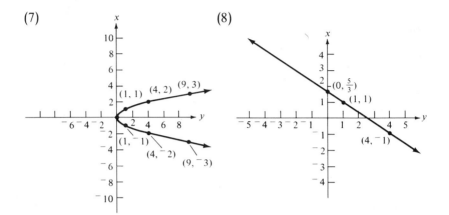 (8)

(9) 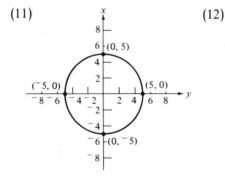 (10)

(11) 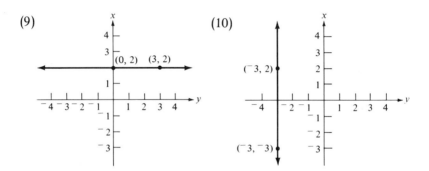 (12)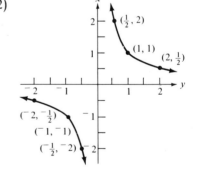

3. (1), (3), (4), (7), (8), (10), (12)

SECTION 4.4

1.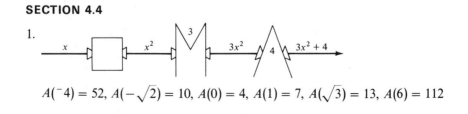

$A(^-4) = 52$, $A(-\sqrt{2}) = 10$, $A(0) = 4$, $A(1) = 7$, $A(\sqrt{3}) = 13$, $A(6) = 112$

3.

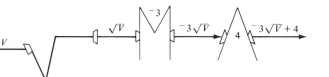

$D(0) = 4$, $D(1) = 1$, $D(3) = {}^-3\sqrt{3} + 4$, $D(4) = {}^-2$, $D(8) = {}^-6\sqrt{2} + 4$, $D(16) = {}^-8$, $D(24) = {}^-6\sqrt{6} + 4$, $D({}^-2)$ does not exist since $\sqrt{{}^-2} \notin Re$. Similar comments for $D({}^-3/2)$ and $D({}^-5)$.

5.

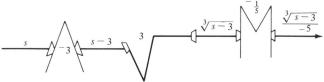

$F({}^-3) = \sqrt[3]{{}^-6}/{}^-5$, $F({}^-2) = \sqrt[3]{{}^-5}/{}^-5$, $F(0) = \sqrt[3]{{}^-3}/{}^-5$, $F(3) = 0$, $F(5) = \sqrt[3]{2}/{}^-5$, $F(24) = \sqrt[3]{21}/{}^-5$, $F(30) = {}^-3/5$

7. $b < 2/3$ 9. $d > 4$

11.

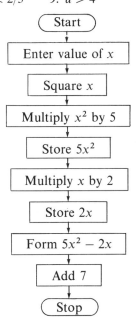

Start

Enter value of x

Square x

Multiply x^2 by 5

Store $5x^2$

Multiply x by 2

Store $2x$

Form $5x^2 - 2x$

Add 7

Stop

13.

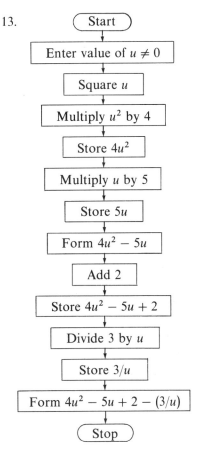

Start

Enter value of $u \neq 0$

Square u

Multiply u^2 by 4

Store $4u^2$

Multiply u by 5

Store $5u$

Form $4u^2 - 5u$

Add 2

Store $4u^2 - 5u + 2$

Divide 3 by u

Store $3/u$

Form $4u^2 - 5u + 2 - (3/u)$

Stop

15.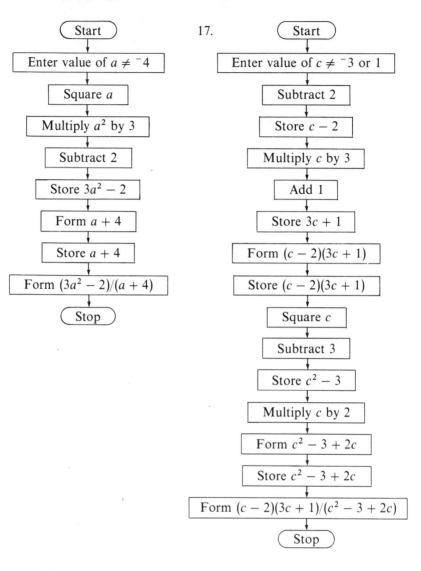

17.

SECTION 4.5

1. b. Yes, $f: A \rightarrow B$ and $R_f = B$. c. No, b would be paired with both 2 and 3. d. $f(1) = a, f(3) = b, f(5) = d$.
3. h is not a mapping from E into F since $h(2) = \pm 2$.
5. Yes 7. a. Yes b. No, 73 is paired with both 1 and 3.

SECTION 4.6

1. b. $f(a) = 1, f(b) = 2, f(c) = 3, f(d) = 2, f(e) = 1$
c. Yes, $f: P \rightarrow Q$ and $R_f = Q$. d. No, $1 \in R_f$ is paired with both a, $e \in D_f$. e. No, f is not 1 : 1.

3. b. *Re*   c. The set of all nonnegative real numbers
   d. No, $h: A \to B$ but $R_h \neq B$ e. No, $2 \in R_h$ is paired with both -2,
   $2 \in D_h$.   f. No, $h: A \to B$ is not $1:1$.   g. (1) 3   (2) 1.5   (3) 0   (4) 2.1
   (5) 3/2   (6) x^2, if $x \in Re$ (7) $3x$, if $x \geq 0$, or ^-3x, if $x < 0$ (8) $x + 2$,
   if $x \geq ^-2$, or $^-(x + 2)$, if $x < ^-2$ (9) $\sqrt{x}$, if $x \geq 0$ (10) x^2, if $x \in Re$
9. a. 5   b. 0   c. $^-2$ d. 2

SECTION 4.7

1. b. 54°F   c. 51° − 54°F   d.

$$T$$

0100	
0400	47°
0700	48°
1000	49°
1200	51°
1400	53°
1500	54°
2300	

e. $T = \{(0100, 47°), (0400, 48°), (0700, 51°), (1000, 53°), (1200, 53°),$
   $(1400, 53°), (1500, 54°), (2300, 49°)\}$
f. No, $53° \in R_T$ is paired with both 1000 and 1200 in D_T.
g. °F

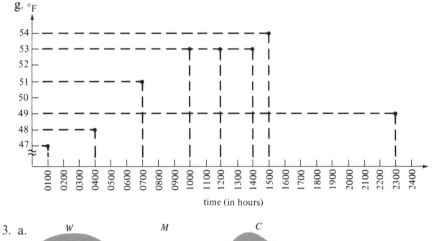

time (in hours)

3. a.

$W =$ set of weights, $C =$ set of costs, first class postage, $M: W \to C$
such that $M(\alpha) =$ cost for first class postage mailed in United States
having weight $\in W$.

b. $M = \{(w, c)|c = M(w) = $ cost for first class postage for letters mailed in the United States having weight $\in W\}$.   c. No, since $13\cancel{c} \in R_M$ is paired with both $\frac{1}{4}$ oz, $\frac{1}{2}$ oz $\in D_M$.   d. (1) 39$\cancel{c}$,   (2) 13$\cancel{c}$,   (3)78$\cancel{c}$, (4) 65$\cancel{c}$,   (5) \$2.08,   (6) \$4.55,   (7) \$6.89,   (8) \$33.28,   (9) \$52.26, (10 52$\cancel{c}$

5. a. No, E has more elements than G; hence, a one-to-one correspondence between their elements cannot be established.
b. Yes, let each element in G be paired with its English transliteration in E as illustrated.
c. $h: G \xrightarrow{\text{onto}} E$ because $h: G \to E$ but $R_h \subset E$.
d. No, D_K would have more elements than R_K.   e. No. See (d).
f. No. See (d) and (e).

SECTION 4.8

1. a. 1   b. $^-$10.2   c. 41.5   d. 13   e. 14.3   f. $\sqrt{2} - \sqrt{3}$ g. $-\sqrt{5}$ h. $\pi + e$
i. $17\sqrt{2} - 9$ j. 13.155   k. 54   l. $^-$133   m. 2.76   n. $^-$36.92   o. 79.38
p. $\sqrt{6}$ q. 1/3   r. 14/27   s. 56/23   t. 76/37

3. No, because we cannot always take an ordered pair of real numbers, divide the first component by the second component, and have a quotient which is a real number; division by 0 is not defined.

REVIEW EXERCISES FOR CHAPTER 4

1. Refer to the following definitions: a. 4.15   b. 4.8   c. 4.8   d. 4.16
e. 4.17   f. 4.18   g. 4.1   h. 4.1   i. 4.3   j. 4.11   k. 4.19

3. a. Yes, the domain of the function would be the set of the first 27 natural numbers and paired with each of these would be the name of an individual student.   b. Yes, each name of a student is paired with a unique natural number from 1 to 27.   c. Yes, each student is assigned one and only one grade.   d. No, since the grade of A, for instance, is assigned to students 1, 6, 11, 16, 21, and 26.

5. a.

weight	$\xrightarrow{\alpha}$ 100 miles
1–23 lb	\$6.30
24–62 lb	8.40
63–100 lb	9.80

α is not one-to-one since $\$6.30 \in R_\alpha$ is paired with all weights between 1 and 23 lb in D_α.

b.

weight	$\xrightarrow{\beta}$ 500 miles
1–23 lb	\$8.90
24–62 lb	11.10
63–100 lb	13.25

β is not one-to-one since $\$8.90 \in R_\beta$ is paired with all weights between 1 and 23 lb in D_β.

c. (1) $6.30   (2) $11.10   (3) $6.30   (4) $8.90
d. Any package weighing between 63–100 lb may be shipped up to 100 miles for $8.90 or any package weighing between 24–62 lb may be shipped up to 200 miles for the same amount.

CHAPTER 5

SECTION 5.1

1. Cardinal   3. Ordinal   5. Cardinal   7. Cardinal
9. Cardinal   11. Cardinal   13. Ordinal   15. Cardinal
17. Ordinal   19. Ordinal

SECTION 5.2

1. a. 3   b. 5   c. 7   d. 18   e. 15   f. 17   g. 24   h. 18   i. 26   j. 16
3. a. True   b. False   c. False   d. False   e. True   f. True   g. False
   h. False   i. False   j. True

SECTION 5.3

1. a. 5   b. 7   c. 5   d. 6   e. 0   f. 10   g. 11   h. 9   i. 13   j. 13

SECTION 5.4

1. a. 15   b. 8   c. 12   d. 2   e. 0   f. 15   g. 16   h. 10   i. 0   j. 24
3. a. Yes   b. Yes   c. Yes   d. Yes   e. Yes

SECTION 5.5

1. a. Commutative property for addition   b. Associative property for multiplication   c. Additive identity property   d. Commutative property for multiplication   e. Multiplicative identity property   f. Additive identity property or additive inverse property   g. Multiplicative identity property   h. Commutative property for addition (twice)
   i. Associative property for addition   j. Associative property for addition   k. Distributive property of multiplication over addition
   l. Commutative property for multiplication   m. Additive identity property   n. Multiplicative identity property (twice)
3. a. $16 \times 21 = 16 \times (20 + 1) = (16 \times 20) + (16 \times 1) = 320 + 16 = 336$
   b. $18 \times 11 = 18 \times (10 + 1) = (18 \times 10) + (18 \times 1) = 180 + 18 = 198$
   c. $31 \times 17 = (30 + 1) \times 17 = (30 \times 17) + (1 \times 17) = 510 + 17 = 527$
   d. $(16 \times 3) + (16 \times 7) = 16 \times (3 + 7) = 16 \times 10 = 160$
   e. $(6 \times 29) + (4 \times 29) = (6 + 4) \times 29 = 10 \times 29 = 290$

SECTION 5.6

1. a. $<$ b. $=$ c. $>$ d. $<$ e. $<$ f. $=$ g. $>$ h. $<$ i. $>$ j. $=$
3. No, no whole number is greater than itself.
5. Yes, since every whole number is equal to itself.

SECTION 5.7

1. a. $8 - 6 = 2 \in W$ since $8 = 6 + 2$ b. $4 - 2 = 2 \in W$ since $4 = 2 + 2$
   c. $23 - 16 = 7 \in W$ since $23 = 16 + 7$ d. $4 - 0 = 4 \in W$ since $4 = 0 + 4$
   e. $9 - 9 = 0 \in W$ since $9 = 9 + 0$ f. $(2 \times 3) - (1 + 4) = 1 \in W$ since
   $(2 \times 3) = (1 + 4) + 1$ g. $3(4 + 2) - 2(3 + 3) = 6 \in W$ since $3(4 + 2) =$
   $2(3 + 3) + 6$ h. $(5 \times 0) - (0 + 0) = 0 \in W$ since $(5 \times 0) = (0 + 0) + 0$
   i. $(8 + 2) - 7 = 3 \in W$ since $8 + 2 = 7 + 3$ j. $13 - (4 + 5) = 4 \in W$
   since $13 = (4 + 5) + 4$
3. a. $3 \times 9 = 3 \times (10 - 1) = (3 \times 10) - (3 \times 1) = 30 - 3 = 27$
   b. $9 \times 11 = 11 \times 9 = 11 \times (10 - 1) = (11 \times 10) - (11 \times 1) =$
   $110 - 11 = 99$ c. $6 \times 29 = 6 \times (30 - 1) = (6 \times 30) - (6 \times 1) =$
   $180 - 6 = 174$ d. $7 \times 39 = 7 \times (40 - 1) = (7 \times 40) - (7 \times 1) =$
   $280 - 7 = 273$ e. $12 \times 99 = 12 \times (100 - 1) = (12 \times 100) - (12 \times 1) =$
   $1200 - 12 = 1188$
5. a. $a = 2, b = 3, a < b$ b. $a = 14, b = 0, a > b$ c. $a = 1, b = 5, a < b$
   d. $a = 1, b = 9, a < b$ e. $a = 7, b = 17, a < b$ 11. Yes, for $a, b \in S$,
   $a \div b \in S$ $(a = b = 1$ only$)$.

SECTION 5.8

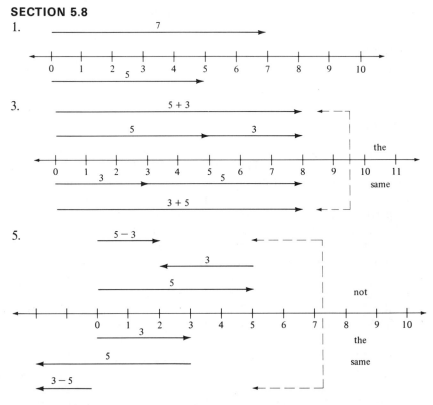

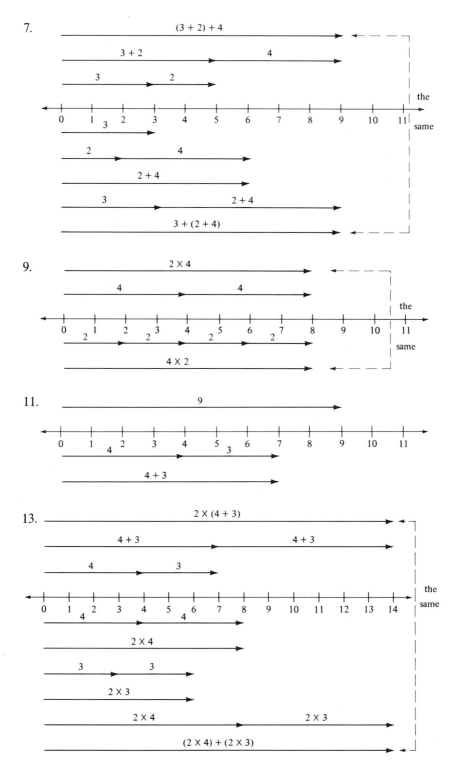

15.

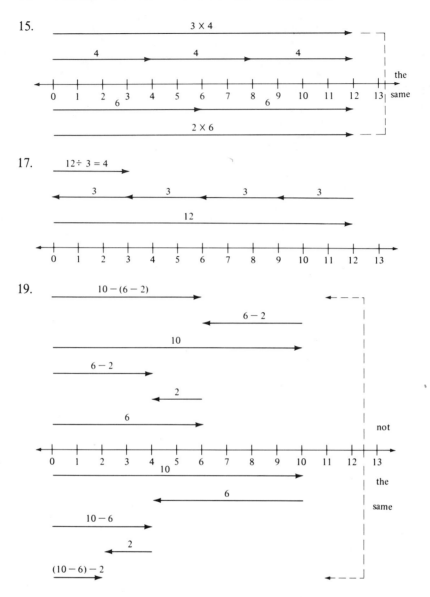

17.

19.

REVIEW EXERCISES FOR CHAPTER 5

1. True; if $a, b \in W$, then $a + b \in W$.   3. True; if $a, b \in W$, then $a \times b \in W$.
5. True; for all $a, b \in W$, $a + b = b + a$.
7. False; $3, 4 \in W$, but $3 - 4 \neq 4 - 3$.
9. False; $3, 4, 5 \in W$, but $(5 - 4) - 3 \neq 5 - (4 - 3)$.
11. False; $8, 4 \in W$ but $8 \div 4 \neq 4 \div 8$.

13. False; 2, 3, 4 ∈ W but 2 + (3 × 4) ≠ (2 + 3) × (2 + 4).

15. False; the statement is known as the trichotomy property.

17. False; let $A = \{a, b, c\}$ and $B = \{p, q, r, s\}$. $n(A) = 3$, $n(B) = 4$ but $n(A \cap B) = 0$.   19. False; if $a = 0$, $0 \div a$ is not defined.

21. True; for all $b \in N$, $0 \div b = 0 \in W$.   23. False; $0 \notin N$.

25. True; see respective definitions.   27. True; for all A, B, $n(A \times B) = n(A) \times n(B)$.   29. True; if $a, b \in W$ and $c \in N$ with $a < b$, then $ca < cb$.

CHAPTER 6

SECTION 6.2

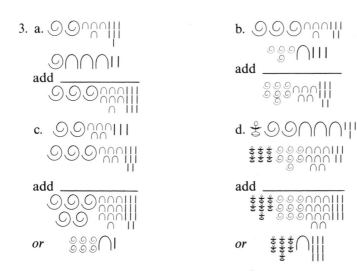

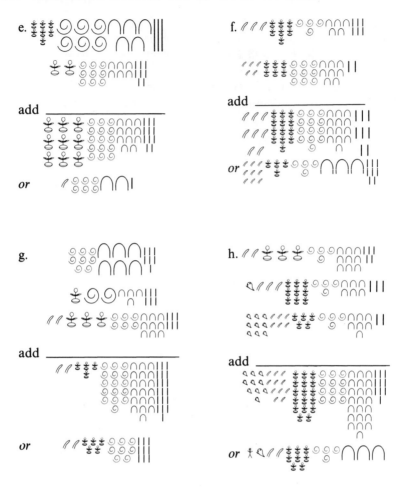

5. a. 1 | ⌒⌒||| (1 × 46)

   2 || ⌒⌒⌒|| (2 × 46)

   3 ||| ⌒⌒⌒|||| (4 × 46)

   4 ||| ⌒⌒⌒⌒|||| (8 × 46)

   5 ∩||| ⌒⌒⌒∩∩∩||| (16 × 46)

   ───────────────────────────────

   1 + 5 ∩||| ⌒⌒⌒∩∩|| (1 × 46) + (16 × 46)
                              or (17 × 46)

b. 1 | (1×53)

2 || (2×53)

3 '|' (4×53)

4 ||| (8×53)

5   (16×53)

---

4 + 5 $(8 \times 53) + (16 \times 53)$
or (24×53)

c. 1 | (1×92)

2 || (2×92)

3 '|' (4×92)

4 ||| (8×92)

5   (16×92)

6   (32×92)

---

1 + 6 $(1 \times 92) + (32 \times 92)$
or (33×92)

d.   1 | (1×123)

2 || (2×123)

3 '|' (4×123)

4 ||| (8×123)

5   (16×123)

6   (32×123)

---

3 + 5 + 6 $(4 \times 123) + (16 \times 123)$
$+ (32 \times 123)$
or (52×123)

e.

1   (1 × 1234)

2   (2 × 1234)

3   (4 × 1234)

4   (8 × 1234)

5   (16 × 1234)

6   (32 × 1234)

7   (64 × 1234)

---

1 + 2 + 3
  + 5 + 7   (1 × 1234)

+ (2 × 1234) + (4 × 1234)
+ (16 × 1234) + (64 × 1234)
or (87 × 1234)

f.

1   (1 × 2468)

2   (2 × 2468)

3   (4 × 2468)

4   (8 × 2468)

5   (16 × 2468)

6   (32 × 2468)

7   (64 × 2468)

---

1 + 2 + 4 + 5
  + 6 + 7   (1 × 2468)

+ (2 × 2468) + (8 × 2468) + (16 × 2468)
+ (32 × 2468) + (64 × 2468)
or (123 × 2468)

SECTION 6.3

1. a. *(cuneiform numeral)*
   b. *(cuneiform numeral)*

   c. *(cuneiform numeral)*
   d. *(cuneiform numeral)*

   e. *(cuneiform numeral)*
   f. *(cuneiform numeral)*

   g. *(cuneiform numeral)*
   h. *(cuneiform numeral)*

   i. *(cuneiform numeral)*
   j. *(cuneiform numeral)*

3. a. *(Egyptian numeral)*
   b. *(Egyptian numeral)*

   c. *(Egyptian numeral)*
   d. *(Egyptian numeral)*

   e. *(Egyptian numeral)*
   f. *(Egyptian numeral)*

   g. *(Egyptian numeral)*
   h. *(Egyptian numeral)*

   i. *(Egyptian numeral)*
   j. *(Egyptian numeral)*

5. a. *(Egyptian numeral)*
   b. *(Egyptian numeral)*

   c. *(Egyptian numeral)*
   d. Not possible

   e. *(Egyptian numeral)*
   f. *(Egyptian numeral)*

SECTION 6.4

1. a. 368   b. 752   c. 831   d. 989   e. 111   f. 5,844   g. 9,579   h. 90,746
   i. 3,428   j. 80,692

3. a. *(cuneiform numeral)*
   b. *(cuneiform numeral)*

   c. *(cuneiform numeral)*
   d. *(cuneiform numeral)*

   e. *(cuneiform numeral)*
   f. *(cuneiform numeral)*

   g. *(cuneiform numeral)*
   h. *(cuneiform numeral)*

   i. *(cuneiform numeral)*
   j. *(cuneiform numeral)*

SECTION 6.5

1. a. 33   b. 57   c. 1,166   d. 649   e. 1,154   f. 2,267   g. 1,749   h. 2,664
   i. 99   j. 1,444   k. 1,666   l. 499   m. 999   n. 6,000   o. 12,550
   p. 9,000,000   q. 1,005,510   r. 600,600,600   s. 2,650,003,747
   t. 3,777,640,708

3. a. CCCLXVII   b. CXLIII   c. MMCDII   d. DCCCXCVIII
   e. MMMCMLIII   f. MMMDLI   g. CCLXI   h. $\overline{\text{XX}}$CDXXIX
   i. $\overline{\text{X}}$CCLX   j. $\overline{\text{VIICLID}}$CLXX

5.

×	I	V	X	L	C	D	M
I	I	V	X	L	C	D	M
V	V	XXV	L	CCL	D	MMD	$\overline{\text{V}}$
X	X	L	C	D	M	$\overline{\text{V}}$	$\overline{\text{X}}$
L	L	CCL	D	MMD	$\overline{\text{V}}$	$\overline{\text{XXV}}$	$\overline{\text{L}}$
C	C	D	M	$\overline{\text{V}}$	$\overline{\text{X}}$	$\overline{\text{L}}$	$\overline{\text{C}}$
D	D	MMD	$\overline{\text{V}}$	$\overline{\text{XXV}}$	$\overline{\text{L}}$	$\overline{\text{CCL}}$	$\overline{\text{D}}$
M	M	$\overline{\text{V}}$	$\overline{\text{X}}$	$\overline{\text{L}}$	$\overline{\text{C}}$	$\overline{\text{D}}$	$\overline{\text{M}}$

SECTION 6.6

1. a., b., c., d., e., f., g., h., i., j.

3. a. XXII   b. CXXVI   c. MMDCXL   d. $\overline{\text{XXXVI}}$DCCCXXII
   e. $\overline{\text{XXXVII}}$CMV   f. $\overline{\text{VII}}$CCI   g. $\overline{\text{LII}}$DLX   h. $\overline{\text{IV}}$DLXXVIII
   i. $\overline{\text{VI}}$CXVI   j. $\overline{\text{XXXVIII}}$CDXVIII

5. a., b., c., d., e., f., g., h., i., j.

SECTION 6.7

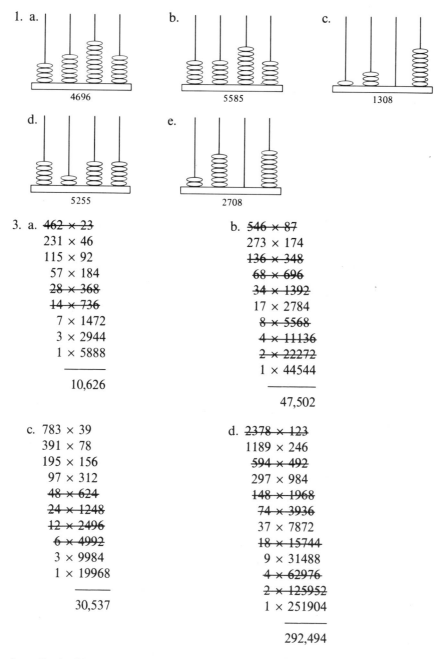

1. a. 4696   b. 5585   c. 1308

 d. 5255   e. 2708

3. a. 462 × 23
      231 × 46
      115 × 92
       57 × 184
       28 × 368
       14 × 736
        7 × 1472
        3 × 2944
        1 × 5888
        ─────
       10,626

 b. 546 × 87
      273 × 174
      136 × 348
       68 × 696
       34 × 1392
       17 × 2784
        8 × 5568
        4 × 11136
        2 × 22272
        1 × 44544
        ─────
       47,502

 c. 783 × 39
      391 × 78
      195 × 156
       97 × 312
       48 × 624
       24 × 1248
       12 × 2496
        6 × 4992
        3 × 9984
        1 × 19968
        ─────
       30,537

 d. 2378 × 123
      1189 × 246
       594 × 492
       297 × 984
       148 × 1968
        74 × 3936
        37 × 7872
        18 × 15744
         9 × 31488
         4 × 62976
         2 × 125952
         1 × 251904
        ─────
       292,494

5. a. 7   b. 14   c. 17   d. 21   e. Yes, since for every $a, b \in W$, $a @ b \in W$.
 f. Yes, for every $a, b \in W$, $a @ b = b @ a$ since $a + b + 1 = b + a + 1$.

g. Yes, for all $a, b, c \in W$, $(a @ b) @ c = a @ (b @ c)$ since $(a + b + 1) + c + 1 = a + (b + c + 1) + 1$.

h. No, there does not exist $e \in W$ such that $a @ e = a$ for every $a \in W$.

SECTION 6.8

1. a. 58   b. 63   c. 612   d. 1225   e. 8261   f. 13,908
3. a. $234 \times (30 + 2) = 7020 + 468 = 7488$
   b. $425 \times (100 + 27) = 42,500 + 11,475 = 53,975$
   c. $6234 \times (200 + 34) = 1,246,800 + 211,956 = 1,458,756$
   d. $5562 \times (400 + 5) = 2,224,800 + 27,810 = 2,252,610$
   e. $27,683 \times (2700 + 4) = 74,744,100 + 110,732 = 74,854,832$
5. a. 237   b. 359   c. 515   d. 623   e. 86   f. 230

SECTION 6.9

1. a. 1, 10, 11, 100, 101, 110, 111, 1000, 1001, 1010, 1011, 1100, 1101, 1110, 1111, 10000, 10001, 10010, 10011, 10100
   b. 1, 2, 3, 4, 10, 11, 12, 13, 14, 20, 21, 22, 23, 24, 30, 31, 32, 33, 34, 40
   c. 1, 2, 3, 4, 5, 6, 7, 10, 11, 12, 13, 14, 15, 16, 17, 20, 21, 22, 23, 24
   d. 1, 2, 3, 4, 5, 6, 7, 8, 10, 11, 12, 13, 14, 15, 16, 17, 18, 20, 21, 22
   e. 1, 2, 3, 4, 5, 6, 7, 8, 9, T, 10, 11, 12, 13, 14, 15, 16, 17, 18, 19
   f. 1, 2, 3, 4, 5, 6, 7, 8, 9, T, E, 10, 11, 12, 13, 14, 15, 16, 17, 18
3. a. $331_{five} R3_{five}$ b. $1414_{five} R3_{five}$ c. $212_{five} R103_{five}$
   d. $21012_{five} R102_{five}$
5. a. 100101_{two} b. 80748_{nine} c. 290344_{twelve} d. 6240_{seven} e. 22223_{four}
7. a. $1000_{two} R10_{two}$ b. $224_{six} R55_{six}$ c. $421_{seven} R153_{seven}$
   d. $134_{eight} R763_{eight}$ e. $2035_{nine} R544_{nine}$ f. $487TE3_{twelve} RT8_{twelve}$
9. a. 210030_{four} b. 40536_{seven} c. 1002020_{three} d. 7622_{twelve} e. 34025_{six}
   f. 1111110001100_{two} g. 112044_{nine} h. 1615_{eight}

REVIEW EXERCISES FOR CHAPTER 6

1. a.   b.

   c.   d.

   e.   f.

   g.   h.

   i.   j.

3. a. CCLXII   b. CCCLXXV   c. DCXXIII   d. MXV   e. $\overline{\text{XIII}}$CCXXXIV
   f. $\overline{\text{XXIII}}$CMLXII   g. $\overline{\text{XXXVII}}$CDLXII   h. $\overline{\text{CLXIX}}$DCCLIV
   i. $\overline{\text{CCCXLII}}$CXXIII   j. $\overline{\text{MXXIII}}$XI

5. a. 112   b. 1,121   c. 100,221   d. 11,424   e. 2,726   f. 1,101,653
7. a. 1,665   b. 2,440   c. 2,754   d. 479   e. 11,004   f. 15,200,044

9. a.

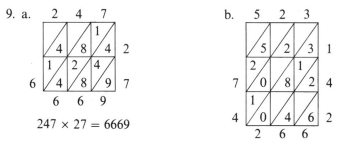

247 × 27 = 6669

523 × 142 = 74,266

11. a. 66   b. 373   c. 9161   d. 19   e. 147   f. 2327
13. a. 123   b. 1,111   c. 972   d. 1,025
15. a. 15,938   b. 2,455,434   c. 871   d. 67,309

CHAPTER 7

SECTION 7.1

1. a. $^-4$ b. 7   c. $^-6$ d. 9   e. 0   f. ^-a g. b h. $^-(2+3)$ i. $^-(4-7)$
   j. $^-(3 \times 6)$ k. $^-(ab)$ l. $^-(a+b)$ m. $^-(b-2)$ n. $^-(3+ ^-3)$
   o. $^-(16 \div 4)$ p. $^-[^-16 \div (8 \div ^-2)]$
3. There isn't a largest integer.   5. 0   7. 1   9. 0

SECTION 7.2

1. a. 9   b. $^-5$ c. $^-2$ d. 2   e. 6   f. 0   g. 0   h. $^-8$ i. $^-5$ j. $^-6$ k. 1
   l. $^-2$ m. $^-3$ n. $^-6$ o. $^-7$
3. a. 4   b. 7   c. $^-5$ d. 0   e. $^-6$ f. 0   g. 5   h. $^-6$
5. $152

SECTION 7.3

1. a. 12   b. $^-8$ c. 15   d. 0   e. 0   f. 0   g. 72   h. 24   i. $^-18$ j. 60
   k. $^-48$ l. 6   m. 12   n. 0   o. $^-18$
3. a. $^-3$ b. $^-4$ c. $^-3$ d. 0   e. $^-30$ f. $^-3$ g. $^-5$ h. 5
7. Yes, let $a = ^-2$ and $b = 3$. Then $(^-a)(^-b) = (2)(^-3) = ^-6$.

SECTION 7.4

1. a. $^-5, ^-3, ^-2, ^-1, 0, 4, 5$ b. $^-5, ^-3, ^-2, 0, 1, 2, 4$
   c. $^-105, ^-103, ^-101, 98, 100, 102$ d. $^-28, ^-27, ^-26, ^-17, 7, 13, 23$
   e. $^-3, 0, 2, (^-2)^2, 5, (3)^2$

3. a. >   b. <   c. =   d. >   e. >   f. <   g. <   h. >
5. a. $x < {}^-1$ b. $y \geq {}^-9$ c. $u \in I$ d. $v \leq {}^-2$

SECTION 7.5

1. a. $5 + ({}^-2)$ b. ${}^-9 + ({}^-5)$ c. ${}^-8 + (5)$ d. $7 + (9)$ e. $13 + ({}^-17)$
   f. $16 + ({}^-11)$ g. $b + ({}^-a)$ h. ${}^-a + ({}^-b)$ i. $a + (b)$ j. ${}^-a + (b)$
3. a. 4   b. ${}^-4$ c. 7   d. 0   e. ${}^-2$ f. 6   g. ${}^-2$ h. ${}^-5$ i. 2   j. 190
5. a. $(a + b) - c = (5 + 4) - 3 = 9 - 3 = 6$;
   $a + (b - c) = 5 + (4 - 3) = 5 + 1 = 6$
   b. $(a + b) - c = (4 + {}^-3) - {}^-2 = 1 - {}^-2 = 3$;
   $a + (b - c) = 4 + ({}^-3 - {}^-2) = 4 + {}^-1 = 3$
   c. $(a + b) - c = (0 + 2) - {}^-1 = 2 + 1 = 3$;
   $a + (b - c) = 0 + (2 - {}^-1) = 0 + 3 = 3$
   d. $(a + b) - c = ({}^-3 + 5) - {}^-7 = 2 + 7 = 9$;
   $a + (b - c) = {}^-3 + (5 - {}^-7) = {}^-3 + 12 = {}^.9$
   e. $(a + b) - c = (9 + {}^-6) - 5 = 3 - 5 = {}^-2$;
   $a + (b - c) = 9 + ({}^-6 - 5) = 9 + {}^-11 = {}^-2$
7. a. No, $3 - 2 \neq 2 - 3$; $2, 3 \in I$.
   b. No, $(6 - 5) - 4 \neq 6 - (5 - 4)$; $4, 5, 6 \in I$.
   c. No, $16 \div 8 \neq 8 \div 16$; $8, 16 \in I$.
   d. No, $(16 \div 8) \div 4 \neq 16 \div (8 \div 4)$; $4, 8, 16 \in I$.
   e. No, ${}^-5 - {}^-7 = 2$; ${}^-5, {}^-7 \in I_n, 2 \notin I_n$.
   f. No, ${}^-16 \div {}^-8 = 2$; ${}^-8, {}^-16 \in I_n, 2 \notin I_n$.

SECTION 7.6

1. a.

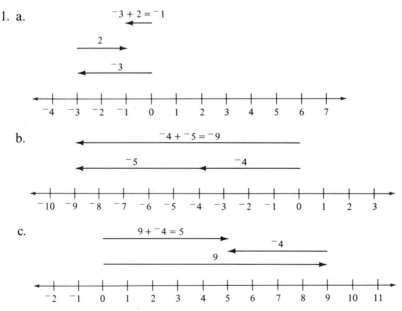

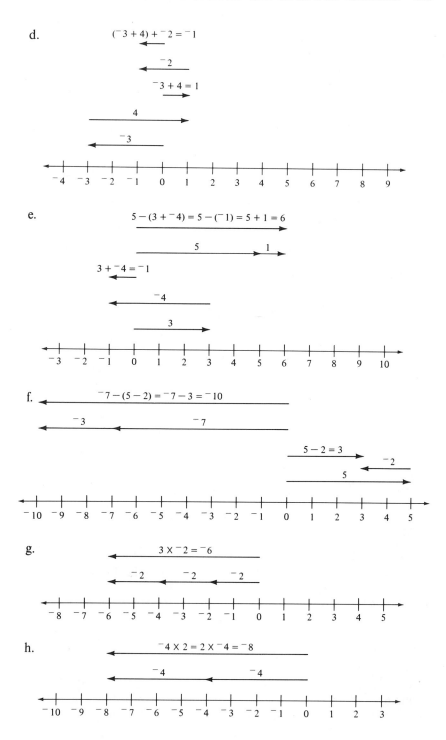

d.

$$(^-3 + 4) + ^-2 = ^-1$$

$^-2$

$$^-3 + 4 = 1$$

4

$^-3$

e.

$$5 - (3 + ^-4) = 5 - (^-1) = 5 + 1 = 6$$

5   1

$$3 + ^-4 = ^-1$$

$^-4$

3

f.

$$^-7 - (5 - 2) = ^-7 - 3 = ^-10$$

$^-3$ $^-7$

$$5 - 2 = 3$$

$^-2$

5

g.

$$3 \times ^-2 = ^-6$$

$^-2$ $^-2$ $^-2$

h.

$$^-4 \times 2 = 2 \times ^-4 = ^-8$$

$^-4$ $^-4$

i.

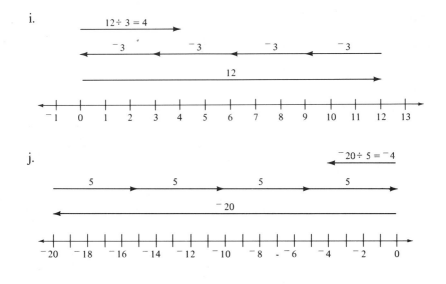

REVIEW EXERCISES FOR CHAPTER 7

1. a. 1 b. ⁻7 c. ⁻2 d. 2 e. ⁻1 f. 5 g. ⁻12 h. 20 i. 24 j. ⁻6
   k. ⁻2 l. 4 m. 5 n. 0 o. 4 p. 6
3. a. > b. > c. = d. > e. = f. >, < g. > h. >
7. a. There isn't any. b. 0 c. 6 d. Closure under subtraction and each
   integer has an additive inverse. e. 1 f. If $a, b \in I$, then $a < b$ or
   $a = b$ or $a > b$. g. 0 h. Not defined (impossible) i. Not defined
   (indeterminate) j. No, since 3, 6 $\in I$ but 3 ÷ 6 $\notin I$.

CHAPTER 8

SECTION 8.1

1. For "a divides b", a is a factor of b and b is a multiple of a. For "a is
   divided by b", b is a factor of a and a is a multiple of b.
3. a. 1, 2, 4, 8 b. 8, 64, 512 c. ⁻16, ⁻40, ⁻72
5. a. 1, 7, 63, 82 b. ⁻2, ⁻13, ⁻131, ⁻499 c. No, 0 $\notin N$. (A divisor is a
   natural number.)
7. a. 2|4, 4|24, and 2|24 b. 3|9, 9|72, and 3|72 c. 5|25, 25|⁻225,
   and 5|⁻225 d. ⁻2|⁻14, ⁻14|98, and ⁻2|98 e. ⁻4|20, 20|⁻600,
   and ⁻4|⁻600
9. a. 3|12, 3|15, and 3|⁻3 b. ⁻2|24, ⁻2|⁻8, and ⁻2|32 c. 6|⁻12, 6|102,
   and 6|⁻114 d. ⁻8|⁻24, ⁻8|⁻32, and ⁻8|8 e. 4|0, 4|⁻48, and 4|48
11. a. Odd b. Odd c. Even d. Odd e. Even f. Even

SECTION 8.2

1. (For each of these, the rightmost digit must be 0, 2, 4, 6, or 8.) a. No
   b. Yes   c. No   d. Yes   e. No   f. Yes   g. No   h. Yes   i. Yes   j. No
   k. Yes   l. No   m. Yes   n. Yes   o. No   p. No   q. Yes   r. Yes   s. No
   t. Yes   u. No   v. Yes   w. No   x. No
3. a. No, $4 \nmid 21$.   b. No, $4 \nmid 46$.   c. No, $4 \nmid 69$.   d. Yes, $4 \mid 28$.
   e. No, $4 \nmid 37$.   f. No, $4 \nmid 62$.   g. No, $4 \nmid 01$.   h. Yes, $4 \mid 00$.   i. Yes, $4 \mid 16$.
   j. No, $4 \nmid 21$.   k. No, $4 \nmid 26$.   l. No, $4 \nmid 69$.   m. Yes, $4 \mid 72$.
   n. Yes, $4 \mid 64$.   o. No, $4 \nmid 17$.   p. No, $4 \nmid 09$.   q. Yes, $4 \mid 76$.   r. Yes, $4 \mid 92$.
   s. No, $4 \nmid 69$.   t. Yes, $4 \mid 68$.   u. No, $4 \nmid 71$.   v. No, $4 \nmid 62$.
   w. No, $4 \nmid 87$.   x. No, $4 \nmid 39$.
5. a. No, 21 is not even.   b. No, $3 \nmid 46$.   c. No, 69 is not even.
   d. No, $3 \nmid 128$.   e. No, 237 is not even.   f. Yes, 462 is even and $3 \mid 462$.
   g. No, 701 is not even.   h. No, $3 \nmid 800$.   i. No, $3 \nmid 916$.   j. No, 1121 is
   not even.   k. Yes, 1326 is even and $3 \mid 1326$.   l. No, 4669 is not even.
   m. Yes, 6372 is even and $3 \mid 6372$.   n. No, $3 \nmid 9764$.   o. No, 11217 is
   not even.   p. No, 15609 is not even.   q. No, $3 \nmid 17876$.   r. No,
   $3 \nmid 20692$.   s. No, 86769 is not even.   t. Yes, 123468 is even and
   $3 \mid 123468$.   u. No, 376871 is not even.   v. No, $3 \nmid 999862$.
   w. No, 1605987 is not even.   x. No, 716890239 is not even.
7. a. No, $8 \nmid 021$.   b. No, $8 \nmid 046$.   c. No, $8 \nmid 069$.   d. Yes, $8 \mid 128$.
   e. No, $8 \nmid 237$.   f. No, $8 \nmid 462$.   g. No, $8 \nmid 701$.   h. Yes, $8 \mid 800$.
   i. No, $8 \nmid 916$.   j. No, $8 \nmid 121$.   k. No, $8 \nmid 326$.   l. No, $8 \nmid 669$.
   m. No, $8 \nmid 372$.   n. No, $8 \nmid 764$.   o. No, $8 \nmid 217$.   p. No, $8 \nmid 609$.
   q. No, $8 \nmid 876$.   r. No, $8 \nmid 692$.   s. No, $8 \nmid 769$.   t. No, $8 \nmid 468$.
   u. No, $8 \nmid 871$.   v. No, $8 \nmid 862$.   w. No, $8 \nmid 987$.   x. No, $8 \nmid 239$.
9. (For each of these, the rightmost digit must be 0.)   a. No   b. No
   c. No   d. No   e. No   f. No   g. No   h. Yes   i. No   j. No   k. No
   l. No   m. No   n. No   o. No   p. No   q. No   r. No   s. No   t. No
   u. No   v. No   w. No   x. No
11. No, $3 \mid 21$ but $9 \nmid 21$. If a number is divisible by 3 and the quotient
    obtained is also divisible by 3, then the number itself is divisible by 9.
13. No, $4 \mid 12$ but $8 \nmid 12$. If a number is divisible by 4 and the quotient
    obtained is even, then the number itself is also divisible by 8.
15. a. 0, 2, 4, 6, 8   b. 3, 6, 9   c. None   d. Any digit   e. 1, 4, 7   f. 1, 8
    g. 2, 6   h. 5   i. 0   j. 1

SECTION 8.3

1. a. Composite   b. Prime   c. Composite   d. Prime   e. Prime   f. Prime
   g. Composite   h. Neither   i. Prime   j. Composite   k. Composite
   l. Composite   m. Prime   n. Neither   o. Composite   p. Prime
   q. Composite   r. Composite

3. 3, 5, 7, 11, 13, 17, 19, 23, 29
5. 2, 3, 5, 7, 11, 13, 17, 19, 23, 29, 31, 37, 41, 43, 47, 53, 59, 61, 67, 71, 73,
   79, 83, 97, 101, 103, 107, 109, 113, 127, 131, 137, 139, 149, 151, 157, 163,
   167, 173, 179, 181, 191, 193, 197, 199
7. 3 and 5, 5 and 7, 11 and 13, 17 and 19, 29 and 31, 41 and 43,
   59 and 61, 71 and 73
9. a. 1, 2, 4, 17, 34   b. 1, 2, 4, 31, 62   c. 1, 2, 3, 5, 6, 7, 10, 14, 15, 21, 30,
   35, 42, 70, 105   d. 1, 5, 61   e. 1, 2, 4, 5, 8, 10, 16, 20, 32, 40, 64, 80, 160
   f. 1, 2, 17, 43, 731   g. 1, 2, 4, 5, 8, 10, 20, 40, 59, 118, 236, 295, 472, 590,
   1180   h. 1, 2, 5, 10, 409, 818, 2045   i. 1, 43, 131   j. 1, 2, 19, 38, 163,
   326, 3097   k. 1, 5, 1447   l. 1, 3, 9, 31, 93, 279, 961, 2883
11. Proper divisors of 1184 are 1, 2, 4, 8, 16, 37, 74, 148, 296, and 592;
    their sum is 1210. Proper divisors of 1210 are 1, 2, 5, 10, 11, 22, 55,
    110, 121, 242, and 605; their sum is 1184.

SECTION 8.4

1. a. 2   b. 13   c. 6   d. 105   e. 81   f. 5   g. 22   h. 8   i. 1   j. 1
3. a. 2   b. 1   c. 24   d. 1   e. 1   f. 25   g. 2   h. 1   i. 1   j. 1
5. a. $^-5$, $^-1$, 1, 5   b. $^-15$, $^-5$, $^-3$, $^-1$, 1, 3, 5, 15
   c. $\{^-5, ^-1, 1, 5\} \subset \{^-15, ^-5, ^-3, ^-1, 1, 3, 5, 15\}$ d. 5   9. 1

SECTION 8.5

1. a. 156   b. 546   c. 3570   d. 6100   e. 1134   f. 1056   g. 70   h. 1734
   i. 17640   j. 34608
3. See Exercise 1.
5. a. $A = \{5, 10, 15, 20, 25, 30, 35, 40, 45, 50, 55, 60, 65, 70, 75\}$
   b. $B = \{15, 30, 45, 60, 75\}$ c. $B \subset A$ d. 15   e. 5
   f. l.c.m.(5, 15) = 15; g.c.d.(5, 15) = 5; l.c.m.(5, 15) × g.c.d.(5, 15) = 15 × 5

SECTION 8.6

1. a. False   b. True   c. True   d. False   e. True   f. True   g. False
   h. True   i. True   j. True
3. a. 3   b. 3   c. 1   d. 8   e. 5   f. 5   g. 1   h. 3   i. 3   j. 2

REVIEW EXERCISES FOR CHAPTER 8

1. a. 1, 2, 3, 4, 6, 9, 12, 18, 36   b. $^-1$, $^-3$, $^-5$, $^-9$, $^-15$, $^-45$
3. a. Odd   b. Odd   c. Even   d. Odd   e. Even   f. Even   g. Odd   h. Odd
5. a. Prime   b. Composite   c. Composite   d. Composite   e. Composite
   f. Composite   g. Prime   h. Composite   i. Neither   j. Neither
   k. Prime   l. Prime

7. c. 6   9. a. 156   b. 221   c. 5200   d. 4042   e. 833   f. 5590

11. a. 672   b. 1050   c. 324   d. 750   e. 660   f. 1890

13. a. False   b. True   c. True   d. False   e. True   f. False

CHAPTER 9

SECTION 9.1

1. b, d, e, f, g, i, j

SECTION 9.2

1. See Definition 9.3.   3. a, c, d, e, g, h   5. a. True   b. True
   c. False   d. True

SECTION 9.3

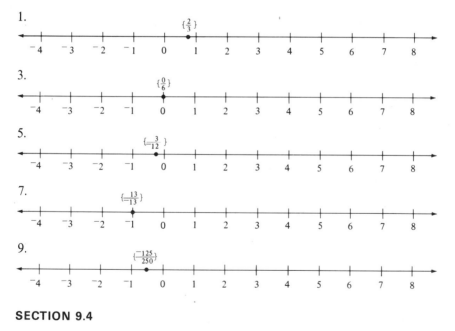

1.

3.

5.

7.

9.

SECTION 9.4

1. a. 23/20   b. 17/($^-$35)   c. $^-$29/117   d. 57/($^-$77)   e. 2/5   f. $^-$11/57
   g. 49/60   h. 3/35   i. 4/21   j. $^-$19/36

3. a. $\left(\dfrac{1}{2} + \dfrac{1}{3}\right) + \dfrac{1}{4} = \dfrac{(1)(3) + (2)(1)}{(2)(3)} + \dfrac{1}{4} = \dfrac{5}{6} + \dfrac{1}{4}$

   $= \dfrac{(5)(4) + (6)(1)}{(6)(4)} = \dfrac{26}{24} = \dfrac{13}{12}$

$$\frac{1}{2} + \left(\frac{1}{3} + \frac{1}{4}\right) = \frac{1}{2} + \frac{(1)(4) + (3)(1)}{(3)(4)} = \frac{1}{2} + \frac{7}{12}$$

$$= \frac{(1)(12) + (2)(7)}{(2)(12)} = \frac{26}{24} = \frac{13}{12}$$

b. $\dfrac{^-2}{5} + \left(\dfrac{1}{4} + \dfrac{^-2}{7}\right) = \dfrac{^-2}{5} + \dfrac{(1)(7) + (4)(^-2)}{(4)(7)} = \dfrac{^-2}{5} + \dfrac{^-1}{28}$

$$= \frac{(^-2)(28) + (5)(^-1)}{(5)(28)} = \frac{^-61}{140}$$

$\left(\dfrac{^-2}{5} + \dfrac{1}{4}\right) + \dfrac{^-2}{7} = \dfrac{(^-2)(4) + (5)(1)}{(5)(4)} + \dfrac{^-2}{7} = \dfrac{^-3}{20} + \dfrac{^-2}{7}$

$$= \frac{(^-3)(7) + (20)(^-2)}{(20)(7)} = \frac{^-61}{140}$$

c. $\left(\dfrac{3}{10} + \dfrac{0}{4}\right) + \dfrac{^-1}{2} = \dfrac{(3)(4) + (10)(0)}{(10)(4)} + \dfrac{^-1}{2} = \dfrac{12}{40} + \dfrac{^-1}{2}$

$$= \frac{(12)(2) + (40)(^-1)}{(40)(2)} = \frac{^-16}{80} = \frac{^-1}{5}$$

$\dfrac{3}{10} + \left(\dfrac{0}{4} + \dfrac{^-1}{2}\right) = \dfrac{3}{10} + \dfrac{(0)(2) + (4)(^-1)}{(4)(2)} = \dfrac{3}{10} + \dfrac{^-4}{8}$

$$= \frac{(3)(8) + (10)(^-4)}{(10)(8)} = \frac{^-16}{80} = \frac{^-1}{5}$$

5. a. $1\frac{11}{12}$ b. $3\frac{1}{12}$ c. $3\frac{2}{17}$ d. 7   e. $11\frac{14}{17}$ f. $27\frac{17}{25}$ g. $^-2\frac{5}{11}$ h. $^-3\frac{11}{26}$ i. $^-3\frac{25}{57}$

SECTION 9.5

1. a. 8/21   b. 2/35   c. $^-$20/117   d. $^-$6/77   e. 4/15   f. 0/1   g. $^-$1/15
   h. 9/35   i. $^-$3/175   j. $^-$1/6

3. a. $\dfrac{2}{3} \times \dfrac{4}{5} = \dfrac{(2)(4)}{(3)(5)} = \dfrac{(4)(2)}{(5)(3)} = \dfrac{4}{5} \times \dfrac{2}{3}$

b. $\dfrac{6}{7} \times \dfrac{0}{3} = \dfrac{(6)(0)}{(7)(3)} = \dfrac{(0)(6)}{(3)(7)} = \dfrac{0}{3} \times \dfrac{6}{7}$

c. $\dfrac{7}{11} \times \dfrac{^-1}{2} = \dfrac{(7)(^-1)}{(11)(2)} = \dfrac{(^-1)(7)}{(2)(11)} = \dfrac{^-1}{2} \times \dfrac{7}{11}$

d. $\dfrac{^-2}{5} \times \dfrac{^-3}{4} = \dfrac{(^-2)(^-3)}{(5)(4)} = \dfrac{(^-3)(^-2)}{(4)(5)} = \dfrac{^-3}{4} \times \dfrac{^-2}{5}$

5. a. $\dfrac{2}{3} \times \left(\dfrac{1}{6} + \dfrac{2}{5}\right) = \dfrac{2}{3} \times \dfrac{17}{30} = \dfrac{17}{45}; \left(\dfrac{2}{3} \times \dfrac{1}{6}\right) + \left(\dfrac{2}{3} \times \dfrac{2}{5}\right) = \dfrac{1}{9} + \dfrac{4}{15} = \dfrac{51}{135} = \dfrac{17}{45}$

b. $\dfrac{^-1}{6} \times \left(\dfrac{2}{3} + \dfrac{^-3}{4}\right) = \dfrac{^-1}{6} \times \dfrac{^-1}{12} = \dfrac{1}{72}; \left(\dfrac{1}{6} \times \dfrac{2}{3}\right) + \left(\dfrac{^-1}{6} \times \dfrac{^-3}{4}\right) = \dfrac{^-1}{9} + \dfrac{1}{8} = \dfrac{1}{72}$

c. $\dfrac{^-3}{4} \times \left(\dfrac{^-1}{2} + \dfrac{^-2}{3}\right) = \dfrac{^-3}{4} \times \dfrac{^-7}{6} = \dfrac{7}{8}; \left(\dfrac{^-3}{4} \times \dfrac{^-1}{2}\right) + \left(\dfrac{^-3}{4} \times \dfrac{^-2}{3}\right) = \dfrac{3}{8} + \dfrac{1}{2} = \dfrac{7}{8}$

7. a. 0/1   b. All integers, excepting 1 and ⁻1.   c. All whole numbers except 1.   d. All natural numbers except 1.

SECTION 9.6

1. a. 5/12   b. ⁻1/14   c. ⁻26/15   d. 2/9   e. ⁻5/7   f. 0/1   g. 6/5   h. ⁻4/45
i. 0/1   j. 13/21
3. a. 12/5   b. 14/9   c. ⁻6/7   d. 5/4   e. 0/1   f. 1/1   g. 1/1   h. 121/441
i. 78/55   j. ⁻34/117
5. a. No solution   b. 2   c. No solution   d. No solution   e. 5
f. No solution   g. No solution   h. No solution   i. No solution   j. 2
7. a. 1/2   b. 2/1   c. ⁻3/1   d. ⁻1/1   e. 5/1   f. 5/2   g. 7/2   h. ⁻1/4
i. 5/2   j. 2/1

9. a. $\left(\dfrac{a}{b} + \dfrac{c}{d}\right) - \dfrac{e}{f} = \left(\dfrac{1}{2} + \dfrac{1}{3}\right) - \dfrac{1}{4} = \dfrac{5}{6} - \dfrac{1}{4} = \dfrac{7}{12}$

$\dfrac{a}{b} + \left(\dfrac{c}{d} - \dfrac{e}{f}\right) = \dfrac{1}{2} + \left(\dfrac{1}{3} - \dfrac{1}{4}\right) = \dfrac{1}{2} + \dfrac{1}{12} = \dfrac{7}{12}$

b. $\left(\dfrac{a}{b} + \dfrac{c}{d}\right) - \dfrac{e}{f} = \left(\dfrac{^-2}{5} + \dfrac{^-3}{4}\right) - \dfrac{^-1}{7} = \dfrac{^-23}{20} - \dfrac{^-1}{7} = \dfrac{^-141}{140}$

$\dfrac{a}{b} + \left(\dfrac{c}{d} - \dfrac{e}{f}\right) = \dfrac{^-2}{5} + \left(\dfrac{^-3}{4} - \dfrac{^-1}{7}\right) = \dfrac{^-2}{5} + \dfrac{^-17}{28} = \dfrac{^-141}{140}$

c. $\left(\dfrac{a}{b} + \dfrac{c}{d}\right) - \dfrac{e}{f} = \left(\dfrac{2}{11} + \dfrac{^-1}{9}\right) - \dfrac{^-3}{7} = \dfrac{7}{99} - \dfrac{^-3}{7} = \dfrac{346}{693}$

$\dfrac{a}{b} + \left(\dfrac{c}{d} - \dfrac{e}{f}\right) = \dfrac{2}{11} + \left(\dfrac{^-1}{9} - \dfrac{^-3}{7}\right) = \dfrac{2}{11} + \dfrac{20}{63} = \dfrac{346}{693}$

d. $\left(\dfrac{a}{b} + \dfrac{c}{d}\right) - \dfrac{e}{f} = \left(\dfrac{4}{7} + \dfrac{0}{2}\right) - \dfrac{0}{^-3} = \dfrac{4}{7} - \dfrac{0}{^-3} = \dfrac{4}{7}$

$\dfrac{a}{b} + \left(\dfrac{c}{d} - \dfrac{e}{f}\right) = \dfrac{4}{7} + \left(\dfrac{0}{2} - \dfrac{0}{^-3}\right) = \dfrac{4}{7} + \dfrac{0}{1} = \dfrac{4}{7}$

SECTION 9.7

1. a. True b. False c. True d. False e. False f. True g. True
   h. False i. True j. True
3. a. > b. < c. = d. < e. < f. = g. > h. < i. > j. >

7. a. True b. False, $\dfrac{2}{3} + \dfrac{0}{1} \geq \dfrac{2}{3} + \dfrac{0}{1}$ but $\dfrac{2}{3} \not> \dfrac{2}{3}$.

   c. False, $\dfrac{1}{2} < \dfrac{2}{3}$ but $\dfrac{^-1}{2} \times \dfrac{1}{2} > \dfrac{^-1}{2} \times \dfrac{2}{3}$. d. True e. False, $p = 0$.

REVIEW EXERCISES FOR CHAPTER 9

1. a. $^-2/21$ b. $29/28$ c. $20/99$ d. $39/28$ e. $186/455$ f. $57/28$
   g. $202/231$ h. $^-199/99$ i. $49/30$ j. $63/5$
3. a. $x < {}^-1$ b. $x \neq 0$ c. $x \geq 7$ d. $x > 1$ e. $x \leq 0$ f. $x < 0$
5. a. $3/2$ b. $^-4/1$ c. $5/3$ d. $^-4/1$ e. $7/30$ f. Does not exist
7. a. $2/3, 2/5, 1/4, 1/10$ b. $^-1/9, {}^-1/5, {}^-2/7, 3/({}^-4)$
   c. $0/({}^-4), {}^-2/3, 4/({}^-5), {}^-11/9$ d. $12/25, {}^-6/125, {}^-1/5, {}^-201/625$
9. a. False b. False c. False d. False e. False f. True g. False
   h. True i. False j. True

CHAPTER 10

SECTION 10.2

1. a. Irrational b. Rational c. Irrational d. Irrational e. Irrational
   f. Irrational g. Rational h. Irrational i. Rational j. Irrational
   k. Irrational l. Rational m. Irrational n. Rational o. Rational
3. a. $2 \in Q$ b. $3 \in Q$ c. $11 \in Q$ d. $19 \in Q$

SECTION 10.2

1. a. .2 b. 1.3 c. 2.3 d. 3.4 e. 4.1 f. 5.3 g. 7.2 h. 8.9 i. 10.1
   j. 23.1 k. 46.9 l. 61.3 m. 82.9 n. 99.9 o. 102.3
3. a. $13/10$ b. $112/10$ c. $239/10$ d. $213/100$ e. $4619/100$ f. $5727/100$
   g. $14123/1000$ h. $23296/1000$ i. $37307/1000$ j. $40029/1000$
   k. $57167/1000$ l. $70002/1000$ 5. b, d, e, g, h, i

SECTION 10.4

1. a. 25.406 b. 12.68 c. 16.327 d. 125.901 e. 165.4 f. ⁻268.533
g. 149.129 h. 449.06 i. ⁻29.537 j. 349.519
3. a. 13.6 b. 2.01 c. 23.2 d. 46.2 e. 1.07 f. 14.4 g. 4.2 h. 4.23
i. 13,700 j. .7
5. a. .49 b. .67 c. .191 d. .8768 e. .999 f. 1.21 g. 1.324 h. .003
i. .0019 j. .00019 k. .00001 l. 12.47 m. 23.178 n. .00999
7. $9,900   9. $19,019

SECTION 10.5

1. a. $.\overline{6}$ b. $.25$ c. $.\overline{285714}$ d. $2.\overline{72}$ e. $.\overline{2352941176470588}$
f. $.125$ g. $.\overline{2}$ h. $.\overline{230769}$ i. $.\overline{45}$ j. $.\overline{571428}$ k. $.\overline{105263157894736842}$
l. $.8\overline{3}$ m. $.75$ n. $.6875$ o. $.8\overline{1}$ p. $.5$ q. $.\overline{441860465116279069767}$
r. $.5$
3. a. 1/9 b. 23/99 c. 145/999 d. 104/333 e. 532/3333 f. 2021/9999
g. 397/9990 h. 13/5550 i. 23/330000 j. 41111/333000
k. 118739/49500 l. 28465/1998
5. a. No b. No c. No, it is neither a repeating nor terminating decimal.

SECTION 10.6

1. a. 201.044_{seven} b. 11.716_{eight} c. $6994E.3E2_{twelve}$ d. 30.031_{five}
e. 11110.0001_{two}

SECTION 10.7

1. a. Irrational b. Irrational c. Irrational d. Rational e. Rational
f. Irrational g. Irrational h. Irrational i. Rational j. Rational
k. Rational l. Rational m. Irrational n. Rational o. Irrational
3. a. 1.41 b. 1.73 c. 2.24 d. 5.20 e. 5.57 f. 6.56 g. 10.05
h. 14.32 i. 17.32

SECTION 10.8

1. a. True b. False c. False d. True e. True f. False g. True
h. False i. True j. True
3. Any real number less than or equal to: a. ⁻3 b. $⁻2\frac{1}{2}$ c. ⁻2 d. 0
e. ⁻5 f. 0 g. 1 h. ⁻1 i. 4 j. ⁻2

REVIEW EXERCISES FOR CHAPTER 10

1. Addition, subtraction, multiplication   3. Addition, multiplication
5. 1   7. 0
9. A rational real number can be expressed as either a terminating or repeating decimal. An irrational real number cannot be expressed as either a terminating or repeating decimal.
11. No. Let $S = \{^-1, 2, 5, 7\}$. There is no element in S between 2 and 5.
15. No, $\sqrt{3} \cdot \sqrt{3} = 3 \in Q.$ 17. About 1570
19. By Theorem 10.2 a nonrepeating, nonterminating decimal since $23 = 1 \times 23$.

CHAPTER 11

SECTION 11.1

1. a.
	A	B
R	6	4
S	4	7
T	8	6

b.
	R	S	T
A	6	4	8
B	4	7	6

3. a. $\begin{bmatrix} 0 & 0 \\ 0 & 0 \\ 0 & 0 \end{bmatrix}$ b. $\begin{bmatrix} 0 & 0 & 0 \\ 0 & 0 & 0 \\ 0 & 0 & 0 \end{bmatrix}$

5. a. Yes; 1×4 b. Yes; 3×1 7. a. 2 b. 0 c. Does not exist
9. a. $c_{11} = 3, c_{22} = 2, c_{33} = 1$ b. $c_{31} = 0, c_{22} = 2, c_{13} = 1$
11. a. 1 b. c c. 57

SECTION 11.2

1. No, A is of order 3×2 and B is of order 2×3.
5. No, their orders are different.

SECTION 11.3

1. a. $\begin{bmatrix} 1 & 2 \\ -3 & 2 \end{bmatrix}$ b. $\begin{bmatrix} 2 & 0 \\ 3 & -1 \end{bmatrix}$ c. $\begin{bmatrix} 3 & 0 \\ -3 & 2 \end{bmatrix}$ d. $\begin{bmatrix} 3 & -1 \\ 2 & 0 \end{bmatrix}$

e. $\begin{bmatrix} 6 & -1 \\ -1 & 2 \end{bmatrix}$ f. $\begin{bmatrix} -2 & 3 \\ 1 & 0 \end{bmatrix}$ g. $\begin{bmatrix} 1 & 3 \\ -2 & 2 \end{bmatrix}$

3. a. $(A + B) + C = \left(\begin{bmatrix} 2 & 0 \\ -3 & 1 \end{bmatrix} + \begin{bmatrix} -1 & 2 \\ 0 & 1 \end{bmatrix}\right) + \begin{bmatrix} 3 & -1 \\ 2 & 0 \end{bmatrix}$

$= \begin{bmatrix} 1 & 2 \\ -3 & 2 \end{bmatrix} + \begin{bmatrix} 3 & -1 \\ 2 & 0 \end{bmatrix} = \begin{bmatrix} 4 & 1 \\ -1 & 2 \end{bmatrix};$

$A + (B + C) = \begin{bmatrix} 2 & 0 \\ -3 & 1 \end{bmatrix} + \left(\begin{bmatrix} -1 & 2 \\ 0 & 1 \end{bmatrix} + \begin{bmatrix} 3 & -1 \\ 2 & 0 \end{bmatrix}\right)$

$= \begin{bmatrix} 2 & 0 \\ -3 & 1 \end{bmatrix} + \begin{bmatrix} 2 & 1 \\ 2 & 1 \end{bmatrix} = \begin{bmatrix} 4 & 1 \\ -1 & 2 \end{bmatrix}$

b. $(B + D) + E = \left(\begin{bmatrix} -1 & 2 \\ 0 & 1 \end{bmatrix} + \begin{bmatrix} -1 & 1 \\ 1 & -1 \end{bmatrix}\right) + \begin{bmatrix} 1 & 0 \\ 0 & 1 \end{bmatrix}$

$= \begin{bmatrix} -2 & 3 \\ 1 & 0 \end{bmatrix} + \begin{bmatrix} 1 & 0 \\ 0 & 1 \end{bmatrix} = \begin{bmatrix} -1 & 3 \\ 1 & 1 \end{bmatrix}$

$B + (D + E) = \begin{bmatrix} -1 & 2 \\ 0 & 1 \end{bmatrix} + \left(\begin{bmatrix} -1 & 1 \\ 1 & -1 \end{bmatrix} + \begin{bmatrix} 1 & 0 \\ 0 & 1 \end{bmatrix}\right)$

$= \begin{bmatrix} -1 & 2 \\ 0 & 1 \end{bmatrix} + \begin{bmatrix} 0 & 1 \\ 1 & 0 \end{bmatrix} = \begin{bmatrix} -1 & 3 \\ 1 & 1 \end{bmatrix}$

c. $(C + E) + F = \left(\begin{bmatrix} 3 & -1 \\ 2 & 0 \end{bmatrix} + \begin{bmatrix} 1 & 0 \\ 0 & 1 \end{bmatrix}\right) + \begin{bmatrix} 0 & 0 \\ 0 & 0 \end{bmatrix}$

$= \begin{bmatrix} 4 & -1 \\ 2 & 1 \end{bmatrix} + \begin{bmatrix} 0 & 0 \\ 0 & 0 \end{bmatrix} = \begin{bmatrix} 4 & -1 \\ 2 & 1 \end{bmatrix}$

$C + (E + F) = \begin{bmatrix} 3 & -1 \\ 2 & 0 \end{bmatrix} + \left(\begin{bmatrix} 1 & 0 \\ 0 & 1 \end{bmatrix} + \begin{bmatrix} 0 & 0 \\ 0 & 0 \end{bmatrix}\right)$

$= \begin{bmatrix} 3 & -1 \\ 2 & 0 \end{bmatrix} + \begin{bmatrix} 1 & 0 \\ 0 & 1 \end{bmatrix} = \begin{bmatrix} 4 & -1 \\ 2 & 1 \end{bmatrix}$

d. $(A + C) + E = \left(\begin{bmatrix} 2 & 0 \\ -3 & 1 \end{bmatrix} + \begin{bmatrix} 3 & -1 \\ 2 & 0 \end{bmatrix}\right) + \begin{bmatrix} 1 & 0 \\ 0 & 1 \end{bmatrix}$

$= \begin{bmatrix} 5 & -1 \\ -1 & 1 \end{bmatrix} + \begin{bmatrix} 1 & 0 \\ 0 & 1 \end{bmatrix} = \begin{bmatrix} 6 & -1 \\ -1 & 2 \end{bmatrix}$

$A + (C + E) = \begin{bmatrix} 2 & 0 \\ -3 & 1 \end{bmatrix} + \left(\begin{bmatrix} 3 & -1 \\ 2 & 0 \end{bmatrix} + \begin{bmatrix} 1 & 0 \\ 0 & 1 \end{bmatrix}\right)$

$= \begin{bmatrix} 2 & 0 \\ -3 & 1 \end{bmatrix} + \begin{bmatrix} 4 & -1 \\ 2 & 1 \end{bmatrix} = \begin{bmatrix} 6 & -1 \\ -1 & 2 \end{bmatrix}$

5. $T_{m \times n}$

7.

A	B	C	
431	350	336	Skilled labor (in hours)
535	380	489	Semi-skilled labor (in hours)
523	474	613	Unskilled labor (in hours)

SECTION 11.4

1. a. $\begin{bmatrix} 4 & 4 & 2 \\ -2 & 0 & 4 \end{bmatrix}$ b. $\begin{bmatrix} -3 & 0 & -12 \\ 0 & 6 & -9 \end{bmatrix}$ c. $\begin{bmatrix} 0 & 1 & -1/2 \\ 3/2 & 1/2 & 1 \end{bmatrix}$

d. $\begin{bmatrix} -3 & 1 & -1 \\ -1 & 0 & -2 \end{bmatrix}$ e. $\begin{bmatrix} 7 & 4 & 14 \\ -2 & -6 & 13 \end{bmatrix}$ f. $\begin{bmatrix} 3 & -8 & 16 \\ -12 & -10 & 1 \end{bmatrix}$

g. $\begin{bmatrix} 15 & -1 & 3 \\ 11 & 2 & 14 \end{bmatrix}$ h. $\begin{bmatrix} 8 & -8 & 2 \\ 6 & 0 & 4 \end{bmatrix}$ i. $\begin{bmatrix} -2 & 5 & -11 \\ -3 & 6 & -7 \end{bmatrix}$

j. $\begin{bmatrix} -9 & 10 & -18 \\ 10 & 10 & -5 \end{bmatrix}$

3. a. $x = 4, y = 1, u = 3, v = 0$ b. $u < 3, x > 2, y < 1$

SECTION 11.5

1. a. Not conformable for matrix multiplication   b. 3×1 c. 4×4
d. 3×4 e. Not conformable for matrix addition   f. 4×1 g. 3×1
h. Not conformable for matrix addition   i. 4×4 j. 4×1 k. 3×3
l. 3×1

3. $\begin{bmatrix} 2 & -3 \\ 1 & 2 \end{bmatrix} \begin{bmatrix} x \\ y \end{bmatrix} = \begin{bmatrix} 1 \\ -3 \end{bmatrix}$

5. $(A + B)^2 = \left(\begin{bmatrix} 1 & 0 \\ 2 & -1 \end{bmatrix} + \begin{bmatrix} -1 & 1 \\ 1 & 2 \end{bmatrix} \right)^2 = \begin{bmatrix} 0 & 1 \\ 3 & 1 \end{bmatrix}^2 = \begin{bmatrix} 3 & 1 \\ 3 & 4 \end{bmatrix}$

$A^2 + 2AB + B^2 = \begin{bmatrix} 1 & 0 \\ 2 & -1 \end{bmatrix}^2 + 2 \begin{bmatrix} 1 & 0 \\ 2 & -1 \end{bmatrix} \begin{bmatrix} -1 & 1 \\ 1 & 2 \end{bmatrix} + \begin{bmatrix} -1 & 1 \\ 1 & 2 \end{bmatrix}^2$

$= \begin{bmatrix} 1 & 0 \\ 0 & 1 \end{bmatrix} + 2 \begin{bmatrix} -1 & 1 \\ -3 & 0 \end{bmatrix} + \begin{bmatrix} 2 & 1 \\ 1 & 5 \end{bmatrix}$

$= \begin{bmatrix} 1 & 0 \\ 0 & 1 \end{bmatrix} + \begin{bmatrix} -2 & 2 \\ -6 & 0 \end{bmatrix} + \begin{bmatrix} 2 & 1 \\ 1 & 5 \end{bmatrix} = \begin{bmatrix} 1 & 3 \\ -5 & 6 \end{bmatrix} \neq \begin{bmatrix} 3 & 1 \\ 3 & 4 \end{bmatrix}$

7. a. $A(B + C) = \begin{bmatrix} 1 & 2 & ^-1 \\ 0 & 1 & 2 \\ 3 & 2 & 1 \end{bmatrix}\left(\begin{bmatrix} 1 & 0 \\ ^-1 & 2 \\ 2 & 1 \end{bmatrix} + \begin{bmatrix} 2 & ^-1 \\ 0 & 1 \\ 1 & 3 \end{bmatrix}\right)$

$= \begin{bmatrix} 1 & 2 & ^-1 \\ 0 & 1 & 2 \\ 3 & 2 & 1 \end{bmatrix}\begin{bmatrix} 3 & ^-1 \\ ^-1 & 3 \\ 3 & 4 \end{bmatrix} = \begin{bmatrix} ^-2 & 1 \\ 5 & 11 \\ 10 & 7 \end{bmatrix}$

$AB + AC = \left(\begin{bmatrix} 1 & 2 & ^-1 \\ 0 & 1 & 2 \\ 3 & 2 & 1 \end{bmatrix}\begin{bmatrix} 1 & 0 \\ ^-1 & 2 \\ 2 & 1 \end{bmatrix}\right) + \left(\begin{bmatrix} 1 & 2 & ^-1 \\ 0 & 1 & 2 \\ 3 & 2 & 1 \end{bmatrix}\begin{bmatrix} 2 & ^-1 \\ 0 & 1 \\ 1 & 3 \end{bmatrix}\right)$

$= \begin{bmatrix} ^-3 & 3 \\ 3 & 4 \\ 3 & 5 \end{bmatrix} + \begin{bmatrix} 1 & ^-2 \\ 2 & 7 \\ 7 & 2 \end{bmatrix} = \begin{bmatrix} ^-2 & 1 \\ 5 & 11 \\ 10 & 7 \end{bmatrix}$

b. $A(B + C) = \begin{bmatrix} 1 & 2 \\ 3 & 4 \end{bmatrix}\left(\begin{bmatrix} 1 & 2 & 3 \\ 3 & 2 & ^-1 \end{bmatrix} + \begin{bmatrix} 4 & 0 & 2 \\ ^-1 & 2 & 3 \end{bmatrix}\right)$

$= \begin{bmatrix} 1 & 2 \\ 3 & 4 \end{bmatrix}\begin{bmatrix} 5 & 2 & 5 \\ 2 & 4 & 2 \end{bmatrix} = \begin{bmatrix} 9 & 10 & 9 \\ 23 & 22 & 23 \end{bmatrix}$

$AB + AC = \left(\begin{bmatrix} 1 & 2 \\ 3 & 4 \end{bmatrix}\begin{bmatrix} 1 & 2 & 3 \\ 3 & 2 & ^-1 \end{bmatrix}\right) + \left(\begin{bmatrix} 1 & 2 \\ 3 & 4 \end{bmatrix}\begin{bmatrix} 4 & 0 & 2 \\ ^-1 & 2 & 3 \end{bmatrix}\right)$

$= \begin{bmatrix} 7 & 6 & 1 \\ 15 & 14 & 5 \end{bmatrix} + \begin{bmatrix} 2 & 4 & 8 \\ 8 & 8 & 18 \end{bmatrix} = \begin{bmatrix} 9 & 10 & 9 \\ 23 & 22 & 23 \end{bmatrix}$

SECTION 11.6

1. a. $\begin{bmatrix} 1 & 2 \\ 2 & ^-1 \end{bmatrix}$ b. $\begin{bmatrix} 4 & 1 \\ 2 & 1 \end{bmatrix}$ c. $\begin{bmatrix} 1 & ^-1 & 2 \\ 2 & 1 & ^-2 \\ 3 & 2 & ^-4 \end{bmatrix}$ d. $\begin{bmatrix} 2 & 1 & ^-3 \\ 1 & ^-3 & 1 \\ 2 & 1 & 2 \\ 3 & ^-2 & ^-1 \end{bmatrix}$

3. a. Interchanging rows b. Adding ($^-1$) times row 1 to row 2
   c. Multiplying row 1 by $\frac{1}{2}$ and row 2 by ($^-2$)
   d. Interchanging rows 1 and 3 and multiplying row 2 by ($^-3$)

SECTION 11.7

1. $\begin{bmatrix} ^-2 & 1 \\ 3/2 & ^-1/2 \end{bmatrix}$ 3. $\begin{bmatrix} 1 & 0 \\ 0 & 1 \end{bmatrix}$ 5. $\begin{bmatrix} 1 & 0 & 0 \\ 0 & 1 & 0 \\ 0 & 0 & 1 \end{bmatrix}$

7. $\begin{bmatrix} 1/7 & 2/7 & 0 \\ ^-1/7 & ^-2/7 & 1 \\ 2/7 & ^-3/7 & 0 \end{bmatrix}$ 9. $\begin{bmatrix} 0 & 2/3 & ^-1/3 \\ 1/3 & ^-10/9 & 2/9 \\ 0 & 1/3 & 1/3 \end{bmatrix}$

REVIEW EXERCISES FOR CHAPTER 11

1. a. $\begin{bmatrix} 3 & ^-1 \\ 4 & 4 \end{bmatrix}$ b. $\begin{bmatrix} 4 & 5 & 4 \\ 10 & 9 & 8 \end{bmatrix}$ c. $\begin{bmatrix} ^-10 & ^-1 \\ ^-9 & ^-2 \\ ^-9 & ^-8 \end{bmatrix}$

d. 2E and F are not conformable for matrix multiplication.
e. D and F are not conformable for matrix addition.

f. $\begin{bmatrix} ^-10 & ^-37 & ^-24 \\ ^-26 & ^-29 & ^-24 \end{bmatrix}$ g. $\begin{bmatrix} 12 & 4 & ^-2 & 16 \\ 20 & 4 & 2 & 12 \\ 32 & 8 & 0 & 28 \end{bmatrix}$

h. $\begin{bmatrix} ^-42 & ^-5 & ^-11 & ^-14 \\ 0 & 2 & ^-4 & 5 \end{bmatrix}$ $\begin{bmatrix} 9 & ^-4 & ^-17 \\ 15 & ^-5 & ^-30 \\ 21 & ^-3 & ^-46 \end{bmatrix}$

j. GD and H are not conformable for matrix multiplication.
3. Yes, the sum of any two matrices of order $m \times n$ is always a matrix of order $m \times n$.   5. The zero matrix of order 4×3.
7. No, the zero matrix of order 2 does not have a multiplicative inverse.
9. a. $u = \frac{2}{3}, v = \frac{4}{3}, x = 6, y = 3$ b. $u = \frac{1}{2}, v = 1, x = 3, y = 5$
 c. $x \geq \frac{3}{2}, y \geq 3$ d. $u \geq \frac{2}{3}, v \geq \frac{1}{3}, x \leq \frac{5}{2}, y \leq ^-3$
11. $x = ^-23/15, y = ^-4/5, z = 5/3$

CHAPTER 12

SECTION 12.1

1. No, a mathematical system must have at least one operation defined on the elements of the nonempty set.   3. a, e, h, j
5. a. Yes, for all $a, b \in Re, a * b = a + b + 1 \in Re$.
 b. Yes, for all $a, b \in Re, a * b = a + b + 1 = b + a + 1 = b * a$.
 c. Yes, for all $a, b, c \in Re, (a * b) * c = (a * b) + c + 1 =$
 $(a + b + 1) + c + 1 = a + (b + c + 1) + 1 = a * (b * c)$.
 d. (1) 6   (2) 8   (3) 7/4   (4) 7   (5) 10

SECTION 12.2

1. a. A group   b. Not a group; property 2   c. Not a group; properties 3 and 4   d. A group   e. Not a group; properties 1 and 3
 f. A group   g. Not a group; property 1   h. Not a group; property 4
3. f

SECTION 12.3

1. a. Not a field; property 10   b. A field   c. Not a field; property 1
d. Not a field; properties 9 and 10   e. Not a field; property 7
f. A field   g. Not a field; property 10

SECTION 12.4

1. a. A ring   b. A ring   c. Not a ring; property 1   d. A ring   e. A ring
f. A ring   g. A ring

SECTION 12.5

1. a. 5   b. 10   c. 11   d. 12   e. 3   f. 7   g. 2   h. 12   i. 11   j. 10

3.

$\oplus$	1	2	3	4	5
1	2	3	4	5	1
2	3	4	5	1	2
3	4	5	1	2	3
4	5	1	2	3	4
5	1	2	3	4	5

$\otimes$	1	2	3	4	5
1	1	2	3	4	5
2	2	4	1	3	5
3	3	1	4	2	5
4	4	3	2	1	5
5	5	5	5	5	5

a. 2   b. 2   c. 3   d. 3   e. 2   f. 3   g. No solution   h. 1   i. 3   j. 1
5. a. 9:30 A.M.   b. 10:40 A.M.   c. 2:00 P.M.   d. 4:10 P.M.   e. 0:01 A.M.
f. 1:45 P.M.   g. 10:15 P.M.   h. 12:00 P.M.   i. 8:00 A.M.   j. 3:50 A.M.
7. a. December b. September   c. April   d. January   e. None   f. May
g. August   h. July   i. April   j. July

SECTION 12.6

1. a. True   b. True   c. False   d. True   e. False   f. False   g. True
3. Infinitely many
5. a. $3 \times 2 \equiv 3 \times 7 \pmod 5$ since $5 \mid {}^-15$; g.c.d. $(3, 5) = 1$;
$2 \equiv 7 \pmod 5$ since $5 \mid {}^-5$.
b. $4 \times 3 \equiv 4 \times 10 \pmod 7$ since $7 \mid {}^-28$ g.c.d. $(4, 7) = 1$;
$3 \equiv 10 \pmod 7$ since $7 \mid {}^-7$.
c. $7 \times 2 \equiv 7 \times {}^-4 \pmod 6$ since $6 \mid 42$; g.c.d. $(7, 6) = 1$;
$2 \equiv {}^-4 \pmod 6$ since $6 \mid 6$.
d. $8 \times 5 \equiv 8 \times 2 \pmod 3$ since $3 \mid 24$; g.c.d. $(8, 3) = 1$;
$5 \equiv 2 \pmod 3$ since $3 \mid 3$.
7. a. Probably correct   b. Probably correct   c. Probably correct
d. Probably correct   e. Definitely incorrect   f. Probably correct
g. Probably correct   h. Probably correct   i. Definitely incorrect
j. Probably correct

9. a. 701   b. 11,797   c. 39,273   d. 174,986   e. 10,038,132   f. 3,582
   g. 4,132   h. 41,993   i. 48,378   j. 17,250,835

SECTION 12.7

1. H 3. R_2 5. D_1 7. I 9. H 11. D_2 13. D_2
15. R_2 17. D_2 19. I

REVIEW EXERCISES FOR CHAPTER 12

1. a. False   b. False   c. True   d. False   e. True   f. True   g. True
   h. False   i. False   j. False
3. Yes, Definition 12.6 is satisfied.
5. Yes, Definition 12.8 is satisfied if property 10 is replaced by the
cancellation property for multiplication.

CHAPTER 13

SECTION 13.1

1. a. Line; $\overleftrightarrow{AB}, \overleftrightarrow{AC}, \overleftrightarrow{BC}$ b. Ray; $\overrightarrow{PQ}, \overrightarrow{PR}$ c. Line segment; $\overline{CD}, \overline{DC}$
   d. Line; $l, \overleftrightarrow{ST}$ e. Ray; $\overrightarrow{CB}, \overrightarrow{CA}$ f. Line segment; $\overline{UV}, \overline{VU}$
3. Infinitely many   5. 1
9. a. False   b. False   c. False   d. True   e. True   f. True   g. True
   h. True   i. True   j. False

SECTION 13.2

1. False   3. True   5. True   7. True   9. True

SECTION 13.3

1. a. $\angle AOB, \angle BOA$ b. $\angle PRQ, \angle QRP$ c. $\angle RST, \angle TSR$
   d. $\angle ACD, \angle C$ e. $\angle V, \angle WVU$ f. $\angle PTF, \angle T$
3. a. $\angle AOB$ and $\angle BOC$, $\angle BOC$ and $\angle COD$, $\angle COD$ and $\angle AOD$,
   $\angle AOD$ and $\angle AOB$
   b. $\angle JGH$ and $\angle HGE$, $\angle HGE$ and $\angle CGE$, $\angle CGE$ and $\angle JGC$,
   $\angle JGC$ and $\angle JGH$, $\angle GEF$ and $\angle FED$, $\angle FED$ and $\angle DEC$,
   $\angle DEC$ and $\angle CEG$, $\angle CEG$ and $\angle GEF$, $\angle GBE$ and $\angle EBC$,
   $\angle EBC$ and $\angle CBA$, $\angle CBA$ and $\angle ABG$, $\angle ABG$ and $\angle GBE$
   c. $\angle MOV$ and $\angle UOP$, $\angle UOP$ and $\angle NOP$, $\angle NOP$ and $\angle MON$,
   $\angle MON$ and $\angle MOU$, $\angle SUT$ and $\angle TUL$, $\angle TUL$ and $\angle LUO$,
   $\angle LUO$ and $\angle OUS$, $\angle OUS$ and $\angle SUT$, $\angle ULD$ and $\angle DLE$,

∠ *DLE* and ∠ *ELP*, ∠ *ELP* and ∠ *PLU*, ∠ *PLU* and ∠ *ULD*,
∠ *OPL* and ∠ *LPR*, ∠ *LPR* and ∠ *RPQ*, ∠ *RPQ* and ∠ *QPO*,
∠ *QPO* and ∠ *OPL*
d. ∠ *GEF* and ∠ *FEH*, ∠ *FEH* and ∠ *HEB*, ∠ *HEB* and ∠ *BEG*,
∠ *BEG* and ∠ *GEF*, ∠ *ABE* and ∠ *EBC*, ∠ *EBC* and ∠ *CBD*,
∠ *CBD* and ∠ *DBA*, ∠ *DBA* and ∠ *ABE*

SECTION 13.4

1. a. A curve   b. A curve   c. Not a curve   d. A curve   e. A curve
   f. Not a curve   g. Not a curve   h. A curve   i. A curve   j. A curve
   k. A curve   l. A curve   m. A curve   n. A curve
3. a. Not closed   b. Closed   c. Not closed   d. Closed   e. Not closed
   f. Closed   g. Not closed   h. Closed   i. Closed   j. Not closed
   k. Closed   l. Not closed   m. Closed   n. Closed   o. Closed   p. Closed
   q. Closed   r. Not closed   s. Closed   t. Not closed
5. a. False   b. False   c. True   d. False   e. True   f. False   g. True
   h. False

SECTION 13.5

1. a. Broken line   b. Broken line   c. Not a broken line
   d. Not a broken line   e. Broken line   f. Not a broken line
   g. Broken line   h. Not a broken line   i. Broken line   j. Broken line
   k. Broken line   l. Broken line   m. Broken line   n. Not a broken line
   o. Broken line   p. Broken line   q. Broken line   r. Not a broken line
3. a, b, g, l, m, o   5. 3   7. Yes   9. Yes
11. Yes, into 3 subsets—the polygon, its interior, and its exterior.
13. Yes   15.

SECTION 13.6

1. a. 8   b. 5   c. 6   d. 7   e. 5   f. 4   g. 8   h. 9   i. 9   j. 6   k. 9   l. 8
   m. 6   n. 13   o. 4
3. a. True   b. False   c. False   d. False   e. False   f. True   g. True
   h. True   i. False   j. False
5. a. 22.2°   b. 39.1°   c. 56.4°   d. 102.5°   e. 68.7°   f. 100.9°   g. 86.6°
   h. 57.3°   i. 180°
7. a. False   b. True   c. True   d. False   e. True   f. True   g. True
   h. False   i. True   j. False

SECTION 13.7

1. a. Congruent   b. Not congruent   c. Congruent   d. Not congruent
   e. Congruent   f. Not congruent   g. Not congruent   h. Not congruent
3. a. Acute   b. Obtuse   c. Equilangular   d. Acute   e. Obtuse
   f. Equilangular   g. Right   h. Obtuse

SECTION 13.8

1. a. 12 sq in   b. 10.5 sq cm   c. 6 sq ft   d. 20 sq mm   e. 17.5 sq yd
   f. 14 sq units   g. 16.4 sq in   h. 30 sq m   i. Not a triangular region
3. a. 25 cm   b. 14.4 in   c. 30 ft   d. 28 yd   e. $28 + 2\sqrt{5}$ m
   f. $44 + 2\sqrt{5}$ units   g. $40 + \sqrt{34}$ units   h. 27 units   i. 43.6 units
5. a, c, d, e, f, h

SECTION 13.9

1. a. Similar   b. Similar   c. Similar   d. Not similar   e. Similar
   f. Not similar   3. 75°   5. $m(\overline{UV}) = \frac{1}{2}m(\overline{ST})$ 7. 13.5 units
9. 85,000 inches or approximately 1.34 miles   11. Yes

SECTION 13.10

1. a. False   b. True   c. False   d. False   e. True   f. True   g. False
   h. True   3. No

SECTION 13.11

1. a. 334 sq cm   b. 180 sq cm   c. 385 cu cm   d. 92 cm
3. a. 492.66 sq cm   b. 623.16 sq cm   c. 776.475 cu cm   d. 130.4 cm

REVIEW EXERCISES FOR CHAPTER 13

1. a. False   b. True   c. True   d. False   e. False   f. True   g. True
   h. True   i. False   j. False   k. True   l. False   m. True   n. True
   o. True   p. True   q. False   r. False   s. True   t. False   u. False
   v. False   w. False   x. True   y. True   z. False

CHAPTER 14

SECTION 14.1

1.

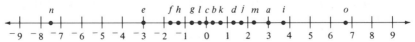

SECTION 14.2

1. a. 5 c. 3 e. $2 - \sqrt{2}$ f. 1/6 h. 5 i. 8 j. $\frac{11}{14}$
3. a. Horizontal; 2 b. Vertical; 11 c. Horizontal; $2\sqrt{3}$ e. Horizontal; 2
   g. Vertical; 2 h. Vertical; 2 i. Horizontal; 11
   j. Vertical; $^-1 + \sqrt{5} + \sqrt{2}$
5. a. $(^-3/2, 3)$ b. $(3, {}^-3/2)$ c. $(3, 5/2)$ d. $(3/2, 11/2)$ e. $(0, 1/2)$
   f. $((\sqrt{2} - 2)/2, {}^-3/2)$ g. $(4, {}^-3/2)$ h. $(5/2, 5/2)$ i. $(^-3/2, (\sqrt{5} - 1)/2)$
   j. $(^-5/2, {}^-2)$ k. $(13/2, {}^-7)$ l. $(3/8, 3/8)$ m. $(11/30, 25/126)$
   n. $(^-1/4, 1/4)$ o. $(^-1/2, 3/2)$

SECTION 14.3

1. a. $^-2/3$ b. $^-3/2$ c. Not defined d. 1/11 e. Not defined
   f. $\sqrt{3}/(2 + \sqrt{2})$ g. $^-3/8$ h. 1 i. $(\sqrt{5} + 1)/(^-7)$ j. 2
3. a. $y = {}^-3x + 2$ b. $2y = x$ c. $3y = 6x - 1$ d. $y = {}^-4$ e. $3y = {}^-2x + 3$
   f. $y = 2cx + 3p$ g. $y = 0$ h. $63y = {}^-9x + 14$
5. a. $2y = x - 6$ b. $2y = {}^-x + 3$ c. $x = 3$ d. $y = 2x - 7$
   e. $y = {}^-2x + 4$ f. $y = 4$ g. $y = {}^-3$ h. $x = 2$ i. $2x - 3y = {}^-6$
   j. $x = 2$
7. a. $3y = 2x - 19$ b. $2y = 3x + 6$ c. $3y = 2x - 5$
9. a. $3x - y - 4 = 0$ b. $5x + y - 1 = 0$ c. $3x - 2y + 2 = 0$
   d. $3x - y - 7 = 0$ e. $x - 2y - 3 = 0$ f. $4x - y - 2 = 0$
   g. $2x - y - 6 = 0$ h. $2x - 3y - 1 = 0$ i. $4x + 3y - 2 = 0$
   j. $5x + 2y - 3 = 0$
11. c, d, f

SECTION 14.4

1. a. $(x - 2)^2 + (y + 3)^2 = 9$ b. $(s + 4)^2 + (t - 1)^2 = 4$
   c. $u^2 + (v + 3)^2 = 3$ d. $(y - \sqrt{2})^2 + (z - 1)^2 = \frac{4}{9}$
   e. $(p - 2)^2 + q^2 = \frac{9}{49}$ f. $(m + 3)^2 + (n - \sqrt{5})^2 = 16$
3. a. Circle b. Circle c. Point d. No graph e. Circle f. Circle
5. a. 32π units b. 8π units c. $2\pi\sqrt{3}$ units f. 2π units

SECTION 14.5

1. a. $x = {}^-1, y = {}^-1$ b. $x = \frac{3}{5}, y = \frac{6}{5}$ c. $x = \frac{3}{4}, y = \frac{1}{8}$
   d. $x = 8/7, y = {}^-3/7$ e. $x = 0, y = {}^-1$ f. $x = \frac{20}{17}, y = \frac{15}{17}$
3. a. Two b. Two c. None d. Two e. One f. Two

REVIEW EXERCISES FOR CHAPTER 14

1. a. $\sqrt{34}$ b. 4 c. $\sqrt{10}$ d. $\sqrt{2}$ e. 3 f. $2\sqrt{2}$ g. $2\sqrt{13}$ h. $2\sqrt{13}$
   i. $\sqrt{178}$ j. $\sqrt{58}$

3. a. $y = 2x + 3$ b. $5x + 3y = 1$ c. $4x - 3y = 12$ d. $y = 2x$ e. $x = {}^-2$
5. $(x + 2)^2 + (y - 3)^2 = 9$ 7. $C(2, {}^-3), r = 2$

CHAPTER 15

SECTION 15.1

1. a. $x = 9$ b. $y = 2$ c. $u = {}^-2$ d. $w = {}^-9/5$ e. $p = {}^-2/9$ f. $t = 7/16$
   g. $x = {}^-31/82$ h. $y = 18/53$ i. $u = 75/82$ j. $m = 5$ k. $x = {}^-60/7$
   l. $t = 16/27$ m. $r = 123/12$ n. $x = 76/17$ o. $m = {}^-82/11$ p. $w = 9$
   q. $x = 29$ r. $y = 3$ s. $u = {}^-25/8$ t. $s = 19/37$
3. 9 and 11   5. 7 in. and 17 in.   7. 19 pennies, 12 dimes, 39 quarters
9. a. $^-35°$ b. $24\frac{4}{9}°$ c. $^-17\frac{7}{9}°$ d. $0°$ e. $32\frac{2}{9}°$ 11. $7\frac{2}{7}$ ft
13. 37.5 mph   15. $5\frac{5}{7}$ hr   17. 1.82 sec   19. 19,700   21. $14.60
23. 1525 reserved and 3205 general admission

SECTION 15.2

1.

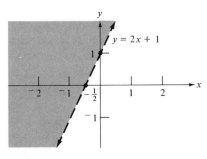

3.

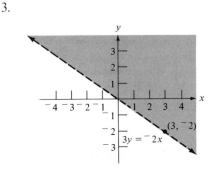

5.

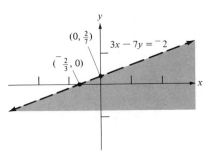

7.

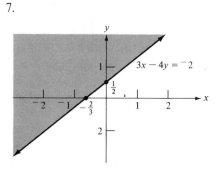

9.

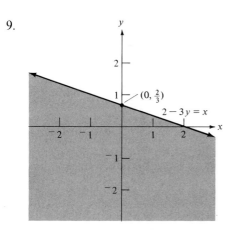

SECTION 15.3

1.

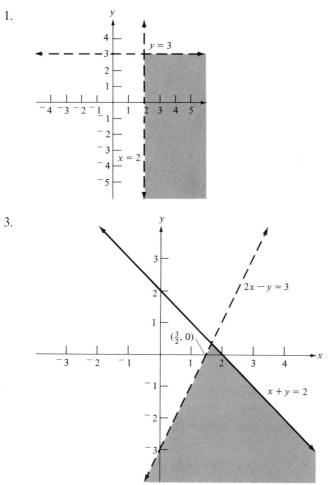

3.

5.

7.

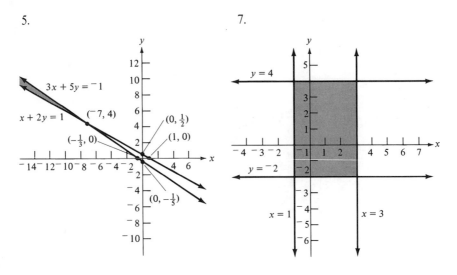

9.

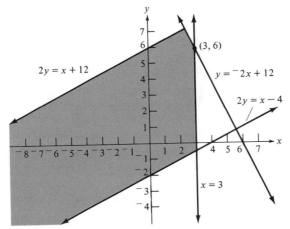

SECTION 15.4

1. a.

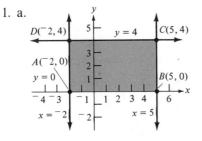

b.

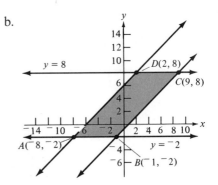

c.

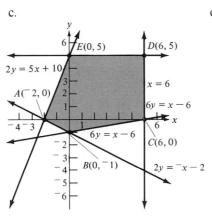

d.

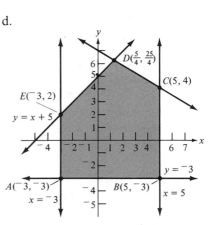

e.

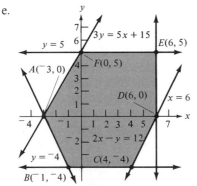

f.

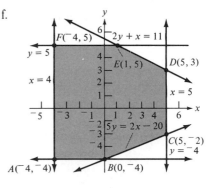

3. a. $A: F(^-2, 0) = ^-8$
   $B: F(5, 0) = 13$
   $C: F(5, 4) = 13$
   $D: F(^-2, 4) = ^-8$

   c. $A: F(^-2, 0) = ^-8$
   $B: F(0, ^-1) = ^-3$
   $C: F(6, 0) = 16$
   $D: F(6, 5) = 21$
   $E: F(0, 5) = 3$

   e. $A: F(^-3, 0) = ^-11$
   $B: F(^-1, ^-4) = ^-9$
   $C: F(4, ^-4) = 6$
   $D: F(6, 0) = 16$
   $E: F(6, 5) = 21$
   $F: F(0, 5) = 3$

   b. $A: F(^-8, ^-2) = ^-28$
   $B: F(^-1, ^-2) = ^-7$
   $C: F(9, 8) = 33$
   $D: F(2, 8) = 12$

   d. $A: F(^-3, ^-3) = ^-14$
   $B: F(5, ^-3) = 10$
   $C: F(5, 4) = 17$
   $D: F(5/4, 25/4) = 8$
   $E: F(^-3, 2) = ^-9$

   f. $A: F(^-4, ^-4) = ^-18$
   $B: F(0, ^-4) = ^-6$
   $C: F(5, ^-2) = 11$
   $D: F(5, 3) = 16$
   $E: F(1, 5) = 6$
   $F: F(^-4, 5) = ^-9$

5. a. $^-8$ at A and D b. $^-28$ at A c. $^-8$ at A d. $^-14$ at A
   e. $^-11$ at A f. $^-18$ at A

7. a. $A: H(^-2, 0) = 3$
   $B: H(5, 0) = 10$ (max)
   $C: H(5, 4) = 6$
   $D: H(^-2, 4) = ^-1$ (min)

   b. $A: H(^-8, ^-2) = ^-1$ (min)
   $B: H(^-1, ^-2) = 6$ (max)
   $C: H(9, 8) = 6$ (max)
   $D: H(2, 8) = ^-1$ (min)

   c. $A: H(^-2, 0) = 3$
   $B: H(0, ^-1) = 6$
   $C: H(6, 0) = 11$ (max)
   $D: H(6, 5) = 6$
   $E: H(0, 5) = 0$ (min)

   d. $A: H(^-3, ^-3) = 5$
   $B: H(5, ^-3) = 13$ (max)
   $C: H(5, 4) = 6$
   $D: H(5/4, 25/4) = 0$ (min)
   $E: H(^-3, 2) = 0$ (min)

   e. $A: H(^-3, 0) = 2$
   $B: H(^-1, ^-4) = 8$
   $C: H(4, ^-4) = 13$ (max)
   $D: H(6, 0) = 11$
   $E: H(6, 5) = 6$
   $F: H(0, 5) = 0$ (min)

   f. $A: H(^-4, ^-4) = 5$
   $B: H(0, ^-4) = 9$
   $C: H(5, ^-2) = 12$ (max)
   $D: H(5, 3) = 7$
   $E: H(1, 5) = 1$
   $F: H(^-4, 5) = ^-4$ (min)

REVIEW EXERCISES FOR CHAPTER 15

1. a. $x = ^-5$ b. $y = \frac{11}{2}$ c. $x = 24$ d. $u = \frac{3}{10}$ e. $t = \frac{1}{13}$ f. $v = \frac{37}{5}$

3. a.
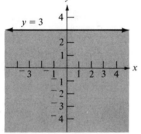

b.
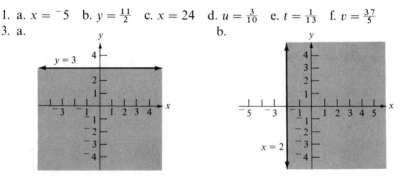

c.
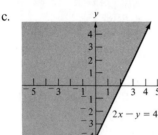

d.

e.
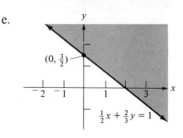

f.
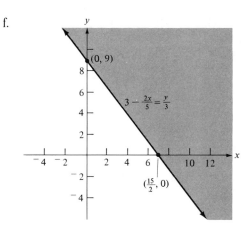

CHAPTER 16

SECTION 16.1

1. a. $S = \{1, 2, 3, 4\}$ b. $S = \{G, R, V, Y\}$ c. $S = \{A, B, C, D, E, F\}$
   d. $S = \{I, II, III\}$
3. a. $S = \{(1, G), (1, R), (1, V), (1, Y), (2, G), (2, R), (2, V), (2, Y), (3, G),$
   $(3, R), (3, V), (3, Y), (4, G), (4, R), (4, V), (4, Y)\}$
   b. $S = \{(1, A), (1, B), (1, C), (1, D), (1, E), (1, F), (2, A), (2, B), (2, C),$
   $(2, D), (2, E), (2, F), (3, A), (3, B), (3, C), (3, D), (3, E), (3, F), (4, A),$
   $(4, B), (4, C), (4, D), (4, E), (4, F)\}$
   c. $S = \{(1, I), (1, II), (1, III), (2, I), (2, II), (2, III), (3, I), (3, II), (3, III),$
   $(4, I), (4, II), (4, III)\}$
   d. $S = \{(G, A), (G, B), (G, C), (G, D), (G, E), (G, F), (R, A), (R, B),$
   $(R, C), (R, D), (R, E), (R, F), (V, A), (V, B), (V, C), (V, D), (V, E),$
   $(V, F), (Y, A), (Y, B), (Y, C), (Y, D), (Y, E), (Y, F)\}$
   e. $S = \{(G, I), (G, II), (G, III), (R, I), (R, II), (R, III), (V, I), (V, II),$
   $(V, III), (Y, I), (Y, II), (Y, III)\}$
   f. $S = \{(A, I), (A, II), (A, III), (B, I), (B, II), (B, III), (C, I), (C, II),$
   $(C, III), (D, I), (D, II), (D, III), (E, I), (E, II), (E, III), (F, I), (F, II),$
   $(F, III)\}$
5. a. $\{HHH, HHT, HTH, HTT, THH, THT, TTH, TTT\}$
   b. $\{HHH, HHT, HTH, HTT\}$ c. $\{HHT, HTH, THH\}$
   d. $\{HHT, HTH, HTT, THH, THT, TTH, TTT\}$
   e. $\{HHH, HTH, THT, TTT\}$ f. $\{HHH\}$ g. $\{THH, THT, TTH, TTT\}$
   h. $\{HHT, HTT, THH, TTH\}$ i. No, $A \cap B = \{HHT, HTH\} \neq \emptyset$.
   j. No, $D \cap E = \{HHH\} \neq \emptyset$.
7. $\{H1, H2, H3, H4, H5, H6, H7, H8, T1, T2, T3, T4, T5, T6, T7, T8\}$
9. a. $\{GGG, GGB, GBG, GBB, BGG, BGB, BBG, BBB\}$
   b. $\{GGG, GGB, BGG, BGB\}$ c. $\{GBB, BGB, BBG\}$ d. $\{GGB, BGG\}$

e. $\{GGB\}$ f. $\{GGG, BBB\}$ g. $\{GBB, BGB, BBG, BBB\}$
h. $\{GGG, GGB, GBG, BGG\}$
i. $\{GGG, GGB, GBG, GBB, BGG, BGB, BBG\}$
j. $\{GGG, BGB, BGB, BBB\}$

SECTION 16.2

1. a. 1/6 b. 5/36 c. 13/18 d. 5/6
3. a. 8/15 b. 7/15 c. 4/5 d. 11/15
5. a. 1/2 b. 3/8 c. 15/16 d. 11/16 e. 1/8 f. 3/8 g. 11/16

SECTION 16.3

1. Let A be the event "tail on the first toss" and B be the event "head on the second toss." Then $p(A) = 1/2$, $p(B) = 1/2$, $p(A \cap B) = 1/4 = p(A) \cdot p(B)$.
3. a. 0.80 b. 0.65 c. 0.07 d. 0.48 e. 0.52 f. 0.93
5. a. $p(A) = 1/2$, $p(B) = 1/2$; $p(A \cap B) = 1/4 = p(A) \cdot p(B)$
   b. $p(A) = 1/2$, $p(C) = 1/2$; $p(A \cap C) = 1/4 = p(A) \cdot p(C)$
   c. $p(B) = 1/2$, $p(C) = 1/2$; $p(B \cap C) = 1/4 = p(B) \cdot p(C)$
   d. $p(A \cap B \cap C) = 1/4$, $p(A) = 1/2$, $p(B) = 1/2$, $p(C) = 1/2$; $1/4 \neq 1/2 \times 1/2 \times 1/2 = 1/8$
7. a. 0.70 b. 0.72 c. 0.30 d. 0.89   9. 0.33

SECTION 16.4

1. 1/3   3. 1/7   5. 5/8   7. a. 45 b. 67/76 c. 9/40 d. 6/25

SECTION 16.5

1. No, favors the seller.   3. No, favors the man.
5. a. $3\frac{1}{2}$ cents b. $^-3\frac{1}{2}$ cents

REVIEW EXERCISES FOR CHAPTER 16

1. Yes, $A \cap B = \varnothing$.   3. Yes, $A \cap B = \varnothing$.
5. The probabilities add up to only 0.95.   7. a, d
9. a. 0.833 b. 0.86 c. 0.14   11. a. 0.86 b. 0.43 c. 0.36
13. a. 59/150 b. 107/150 c. 43/150   15. 5/7   17. 5/9
19. a. $^-\$.71\frac{3}{7}$ b. No, the expectation is not 0. c. The "house."

CHAPTER 17

SECTION 17.1

1. False   3. False   5. False   7. True   9. False

SECTION 17.2

1. a. 10   b. $\frac{73}{12}$ c. $\frac{43}{90}$ d. $\frac{137}{60}$ e. $\frac{11}{3}$ f. 40
3. a. 3   b. 9.5   c. 17   d. 91   e. 98   f. 36
5. There were probably a few highly paid members whose salaries distorted the mean.

SECTION 17.3

1. a.

X	f	X	f	X	f	X	f
98	1	83	2	69	1	55	1
97	1	82	1	68	2	54	0
96	2	81	1	67	0	53	0
95	0	80	1	66	0	52	0
94	0	79	3	65	1	51	1
93	0	78	0	64	0	50	0
92	2	77	4	63	2	49	1
91	1	76	0	62	0	48	0
90	4	75	2	61	1	47	0
89	2	74	2	60	1	46	0
88	2	73	1	59	1	45	0
87	1	72	1	58	1	44	0
86	1	71	3	57	1	43	0
85	2	70	0	56	0	42	1
84	5						

b.

Class interval	f	Class interval	f
98–101	1	66–69	3
94–97	3	62–65	3
90–93	7	58–61	4
86–89	6	54–57	2
82–85	10	50–53	1
78–81	5	46–49	1
74–77	8	42–45	1
70–73	5		$N = 60$

c. 41.5–45.5   d. 97.5–101.5
e.

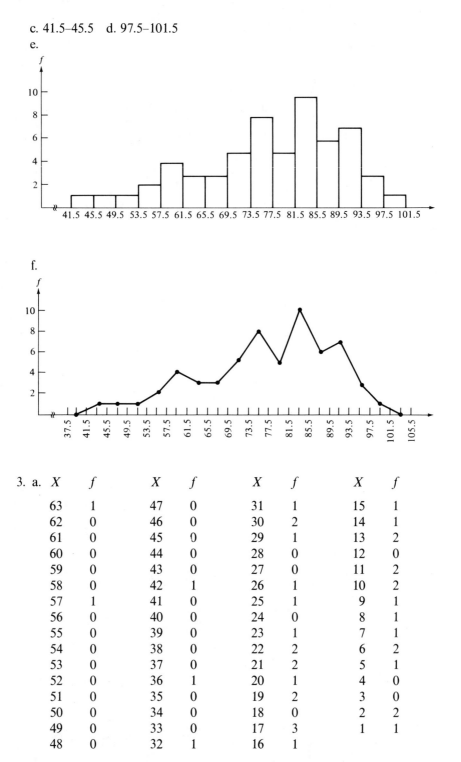

f.

3. a.

X	f	X	f	X	f	X	f
63	1	47	0	31	1	15	1
62	0	46	0	30	2	14	1
61	0	45	0	29	1	13	2
60	0	44	0	28	0	12	0
59	0	43	0	27	0	11	2
58	0	42	1	26	1	10	2
57	1	41	0	25	1	9	1
56	0	40	0	24	0	8	1
55	0	39	0	23	1	7	1
54	0	38	0	22	2	6	2
53	0	37	0	21	2	5	1
52	0	36	1	20	1	4	0
51	0	35	0	19	2	3	0
50	0	34	0	18	0	2	2
49	0	33	0	17	3	1	1
48	0	32	1	16	1		

b. *Class interval* *f*   *Class interval* *f*

62–65	1		26–29	2
58–61	0		22–25	4
54–57	1		18–21	5
50–53	0		14–17	6
46–49	0		10–13	6
42–45	1		6–9	5
38–41	0		2–5	3
34–37	1		⁻2–1	1
30–33	4			

$$N = 40$$

c. ⁻2.5–1.5

d.

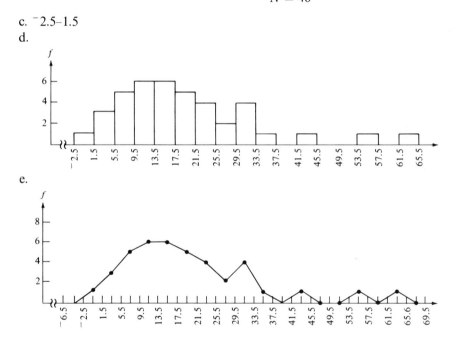

e.

SECTION 17.4

1. a. 27   b. 3   c. $3\frac{5}{9}$ d. 3
3. a. $12,157.40   b. $14,999.50   c. $17,904.26   d. $22,922.58
   e. $16,466.17
5. a. $11,999.50   b. $14,874.50   c. $17,691.73   d. $22,499.50
   e. $15,874.50
7. a. No, $4 would be the average value per sale only if equal numbers of items were sold at the varying rates.   b. $3.84

SECTION 17.5

1. a. $6\frac{4}{9}$ b. 6   c. 2   d. 10   e. $3\frac{49}{81}$ f. $s^2 = 15.80$, $s = 3.7$
   g. $s^2 = 15.80$, $s = 3.7$ 3. a. 35.89   b. 5.99   5. a. (2)   b. (4)

SECTION 17.6

1. a. P_{90} b. P_{70} c. P_{75} d. P_{50} e. P_{77} f. P_{72} g. P_{87} h. P_{84} i. P_{74}
   j. P_{32}
3. a. 5   b. 8   c. 11   d. 16   e. 19   f. 25   g. 31   h. 31   i. 36   j. 38   k. 40
   l. 43   m. 57   n. 55   o. 32   p. 20
5. a. 88.4   b. 65   c. 75.2   d. 79.4   e. 85.9   f. 47.8   g. 82.4   h. 42.5
   i. 95.4   j. 83.9

REVIEW EXERCISES FOR CHAPTER 17

1. False   3. True   5. True   7. True   9. True   11. False
13. False   15. False   17. True   19. False   21. False
23. True   25. True

INDEX